NUTRITION AND HEALTH

Adrianne Bendich, Ph.D., FACN, FASN

Connie W. Bales, Ph.D., R.D., SERIES EDITORS

More information about this series at http://www.springer.com/series/7659

Nathan S. Bryan • Joseph Loscalzo
Editors

Nitrite and Nitrate in Human Health and Disease

Second Edition

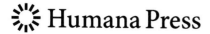 Humana Press

Editors
Nathan S. Bryan
Department of Molecular and Human Genetics
Baylor College of Medicine,
Houston, TX, USA

Joseph Loscalzo
Department of Medicine
Harvard Medical School
Boston, MA, USA

Department of Medicine
Brigham and Women's Hospital
Boston, MA, USA

Nutrition and Health
ISBN 978-3-319-83463-4   ISBN 978-3-319-46189-2   (eBook)
DOI 10.1007/978-3-319-46189-2

Printed on acid-free paper

This Humana Press imprint is published by Springer Nature
The registered company is Springer International Publishing AG
The registered company address is: Gewerbestrasse 11, 6330 Cham, Switzerland

Foreword

The short-lived, free radical molecule nitric oxide (NO) has emerged as one of the most versatile cell signaling transmitters produced by mammalian biological systems. NO, identified as "endothelium-derived relaxing factor" and proclaimed "Molecule of the Year" in 1992, functions critically in physiology, neuroscience, and immunology. The vascular effects of NO alone include vasodilatation, inhibition of platelet aggregation and leukocyte adhesion to the endothelium, scavenging of superoxide anions, and inhibition of smooth muscle cell hyperplasia. Early studies on NO stemmed from work with nitroglycerin in an attempt to elucidate the mechanism through which it relieved pain due to angina pectoris. It was discovered that the formation of NO from nitroglycerin accounts for its therapeutic efficacy for angina by dilating constricted and diseased blood vessels in the heart. Not surprisingly, some of the most prevalent diseases result, at least in part, from decreased NO availability, for example, hypertension, atherosclerosis, diabetes mellitus, and hypercholesterolemia.

The discovery of the formation of NO from the semi-essential amino acid L-arginine through one of three isoforms of nitric oxide synthase provided a key therapeutic target, which is still the focus of much research today. Dietary supplementation of L-arginine has been shown to enhance NO production in healthy individuals (despite already saturated extracellular concentrations), and this may both provide cardiovascular protective effects and enhance athletic performance. Indeed, endothelial dysfunction, an early sign of cardiovascular disease, has been reversed through enhanced NO production. This observation leads us to believe that intervention through the NO pathway is a viable route for treatment and prevention of vascular dysfunction.

Recently, the oxidative "waste" products of nitric oxide, nitrite and nitrate, have been evaluated in a new context, due to their ability to form NO independent of nitric oxide synthase enzymes, through reductive electron exchanges. Since nitrate (as well as nitrite) is primarily ingested in the form of fruits and vegetables, which have been known for some time to protect against diseases from atherosclerosis to cancer, a new paradigm has emerged regarding the role of these once feared nitrogen oxides. Both public and scientific perception of nitrite and nitrate still revolve around fears of nitrosamine formation and carcinogenesis. What has not been considered, however, is the fact that consumption of antioxidants with nitrite and nitrite (both significant components of fruits and vegetables) inhibits the formation of nitrosamines in the gastric milieu. Furthermore, a human nitrogen cycle consisting of commensal bacteria in the oral cavity, which serve a reductive role in the conversion of approximately 20 % of ingested nitrate to nitrite, now appears to provide a significant NOS-independent source of NO generation.

This body of work may have revolutionary implications in terms of developing strategies to combat heart disease and many other contemporary diseases associated with a NO deficiency. Furthermore we may finally have an explanation for the many known and undisputed benefits of the Mediterranean diet. Perhaps now we should consider nitrite and nitrate as the bioactive food components that account for the protective benefits of certain foods and diets. Numerous clinical trials of supplementation with various antioxidants borrowed from heart-healthy diets, such as those typical of Mediterranean countries, have consistently failed to replicate the protective effects of the foods themselves. Consistently absent, but the primary human source, is dietary nitrate and nitrite. Recent work has shown various cardioprotective effects from modest supplementation of nitrite and nitrate. Nitrite, in particular, has been shown to prevent hypercholesterolemic microvascular inflammation and protect against injury from ischemic events.

The broader context of research regarding nitrate, nitrite, and nitric oxide suggests these simple nitrogen oxides serve as a critical dietary component for protection against various chronic diseases. Currently, heart disease and cancer lead the nation in cause of deaths. Concurrently, the dietary patterns of the West have transitioned towards heavily processed foods and lack significant quantities of fruits and vegetables. The explanations have been varied but overlook simple molecules known to play critical roles in multiple organ systems through the chemical messenger NO. The dietary contributions to normal NO homeostasis would not only help explain significantly lower rates of cardiovascular disease in those who regularly consume fruits and vegetables but also arm scientists and physicians with a relatively simple and inexpensive therapeutic intervention.

This text effectively overviews the important role nitrite and nitrate play in biological systems and NO homeostasis. A risk benefit analysis has shown nitrite and nitrate present no danger when consumed in modest quantities and preferably with antioxidants. In fact, research appears to suggest nitrite acts as a redundant NO reservoir when NOS activity is insufficient or stress requires a secondary source. The future use of nitrite/nitrate in dietary considerations will likely have a significant impact on current public health policy. This book brings the NO-story full circle and presents novel thought on the future treatment for many of the country's most pressing health issues. This is a relatively new area of nitric oxide research but a very exciting one. The L-arginine pathway for NO synthesis may turn out to be only part of the story. The symbiosis between humans and the bacteria that reside in and on our body may be just as important in terms of utilizing nitrate and nitrite to make NO under conditions when the oxidation of L-arginine is dysfunctional. Drs. Nathan S. Bryan and Joseph Loscalzo have assembled the world's experts to present a first of its kind, comprehensive work on nitrite and nitrate in human health and disease, carefully examining the context for a risk benefit assessment.

Louis J. Ignarro

Preface

Our major objective and driving force in developing the second edition of *Nitrite and Nitrate in Human Health and Disease* is to consolidate and update all the key research and new knowledge in one volume in order to establish a framework based on the totality of evidence for nitrite and nitrate and the effects of these two anions on human health and disease. Since the publication of the first edition in 2011, more evidence has been published in human trials demonstrating the safety and efficacy of nitrite and nitrate in human disease, but also more insight has been gained into mechanisms of action. These new data are now included in the second edition. Although the biomedical science community is excited and optimistic about the potential for developing new therapeutics and perhaps regimens of disease prevention based on their ability to generate nitric oxide under appropriate conditions, epidemiologists, nutritionists, and cancer biologists have historically had cause for concern owing to the inherent nitrosative chemistry of nitrite and NO that could form potentially carcinogenic N-nitrosamines. The new data on nitrite and nitrate have even shone new light on this minimal potential risk. As a result, a risk benefit analysis for nitrite and nitrate can now be evaluated where it appears that at the right doses and routes of administration, the potential benefits far outweigh any potential risks. This is the second edition of the first book in the Springer/Humana Nutrition Series dedicated to understanding the nutritional aspects of nitrite and nitrate for human health. It is our intent to deliver a comprehensive review of nitrite and nitrate, from basic biochemistry to the complex physiology and metabolism of these two naturally occurring molecules in the human body.

Overall, the book contains well-organized and well-referenced chapters by respected scientists and physicians that covers the rich history of nitrite and nitrate, sources of exposure and physiological effects when consumed through foods containing nitrite and nitrate. The first portion of the book describes the biochemistry, metabolism, and physiology of nitrite and nitrate, and how these molecules are incorporated into the foods we eat and subsequently how they are systematically metabolized to bioactive nitric oxide. This biochemistry involves the environmental processes of nitrogen fixation and the presence of a human nitrogen cycle involving symbiotic bacteria that reside in and on the human body. The book then shifts focus to the sources of exposure to nitrite and nitrate, both environmental and dietary, as a means to quantify exposure estimates and what this may mean for human health. We also discuss the epidemiology and dietary effects on the nitric oxide pathway. This portion of the book also examines systems in nature which this pathway is exploited, including the breast milk of nursing mothers. Finally, the last section of the book discusses nitric oxide-based therapeutics and how nitrite and nitrate biochemistry can be harnessed safely and efficaciously to improve human health through the production of nitric oxide. We end with a summary of the collective body of knowledge presented in the book and what we might expect going forward. Each chapter begins with an abstract and Key Points that outline the concepts presented to assist the reader in understanding the fundamental principles presented in each chapter.

It is undisputed in the biomedical community that NO is one of the most important molecules the human body produces. If NO is such an important molecule in practically every organ system in our body, why, then, is there only a singular pathway for its production, i.e., the complex oxidation of L-arginine? Phosphorylation is another fundamental cellular process that is just as important in cell signaling, with over 500 recognized kinases and many phosphatases to regulate this biochemical process; by contrast, there is only one class of enzymes, the nitric oxide synthases, to produce NO. Most physiological systems are rich in redundancy, allowing backup systems to support the primary system. The provision of nitrate and nitrite as sources of NO may then be viewed as a system of redundancy. After all, a one-electron reduction is energetically and kinetically favorable to a five-electron oxidation.

According to the World Health Organization, cardiovascular disease is the number one killer of both men and women in the United States. These deaths represent a staggering 40 % of all deaths. Close to one million people die each year and more than six million are hospitalized due to cardiovascular disease. Therefore, developing new strategies to correct NO insufficiency and replete NO availability is of paramount importance and could potentially save millions of lives worldwide and lessen the burden on the health care system. We now appreciate that reduced or insufficient NO production or activity is a hallmark of a number of disorders, including many complex, chronic cardiovascular diseases and even Alzheimer's and diabetes mellitus. Therefore, developing new strategies to restore and replete bioactive NO is of paramount importance and could potentially lessen the burden of disease for society. Thus, understanding the biological activity of nitrite and nitrate may not only lead to novel treatments for disease but may lead to strategies to prevent disease development or progression and even the physiological basis for the benefits of certain diets such as the Mediterranean diet. To achieve this laudable goal, we must first establish the context for potential benefit while preventing unwanted risks or harm. We hope the information provided in this text will begin to help define that context, be a source of valuable information, and be useful for anyone who wants the most important and updated information about nitrate and nitrite.

We have invited the world's leading experts to share their research and perspectives, which we hope will help define the context for benefits vs. any potential risks associated with nitrite and nitrate, through either dietary ingestion or therapeutic dosing. This diverse collection of authors includes muscle biologists, physiologists, physicians, epidemiologist, cancer biologist, registered dietician, chemist, and public health experts from five countries around the world in both academia and government. This approach will provide a fair and balanced view of nitric oxide biochemistry, and nitrite and nitrate biochemistry in physiology and in the food sciences. As a result, we are indebted to these many individuals. We feel extremely honored and grateful to have many of the world's experts contribute their knowledge and perspective. We realize the time and dedication it takes to compose a book chapter on the latest body of knowledge, so we appreciate these authors' taking the time to help develop this volume. Without their creative contributions, this book would not have been possible. We also sincerely thank Springer-Humana Press for including this body of work in the Nutrition and Health Series with Dr. Adrianne Bendich as Series Editor. The Editors also wish to thank Stephanie Tribuna for her time and effort in organizing and managing this project with the Editors. This is an exciting time in NO and nitrite-based research. This has been—and we predict will continue to be—an area of intense research in the future. It is our hope that the information contained here will educate and inform scientists, physicians, health care professionals, nutritionists, dieticians, and even the general public on the effects of nitrite and nitrate in human health and disease.

Houston, TX   Nathan S. Bryan
Boston, MA   Joseph Loscalzo

Series Editor Page

The great success of the Nutrition and Health Series is the result of the consistent overriding mission of providing health professionals with texts that are essential because each includes (1) a synthesis of the state of the science, (2) timely, in-depth reviews by the leading researchers and clinicians in their respective fields, (3) extensive, up-to-date fully annotated reference lists, (4) a detailed index, (5) relevant tables and figures, (6) identification of paradigm shifts and the consequences, (7) virtually no overlap of information between chapters, but targeted, inter-chapter referrals, (8) suggestions of areas for future research, and (9) balanced, data-driven answers to patients as well as health professionals questions which are based upon the totality of evidence rather than the findings of any single study.

The series volumes are not the outcome of a symposium. Rather, the editors have the potential to examine a chosen area with a broad perspective, both in subject matter and in the choice of chapter authors. The international perspective, especially with regard to public health initiatives, is emphasized where appropriate. The editors, whose trainings are both research and practice oriented, have the opportunity to develop a primary objective for their book; define the scope and focus, and then invite the leading authorities from around the world to be part of their initiative. The authors are encouraged to provide an overview of the field, discuss their own research, and relate the research findings to potential human health consequences. Because each book is developed *de novo*, the chapters are coordinated so that the resulting volume imparts greater knowledge than the sum of the information contained in the individual chapters.

"**Nitrite and Nitrate in Human Health and Disease, 2nd Edition,**" edited by **Nathan S. Bryan, Ph.D., and Joseph Loscalzo, M.D., Ph.D.,** is a welcome addition to the Nutrition and Health Series and fully exemplifies the Series' goals. This unique volume is a very timely update of the first edition that was published in 2011. Over the past 6 years, there has been a significant increase in the understanding of the huge role of nitrate and nitrite in health and disease. This comprehensive review of the science behind active nitrogen-containing molecules and their effects on humans is of great importance to the nutrition community as well as for health professionals who have to answer client questions about this expanding area of clinical research. The editors strive to inform scientists, physicians, related health care professionals, nutritionists, dieticians, advanced students, and those who have an interest in objective, data-driven analyses of the health effects of nitrite and nitrate. Thus, with these critical issues in mind, the editors have decided to not only update the relevant chapters from the first edition but also add nine new, timely chapters that are of great value to clinicians, nutritional health professionals, and advanced students. The importance of this updated volume is captured in the Foreword, written by Dr. Louis Ignarro, 1998 recipient of the Nobel Prize in Medicine, who writes: "This book brings the NO-story full circle and presents novel thought on the future treatment for many of the country's most pressing health issues. This is a relatively new area of nitric oxide research but a very exciting one."

The internationally recognized clinical researchers who serve as the editors of this as well as the first edition of *Nitrite and Nitrate in Human Health and Disease* are highly qualified in nitrate and nitrite investigations. Dr. Nathan S. Bryan, Ph.D., is currently Adjunct Assistant Professor at Baylor College of Medicine in the Department of Molecular and Human Genetics in Houston, TX. He is a pioneer in recognizing the important effects of dietary nitrite and nitrate and elucidating the novel metabolism of nitrate and nitrite in humans. Dr. Bryan has been involved in nitric oxide research for more than 16 years and has made many seminal discoveries in the field that have resulted in seven patents. Specifically, Dr. Bryan was the first to describe nitrite and nitrate as indispensable nutrients required for optimal cardiovascular health. Dr. Bryan's lab was the first to demonstrate and discovered an endocrine function of nitric oxide via the formation of S-nitrosoglutathione and inorganic nitrite. Through the drug discovery program in natural products chemistry, Dr. Bryan discovered unique compositions of matter that can be used to safely and effectively generate and restore nitric oxide in humans. This technology is now validated in many published clinical trials. These discoveries and findings have unveiled many beneficial effects of nitrite in the treatment and prevention of human disease and may provide the basis for new preventive or therapeutic strategies in diseases associated with NO insufficiency and new guidelines for optimal health. He is also a successful entrepreneur who has commercialized his nitric oxide technology through the formation of Human to the Power of N (formerly Neogenis Labs, Inc.) where he is Co-founder and Chief Science Officer. Dr. Bryan's most recent work has focused on the role of oral nitrate reducing bacteria in control of endogenous nitric oxide production and hypertension. He has published a number of highly cited papers and authored or edited six books. Dr. Bryan is a member of the Society for Free Radical Biology and Medicine, American Physiological Society, Nitric Oxide Society, American Heart Association, and the American Association for the Advancement of Science.

Joseph Loscalzo, M.D., Ph.D., is the Hersey Professor of the Theory and Practice of Medicine at Harvard Medical School and serves as Chairman of the Department of Medicine and Physician-in-Chief at Brigham and Women's Hospital in Boston, Massachusetts. Dr. Loscalzo is recognized as an outstanding cardiovascular scientist, clinician, and teacher. Dr. Loscalzo is one of the early pioneers in the nitric oxide field with many of his seminal publications and discoveries dating from the early 1980s. He was the first to demonstrate the formation of S-nitrosoproteins and to elucidate the anti-thrombotic and antiplatelet mechanisms of NO, as well as many other contributions to this field. He has received many awards, including the Clinician-Scientist Award, the Distinguished Scientist Award, the Research Achievement Award, and the Paul Dudley White Award from the American Heart Association; a Research Career Development Award, a Specialized Center of Research in Ischemic Heart Disease Award, and a MERIT Award from the National Institutes of Health; the George W. Thorn Award for Excellence in Teaching at Brigham and Women's Hospital, and Educator of the Year Award in Clinical Medicine from Boston University; the Glaxo Cardiovascular Research Award, and the Outstanding Investigator Prize from the International Society for Heart Research; and election to the American Society for Clinical Investigation, the Association of American Physicians, and the Institute of Medicine of the National Academy of Sciences. He has served on several NIH study sections and editorial boards and has chaired the Gordon Conference on Thrombolysis. He served as an associate editor of the *New England Journal of Medicine* for 9 years, Editor-in-Chief of *Circulation* for 12 years, Chair of the Cardiovascular Board of the American Board of Internal Medicine, Chair of the Research Committee of the American Heart Association, Chair of the Scientific Board of the Stanley J. Sarnoff Society of Fellows for Research in the Cardiovascular Sciences, and Chair of the Board of Scientific Counselors of the National Heart, Lung, and Blood Institute of the National Institutes of Health. He is a former member of the Advisory Council of the National Heart, Lung, and Blood Institute and a former member of the Council of Councils of the National Institutes of Health. He is currently Director of the NIH-funded Center for Accelerated Innovation (the Boston Biomedical Innovation Center), the NIH-funded Harvard Undiagnosed Disease Network program, and a senior editor of *Harrison's Principles of Internal Medicine*. His most recent work has established the field of network medicine, a paradigm-changing discipline that seeks to redefine disease and therapeutics from an integrated perspective using systems biology and network science. Moreover, these two eminent editors have chosen the most respected and knowledgeable authors for the volume's 22 insightful chapters. Thus, this volume addresses the critical issues of the roles of nitrite and nitrate in human health and disease with objective, data-driven chapters that reflect the goals of clinical nutrition today. The chapters contain Key Points as well as integrated Conclusions as it is the intention of the editors that this volume serve as an important resource for educators and students.

This comprehensive volume contains 22 chapters that are organized in three broad parts: Biochemistry, Molecular Biology, Metabolism and Physiology; Food and Environmental Exposures to Nitrite and Nitrate; and Nitrite and Nitrate in Therapeutics and Disease.

Part I: Biochemistry, Molecular Biology, Metabolism and Physiology

The six chapters that comprise the first, introductory part of the volume provide a grounding in the chemistry, biochemistry, and functions of the nitrogen oxides, nitrite and nitrate. The first chapter, written by the volume's editors, describes the discovery of the gas, nitric oxide (NO), and its formation in the body. We learn that the L-arginine-NOS (nitric oxide synthase) pathway was the first NO production system discovered and is responsible for about 50 % of endogenous nitrate and nitrite in humans. Nitrite can also form directly in blood and tissues via NO autoxidation, which involves the reaction of two molecules of NO with oxygen, a reaction that is catalyzed by the plasma protein ceruloplasmin. Nitrate is by far the dominant final NO oxidation product; the levels in blood and tissues exceed those of nitrite by at least two orders of magnitude. The half-life of nitrite is about 110 s when added to venous blood. Nitrate, by contrast, has a circulating half-life of 5–8 h. The interactions between oral bacteria, stomach acid, and dietary antioxidants mainly from fruits and vegetables are reviewed. The importance of NO for cardiovascular health is also discussed. The second chapter examines the natural nitrogen cycle and its role in provision of sources of nitrite and nitrate in our diets. Nitrogen (N_2) gas makes up approximately 78 % of the atmosphere. It is highly inert, being bound together by a triple bond that requires considerable energy to break, and higher organisms, including humans, cannot convert nitrogen to biologically active forms. The conversion of nitrogen from inert N_2 to biologically available nitrogen (called fixation) requires multistep enzymatic pathways; vast amounts of energy in the case of lightning; or a combination of heat, pressure, and metallic catalysts. The conversion of atmospheric nitrogen to a fixed, biologically available nitrogen species is only accomplished by prokaryotes, eubacteria, and archaea. These organisms are known as diazotrophs and they have three modes of survival: free-living, symbiotic (within plant roots), and those that are associated mainly with plant roots. The formation and use of fixed nitrogen by plants is described. In humans, nitrogen is principally ingested as protein. The average US young adult consumes approximately 90 g of protein daily which equates to about 14.5 g of nitrogen, nearly all of which is excreted in the urine as urea, creatinine, uric acid, and ammonia. Only a tiny proportion of this nitrogen is converted to nitric oxide via the nitric oxide synthase pathway and some is also generated from L-arginine. The importance of the oral commensal bacteria in salivary glands in the generation of nitrite is reviewed.

Chapter 3, coauthored by the volume's coeditor, discusses the importance of human oral nitrate reducing bacteria as humans do not have the enzymes to reduce nitrate to nitrite which is then swallowed and further reduced to NO in the acidic stomach. The chapter reviews the enterosalivary nitrate-nitrite-nitric oxide pathway that represents a potential symbiotic relationship between oral bacteria and their human hosts. The oral commensal bacteria provide an important metabolic function in human physiology by contributing an NOS-independent source of NO. The chapter stresses the importance of these bacteria in maintaining adequate concentrations of NO that are associated with reduced risk of cardiovascular disease. Conversely, the chapter explores the consequences of disruption of nitrite and NO production in the oral cavity that may contribute to the oral-systemic link between oral hygiene and cardiovascular risk and disease. The new emerging area of epigenetics and how these changes to DNA's environment can affect NO production are the focus of Chap. 4 that is coedited by one of the volume's editors. Epigenetics involves chromatin-based modifications to DNA or histone proteins that influence patterns of gene expression. NOS genes and other genes involved in NO production are targets for epigenetic modification. Of relevance, prenatal exposure to pharmacological agents or special diets can promote epigenetic changes that affect gene expression and physiology in offspring. Maternal diet can affect epigenetic regulation of gene expression and modulate disease outcomes. This detailed chapter, containing over 100 relevant references and six helpful figures, reviews numerous targets for epigenetic alterations in methylation reactions as well as changes in NOS expression. There is a discussion of the direct and indirect epigenetic mechanisms that regulate NOS gene expression and NO production.

Chapter 5 examines the role of mitochondria and the link to nitrogen-containing molecules. Nitrite has been shown to be an endogenous signaling molecule that regulates mitochondrial number, morphology, and function. The chapter, containing over 125 references, provides an in-depth examination of the functions of the mitochondria as these have expanded greatly in the past decade. Included are reviews of the importance of the mitochondrion as a physiological subcellular target for nitrite; regulation of mitochondrial respiration, reactive oxygen species generation and apoptotic signaling, as well as nitrite's control over mitochondrial number and dynamics. The last chapter in this part, Chap. 6, outlines the major endogenous and exogenous sources of exposure to nitrogen oxides. Nitrate and nitrite are found in the human diet, are present in drinking water, and in many vegetables as well as part of food preservation systems in cured meats. Exposure is dependent on geographic location and diet choices. Endogenously produced nitrate and nitrite are derived from nitric oxide metabolism and approximately 25 % of plasma nitrate is retained in the body via an enterosalivary pathway discussed in detail in Chap. 3. The enterosalivary recycling involves the oral reduction of nitrate to nitrite by commensal bacteria. There is a discussion of the discovery of the endogenous formation of NO that is synthesized by mammalian cells from L-arginine: NO was shown to be a potent vasodilator, inhibitor of platelet aggregation, and active species of nitroglycerin. The importance of maintaining a safe level of nitrate and nitrite in drinking water, especially in areas of high use in plant fertilizers, is reviewed. Also outlined are the major environmental exposures to nitrogen oxides including cigarette smoke and air pollution and the potential exposure to nitrosamines.

Part II: Food and Environmental Exposures to Nitrite

Part II contains six chapters that examine the sources of nitrite and nitrate from food and environmental exposure and their biological impacts. Chapter 7 reviews the long history of use of salting meats as a means of preservation and indicates that, in ancient times, salt was obtained from crystalline deposits by mining directly from the earth, or evaporating water from brine pools or seawater. As a consequence, it often contained natural contaminants such as sodium or potassium nitrate (saltpeter or "niter") that contributed directly to the curing reaction and the preservation process. The contaminants were actually the primary components in curing reactions. Importantly, nitrite inhibits the growth of a number of aerobic and anaerobic microorganisms and especially suppresses the growth of *Clostridium botulinum* spores. Nitrite in combination with salt and other curing factors may also control the growth of other pathogens such as *Bacillus cereus*, *Staphylococcus aureus*, and *Clostridium perfringens*. Nitrite also reacts with myoglobin to produce the reddish-pink cured meat color, inhibits lipid oxidation (and reduces oxidative rancidity), and contributes to the cured flavor of meat products.

The chapter includes an overview of the regulatory limits of nitrite and nitrate in cured foods as established by the United States Department of Agriculture—Food Safety Inspection Service (USDA-FSIS) or Food and Drug Administration (FDA), the scientific research to determine the limit of consumption based upon the formation of nitrosamines that are considered as carcinogenic agents, and tables that describe the concentrations of naturally occurring nitrates in selected vegetables and an extensive discussion of the regulatory history of establishing safe levels of nitrate/nitrite in municipal water supplies in the United States and many other countries.

Chapter 8 provides an overview of the field of nutritional epidemiology and specifically examines the critical issues involved in determining dietary intakes of nitrogen-containing compounds in food and water. Nitrite and nitrite oxide are biologically active compounds involved in vasodilatation, inhibition of endothelial inflammation, and platelet aggregation. Nitrates and nitrites in the diet are the relevant compounds for studying the nutritional components of cardiovascular disease. However, compared to other nutrients, quantifying nitrate and nitrite intake is challenging. There are large within-food variation in nitrate and nitrite concentrations and large geographical and seasonal differences in the nitrate content in drinking water. Conclusions include that to date, the epidemiologic evidence for the relationship between dietary nitrate and nitrite and bladder or gastric cancers is limited and weak. There is no direct epidemiological evidence relating dietary nitrite and nitrate to risk of cardiovascular disease. However, the primary source of dietary nitrate, green leafy vegetables, has been associated with lower risk of cardiovascular disease. Processed meats are important contributors of nitrite intake. However, the increased risks for cardiovascular disease are also observed for red meat to which no preservatives are added. A recent finding links dietary nitrate with a reduced risk of a particular type of glaucoma, and this finding needs to be confirmed.

Although the nutritional epidemiology does not find a strong link between dietary sources of nitrite/nitrate and reduction in cardiovascular disease risk, Chaps. 9 and 10 examine the physiological effects at the cellular level of NO and how these effects could affect cardiovascular disease risk. NO relaxes the underlying vascular smooth muscle to improve vascular compliance and to reduce vascular resistance. In addition, endothelium-derived NO inhibits platelet adhesion and aggregation, suppresses leukocyte adhesion and vascular inflammation, and limits the proliferation of the underlying vascular smooth muscle cells. The chapter objectively reviews the data indicating that short-term effects of NO may not result in overall reduction in cardiac vessel disease or its consequences. The list of substances examined include, but are not limited to, arginine, B vitamins, antioxidants, fiber, vegetable protein, lipids, soy, and phytoestrogens and several diets including the DASH diet and the Mediterranean Diet, and the chapter includes over 150 relevant references. Chapter 10 reviews the importance of NO for optimal vascular functions and suggests that disturbed NO homeostasis is a hallmark of the development of chronic cardiovascular disease and affects the level of risk for acute ischemic events such as an acute myocardial infarction. Additionally, the flavonoids that are often found in nitrate-containing foods are reviewed as important bioactive molecules that enhance NO vascular activity. Several of the key dietary components that have high flavonoid content are green tea, red wine, and products made from cocoa beans; these are described in detail.

In addition to plants, human milk contains relatively high concentrations of nitrate and nitrite. Chapter 11, coauthored by one of the volume's editors, examines the importance of these molecules in infant development. The chapter reviews the evidence that these anions are necessary for growth and development as well as immunological support in the newborn infant. Emerging data suggest that the difference between the nitrite and nitrate content of breast milk compared to formula milk may account for some of the health factors that have been shown to differ in breast-fed versus formula-fed infants. The chapter includes a detailed explanation of "blue-baby syndrome" that is linked to excessive nitrate in ground water, and not from breast feeding. Methemoglobinemia, or blue-baby syndrome, has been seen in infants less than 6 months of age that are exposed to excess nitrates in bacterially contaminated well water. Infants who are younger than 6 months old are particularly susceptible to nitrate-induced methemoglobinemia because of the lower than needed stomach acid production, large concentration of nitrite-reducing bacteria, the relatively easy oxidation of fetal hemoglobin, and immaturity of the methemoglobin reductase system. There is a detailed description of the nutritional and physiological importance of colostrum as well as breast milk and the changes in composition that occur during 6 months of breastfeeding. The final chapter in this part, Chap. 12, describes the dilemma currently that has arisen because of the known adverse effects of overexposure to environmental nitrate pollution in the face of relatively new data on the importance of nitrite and nitrate and NO for cardiovascular function, immune function, and other clinical areas that are discussed in detail in subsequent chapters. The authors suggest that there be a reappraisal of all of the clinical data to better understand the risk/benefit ratio and provide data-driven guidance to consumers and health professionals.

Part III: Nitrite and Nitrate in Therapeutics and Disease

Part III addresses the recent uses of nitrogen-containing molecules as therapeutics and reviews the association of NO status and disease risk in ten clinically focused chapters. Chapter 13 provides an overview of the section and is authored by one of the volume's coeditors. The authors enumerate some of the major disease conditions that have been researched and published in over 140,000 publications in the last 12 years. NO has been shown to be involved in and affect neurotransmission, memory, stroke, glaucoma and neural degeneration, pulmonary hypertension, penile erection, angiogenesis, wound healing, atherogenesis, inflammation such as seen in rheumatoid arthritis, nephritis, colitis, other autoimmune diseases, destruction of pathogens and tumors, asthma, tissue transplantation, septic shock, platelet aggregation and blood coagulation, sickle cell disease, gastrointestinal motility, hormone secretion, gene regulation, hemoglobin delivery of oxygen, stem cell proliferation and differentiation, and bronchodilation. The chapter focuses on the cardiovascular and nervous systems and the functions of NO in the immune system. There are detailed discussions of the biochemistry of the three enzymes involved in the synthesis of NO within cells: neuronal NOS, inducible NOS, and endothelial NOS and the reactions that follow the formation of NO within cells.

Chapters 14–17 are practice-oriented chapters that examine the therapeutic usefulness of nitrogen-containing molecules in patients with cardiovascular disease (CVD) and/or pulmonary diseases that require dilation of vessels and organs such as the lungs and heart. Chapter 14, containing more than 100 useful references, describes the use of inhaled NO as a therapeutic agent for the treatment of pulmonary vasodilation. There is an historic review of the use of nitric oxide-generating drugs, including nitroglycerin and sodium nitroprusside, that have been used by clinicians to treat patients with a broad spectrum of cardiovascular diseases including angina pectoris and systemic hypertension. In addition to reviewing the animal model data on the development of inhaled NO for a number of clinical indications, the chapter highlights the use of inhaled NO to treat hypoxia in the newborn, to prevent the development of bronchopulmonary dysplasia in premature infants, and several other new therapies. One example is the use of inhaled NO in adults and children undergoing cardiac surgery and cardiac transplantation to treat pulmonary hypertension and improve cardiac output in the perioperative period. Of great value to clinicians, the safe use of inhaled NO is clearly described. Chapter 15 provides detailed explanations of the biochemical reactions and therapeutic value of the vasodilation that result from using NO-generating therapies. The chapter, containing over 150 relevant references, also explains the negative phenomenon of nitrate tolerance and the induction of endothelial dysfunction: these side effects are described and new data, including epigenetic effects of organic nitrates, as well as methods to avoid the appearance of these adverse effects, are outlined. Chapter 16 examines the potential for nitrite to improve recovery from ischemia-reperfusion injury. This injury, which occurs following a myocardial infarct, ischemic stroke, or other cause of blood flow disruption, is characterized by numerous detrimental cellular events including oxidative damage to proteins and lipids, enzyme release, and inflammatory responses, which ultimately lead to necrosis and apoptosis of affected tissues. The chapter includes an extensive review and tabulation of the clinical studies that have used low dose nitrite or NO donor molecules in patients with ischemia/reperfusion injuries. Chapter 17 reviews the importance of nitrogen-containing molecules for hypertensive patients. NO is unique in its capacity to maintain blood pressure homeostasis because its vasodilatory action affects both arterial and venous sides of the circulation. The chapter includes an historical perspective on the use of inorganic nitrite as a vasodilator and the recent epidemiological data linking consumption of green leafy vegetables (natural sources of nitrate) with reduced risk of hypertension has encouraged the development of small clinical studies using beet juice compared to juice that does not contain nitrate in normal and hypertensive patients. Preliminary data are promising, but further work is warranted. Also discussed are the potentially detrimental effects of antimicrobial mouth washes that adversely affect oral bacteria involved in the generation of nitrite from food nitrate. In total, these four clinically interrelated chapters provide the reader with over 650 targeted references as well as 16 valuable tables and figures.

The unique chapter, Chap. 18, examines the historical significance of Chinese and other traditional herbal medicines and supplements that have been in use for centuries to treat cardiovascular and other diseases. A number of these herbs have the potential to increase intakes of nitrite and nitrate directly and may also provide compounds that can alter synthesis of nitrogen-containing molecules. The author reminds us that plants require nitrogen for growth. In addition, we learn that many herbal medicines are made from the roots and leaves of certain plants that contain abundant nitrate/nitrite or related compounds. In these plants, NO can be produced by several routes via plant L-arginine-dependent or independent NO-synthesizing enzymes; by the action of plasma membrane-bound nitrate/nitrite reductase; as an end product of the mitochondrial electron transport chain; and through nonenzymatic reactions as the roots of certain plants may contain high levels of unstable nitro compounds, which may release NO spontaneously. Moreover, plants exposed to nitro-types of fertilizers typically generate more NO than those grown in the wild. The recent interest in the plants and plant parts used to affect cardiovascular symptoms is reviewed and findings are tabulated as well as illustrated in informative figures.

The final four chapters in this comprehensive volume examine the effects of aging on NO availability, the role of nitrite and nitrate in exercise, and the association of NO status and the risk of cancer, followed by a critically important chapter by the editors on future research areas. Chapter 19 provides the rationale for the decline in NO with aging and evidence that enhancement of NO status during the senior years may reduce the risk of chronic, noninfectious diseases of aging. NO bioavailability decreases with age. The reasons for the age-associated decrease in NO bioavailability are multifactorial, but evidence suggests that an age-related increase in oxidative stress is the primary cause. Oxidative stress is defined as an imbalance in reactive oxygen species relative to antioxidant defenses and results in the destruction in cellular concentrations of NO. Additionally, chronic, low-grade inflammation also develops with advancing age and this also reinforces oxidative stress; low-grade inflammation has also been seen in obesity, diabetes, rheumatoid arthritis, and certain other diseases. The well-accepted function of NO as a vasodilator is especially important in the aging population. The authors remind us that NO is produced within the vascular endothelium and diffuses to the vascular smooth muscle, providing a powerful vasodilatory signal that adjusts the diameter of the vessel to accommodate changes in blood flow. In aging as well as pathological states in which NO bioavailability is low, this signaling pattern is disrupted, creating an environment conducive to endothelial dysfunction and stiffening of the large elastic arteries, two primary contributors to the increased risk of CVD in middle-aged and older adults. There are also data on cognitive and muscle function that are adversely affected by lower concentrations of NO seen with aging. Short-term clinical studies using diet and/or supplements are reviewed and the chapter includes over 180 clinically relevant references.

Chapter 20 examines the physiology of muscle function during exercise and the role of NO in muscle function. We learn that nitrite is a precursor of nitric oxide synthesis in hypoxic and acidic conditions. Skeletal muscles become acidic and hypoxic during exercise, increasing the potential for NO synthesis in the presence of nitrite. In fact, nitrate supplementation has been shown to improve exercise economy and endurance performance in moderately trained, but not well-trained, subjects and improve short-duration, high-intensity exercise performance. The chapter, containing over 130 relevant references, summarizes nitrate supplementation procedures, often with beet juice, exercise settings, and the current status of supplementation studies in several sports and exercise routines. Chapter 21 critically reviews the early in vitro and laboratory animal studies using very high doses of nitrite that linked their formation into carcinogenic nitrosamines and increased the risk of gastric cancer. Newer, more controlled laboratory studies have failed to replicate the earlier findings and human epidemiological studies have also not consistently found a positive association between dietary intake of nitrite/nitrate and cancer risk. Additionally, we have learned that vegetables are the major source of dietary nitrogen-containing compounds and that cured meats now have antioxidants added to prevent nitrosamine formation. The chapter carefully reviews the potential points of error that can be found in epidemiological studies, such as when the exposure to the potential carcinogen is not known—this is the case of nitrate, nitrite, and other compounds that can be involved in nitrosamine synthesis. It is difficult to accurately determine exposure because of growing conditions and environmental stressors to plants; also, it appears that nitrosamines may be carcinogenic in animal models but may not be carcinogenic in humans. This balanced, data-based chapter corroborates the independently arrived at similar conclusions found in Chap. 8. The authors of both chapters confirm the lack of consistent data linking dietary intake of nitrite and nitrate with increased risk of gastric cancer. The authors provide excellent examples of the importance of objective evaluation of epidemiological studies.

The final chapter, written by the volume's excellent editors, objectively identifies the key areas where future research is needed in order to move clinical research in the functions of nitrogen-containing compounds forward. Two key areas identified are the determination of optimal ranges of intake and the establishment of national government—recommended dietary guidelines that are associated with maximum benefit and assurance of reduced risk of overexposure. Once this information is accepted by the clinical nutrition community and related health professionals, future research using safe and effective doses of nitrite and nitrate in long-term clinical trials will become possible.

Conclusions

The above description of the volume's 22 chapters attests to the depth of information provided by the 39 well-recognized and respected chapter authors. Each chapter includes complete definitions of terms with the abbreviations fully defined and consistent use of terms between chapters. Key features of this comprehensive volume include over 60 detailed tables and informative figures, an extensive, detailed index and more than 2200 up-to-date references that provide readers with excellent sources of worthwhile information that is of great value to clinicians, health researchers and providers, government food and nutrition policy leaders as well as graduate and medical students.

In conclusion, "**Nitrite and Nitrate in Human Health and Disease, 2nd Edition**," edited by **Nathan S. Bryan, Ph.D., and Joseph Loscalzo, M.D., Ph.D.,** provides health professionals in many areas of research and practice with the most up-to-date, well-referenced volume on the importance of nitrite and nitrate, precursors of NO as key molecules in the maintenance of cardiovascular health and optimal functioning of the nervous and immune systems and other critical cells and tissues in the body. The value of this volume is captured in the Foreword, written by Dr. Louis Ignarro who was awarded the Nobel Prize in Medicine for his seminal research on the effects of NO in blood vessels.

The chapters review the role of diet, food, salivary bacterial nitrite synthesis, essential and nonessential nutrients, water, and other components of the diet in maintaining overall health as these provide sources of nitrite and nitrate as well as complementary molecules, such as antioxidants, that enhance the actions of NO with cells, mitochondria, and other cellular components. There is an overriding goal of the editors to evaluate the multiple effects of NO, within populations that suffer from undernutrition to obesity throughout the lifecycle, so that targeted progress can be made in improving the health of those at greatest risk of deficits of dietary and supplemental sources of nitrogen-containing compounds. Areas emphasized include new research on the importance of NO for maternal and infant health, especially in preterm infants, adolescent health and exercise strategies, and the health of older individuals at risk of increased oxidative stress. The importance of optimal dietary intakes of vegetables that are excellent sources of nitrate and the safe use of supplements (such as arginine and other compound that increase circulating levels of NO) are reviewed in detail in populations with increased risk of noncommunicable diseases, especially cardiovascular disease, diabetes, and obesity. The broad base of knowledge in this volume is evidenced by the contrasting, well-referenced chapters that present differing perspectives on the historic use of certain products versus the reliance on randomized, controlled clinical studies. Unique chapters review the implications of applying an evidence-based medicine approach to the use of NO-enhancing therapies, and in contrast, there is an equally informative chapter that describes the use of Chinese herbal medicine products for the same purpose. The volume serves the reader as the benchmark in this complex area of interrelationships between molecules that, at high concentrations, are air pollutants and perhaps carcinogens, and yet at low doses, within our bodies, are critically important in maintaining the movement of blood throughout the body. The editors, Dr. Nathan S. Bryan, Ph.D., and Dr. Joseph Loscalzo, M.D., Ph.D., and the excellent chapter authors are applauded for their efforts to develop the most authoritative and unique resource on the importance of nitrite and nitrate in health and disease, and this excellent text is a very welcome addition to the Nutrition and Health Series.

Morristown, NJ, USA

Adrianne Bendich, Ph.D., FACN, FASN
Series Editor

About the Series Editors

Adrianne Bendich, Ph.D., F.A.S.N., F.A.C.N. has served as the "Nutrition and Health" Series Editor for 20 years and has provided leadership and guidance to more than 200 editors that have developed the 70+ well-respected and highly recommended volumes in the Series.

In addition to "Nitrite and Nitrate in Human Health and Disease, 2nd Edition," edited by Nathan S. Bryan, Ph.D., and Joseph Loscalzo, M.D., Ph.D., major new editions published from 2012 to 2017 include:

1. Nutrition and Health in a Developing World, 3rd Edition edited by Dr. Saskia de Pee, PhD, Dr. Douglas Taren, PhD, and Dr. Martin W Bloem, MD, PhD, 2017
2. Nutrition in Lifestyle Medicine edited by James M. Rippe, MD, 2016
3. L-Arginine in Clinical Nutrition edited by Vinood B. Patel, Victor R. Preedy, and Rajkumar Rajendram, 2016
4. Mediterranean Diet: Impact on Health and Disease edited by Donato F. Romagnolo, Ph.D., and Ornella Selmin, Ph.D., 2016
5. Nutrition Support for the Critically Ill edited by David S. Seres, MD, and Charles W. Van Way, III, MD, 2016
6. Nutrition in Cystic Fibrosis: A Guide for Clinicians edited by Elizabeth H. Yen, M.D., and Amanda R. Leonard, MPH, RD, CDE, 2016
7. Preventive Nutrition: The Comprehensive Guide For Health Professionals, Fifth Edition, edited by Adrianne Bendich, Ph.D., and Richard J. Deckelbaum, M.D., 2016
8. Glutamine in Clinical Nutrition edited by Rajkumar Rajendram, Victor R. Preedy, and Vinood B. Patel, 2015
9. Nutrition and Bone Health, Second Edition, edited by Michael F. Holick and Jeri W. Nieves, 2015

Earlier books included Vitamin D, Second Edition, edited by Dr. Michael Holick; "Dietary Components and Immune Function" edited by Dr. Ronald Ross Watson, Dr. Sherma Zibadi, and Dr. Victor R. Preedy; "Bioactive Compounds and Cancer" edited by Dr. John A. Milner and Dr. Donato F. Romagnolo; "Modern Dietary Fat Intakes in Disease Promotion" edited by Dr. Fabien De Meester, Dr. Sherma Zibadi, and Dr. Ronald Ross Watson; "Iron Deficiency and Overload" edited by Dr. Shlomo Yehuda and Dr. David Mostofsky; "Nutrition Guide for Physicians" edited by Dr. Edward Wilson, Dr. George A. Bray, Dr. Norman Temple, and Dr. Mary Struble; "Nutrition and

Metabolism" edited by Dr. Christos Mantzoros and "Fluid and Electrolytes in Pediatrics" edited by Leonard Feld and Dr. Frederick Kaskel. Recent volumes include "Handbook of Drug-Nutrient Interactions" edited by Dr. Joseph Boullata and Dr. Vincent Armenti; "Probiotics in Pediatric Medicine" edited by Dr. Sonia Michail and Dr. Philip Sherman; "Handbook of Nutrition and Pregnancy" edited by Dr. Carol Lammi-Keefe, Dr. Sarah Couch, and Dr. Elliot Philipson; "Nutrition and Rheumatic Disease" edited by Dr. Laura Coleman; "Nutrition and Kidney Disease" edited by Dr. Laura Byham-Gray, Dr. Jerrilynn Burrowes, and Dr. Glenn Chertow; "Nutrition and Health in Developing Countries" edited by Dr. Richard Semba and Dr. Martin Bloem; "Calcium in Human Health" edited by Dr. Robert Heaney and Dr. Connie Weaver; and "Nutrition and Bone Health" edited by Dr. Michael Holick and Dr. Bess Dawson-Hughes.

Dr. Bendich is President of Consultants in Consumer Healthcare LLC and is the editor of ten books including "Preventive Nutrition: The Comprehensive Guide for Health Professionals, Fifth Edition," coedited with Dr. Richard Deckelbaum (www.springer.com/series/7659). Dr. Bendich serves on the Editorial Boards of the *Journal of Nutrition in Gerontology and Geriatrics* and *Antioxidants* and has served as Associate Editor for *Nutrition*, the International Journal; served on the Editorial Board of the *Journal of Women's Health and Gender-Based Medicine* and served on the Board of Directors of the American College of Nutrition.

Dr. Bendich was Director of Medical Affairs at GlaxoSmithKline (GSK) Consumer Healthcare and provided medical leadership for many well-known brands including TUMS and Os-Cal. Dr. Bendich had primary responsibility for GSK's support for the Women's Health Initiative (WHI) intervention study. Prior to joining GSK, Dr. Bendich was at Roche Vitamins Inc. and was involved with the groundbreaking clinical studies showing that folic acid-containing multivitamins significantly reduced major classes of birth defects. Dr. Bendich has coauthored over 100 major clinical research studies in the area of preventive nutrition. She is recognized as a leading authority on antioxidants, nutrition and immunity, pregnancy outcomes, vitamin safety, and the cost-effectiveness of vitamin/mineral supplementation.

Dr. Bendich received the Roche Research Award, is a *Tribute to Women and Industry* Awardee, and was a recipient of the Burroughs Wellcome Visiting Professorship in Basic Medical Sciences. Dr. Bendich was given the Council for Responsible Nutrition (CRN) Apple Award in recognition of her many contributions to the scientific understanding of dietary supplements. In 2012, she was recognized for her contributions to the field of clinical nutrition by the American Society for Nutrition and was elected a Fellow of ASN. Dr. Bendich is Adjunct Professor at Rutgers University. She is listed in Who's Who in American Women.

Connie W. Bales, Ph.D., R.D. is a Professor of Medicine in the Division of Geriatrics, Department of Medicine, at the Duke School of Medicine and Senior Fellow in the Center for the Study of Aging and Human Development at Duke University Medical Center. She is also Associate Director for

Education/Evaluation of the Geriatrics Research, Education, and Clinical Center at the Durham VA Medical Center. Dr. Bales is a well-recognized expert in the field of nutrition, chronic disease, function, and aging. Over the past two decades her laboratory at Duke has explored many different aspects of diet and activity as determinants of health during the latter half of the adult life course. Her current research focuses primarily on the impact of protein enhanced meals on muscle quality, function, and other health indicators during obesity reduction in older adults with functional limitations. Dr. Bales has served on NIH and USDA grant review panels and is a member of the American Society for Nutrition's Medical Nutrition Council. Dr. Bales has edited three editions of the *Handbook of Clinical Nutrition in Aging* and is Editor-in-Chief of the *Journal of Nutrition in Gerontology and Geriatrics*.

About Volume Editors

Nathan S. Bryan, Ph.D. earned his undergraduate Bachelor of Science degree in Biochemistry from the University of Texas at Austin and his doctoral degree from Louisiana State University School of Medicine in Shreveport where he was the recipient of the Dean's Award for Excellence in Research. He pursued his postdoctoral training as a Kirschstein Fellow at Boston University School of Medicine in the Whitaker Cardiovascular Institute.

After a 2-year postdoctoral fellowship, in 2006 Dr. Bryan was recruited to join the faculty at the University of Texas Health Science Center at Houston by Ferid Murad, M.D., Ph.D., 1998 Nobel Laureate in Medicine or Physiology. During his tenure as faculty and principle investigator at UT, his research focused on drug discovery through screening natural product libraries for compounds and extracts that could generate authentic NO gas or stimulate NO release from endothelial cells. His 9 years at UT led to several discoveries which have resulted in seven currently issued patents.

Dr. Bryan has been involved in nitric oxide research for more than 16 years and has made many seminal discoveries in the field. Specifically, Dr. Bryan was the first to describe nitrite and nitrate as vitamins and indispensable nutrients required for optimal cardiovascular health. Dr. Bryan's lab was the first to demonstrate and discover an endocrine function of nitric oxide via the formation of S-nitrosoglutathione and inorganic nitrite. Through the drug discovery program in natural product chemistry, Dr. Bryan discovered unique compositions of matter that can be used to safely and effectively generate and restore nitric oxide in humans. This technology is now validated in many published clinical trials. These discoveries and findings have unveiled many beneficial effects of nitrite in the treatment and prevention of human disease and may provide the basis for new preventive or therapeutic strategies in diseases associated with NO insufficiency and new guidelines for optimal health.

He is also a successful entrepreneur who has commercialized his nitric oxide technology through the formation of Human to the Power of N (formerly Neogenis Labs, Inc) where he is Co-founder and Chief Science Officer.

Dr. Bryan's most recent work has focused on the role of oral nitrate reducing bacteria in control of endogenous nitric oxide production and hypertension. He has published a number of highly cited papers and authored or edited six books.

Joseph Loscalzo, M.D., Ph.D. is Hersey Professor of the Theory and Practice of Medicine at Harvard Medical School, Chairman of the Department of Medicine, and Physician-in-Chief at Brigham and Women's Hospital. Dr. Loscalzo received his A.B. degree, *summa cum laude*, his Ph.D. in biochemistry, and his M.D. from the University of Pennsylvania. His clinical training was completed at Brigham and Women's Hospital and Harvard Medical School, where he served as Resident and Chief Resident in medicine and Fellow in cardiovascular medicine.

After completing his training, Dr. Loscalzo joined the Harvard faculty and staff at Brigham and Women's Hospital in 1984. He rose to the rank of Associate Professor of Medicine, Chief of Cardiology at the West Roxbury Veterans Administration Medical Center, and Director of the Center for Research in Thrombolysis at Brigham and Women's Hospital. He joined the faculty of Boston University in 1994, first as Chief of Cardiology and, in 1997, Wade Professor and Chair of Medicine, Professor of Biochemistry, and Director of the Whitaker Cardiovascular Institute. He returned to Harvard and Brigham and Women's Hospital in 2005.

Dr. Loscalzo is recognized as an outstanding cardiovascular scientist, clinician, and teacher. He has received many awards, including the Clinician-Scientist Award, the Distinguished Scientist Award, the Research Achievement Award, and the Paul Dudley White Award from the American Heart Association; a Research Career Development Award, a Specialized Center of Research in Ischemic Heart Disease Award, and a MERIT Award from the National Institutes of Health; the George W. Thorn Award for Excellence in Teaching at Brigham and Women's Hospital, the Educator of the Year Award in Clinical Medicine from Boston University, and the William Silen Lifetime Achievement in Mentorship Award from Harvard Medical School; the Glaxo Cardiovascular Research Award, and the Outstanding Investigator Prize from the International Society for Heart Research; election to fellowship in the American Association for the Advancement of Science, and in the American Academy of Arts and Sciences; and election to the American Society for Clinical Investigation, the Association of American Physicians, and the Institute of Medicine of the National Academy of Sciences (National Academy of Medicine). He has served on several NIH study sections and editorial boards and has chaired the Gordon Conference on Thrombolysis. He served as an associate editor of the *New England*

Journal of Medicine for 9 years, Editor-in-Chief of *Circulation* for 12 years, Chair of the Cardiovascular Board of the American Board of Internal Medicine, Chair of the Research Committee of the American Heart Association, Chair of the Scientific Board of the Stanley J. Sarnoff Society of Fellows for Research in the Cardiovascular Sciences, and Chair of the Board of Scientific Counselors of the National Heart, Lung, and Blood Institute of the National Institutes of Health. He is a former member of the Advisory Council of the National Heart, Lung, and Blood Institute and a former member of the Council of Councils of the National Institutes of Health. He is currently Director of the NIH-funded Center for Accelerated Innovation (the Boston Biomedical Innovation Center), the NIH-funded Harvard Undiagnosed Disease Network program, and a senior editor of **Harrison's Principles of Internal Medicine**.

Dr. Loscalzo has been a visiting professor at many institutions, holds two honorary degrees, has authored or coauthored more than 800 scientific publications, has authored or edited 41 books, and holds 31 patents for his work in the field of nitric oxide and redox biology. He is also the recipient of many grants from the NIH and industry for his work in the areas of vascular biology, thrombosis, atherosclerosis, and, more recently, systems biology over the past 30 years. He founded three companies, one of which became a publically traded company, and sits on the scientific advisory and corporate boards of several biotechnology and pharma companies. His most recent work has established the field of network medicine, a paradigm-changing discipline that seeks to redefine disease and therapeutics from an integrated perspective using systems biology and network science.

Contents

Contributors

Stephen J. Bailey, Ph.D. Department of Sport and Health Sciences, College of Life and Environmental Sciences, University of Exeter, Exeter, UK

Pamela D. Berens, M.D. Department of Obstetrics, Gynecology, and Reproductive Studies, McGovern Medical School, the University of Texas Health Science Center at Houston, Houston, TX, USA

Nigel Benjamin, M.D. Peninsula Medical School, Exeter, UK

Kenneth D. Bloch, M.D. Department of Anesthesia, Critical Care, and Pain Medicine, Anesthesia Center for Critical Care Research, Cardiovascular Research Center, Massachusetts General Hospital, Harvard Medical School, Boston, MA, USA

Nathan S. Bryan Department of Molecular and Human Genetics, Baylor College of Medicine, Molecular and Human Genetics, Houston, TX, USA

Melissa N. Conley, M.S. School of Biological and Population Health Sciences, Oregon State University, Corvallis, OR, USA

John P. Cooke, M.D., Ph.D. Department of Cardiovascular Sciences, Center for Cardiovascular Regeneration, Houston Methodist Hospital, Houston, TX, USA

Andreas Daiber, Ph.D. Center for Cardiology, University Medical Center Mainz, Mainz, Germany

Allison E. DeVan, Ph.D. Medical College of Wisconsin, Milwaukee, WI, USA

Yong-Jian Geng, M.D., Ph.D. The Center for Cardiovascular Biology and Atherosclerosis Research, University of Texas McGovern School of Medicine at Houston, Houston, TX, USA

Mark Gilchrist, M.D., Ph.D. Peninsula Medical School, Exeter, UK

Danielle A. Guimaraes, Ph.D. Vascular Medicine Institute, University of Pittsburg, Pittsburgh, PA, USA

Diane E. Handy, Ph.D. Cardiovascular Division, Department of Medicine, Brigham and Women's Hospital and Harvard Medical School, Boston, MA, USA

Norman G. Hord, Ph.D., M.P.H., R.D. School of Biological and Population Health Sciences, Oregon State University, Corvallis, OR, USA

Lawrence C. Johnson, M.S. Department of Integrative Physiology, University of Colorado Boulder, Boulder, CO, USA

Andrew M. Jones, Ph.D. Department of Sport and Health Sciences, University of Exeter, Exeter, UK

Jamie N. Justice, Ph.D. Department of Internal Medicine—Geriatrics, Wake Forest School of Medicine, Winston-Salem, NC, USA

Vikas Kapil, M.D. M.R.C.P. Ph.D. William Harvey Research Institute, Centre for Clinical Pharmacology, NIHR Biomedical Research Unit in Cardiovascular Disease at Barts, Queen Mary University of London, London, UK

Barts Heart Centre, Barts BP Centre of Excellence, St Bartholomew's Hospital, Barts Health NHS Trust, West Smithfield, London, UK

Jimmy T. Keeton, B.S., M.S., Ph.D. Department of Nutrition and Food Science, Texas A&M University, College Station, TX, USA

David M. Klurfield, Ph.D. National Program Leader for Human Nutrition, USDA Agricultural Research Service, Beltsville, MD, USA

Martin Lajous, M.D. Centro de Investigación en Salud Poblacional, Instituto Nacional de Salud Pública, Morelos, Mexico

Department of Global Health and Population, Harvard School of Public Health, Boston, MA, USA

Jack R. Lancaster Jr. , Ph.D. Departments of Pharmacology and Chemical Biology, Medicine, and Surgery, University of Pittsburg School of Medicine, Pittsburgh, PA, USA

David J. Lefer, Ph.D. Cardiovascular Center of Excellence and Department of Pharmacology, LSU Health Center Science-New Orleans, New Orleans, LA, USA

Department of Pharmacology and Experimental Therapeutics, Louisiana State University Health Sciences Center, New Orleans, LA, USA

Joseph Loscalzo Department of Medicine, Harvard Medical School, Boston, MA, USA

Department of Medicine, Brigham and Women's Hospital, Boston, MA, USA

Lisa Lohmeyer, M.D. Department of Medicine I: Hematology, Oncology, and Stem Cell Transplantation, University Medical Center Freiburg, Freiburg im Breisgau, Germany

Rajeev Malhotra, M.D., M.S. Cardiology Division, Department of Medicine, Massachusetts General Hospital, Boston, MA, USA

Andrew L. Milkowski, Ph.D. Department of Animal Sciences, Meat Science and Muscle Biology, University of Wisconsin—Madison, Madison, WI, USA

Thomas Munzel, M.D. Center for Cardiology, University Medical Center Mainz, Mainz, Germany

Chelsea L. Organ, B.S. Cardiovascular Center of Excellence and Department of Pharmacology, LSU Health Center Science-New Orleans, New Orleans, LA, USA

Department of Pharmacology and Experimental Therapeutics, Louisiana State University Health Sciences Center, New Orleans, LA, USA

Joseph F. Petrosino, Ph.D. Virology & Microbiology, Baylor College of Medicine, Houston, TX, USA

Tienush Rassaf, M.D., Ph.D. Department of Cardiology, West German Heart and Vessel Center, University Hospital Essen, University of Duisburg-Essen, Essen, Germany

Chris Reyes, B.S. Department of Bioengineering, University of Pittsburgh, Pittsburgh, PA, USA

Douglas R. Seals, Ph.D. Department of Integrative Medicine, University of Colorado Boulder, Boulder, CO, USA

Sruti Shiva, Ph.D. Department of Pharmacology and Chemical Biology, Center for Metabolism and Mitochondrial Medicine (C3M), Vascular Medicine Institute, University of Pittsburgh School of Medicine, Pittsburgh, PA, USA

Andrea U. Steinbicker, M.D., M.P.H. Department of Anesthesiology, Intensive Care and Pain Medicine, University Hospital Muenster, Muenster, Germany

Matthias Totzeck, M.D. Department of Cardiology, West German Heart and Vessel Center, University Hospital Essen, University of Duisburg-Essen, Essen, Germany

Anni Vanhatalo, Ph.D. Department of Sport and Health Sciences, College of Life and Environmental Sciences, University of Exeter, Exeter, UK

Walter C. Willett, M.D., Dr.P.H. Channing Division of Network Medicine, Department of Medicine, Brigham and Women's Hospital, Boston, MA, USA

Department of Nutrition, Harvard T.H. Chan School of Public Health, Boston, MA, USA

Department of Epidemiology, Harvard School of Public Health, Brigham and Women's Hospital and Harvard Medical School, Boston, MA, USA

Wing Tak Wong, Ph.D. Department of Cardiovascular Sciences, Houston Methodist Hospital, Houston, TX, USA

Part I
Biochemistry, Molecular Biology, Metabolism and Physiology

Chapter 1
Introduction

Nathan S. Bryan and Joseph Loscalzo

Key Points

- The reduction of inorganic nitrate to nitrite and nitric oxide is now recognized as a redundant pathway for maintenance of nitric oxide homeostasis in humans.
- Oral nitrate reducing bacteria are required for metabolic activation and reduction of nitrate.
- High nitrate foods, such as green leafy vegetables may exert their known cardioprotective and anticancer effects due to their ability to form nitric oxide.
- Strategies to enhance consumption of high nitrate foods while preventing nitrosative chemistry may be considered a first line of therapy for conditions associated with nitric oxide deficiency.

Keywords Food • Nutrition • Nitric oxide • Nitrite • Nitrate • Diet

Introduction

The discovery of the nitric oxide (NO) pathway in the 1980s represented a critical advance in the understanding of cell signaling and subsequently into major new developments in many clinical areas, including, but not limited to, cardiovascular medicine. This seminal finding was viewed as so fundamentally important that the Nobel Prize in Physiology or Medicine was awarded to its discoverers, Drs. Louis J. Ignarro, Robert Furchgott, and Ferid Murad in 1998, a short 11 years after NO was identified. The Swedish Nobel Assembly sagely noted, "The signal transmission by a gas that is produced by one cell, penetrates through membranes and regulates the function of another cell, represents an entirely new principle for signaling in biological systems." It was shocking to realize that NO, a colorless, odorless gas, was able to perform such important biochemical functions. Dr. Valentin Fuster, then president of the American Heart Association, noted in a 1998 interview that "the

N.S. Bryan, Ph.D.
Department of Molecular and Human Genetics, Baylor College of Medicine, Molecular and Human Genetics,
One Baylor Plaza, BCM225, Houston, TX 77030, USA

J. Loscalzo, M.D., Ph.D. (✉)
Department of Medicine, Harvard Medical School, Boston, MA, USA

Department of Medicine, Brigham and Women's Hospital,
75 Francis Street, TR 1, Rm 210, Boston, MA 02115, USA
e-mail: jloscalzo@parters.org

N.S. Bryan, J. Loscalzo (eds.), *Nitrite and Nitrate in Human Health and Disease*, Nutrition and Health,
DOI 10.1007/978-3-319-46189-2_1, © Springer International Publishing AG 2017

discovery of NO and its function is one of the most important in the history of cardiovascular medicine." More than a decade after the Nobel Prize was awarded for the discovery of NO and after more than 140,000 scientific papers have been published on it, we still do not have a firm grasp on its production and regulation or understand all of its biological functions.

Although one of the simplest biological molecules in nature, NO has found its way into nearly every phase of biology and medicine, ranging from its role as a critical endogenous regulator of blood flow and thrombosis to a principal neurotransmitter mediating penile erectile function, as well as a major pathophysiological mediator of inflammation and host defense. Continuous generation of NO is essential for the integrity of the cardiovascular system, and decreased production and/or bioavailability of NO is central to the development of many cardiovascular disorders [1, 2]. Research has since shown that NO is just as important in other organ systems. Nitric oxide is important for sensoral communication in the nervous system [3] and is a critical molecule used by the immune system to kill invading pathogens, including bacteria [4] and cancer cells. The production of NO from L-arginine is a complex and complicated biochemical process involving a 5-electron oxidation with many cofactors and prosthetic groups. As a result, there are many steps and/or factors that may be altered and affect ultimate NO production. Once produced, NO can be quickly scavenged before it has a chance to perform its actions. It is, therefore, a war of attrition when it comes to producing bioactive NO, and it is a remarkable feat that this short-lived gas is responsible for so many essential cellular activities.

Two Redundant Pathways for Endogenous NO Synthesis

The L-arginine-NOS pathway was the first NO production system discovered and is responsible for about 50 % of endogenous nitrate and nitrite in humans. The major metabolic pathway for NO is the stepwise oxidation to nitrite and nitrate. In plasma, NO is oxidized almost completely to nitrite or nitrate. In blood, this nitrate forms directly from the NO-dioxygenation reaction between NO and oxyhemoglobin: NO reacts with oxyhemoglobin to form nitrate and methemoglobin (Eq. 1.1) [5–7].

$$NO + Fe^{+2} - O_2 \rightarrow NO_3^- + Fe^{+3} \tag{1.1}$$

Nitrite can also form directly in blood and tissues via NO autooxidation, which involves the reaction of two molecules of NO with oxygen, a reaction that is catalyzed by the plasma protein ceruloplasmin [8]. Nitrate is by far the dominant final NO oxidation product; the levels in blood and tissues exceed those of nitrite by at least two orders of magnitude (μM vs. nM) [9]. The half-life of nitrite is about 110 s when added to venous blood. Nitrate, by contrast, has a circulating half-life of 5–8 h. During fasting conditions with a low prior intake of nitrite/nitrate, enzymatic NO formation accounts for the majority of steady-state blood levels of nitrite [10]. On the basis of these observations, it was believed that NO was acutely terminated by oxidation to nitrite and nitrate. Consistent with this view, for the past 20 years, we have used plasma nitrite as a biomarker to reflect acute changes in endogenous NO production. Nitrate, by contrast, is somewhat sensitive to chronic changes in endogenous NO production; however, dietary factors often confound the use of nitrate as a biomarker of NO activity.

In addition to the NOS system, our everyday diet represents an equally important source of nitrate and nitrite. Vegetables are the dominant source of nitrate in the diet (>80 %), and leafy green vegetables contain particularly high levels [11, 12]. The contribution of nitrate from endogenous versus exogenous sources varies greatly depending on diet, physical activity, disease, and medication. Inflammatory conditions with activation of inducible NOS (iNOS) [13] or physical exercise with activation of endothelial NOS (eNOS) [14] will increase endogenous levels of nitrate from NOS-generated NO and subsequent oxidation to nitrate. Conversely, dietary intake of nitrate can increase systemic nitrate levels dramatically. As an example, one serving of beetroot, lettuce, or spinach contains more nitrate than what is generated by the entire NOS system over a day [12, 15]. This nitrate can be recycled to regenerate NO.

Salivary Nitrate Concentration and Reduction

A new paradigm has emerged whereby nitrate and nitrite, instead of being inert oxidation products of NO, can be recycled to reform NO under certain conditions. In 1994, two groups independently presented evidence for the generation of NO in the stomach resulting from the acidic reduction of inorganic nitrite. Benjamin and colleagues [16] demonstrated that the antibacterial effects of acid alone were markedly enhanced by the addition of nitrite, which is present in saliva, whereas Lundberg and colleagues measured high levels of NO in expelled air from the stomach in humans [17]. These levels were abolished after pretreatment with a proton pump inhibitor and markedly increased after ingestion of nitrate, showing the importance of both gastric luminal pH and the conversion of nitrate to nitrite for NO generation in the stomach. These were the first reports of NO synthase-independent formation of NO in vivo. In the classical NO synthase pathway, NO is formed by oxidation of the guanidino nitrogen of L-arginine with molecular oxygen as the electron acceptor. This complex reaction is catalyzed by specific heme-containing enzymes, the NO synthases, and the reaction requires several cofactors. A very simplified schematic of the L-arginine-NO pathway is illustrated in Fig. 1.1. The alternative pathway is fundamentally different: instead of L-arginine, it uses the simple inorganic anions nitrate (NO_3^-) and nitrite (NO_2^-) as substrates in a stepwise reduction process that does not require NO synthase or cofactors.

The biochemical pathway and biological effects of nitrate reduction to nitrite and further to NO in the gastrointestinal tract are now better understood. Oral commensal bacteria are essential for the first step in the nitrate–nitrite–NO pathway since they are responsible for the enzymatic reduction of the

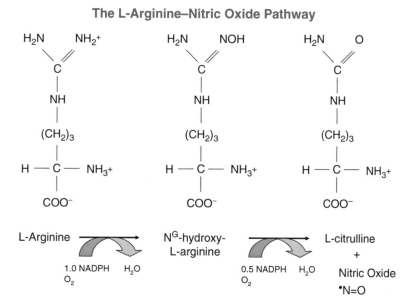

Fig. 1.1 The L-arginine–nitric oxide pathway

higher nitrogen oxide nitrate to nitrite [18]. It was known from the early literature that the salivary glands extract nitrate from plasma, but the reason for this active process was not explained. This process leads to levels of salivary nitrate that are 10–20-fold higher than in plasma. Circulating nitrate both from oxidation of NO and from the diet is actively taken up from blood by the salivary glands and concentrated 10–20-fold in saliva [19]. This concentration of nitrate can result in salivary nitrate levels of several millimolar [20]. Up to 25 % of all circulating nitrate enters this peculiar enterosalivary cycle, while a large portion is eventually excreted via the kidney. In the oral cavity, nitrate reductase enzymes in commensal facultative anaerobic bacteria reduce approximately 20 % of the nitrate to nitrite [19, 21]. These bacteria use nitrate as an alternative terminal electron acceptor to generate ATP in the absence of oxygen. As a result of bacterial enzymatic activity, the levels of nitrite in saliva also become very high. This highly effective bacterial nitrate reduction results in salivary levels of nitrite that are 1000-fold higher than those found in plasma. When nitrite-rich saliva meets the acidic gastric juice, nitrite is protonated to form nitrous acid (HNO_2), which then decomposes to NO and other nitrogen oxides. The presence of antioxidants can facilitate the reduction of nitrite to NO. Salivary-derived nitrite is metabolized to NO and other reactive nitrogen oxides locally in the acidic stomach; a number of physiological roles for gastric NO have been demonstrated [22–26]. Of particular relevance is the fact that much nitrite also survives gastric passage and enters the systemic circulation [27]. This is of great interest since there are now numerous mechanisms described in blood and tissues for nitrite reduction to NO and other bioactive nitrogen oxides [15, 28–30], which will be discussed in other chapters. This pathway becomes extremely important when you consider that nitrite and nitrate are part of our diet, which means that there is now a recognized pathway to affect endogenous NO production based on which foods we eat. The nitrate–nitrite–NO pathway is illustrated in Fig. 1.2. This emerging paradigm of the benefits of dietary sources of nitrite and nitrate may come as a surprise to many since, historically, nitrite and nitrate have been viewed as adversely affecting our food supply and drinking water.

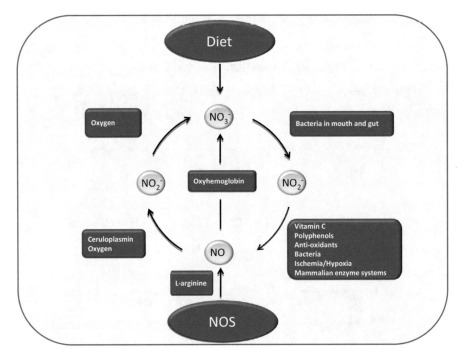

Fig. 1.2 The nitrite–nitrate–NO pathway

Salivary Nitrite Reduction to NO

The levels of NO generated from acidified nitrite in the stomach are in the range of 10–100 ppm, i.e., several orders of magnitude higher than those necessary for dilation of blood vessels [16, 17, 21]. In such high concentrations, NO and its reaction products are toxic to a variety of pathogenic bacteria and other microorganisms, which suggested a role for gastric NO in host defense. Enteropathogens exposed to different combinations of acid and nitrite reveal that *E. coli* and *Candida albicans* species were remarkably resistant when exposed to acid alone but were killed if nitrite was added [16]. Subsequent studies have showed that acidified nitrite inhibits the growth of a variety of enteropathogens, including *Salmonella*, *Shigella*, and *Helicobacter pylori* [31, 32]. More recently, Björne et al. tested the antibacterial effects of authentic human gastric juice and saliva, and, again, this combination was effective when salivary nitrite was high [26]. Beside NO, a variety of other reactive nitrogen intermediates (RNIs) are generated from nitrite under acidic conditions, and it is likely that several of these contribute to the antibacterial effects. Altogether, there are good indications for a role of salivary nitrite in primary host defense against swallowed pathogens. When nitrite is acidified, it yields nitrous acid (HNO$_2$, reaction 1.2), which spontaneously decomposes to NO and other nitrogen oxides (reaction 1.3 and 1.4):

$$NO_2^- + H^+ \leftrightarrow HNO_2 \left(pKa \ 3.2 - 3.4 \right) \tag{1.2}$$

$$2HNO_2 \rightarrow N_2O_3 + H_2O \tag{1.3}$$

$$N_2O_3 \rightarrow NO + NO_2 \tag{1.4}$$

The chemistry of acidified nitrite is very complex and the amount of NO generated from nitrite is dependent not only on pH and nitrite concentrations but also on the presence of other reducing agents (e.g., vitamin C, thiocyanate, polyphenols), proximity to heme groups, proteins, thiols, and oxygen tension [33]. With very high levels of NO generated within the lumen of the stomach, it is not unreasonable to assume that enough NO would reach the gastric mucosa to affect physiological processes. Nitrite-rich saliva, collected after oral ingestion of nitrate, causes an increase in blood flow and mucus production [22]. In addition, feeding rats nitrate in the drinking water for 1 week leads to a sustained increase in gastric mucosal blood flow and a thicker mucus layer [24]. These effects of nitrite are cGMP dependent and associated with NO gas formation. They are also independent of NO synthase and cyclooxygenase activity as evident from experiments using N$^\omega$-nitro-L-arginine methyl ester (L-NAME) and indomethacin [22]. Several studies have found a potent, protective effect of dietary nitrate in animal models of gastric ulcers [23, 25, 34, 35]. Jansson and colleagues found that daily nitrate treatment for 7-days protected rats against ulcers induced by a nonsteroidal anti-inflammatory drug (NSAID) [23]. Miyoshi found gastroprotective effects of dietary nitrate in a stress-induced model of gastric ulcers [25]. Interestingly, if they eliminated the oral microflora, this gastroprotection was lost. Sobko and colleagues showed that in contrast to normally housed rats, gastric NO levels are extremely low in germ-free rats and do not increase after ingestion of nitrate [36]. Together, this demonstrates the central role of bacteria in bioactivation of salivary nitrate to nitrite. The enterosalivary circulation of nitrate and its reduction to nitrite in the mouth by commensal bacteria seem to be a beautiful example of symbiosis. The bacteria receive from the host nitrate, a compound necessary for their respiration, and in return they provide us with nitrite, a substrate needed for the generation of gastroprotective NO. The systemic effects of salivary nitrite are perhaps more impressive therapeutically than the antimicrobial effects seen in the stomach.

Safely Harnessing the Therapeutic Effects of Nitrite and Nitrate

The early implications of dietary nitrite and nitrate causing methemoglobinemia and their propensity to form potentially carcinogenic N-nitrosamines have led to government efforts to quantify exposure rates in humans and establish regulations on the amounts in the food supply and drinking water. Since the early 1970s there have been both government and academic research efforts to understand the toxicology of nitrite, nitrate, and N-nitrosamines. That collective body of science is presented in several of the chapters with the hope that the reader can distill the burden of evidence on the risks and benefits of exposure to sources of nitrite and nitrate.

As we move forward with paradigm shifts and medical discoveries, the scientific community's main objective is to understand mechanisms of disease development to the extent needed to design rational therapies but with the ultimate goal of developing strategies for the *prevention* of human diseases. One could make a strong argument that diet should be a first target for disease prevention. Very little can affect our health more than what we choose to eat and our daily lifestyle habits. The recognition of a nitrate–nitrite–nitric oxide pathway suggests that NO can be modulated by the diet independent of its enzymatic synthesis from L-arginine, e.g., the consumption of nitrite- and nitrate-rich foods, such as fruits, green leafy vegetables, and some meats along with antioxidants. Antioxidants, such as vitamin C and polyphenols, can positively affect NO production from both pathways. First, they help protect the essential cofactors for the NOS pathway, such as tetrahydrobiopterin, from becoming oxidized and, therefore, promote L-arginine conversion to NO. Second, the presence of antioxidants protects NO from being scavenged once it is produced. Third, vitamin C and polyphenols can facilitate the reduction of nitrite to NO in the presence of an electron acceptor, thereby providing a recycling pathway. Lastly, vitamin C and polyphenols are very potent inhibitors of nitrosation reactions. Collectively antioxidants promote beneficial effects of nitrate and nitrite reduction to NO and prevent any unwanted nitrosation reactions.

Regular intake of nitrate- and nitrite-containing foods may ensure that blood and tissue levels of nitrite and NO pools are maintained at a level sufficient to compensate for any disturbances in endogenous NO synthesis from NOS. Since low levels of supplemental nitrite and nitrate have been shown to enhance blood flow, dietary sources of NO metabolites can, therefore, improve blood flow and oxygen delivery, and protect against various cardiovascular disease states or any condition associated with NO insufficiency. As science advances and new discoveries become apparent, it is important to be able to incorporate these new findings into meaningful guidelines that can enhance health, lengthen life, and reduce illness and disability. It takes approximately 17 years for basic scientific discoveries to become standard of care or be fully implemented into public health policy. As Paracelsus exclaimed, "dose makes the poison." There are clear and delineated doses of both nitrite and nitrate that provide indisputable evidence of promoting health and even treating serious medical conditions. Fortunately, these doses fall well below toxic and fatal doses. This provides a sufficient range for the normal dietary guidelines to be established. With such a beneficial toxic-therapeutic ratio, there is sufficient efficacy and safety to establish nutrient recommendations and dietary reference intakes (DRIs). Strong evidence for such has been reported [37]. This view has led to the current belief that a healthy diet focus not only on reducing fat and caloric intake, but on adding foodstuffs promoting nitric oxide bioactivity, which can include foods enriched in nitrite, nitrate, L-arginine, and antioxidants to promote NO production and availability. This pathway is also the current focus of a number of biotechnology and pharmaceutical companies in their attempts to develop NO- and nitrite-based therapeutics.

Conclusion

Many of the chronic conditions that lead to the highly prevalent burden of cardiovascular disorders, including diabetes and obesity, are the result of a dysfunctional endothelium and inability to produce NO and/or maintain NO homeostasis and signaling. Understanding and developing new strategies to restore NO homeostasis will have a profound impact on public health and on the healthcare system. Defining the context for the role of nitrite and nitrate in human health and disease is an essential first step in the process.

References

1. Vita JA, Keaney Jr JF. Endothelial function: a barometer for cardiovascular risk? Circulation. 2002;106(6):640–2.
2. Vita JA, et al. Coronary vasomotor response to acetylcholine relates to risk factors for coronary artery disease. Circulation. 1990;81(2):491–7.
3. Garthwaite J, Charles SL, Chess-Williams R. Endothelium-derived relaxing factor release on activation of NMDA receptors suggests role as intercellular messenger in the brain. Nature. 1988;336(6197):385–8.
4. Stuehr DJ, Marletta MA. Mammalian nitrate biosynthesis: mouse macrophages produce nitrite and nitrate in response to Escherichia coli lipopolysaccharide. Proc Natl Acad Sci U S A. 1985;82(22):7738–42.
5. Doherty DH, et al. Rate of reaction with nitric oxide determines the hypertensive effect of cell-free hemoglobin. Nat Biotechnol. 1998;16(7):672–6.
6. Dou Y, et al. Myoglobin as a model system for designing heme protein based blood substitutes. Biophys Chem. 2002;98(1–2):127–48.
7. Gladwin MT, et al. Relative role of heme nitrosylation and beta-cysteine 93 nitrosation in the transport and metabolism of nitric oxide by hemoglobin in the human circulation. Proc Natl Acad Sci U S A. 2000;97(18):9943–8.
8. Shiva S, et al. Ceruloplasmin is a NO oxidase and nitrite synthase that determines endocrine NO homeostasis. Nat Chem Biol. 2006;2(9):486–93.
9. Moncada S, Higgs A. The L-arginine nitric oxide pathway. N Engl J Med. 1993;329(27):2002–12.
10. Kleinbongard P, et al. Plasma nitrite concentrations reflect the degree of endothelial dysfunction in humans. Free Radic Biol Med. 2006;40(2):295–302.
11. Lundberg JO, Weitzberg E, Gladwin MT. The nitrate-nitrite-nitric oxide pathway in physiology and therapeutics. Nat Rev Drug Discov. 2008;7(2):156–67.
12. Hord NG, Tang Y, Bryan NS. Food sources of nitrates and nitrites: the physiologic context for potential health benefits. Am J Clin Nutr. 2009;90(1):1–10.
13. Crawford JH, et al. Transduction of NO-bioactivity by the red blood cell in sepsis: novel mechanisms of vasodilation during acute inflammatory disease. Blood. 2004;104(5):1375–82.
14. Jungersten L, et al. Both physical fitness and acute exercise regulate nitric oxide formation in healthy humans. J Appl Physiol. 1997;82(3):760–4.
15. Lundberg JO, et al. Nitrate and nitrite in biology, nutrition and therapeutics. Nat Chem Biol. 2009;5(12):865–9.
16. Benjamin N, et al. Stomach NO synthesis. Nature. 1994;368(6471):502.
17. Lundberg JO, et al. Intragastric nitric oxide production in humans: measurements in expelled air. Gut. 1994;35(11):1543–6.
18. Lundberg JO, et al. Nitrate, bacteria and human health. Nat Rev Microbiol. 2004;2(7):593–602.
19. Spiegelhalder B, Eisenbrand G, Preussman R. Influence of dietary nitrate on nitrite content of human saliva: possible relevance to in vivo formation of N-nitroso compounds. Food Cosmet Toxicol. 1976;14:545–8.
20. Govoni M, et al. The increase in plasma nitrite after a dietary nitrate load is markedly attenuated by an antibacterial mouthwash. Nitric Oxide. 2008;19(4):333–7.
21. Duncan C, et al. Chemical generation of nitric oxide in the mouth from the enterosalivary circulation of dietary nitrate [see comments]. Nat Med. 1995;1(6):546–51.
22. Björne HH, et al. Nitrite in saliva increases gastric mucosal blood flow and mucus thickness. J Clin Invest. 2004;113(1):106–14.
23. Jansson EA, et al. Protection from nonsteroidal anti-inflammatory drug (NSAID)-induced gastric ulcers by dietary nitrate. Free Radic Biol Med. 2007;42(4):510–8.
24. Petersson J, et al. Dietary nitrate increases gastric mucosal blood flow and mucosal defense. Am J Physiol Gastrointest Liver Physiol. 2007;292(3):G718–24.

25. Miyoshi M, et al. Dietary nitrate inhibits stress-induced gastric mucosal injury in the rat. Free Radic Res. 2003;37(1):85–90.
26. Björne HH, Weitzberg E, Lundberg JO. Intragastric generation of antimicrobial nitrogen oxides from saliva — physiological and therapeutic considerations. Free Radic Biol Med. 2006;41(9):1404–12.
27. Lundberg JO, Govoni M. Inorganic nitrate is a possible source for systemic generation of nitric oxide. Free Radic Biol Med. 2004;37(3):395–400.
28. Bryan NS. Nitrite in nitric oxide biology: cause or consequence? A systems-based review. Free Radic Biol Med. 2006;41(5):691–701.
29. van Faassen EE, et al. Nitrite as regulator of hypoxic signaling in mammalian physiology. Med Res Rev. 2009;29(5):683–741.
30. Lundberg JO, Weitzberg E. NO generation from nitrite and its role in vascular control. Arterioscler Thromb Vasc Biol. 2005;25(5):915–22.
31. Dykhuizen RS, et al. Antimicrobial effect of acidified nitrite on gut pathogens: importance of dietary nitrate in host defense. Antimicrob Agents Chemother. 1996;40(6):1422–5.
32. Duncan C, et al. Protection against oral and gastrointestinal diseases: importance of dietary nitrate intake, oral nitrate reduction and enterosalivary nitrate circulation. Comp Biochem Physiol A Physiol. 1997;118(4):939–48.
33. Weitzberg E, Lundberg JO. Nonenzymatic nitric oxide production in humans. Nitric Oxide. 1998;2(1):1–7.
34. Larauche M, et al. Protective effect of dietary nitrate on experimental gastritis in rats. Br J Nutr. 2003;89(6):777–86.
35. Larauche M, Bueno L, Fioramonti J. Effect of dietary nitric oxide on gastric mucosal mast cells in absence or presence of an experimental gastritis in rats. Life Sci. 2003;73(12):1505–16.
36. Sobko T, et al. Gastrointestinal nitric oxide generation in germ-free and conventional rats. Am J Physiol Gastrointest Liver Physiol. 2004;287(5):G993–7.
37. Bryan NS, Ivy JL. Inorganic nitrite and nitrate: evidence to support consideration as dietary nutrients. Nutr Res. 2015;35(8):643–54.

Chapter 2
From Atmospheric Nitrogen to Bioactive Nitrogen Oxides

Mark Gilchrist and Nigel Benjamin

Key Points

- Nitrogen is the most abundant element in the atmosphere.
- N_2 is chemically inert.
- Conversion to biologically active, "fixed," nitrogen requires considerable energy input.
- Nitrogen, inorganic nitrogen oxides, and organic nitrogen participate in complex biological cycle.
- The individual steps in the cycle may be used for respiration, anabolic processes, detoxification, or host defense.
- Plants and animals often have symbiotic relationships with specific micro-organisms which catalyze the parts of the cycle which they are unable to do.
- Mammals, including humans, have a complex nitrogen cycle which is, as yet, only partly understood.
- An understanding of the handling of nitrogen by micro-organisms may broaden our understanding of the role of nitrate, nitrite, and nitric oxide in health and disease.

Keywords Nitrogen fixation • Bacteria • Symbiosis • Nitrate reductase • Fertilizers • Denitrification

Introduction

Nitrogen atoms are a constituent part of a vast array of biologically important chemicals from the complex, such as proteins and nucleic acids, to the apparently more simple diatomic nitric oxide. Nitrogen (N_2) gas makes up approximately 78 % of the atmosphere. It is highly inert, being bound together by a triple bond that requires considerable energy to break (bond energy 940 kJ mol^{-1}); higher organisms lack the apparatus to do so. The conversion of nitrogen from inert N_2 to fixed, and, thus, biologically available nitrogen is essential to life on earth and requires dedicated multistep enzymatic pathways; vast amounts of energy in the case of lightning; or a combination of heat, pressure, and metallic catalysts in the Haber–Bosch process. The overview of the biological nitrogen cycle is illustrated in Fig. 2.1.

M. Gilchrist, M.D., Ph.D. • N. Benjamin, M.D. (✉)
Peninsula Medical School, St. Luke's Campus, Exeter, UK
e-mail: Nigel.Benjamin@phnt.swest.nhs.uk

N.S. Bryan, J. Loscalzo (eds.), *Nitrite and Nitrate in Human Health and Disease*, Nutrition and Health, DOI 10.1007/978-3-319-46189-2_2, © Springer International Publishing AG 2017

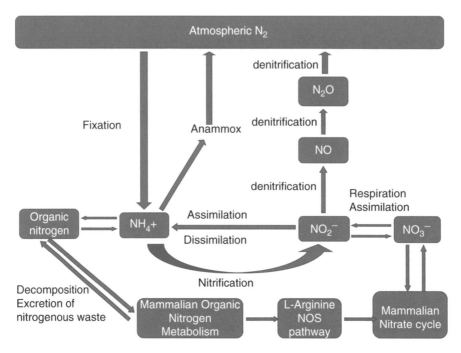

Fig. 2.1 Overview of the biological nitrogen cycle. For details see text

Theories exist that life on earth in its earliest and most primitive stages, when the gaseous constitution of the atmosphere was very different from that found today, synthesized NO to detoxify ozone or oxygen [1] or, indeed, developed NO reductases to remove NO from the cell. Whatever the reason, the chemical properties of NO, NO_2^-, and NO_3^- played a key role in the lives of the most primitive organisms. It is striking that as evolution progressed, higher organisms became utterly dependent on the nitrogen fixing and cycling abilities of their more primitive antecedents. Indeed, the ability to process nitrogen is no less important to familiar plants such as clover and their symbionts to extremophile bacteria.

The surge in knowledge and interest in the microbial/environmental nitrogen cycle is mirrored by the burgeoning interest in the mammalian nitrate cycle and the role of its multifarious reactive nitrogen oxide intermediates in health and disease. The two are, however, rarely considered together.

Nitrate, Agriculture, and Economics

The importance of nitrogen for plant growth was realized in the nineteenth century, first by Boussingault and later by Gilbert and Lawes' Rothamsted experiments [2]. Thus, readily available sources of fixed nitrogen were of considerable economic value. Guano, the nitrogen-rich desiccated droppings of sea birds, proved to be a very valuable export commodity for Peru in the early and mid-1800s. Indeed, such was the economic value of fixed nitrogen that Chile contested a war with Bolivia and Peru from 1879 to 1883 [3]. Bolivia increased taxation on the export of nitrate mined by Chilean companies. In retaliation, Chile seized the Bolivian port of Antofagasta. Bolivia declared war on Chile with Peru joining the Bolivian side in line with a secret treaty. Chile won the conflict that ensued; consequently, the nitrate-rich zones of Bolivia and Peru were ceded to Chile. The redrawn national boundaries also left Bolivia landlocked.

While these sources were economically important at the time, they did not represent a sustainable source of fixed nitrogen. Sustained plant growth requires the continued availability of fixed nitrogen, which can be derived from any of the following sources.

Nitrogen Fixation

Atmospheric nitrogen can be fixed via the following processes

1. Microbial dinitrogen fixation
2. Industrially by the Haber–Bosch process
3. Fossil fuel combustion
4. Lightning

Microbial Dinitrogen Fixations

The conversion of atmospheric nitrogen to a fixed, biologically available nitrogen species is the exclusive preserve of prokaryotes, eubacteria, and archea. Eukaryotes are not capable of this action. Collectively these organisms are known as diazotrophs. They fall into three distinct groups: free living, symbiotic (within plant roots), and those with a loose association with another organism, typically plant roots.

Nitrogen fixation can only occur in anaerobic conditions. Nitrogenase, the enzyme responsible for fixing atmospheric nitrogen, is exquisitely sensitive to oxygen. To maintain anaerobic conditions, a variety of approaches may be adopted: a very high respiration rate to maintain a low internal oxygen tension and/or binding of O_2 to leghemoglobin, a hemoglobin homolog, in the roots of leguminous plants [4].

The symbiotic relationship between rhizobia and leguminous plants is worthy of special consideration. The plant provides rhizobia, which are obligate anaerobes, with an anaerobic environment and sucrose in return for fixed nitrogen. The rhizobia infect plant roots of compatible species. A signaling cascade takes place between the plant and rhizobia resulting in changes in gene transcription in the plant driving root nodule formation and progressive infection of the nodule. Fixed nitrogen in the form of NH_3 is transferred from an infected root cell to an adjacent uninfected cell for incorporation into amides, typically glutamate or asparagine, or ureides, such as allantoin for transport to the upper part of the plant.

Haber–Bosch

Such was the economic importance of fixed nitrogen for agriculture that the development of an industrial process became a prime concern for the chemical industry. In 1910 Fritz Haber began to develop what would ultimately become known as the Haber–Bosch process. This process, still widely used today, involves reacting N_2 and H_2 gases at approximately 500 °C and a pressure of 300 atm over an iron catalyst to yield ammonia.

$$N_2 + 3H_2 \xrightarrow{Fe} 2NH_3$$

Some estimates suggest that 40 % of the plant-derived protein consumed by humans globally is derived from fertilizers utilizing nitrogen fixed by the Haber–Bosch process [5].

Fossil Fuel Combustion

The burning of fossil fuel for energy is responsible for less than 10 % of the fixed biologically available nitrogen deposited on the terrestrial surface [6]. The majority of the nitrogen made biologically available via this route is already fixed. Burning fossil fuels simply releases it from long-term sequestration in geological stores [7]. At sufficiently high temperatures, some de novo nitrogen fixation occurs.

Atmospheric Nitrogen Fixation by Lightning

While nitrogen fixation by lightning is a relatively minor player in the overall turnover of nitrogen within the global cycle, it is worth mentioning as it serves as a potent reminder of the magnitude of the energy investment required to fix nitrogen. The conversion of N_2 to NO and rapid oxidation to NO_2 (nitrogen dioxide) in the upper atmosphere by lightning is notoriously difficult to study and quantify. The primary interest in this process pertains to its importance to stratospheric and tropospheric chemistry and its impact on climate [8], as comparatively little of the nitrogen fixed in this way finds its way into biological systems.

Regardless of the specific mechanism, once fixed, the nitrogen will enter anyone of a variety of biological pathways.

Nitrate Assimilation in Plants and Bacteria

Most plants do not have a symbiont partner to provide them with fixed nitrogen. For those plants and bacteria unable to fix nitrogen independently, nitrate, found in soil, is the preferred source of fixed nitrogen. Nitrate is initially reduced to nitrite and ultimately ammonium before being acted upon by glutamate synthase and incorporated into carbon skeletons as described earlier for transport elsewhere in the plant. Certain plant species reduce nitrate rapidly on entry into the root system. Others transport unmetabolized nitrate to the upper part of the plant.

Nitrification

Decomposition of organic nitrogen-containing compounds such as proteins by a wide variety of bacteria and fungi produces ammonium. This ammonium is oxidized to nitrite and ultimately nitrate in reactions catalyzed by the chemolithoautotrophs nitrosomonas and nitrobacter which use the energy released to assimilate carbon from CO_2.

$$2NH_4^+ + 3O_2 \rightarrow 2NO_2^- + 4H^+ + 2H_2O + \text{energy}$$

$$2NO_2^- + O_2 \rightarrow 2NO_3^- + \text{energy}$$

It is worth noting that under certain circumstances, nitrifiers are capable of denitrification.

Denitrification

Fixed nitrogen is converted back to inert nitrogen gas by micro-organisms, which utilize this multi-step reductive process as an alternative respiratory pathway to oxygen-dependent respiration in anaerobic or near anaerobic conditions. Oxygen respiration is more energetically favorable and is, thus, preferred. It is only when oxygen tension falls to levels that limit the microbe's ability to respire that the enzymes of denitrification are expressed. Conversely, in conditions of rising oxygen tension, denitrification is inhibited.

Some organisms are only capable of catalyzing part of the denitrification pathway. This may represent a defense mechanism in some important human pathogens, such as *Neisseria gonorrhoeae* [9], which may use it as a device for surviving NO attack from the host.

Denitrification is a four-stage process with each step catalyzed by complex multisite metalloenzymes.

$$\underset{\text{Nitrate}}{NO_3^-} \rightarrow \underset{\text{Nitrite}}{NO_2^-} \rightarrow \underset{\text{Nitric oxide}}{NO} \rightarrow \underset{\text{Dinitrogen monoxide}}{N_2O} \rightarrow \underset{\text{Nitrogen}}{N_2}$$

Typically, nitrate reduction by membrane-bound nitrate reductase enzyme (NAR) occurs on the cytoplasmic side. Some organisms express a periplasmic nitrate reductase, NAP. Subsequent steps occur in the periplasm or on the periplasmic side of membrane-bound proteins. Thus, denitrifying bacteria have developed specialized nitrate and nitrite transport mechanisms [10]. The nitrite produced by this process is further reduced to nitric oxide by either a heme or copper containing nitrite reductase, NIR [11]. The NO generated must be rapidly reduced to N_2O as accumulation of NO would be fatal to the organism. Bacterial nitric oxide reductases (NOR) contain the typical components of the main catalytic subunit of heme/copper cytochrome oxidases [12]. The final step in the denitrification pathway is the reduction of N_2O to N_2 by another copper containing enzyme, N_2O reductase; while it represents the terminal step in a complete denitrification pathway, it also represents a separate and independent respiratory process [13].

It is worth noting that NO, N_2O, or N_2 may be released in the denitrification process. The stoichiometry of the gas mix will be affected by the relative activities of the pertinent reductases.

The denitrification process is usually thought to occur in aerobic soil or sediment environments. Recently, it has emerged that a complete denitrification pathway exists in human dental plaque [14], a finding that may have important implications in the mammalian nitrate cycle.

Dissimilatory Reduction of Nitrate to Ammonia

A third route for nitrate reduction exists that provides neither energy nor components for anabolic processes within the cell. Dissimilatory nitrate reduction appears to exist as a method for detoxifying excess nitrite following respiratory reduction of nitrate or balancing an excessive quantity of reductants [15]. It is exclusively an anaerobic process. Under certain conditions such as may be found during photoheterotrophic growth [16], in anaerobic marine sediments, and for some organisms in the human gastrointestinal tract [17] this may be the principal nitrate reduction pathway. A reductant-rich environment has a paucity of electron acceptors. Reduction of NO_2^- to NH_4^+ consumes six electrons compared with the two to three utilized in denitrification [18]. In these circumstances, dissimilatory nitrate reduction to ammonia is a key to maintaining the overall redox balance of the cell.

Anammox

The biological nitrogen cycle was widely held to be complete until the discovery of a pathway for anaerobic ammonium oxidation in 1990 [19]. When a wastewater plant reported higher than expected generation of dinitrogen gas, investigations revealed that certain bacteria were able to oxidize ammonium using nitrite as the electron acceptor to generate energy for growth in anoxic conditions. This process was thought to provide sufficient energy for slow growth only, with a bacterial doubling time of 11 days [20]. It has since emerged that anammox may provide for more rapid growth with a doubling time of 1.8 days, approaching that of ammonium oxidizers [21]:

$$NH_4^+ + NO_2^- \rightarrow N_2 + 2H_2O$$

As biologically available nitrogen is converted back to dinitrogen gas, anammox can be thought of as a form of denitrification. This entire complex pathway is illustrated in Fig. 2.1.

Nitrogen Balance in Mammals

In humans, nitrogen is principally ingested as protein. The average US young adult consumes approximately 90 g of protein daily [22]. This equates to about 14.5 g of nitrogen, nearly all of which is excreted in the urine as urea, creatinine, uric acid, and ammonia. Only a tiny proportion of this nitrogen is converted to nitric oxide via the nitric oxide synthase pathway. In healthy humans, about 1 mmol of nitrate is generated from L-arginine and then nitric oxide oxidation, which represents about one-thousandth of the amount of nitrogen ingested; during illness, such as gastroenteritis, this can increase by as much as eightfold [23].

Ruminants, and other mammals which rely on symbiotic bacteria to metabolize cellulose, can also make use of nonprotein nitrogen sources for protein synthesis. Rumen bacteria can convert urea to ammonia which is used to produce amino acids that can be incorporated into mammalian proteins [24].

L-Arginine Nitric Oxide Synthase Pathway

Three distinct NOS isoforms exist in mammals: inducible, neuronal, and endothelial. The loci of each gives an indication as to the pluripotent effects of nitric oxide in mammals, with roles in host defense, neuronal and other cellular signaling pathways, and vascular control. L-arginine provides organic nitrogen as a substrate which, with $2O_2$ and NADPH as a cofactor, is converted to nitric oxide and L-citrulline. The nitric oxide produced has a very short half-life, being rapidly oxidized in the presence of superoxide or oxyhemoglobin to nitrate, which enters the mammalian nitrate cycle.

Dietary Sources of Nitrate and Nitrite

Certain foods such as green leafy vegetables and beetroot are particularly rich in nitrate. Consumption of a typical Western diet results in the ingestion of approximately 1–2 mmol nitrate per day. Nitrite, because of its antibotulism effect, has been used as preservative and colorant for centuries. In addition, humans are also exposed to biologically active nitrogen oxides from the combustion of fossil fuels or inhalation of tobacco smoke.

Enterosalivary Circulation of Nitrate/Nitrite/NO

The remarkable symbiosis between legumes and nitrogen-fixing rhizobia finds its counterpoint in the relationship between nitrate-reducing bacteria hidden in crypts in mammalian tongues. Nitrate from the diet is rapidly and completely absorbed from the upper gastrointestinal tract. This, along with nitrate derived from the oxidation of NO synthesized by the L-arginine NOS pathway, is actively taken up the salivary glands. The resulting salivary nitrate concentration may be ten times greater than the plasma nitrate concentration. In crypts on the dorsum of the tongue, facultative anaerobes, e.g., *Veillonella* species, utilize nitrate as an alternative electron acceptor [25]. The nitrite released elevates salivary nitrite to levels 1000 times that of plasma in the resting state. In the presence of acid-generating plaque bacteria, some nitrite is chemically reduced to nitric oxide [26]. The remaining salivary nitrite is then swallowed. In the acidic environment of the stomach, some of this nitrite is further reduced to nitric oxide [27] which has an important role in both protection against enteric pathogens and regulation of gastric blood flow and mucous production [28, 29]. Some of the nitrite is absorbed from the stomach with important consequences for mammalian vascular physiology. This pathway is illustrated in Fig. 2.2.

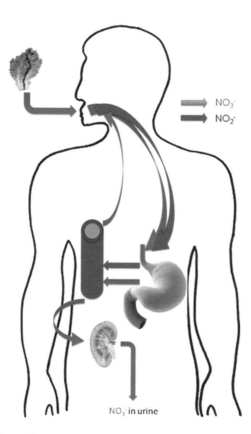

Fig. 2.2 Simplified representation of the enterosalivary circulation of nitrate. Nitrate (represented by the *black arrows*) derived from the diet is swallowed. It is rapidly and completely absorbed in the upper gastrointestinal tract. Approximately 25 % is concentrated in the salivary glands and secreted into the mouth. Here it is reduced to nitrite (represented by the *gray arrows*) by facultative anaerobes on the dorsum of the tongue and swallowed. Some of the nitrite undergoes acidic reduction to nitric oxide in the stomach, with the remainder being absorbed. The fate of nitrite is discussed in depth later. Sixty percent of ingested nitrate is lost in the urine within 48 h

In 1996, it was discovered that nitric oxide is continually released from the surface of normal human skin [30]. Although it was initially thought that nitric oxide synthase would be responsible, inhibition of this enzyme by infusing monomethyl L-arginine into the brachial artery showed that this was not the case. Further studies showed that nitrate is excreted in human sweat and reduced to nitrite by skin bacteria. As normal skin is slightly acidic (pH around 5.5) this nitrite is reduced to nitric oxide. The function of this NO is thought to be to inhibit skin pathogens—particularly fungi [31]—and, intriguingly, when normal saliva is applied to healthy skin, the high concentrations of nitrite considerably increase nitric oxide synthesis, perhaps to protect against infection and encourage wound healing [32].

Breast milk has recently been shown to contain variable amounts of nitrite and nitrate, and it has been suggested that conversion of these anions to more reactive nitrogen oxides may be a factor underlying the protective effect of breastfeeding against infant gastroenteritis [33].

Urinary Nitrate Excretion

Approximately 60 % of the nitrate ingested or endogenously synthesized will be lost in the urine within 48 h [34]. The discovery of complete denitrification pathways in human dental plaque flora [14] may offer a clue as to the fate of at least part of the remainder. Nitrate is freely filtered at the glomerulus. Studies in dogs suggest that as much as 90 % may be reabsorbed by the renal tubules [35]. Nitrite is not found in human urine under normal physiological conditions. Its presence indicates infection with nitrate-reducing organisms. Detection of urinary nitrite is in widespread use as a simple bedside test for diagnosing urinary tract infection.

Conclusion

When considering the biological nitrogen cycle as part of the global nitrogen cycle, mammalian nitrogen cycling is often considered separately from the processes occurring in plants and micro-organisms. Scientific interest in both these spheres has undergone a resurgence over the last 20–30 years. The complexity of the processes and relationships between animals, plants, and micro-organisms is still being unraveled.

References

1. Feelisch M, Martin JF. The early role of nitric oxide in evolution. Trends Ecol Evol. 1995;10(12):496–9.
2. Addiscott T. Nitrate, agriculture and the environment. Oxford: CABI; 2005.
3. Kiernan VG. Foreign interests in the War of the Pacific. Hisp Am Hist Rev. 1955;35(1):14–36.
4. Harutyunyan EH et al. The structure of deoxy- and oxy-leghaemoglobin from Lupin. J Mol Biol. 1995;251(1):104–15.
5. Smil V. Enriching the earth: Fritz Haber, Carl Bosch, and the transformation of world food production. Cambridge: MIT; 2004.
6. Schlesinger WH. On the fate of anthropogenic nitrogen. Proc Natl Acad Sci. 2009;106(1):203–8.
7. Vitousek PM, Aber JD, Howarth RW, Likens GE, Matson PA, Schindler DW, et al. Human alteration of the global nitrogen cycle: sources and consequences. Ecol Appl. 1997;7(3):737–50.
8. Schumann U, Huntrieser H. The global lightning-induced nitrogen oxides source. Atmos Chem Phys. 2007;7(14):3823–907.
9. Barth KR, Isabella VM, Clark VL. Biochemical and genomic analysis of the denitrification pathway within the genus *Neisseria*. Microbiology. 2009;155(12):4093–103.

10. Moir JW, Wood NJ. Nitrate and nitrite transport in bacteria. Cell Mol Life Sci. 2001;58(2):215–24.
11. Cutruzzolà F. Bacterial nitric oxide synthesis. Biochim Biophys Acta. 1999;1411(2–3):231–49.
12. Hendriks J, Oubrie A, Castresana J, Urbani A, Gemeinhardt S, Saraste M. Nitric oxide reductases in bacteria. Biochim Biophys Acta. 2000;1459(23):266–73.
13. Zumft W. Cell biology and molecular basis of denitrification. Microbiol Mol Biol Rev. 1997;61(4):533–616.
14. Schreiber F, Stief P, Gieseke A, Heisterkamp I, Verstraete W, de Beer D, et al. Denitrification in human dental plaque. BMC Biol. 2010;8(1):24.
15. Moreno-Vivián C, Ferguson SJ. Definition and distinction between assimilatory, dissimilatory and respiratory pathways. Mol Microbiol. 1998;29(2):664–6.
16. Berks BC, Ferguson SJ, Moir JWB, Richardson DJ. Enzymes and associated electron transport systems that catalyse the respiratory reduction of nitrogen oxides and oxyanions. Biochim Biophys Acta. 1995;1232(3):97–173.
17. Parham NJ, Gibson GR. Microbes involved in dissimilatory nitrate reduction in the human large intestine. FEMS Microbiol Ecol. 2000;31(1):21–8.
18. Bothe H, Ferguson SJ, Newton WE. Biology of the nitrogen cycle. 1st ed. Amsterdam: Elsevier; 2007.
19. Jetten MSM, Strous M, Pas-Schoonen KT, Schalk J, Dongen UGJM, Graaf AA, et al. The anaerobic oxidation of ammonium. FEMS Microbiol Rev. 1998;22(5):421–37.
20. Strous M, Heijnen JJ, Kuenen JG, Jetten MSM. The sequencing batch reactor as a powerful tool for the study of slowly growing anaerobic ammonium-oxidizing microorganisms. Appl Microbiol Biotechnol. 1998;50(5):589–96.
21. Isaka K, Date Y, Sumino T, Yoshie S, Tsuneda S. Growth characteristic of anaerobic ammonium-oxidizing bacteria in an anaerobic biological filtrated reactor. Appl Microbiol Biotechnol. 2006;70(1):47–52.
22. Fulgoni III VL. Current protein intake in America: analysis of the National Health and Nutrition Examination Survey, 2003–2004. Am J Clin Nutr. 2008;87(5):1554S–7.
23. Forte P, Dykhuizen RS, Milne E, McKenzie A, Smith CC, Benjamin N. Nitric oxide synthesis in patients with infective gastroenteritis. Gut. 1999;45(3):355–61.
24. Huntington GB, Archibeque SL. Practical aspects of urea and ammonia metabolism in ruminants. J Anim Sci. 2000;77(E-Suppl):1–11.
25. Doel JJ, Benjamin N, Hector MP, Rogers M, Allaker RP. Evaluation of bacterial nitrate reduction in the human oral cavity. Eur J Oral Sci. 2005;113(1):14–9.
26. Duncan C, Dougall H, Johnston P, Green S, Brogan R, Leifert C, et al. Chemical generation of nitric oxide in the mouth from the enterosalivary circulation of dietary nitrate [see comment]. Nat Med. 1995;1(6):546–51.
27. Benjamin N, O'Driscoll F, Dougall H, Duncan C, Smith L, Golden M, et al. Stomach NO synthesis [see comment]. Nature. 1994;368(6471):502.
28. Wallace JL, Miller MJ. Nitric oxide in mucosal defense: a little goes a long way. Gastroenterology. 2000;119(2):512–20.
29. Dykhuizen RS, Frazer R, Duncan C, Smith CC, Golden M, Benjamin N, et al. Antimicrobial effect of acidified nitrite on gut pathogens: importance of dietary nitrate in host defense. Antimicrob Agents Chemother. 1996;40(6):1422–5.
30. Weller R, Pattullo S, Smith L, Golden M, Ormerod A, Benjamin N. Nitric oxide is generated on the skin surface by reduction of sweat nitrate. J Invest Dermatol. 1996;107(3):327–31.
31. Weller R, Price RJ, Ormerod AD, Benjamin N, Leifert C. Antimicrobial effect of acidified nitrite on dermatophyte fungi, Candida and bacterial skin pathogens. J Appl Microbiol. 2001;90(4):648–52.
32. Benjamin N, Pattullo S, Weller R, Smith L, Ormerod A. Wound licking and nitric oxide [see comment]. Lancet. 1997;349(9067):1776.
33. Hord NG, Ghannam JS, Garg HK, Berens PD, Bryan NS. Nitrate and nitrite content of human, formula, bovine and soy milks: implications for dietary nitrite and nitrate recommendations. Breastfeed Med. 2011;6(6):393–9.
34. Wagner DA, Schultz DS, Deen WM, Young VR, Tannenbaum SR. Metabolic fate of an oral dose of 15N-labeled nitrate in humans: effect of diet supplementation with ascorbic acid. Cancer Res. 1983;43(4):1921–5.
35. Godfrey M, Majid DS. Renal handling of circulating nitrates in anesthetized dogs. Am J Physiol. 1998;275(1 Pt 2):F68–73.

Chapter 3
Nitrate-Reducing Oral Bacteria: Linking Oral and Systemic Health

Nathan S. Bryan and Joseph F. Petrosino

Key Points

- Inorganic nitrate from the diet can only be reduced by bacteria in and on the human body since humans lack a nitrate reductase gene.
- Oral facultative anaerobic bacteria can perform the two-electron reduction of nitrate to nitrite.
- Humans lacking these specific bacteria become nitric oxide deficient and appear to suffer the clinical consequences of NO deficiency including increased risk of cardiovascular disease.
- Overuse of antibiotics and antiseptic mouthwash eradicates these essential bacteria and therefore may cause undue risk in patients.
- Understanding the identity and how to repopulate these communities may allow for novel therapeutic strategies to restore NO homeostasis in humans.

Keywords Denitrifying bacteria • Dietary nitrate • Nitrite • Periodontitis • Hypertension • Stomach acid

Introduction

The human microbiome is composed of many different bacterial species, which outnumber our human cells ten to one and provide functions that are essential for our survival. The human gastrointestinal tract represents a major habitat for bacterial colonization and has been the focus of much of the human microbiome project. The microbiota of the lower intestinal tract is widely recognized as playing a symbiotic role in maintaining a healthy host physiology [1] by participating in nutrient acquisition and bile acid recycling, among other activities. In contrast, although the role of oral microbiota in disease is well studied, specific contributions to host health are not well defined. A human nitrogen cycle has been identified. This pathway, termed the entero-salivary nitrate–nitrite–nitric oxide

N.S. Bryan, Ph.D. (✉)
Department of Molecular and Human Genetics, Baylor College of Medicine,
One Baylor Plaza, BCM225, Houston, TX 77030, USA
e-mail: Nathan.Bryan@bmc.edu

J.F. Petrosino, Ph.D.
Virolology & Microbiology, Baylor College of Medicine,
One Baylor Plaza, BCM225, Houston, TX 77030, USA

N.S. Bryan, J. Loscalzo (eds.), *Nitrite and Nitrate in Human Health and Disease*, Nutrition and Health,
DOI 10.1007/978-3-319-46189-2_3, © Springer International Publishing AG 2017

pathway, can positively affect nitric oxide (NO) homeostasis and represents a potential symbiotic relationship between oral bacteria and their human hosts [**2**, **3**]. It is now recognized that the oral commensal bacteria provide an important metabolic function in human physiology by contributing a nitric oxide synthase (NOS)-independent source of NO. NO is an essential signaling molecule for diverse physiological and disease-prevention functions. This process is analogous to the environmental nitrogen cycle whereby soil bacteria convert atmospheric nitrogen oxides to usable forms for plant growth. Human nitrate reduction requires the presence of nitrate-reducing bacteria as mammalian cells cannot effectively reduce this anion. The discovery of the NO pathway in the **1980**s represented a critical advance in understanding cardiovascular disease, and today a number of human diseases are characterized by NO insufficiency. There is sufficient and convincing evidence in the literature that these bacterial communities provide the host a source of NO that may be able to overcome insufficient NO production from the endothelium [**2–5**]. The focus of this chapter is to reveal the biochemistry and molecular biology of oral bacterial nitrate reduction thereby providing the host a rescue pathway for conditions of NO insufficiency and affecting diseases associated with NO insufficiency.

Nitrate–Nitrite–Nitric Oxide Pathway

Inorganic nitrite (NO_2^-) and nitrate (NO_3^-) were known predominantly as undesired residues in the food chain or as inert oxidative end products of endogenous NO metabolism. However, from research performed over the past decade, it is now apparent that nitrate and nitrite are physiologically recycled in blood and tissue to form NO and other bioactive nitrogen oxides [6–9]. As a result, they should now be viewed as storage pools for NO-like bioactivity to be acted upon when enzymatic NO production from NOS is insufficient, such as during anaerobic conditions or uncoupling of NOS. Nitrite is an oxidative breakdown product of NO that has been shown to serve as an acute marker of NO flux/formation [10]. Nitrite has recently moved to the forefront of NO biology, as it represents a major storage form of NO in blood and tissue [11]. Nitrite is in steady-state equilibrium with *S*-nitrosothiols [6, 12] and has been shown to activate soluble guanylyl cyclase (sGC) and increase cGMP levels in tissues [6]. Therefore, it is an ideal candidate for restoring both cGMP-dependent and cGMP-independent NO signaling. In addition to the oxidation of NO, nitrite is also derived from reduction of salivary nitrate by commensal bacteria in the mouth and gastrointestinal tract [13, 14], as well as from dietary sources, such as meat, vegetables, and drinking water. Understanding nitrite and NO production pathways in the human body will allow for novel therapeutic strategies for conditions of NO insufficiency.

The bioactivation of nitrate from dietary (mainly green leafy vegetables) or endogenous sources requires its initial reduction to nitrite, and because mammals lack specific and effective nitrate reductase enzymes, this conversion is mainly carried out by commensal bacteria [15]. Dietary nitrate is rapidly absorbed in the upper gastrointestinal tract. In the blood, it mixes with the nitrate formed from the oxidation of endogenous NO produced from the NOS enzymes. After a meal rich in nitrate, the nitrate levels in plasma increase greatly and remain high for a prolonged period of time (the plasma half-life of nitrate is 5–6 h). The nitrite levels in plasma also increase after nitrate ingestion after approximately 90 min [16]. Although much of the nitrate is eventually excreted in the urine, *up to 25 % is actively taken up by the salivary glands and is concentrated up to 20-fold in saliva* [16, 17]. In the mouth, commensal facultative anaerobic bacteria reduce salivary nitrate to nitrite during anaerobic respiration by the action of nitrate reductases [3, 15]. After a dietary nitrate load, the salivary nitrate and nitrite levels can approach 10 and 1–2 mM, respectively [16]. When saliva enters the acidic stomach (1–1.5 L per day), much of the nitrite is rapidly protonated to form nitrous acid (HNO_2; pKa ~3.3), which decomposes further to form NO and other nitrogen oxides [8, 9]. A simplified human nitrogen cycle is illustrated in Fig. 3.1. More recent studies indicate that nitrite does not have to be protonated to be absorbed and is about 98 % bioavailable when swallowed in an aqueous solution [18].

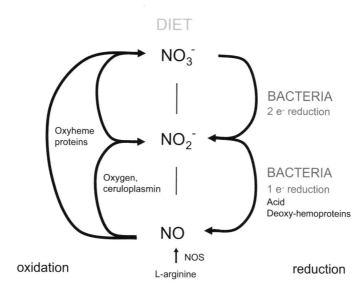

Fig. 3.1 The human nitrogen cycle whereby nitrate is serially reduced to nitrite and NO providing the host with a source of bioactive NO

Nitrite-Based Signaling

The production of nitrite from nitrate-reducing bacteria may have profound implications on the health of the human host. Numerous studies have shown that nitrite produced from bacterial nitrate reduction is an important storage pool for NO in blood and tissues when NOS-mediated NO production is insufficient [19–23]. In various animal models and in humans, dietary nitrate supplementation has shown numerous beneficial effects, including a reduction in blood pressure, protection against ischemia–reperfusion damage, restoration of NO homeostasis with associated cardioprotection, increased vascular regeneration after chronic ischemia, and a reversal of vascular dysfunction in the elderly [24, 25]. Some of these benefits were reduced or completely prevented when the oral microbiota were abolished with an antiseptic mouthwash [24, 26]. Additionally, it was recently shown that in the absence of any dietary modifications, a 7-day period of antiseptic mouthwash treatment to disrupt the oral microbiota reduced both oral and plasma nitrite levels in healthy human volunteers and was associated with a sustained increase in both systolic and diastolic blood pressure [27]. Altogether, these studies firmly establish the role for oral nitrate-reducing bacteria in making a physiologically relevant contribution to host nitrite and thus NO levels, with measureable physiological effects.

Nitrite is now recognized as a signaling molecule that may act independent of NO [6]. Much of the recent focus on nitrite physiology is due to its ability to be reduced to NO during ischemic or hypoxic events [7, 11, 28]. Since nitrate reduction by microbial communities generates nitrite, it is of great importance to be able to recognize specific bacteria in and on the body that are capable of generating nitrite from nitrate. Once nitrite is formed, it can be utilized as a substrate for NO production. Although largely inefficient, there exists a number of nitrite-reducing system in mammals. Nitrite reductase activity in mammalian tissues has been linked to the mitochondrial electron transport system [29, 30], protonation [7], deoxyhemoglobin [31], and xanthine oxidase [25, 32]. Nitrite can also transiently form S-nitrosothiols (RSNOs) under both normoxic and hypoxic conditions [28], and a recent study by Bryan et al. demonstrates that steady-state concentrations of tissue nitrite and S-nitroso species are affected by changes in dietary NOx (nitrite and nitrate) intake [6]. Furthermore, enriching dietary intake of nitrite and nitrate translates into significantly less injury from myocardial infarction [19].

Previous studies demonstrated that nitrite therapy given intravenously prior to reperfusion protects against hepatic and myocardial ischemia/reperfusion (I/R) injury [33]. Additionally, experiments in primates revealed a beneficial effect of long-term application of nitrite on cerebral vasospasm [34]. Moreover, inhalation of nitrite selectively dilates the pulmonary circulation under hypoxic conditions in vivo in sheep [35]. Topical application of nitrite improves skin infections and ulcerations [36]. Furthermore, in the stomach, nitrite-derived NO seems to play an important role in host defense [15] and in regulation of gastric mucosal integrity [37]. All of these studies, along with the observation that nitrite can act as a marker of NOS activity [10], have opened new avenues for the diagnostic and therapeutic applications of nitrite, especially in cardiovascular diseases, using nitrite as a marker as well as an active agent. Oral nitrite has also been shown to reverse L-NAME (NOS inhibitor)-induced hypertension and serve as an alternate source of NO in vivo [38]. In fact, a report by Kleinbongard et al. [39] demonstrates that plasma nitrite levels progressively decrease with increasing cardiovascular risk. Since a substantial portion of steady-state nitrite concentrations in blood and tissue are derived from dietary sources [6], modulation of nitrate intake along with optimal nitrate-reducing microbial communities may provide a first line of defense for conditions associated with NO insufficiency [11]. To support this approach, it has been reported that dietary nitrate reduces blood pressure in healthy volunteers [21, 40], and that the effects are abolished after rinsing with oral antiseptic mouthwash [41]. Therefore, nitrite production through nitrate-reducing bacteria may restore NO homeostasis resulting from endothelial dysfunction and provide benefits to individuals afflicted by a number of diseases that are characterized by NO insufficiency. As a result, this new field has the potential to provide the basis for new preventive or therapeutic strategies and new dietary guidelines for optimal health. From a public health perspective, we may be able to make better recommendations on diet and relevant host–microbe communities to affect dramatically the incidence and severity of a number of symptoms and diseases characterized by NO insufficiency. Development of novel therapeutics or strategies to restore NO homeostasis can have a profound impact on disease prevention [42]. These new discoveries in the oral bacterial microbiome suggest that an effective strategy to promote therapeutically NO production and overcome conditions of NO insufficiency may not solely be to target NOS, but, rather, to focus on understanding specific oral bacterial communities and the optimal conditions for efficient oral nitrate reduction. It appears from early studies that many individuals may not have the optimal microbial communities for maximum nitrate reduction, causing a disruption in a critical NO production pathway. Understanding and harnessing this alternative pathway may prove to be a viable and cost-effective strategy for maintaining NO homeostasis in humans. Because NO signaling affects all organ systems and almost all disease processes described to date, this novel approach to NO regulation has the potential to affect the study and treatment of many diseases across all organ systems.

Bacterial Nitrate Reduction

Salivary nitrate is metabolized to nitrite via a two-electron reduction, a reaction that mammalian cells are unable to perform, during anaerobic respiration by nitrate reductases produced by facultative and obligate anaerobic commensal oral bacteria [3, 15]. Manipulation of the human microbiome as a therapeutic target for disease management is on the near horizon. The oral cavity is an attractive target for probiotic and/or prebiotic therapy because of the ease of access. The potential to restore the oral flora as a means to provide NO production is a completely new paradigm for NO biochemistry and physiology as well as for cardiovascular medicine and dentistry.

As illustrated in Fig. 3.2, identification of specific partial denitrifying bacteria in the oral cavity can provide optimal conditions for nitrate reduction with nitrite accumulation in the saliva. The potential phenotypic correlations of select bacteria within the oral flora will lead to what we anticipate will be

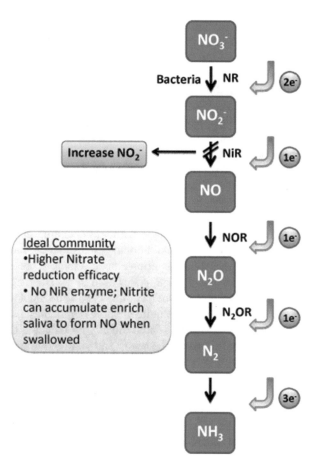

Fig. 3.2 The biological nitrogen cycle. Nitrate is reduced all the way down to elemental nitrogen by a series of enzymatic steps from nitrate reductase (NR), nitrite reductase [43], NO reductase (NOR), nitrous oxide reductase (N_2OR). Identifying bacteria or communities that only reduce nitrate and/or nitrite will allow for nitrite accumulation and more efficient NO production

transformative studies in the broader field of NO regulation in human disease. Hence, understanding the specific conditions of optimal nitrate reduction and the community ecology that best serves the host by providing a source of bioactive nitrite and NO may open up new therapeutic strategies for restoring NO. Although a few nitrate-reducing bacteria have been identified in the oral cavity [44], we have analyzed nitrate reduction by bacterial communities present in tongue scrapings from healthy human volunteers during 4 days of in vitro growth and performed a parallel metagenomic analysis of these samples to identify specific bacteria associated with nitrate reduction. Through 16S rRNA gene pyrosequencing and whole genome shotgun (WGS) sequencing and analysis, we identified specific taxa that likely contribute to nitrate reduction. Biochemical characterization of nitrate and nitrite reduction by four candidate species indicates that complex community interactions contribute to nitrate reduction [45]. This model assessed nitrate reduction over 4 days in biofilm culture in vitro. The bacterial communities had varying potential for nitrate reduction, and this study identified 14 candidate species that were present in communities with the best nitrate reduction activity. The 14 species present at an abundance of at least 0.1 % in the best nitrate-reducing sample and at the highest abundance in this sample compared to the intermediate and worst reducing sample that belonged to the genera of interest and were identified through 16S rRNA gene pyrosequencing and analysis were as follows: *Granulicatella adiacens, Haemophilus parainfluenzae, Actinomyces odontolyticus, Actinomyces viscosus, Actinomyces oris, Neisseria flavescens, Neisseria mucosa, Neisseria sicca, Neisseria subflava, Prevotella melaninogenica, Prevotella salivae, Veillonella dispar, Veillonella parvula, and Veillonella atypica.* Additionally, *Fusobacterium nucleatum* and *Brevibacillus brevis* were designated as species of interest even though they were not at a relative abundance of at least 0.1 % in the WGS best nitrate-reducing sample [45]. Previously the Doel et al., 2005, study isolated and identified five genera of oral nitrate-reducing bacterial taxa on the tongues of healthy individuals: *Veillonella, Actinomyces, Rothia, Staphylococcus,* and *Propionibacterium* [44]. In our investigation, *Veillonella* species were the most abundant group of nitrate reducers isolated from the tongue, followed by *Actinomyces* spp. *Veillonella* was the most abundant nitrate-reducing genus detected in the original tongue scrapings, although *Prevotella, Neisseria,* and *Haemophilus* were all found at a higher abundance than *Actinomyces,* highlighting the higher resolution of our study. This difference in resolution is likely due to our use of a sequencing-based approach, which allowed us to survey the native bacterial environment on the dorsal surface of the tongue without depending on the growth requirement necessary for classic culture-based techniques. Additionally, in our study, the original samples were subsequently grown under anaerobic biofilm conditions that are more representative of the environment in the deep crypts of the tongue than the agar plates used in the Doel, et al. 2005, study. A final key difference between our study and that of Doel, et al. is that while Doel, et al., were specifically targeting active nitrate reducers, we sought to acquire a whole community picture, targeting all species contributing to community nitrate reduction, both directly by reducing nitrate and indirectly by acting as helper species or otherwise increasing the health and vitality of the community. Conversely, we identify bacteria that may further reduce nitrite generated from nitrate reduction by species that contain nitrite reductase enzymes. The presence of these bacteria may not allow for sufficient nitrite accumulation in the saliva, thereby suppressing the nitrate–nitrite–nitric oxide pathway. Metagenomic data are important and informative in that one can determine what a bacterial community is *capable* of doing, yet it is limited in the sense that it cannot inform what the community is *actually* doing, which can vary under different circumstances. Indeed, we observed relatively uniform bacterial communities in terms of pathway abundance across the three samples that underwent WGS analysis; however, our in vitro data clearly demonstrated that nitrate reduction varied widely between these samples. Thus, while each community had the same capacity for nitrate reduction, the true activity clearly differed between these communities. Our study, therefore, highlights that a full understanding of the entero-salivary nitrate–nitrite–NO pathway will require the generation and integration of a complete set of data from metagenomic, metatranscriptomic, metaproteomic, and metametabolomic studies coupled to biochemical functional assays.

Another interesting finding was not only identification of nitrate-reducing bacteria, but also the presence of bacterial species in these communities not genetically capable of nitrate reduction and not previously implicated in oral nitrate reduction. This poses several intriguing questions: [43] do these bacteria act as helper species, enabling nitrate reducers to reduce nitrate more efficiently? [1] do these bacteria "ride along" with nitrate-reducing bacteria, feeding off a metabolic by-product produced by nitrate reducers? and [2] do these species contribute to the health and structure of the biofilm by acting as "scaffold species" to form and maintain the biofilm? This implies that in addition to bacteria that reduce nitrate, other "helper" species may be present to maximize the metabolic efficiency of the community. Furthermore, as not all healthy donors had nitrate-reducing bacteria in their oral cavity, there may be a significant number of individuals who may not have a functioning nitrate–nitrite–NO enterosalivary pathway to support systemic health. The presence or absence of these select bacteria may be a new determinant of nitrite and NO bioavailability in humans and, thus, a new consideration for cardiovascular disease risk. Further studies on multispecies biofilms integrating biochemical, metagenomic, and metatranscriptomic data will answer these important questions and provide more information regarding the community dynamics that contribute to oral nitrate reduction that results in nitrite accumulation.

How to Determine the Presence or Absence of Nitrate-Reducing Bacteria

Since nitrite accumulates in saliva from the reduction of nitrate in the oral cavity, determining salivary nitrite concentrations may offer a simple means to determine the presence or absence of nitrate-reducing bacteria. However, understanding that the use of salivary nitrite as a marker of endogenous NO generation, both from oxidation of NO and from dietary sources, is totally dependent upon the uptake and excretion of nitrate by the salivary glands and subsequent reduction of nitrate to nitrite by lingual bacteria, there are steps in the pathway that can become disrupted and lead to changes in salivary nitrite. Each step is described as follows:

1. Nitrate (from oxidation of NO or from diet) uptake in the salivary glands: It was recently reported that the sialic acid (SA)/H^+ cotransporter is involved in nitrate uptake into salivary glands [46].
2. Nitrate secretion by salivary glands: The volumes of saliva produced vary depending on the type and intensity of stimulation, the largest volumes occurring with cholinergic stimulation.
3. Oral bacterial nitrate reduction: Humans lack a functional nitrate reductase, so salivary nitrate reduction is dependent upon oral commensal nitrate-reducing bacteria.
4. Oral pH: Healthy oral pH is between 6.5 and 7.5. The pKa of nitrite is 3.4 so any condition that lowers the pH in the oral cavity may destabilize nitrite and affect the use of salivary nitrite as a measure of nitrate reduction.

As a result of this human nitrogen cycle and enterosalivary circulation of nitrate (both from diet and from endogenous NO production) and subsequent reduction to nitrite in the mouth, sampling salivary nitrite can be used as an accurate representation of the presence or absence of nitrate-reducing bacteria. Saliva offers a number of advantages as a biological compartment for diagnostics [47]. There is sufficient evidence to show that increased circulating levels of plasma nitrite correlate with changes in blood pressure [21, 48]. Furthermore, circulating and salivary levels of nitrite increase after a nitrate load [49] and eradication of oral bacteria cause a decrease in salivary nitrite and an increase in blood pressure [27]. To the contrary, as NO availability is decreased, both plasma and salivary levels of nitrite decline [50]. Whether there is sufficient NO produced by the vascular endothelium to form nitrite and nitrate in the circulation, which is then concentrated in our salivary glands, or whether there is sufficient nitrate ingested in the diet, this will be reflected as nitrite in the saliva. Understanding the basis and rationale for sampling salivary nitrite as well as recognizing the limitations, this noninvasive diagnostic can be an accurate assessment of total body NO availability and provide physicians and patients new information to cardiovascular risk and NO homeostasis. More research is needed in order to determine if this approach may have clinical utility. Salivary nitrite measurements may offer an indirect measure of the ability of humans to reduce nitrate. Rapid, point of care diagnostics that can determine specific and selective known nitrate-reducing bacteria is preferred but to date, this technology is not available.

Restoring and Repopulation of Nitrate-Reducing Bacteria

The potential to exploit the symbiotic nitrate–nitrite–NO pathway to NO production is profound, particularly because adequate and sustained control of blood pressure is achieved in only about 50 % of treated hypertensive patients, including all classes of antihypertensives [51] not to mention all the other biological effects of NO. As cardiovascular disease remains the top killer in the U.S., accounting for more deaths each year than cancer, designing new diagnostics, treatments, and preventives for diseased and at-risk individuals is essential. Additionally, because NO is an important signaling molecule in various body systems, exploiting the oral microbiome to contribute to NO production and maintain NO homeostasis has the potential to affect human health beyond the cardiovascular system. Being able to repopulate essential nitrate-reducing bacteria in the oral cavity is of immense interest since this may be an effective means to restore and replete NO availability in the human body. In a recent study, we observed significant changes in the relative abundances of taxa present on the tongues of sodium nitrate-supplemented rats compared to nonsupplemented rats, including a significant increase in nitrate-reducing *Haemophilus parainfluenzae* [52]. Additionally, *Granulicatella* and *Aggregatibacter*, which have both been associated with poor oral health [53, 54] in humans, decreased with NaNO$_3$ supplementation. These results suggest that high nitrate diets may induce changes in oral microbiome communities to reduce nitrate to nitrite and NO more efficiently, which could be beneficial both by reducing blood pressure and by inhibiting bacterial species associated with poor oral health. Since nitrite and NO [55, 56] are toxic to many pathogenic bacteria, dietary nitrate supplementation may allow for specific reduction or eradication of pathogenic bacteria and provide an environment for beneficial bacteria to flourish. The results of our study show that the oral microbiome can be modified by dietary nitrate alone.

Conclusion

For the past 30 years, scientists have focused on NO production/regulation at the level of nitric oxide synthase (NOS), through the five-electron oxidation of L-arginine. However, this pathway becomes dysfunctional with age and disease [57]. The notion that this deficiency can be overcome by targeting oral bacteria is profound and revolutionary. Therapeutically, then, perhaps an effective strategy to promote NO production and overcome conditions of NO insufficiency may not be targeted at eNOS, but rather to target specific oral nitrate-reducing bacterial communities. Understanding and harnessing this redundant compensatory pathway may prove to be a viable and cost-effective strategy. Furthermore, new research reveals that this may in fact be the biochemical and physiological link between oral health and cardiovascular disease (CVD) through maintenance of NO production. Because NO signaling affects all organ system and almost all disease processes described to date, this novel approach to NO regulation has the potential to affect the study and treatment of many diseases across all organ systems.

Nitric oxide research has expanded rapidly in the past 30 years, and the roles of NO in physiology and pathology have been extensively studied. The pathways of NO synthesis, signaling, and metabolism in vascular biological systems have been and continue to be a major area of research. As a gas (in the pure state and under standard temperature and pressure) and free radical with an unshared electron, NO participates in various biological processes. Understanding its production, regulation, and molecular targets will be essential for the development of new therapies for various pathological conditions characterized by an unbalanced production and metabolism of NO. An outstanding key question is whether the decreased abundance or absence of nitrate-reducing communities is correlated with a state of NO insufficiency and an increased risk for cardiovascular disease. A recent study indicates that eradicating oral bacteria with antiseptic mouthwash leads to an increase in systemic blood pressure [27]. There is a known correlation between oral health and systemic disease [58]. Importantly, because oral NO production is dependent on oral nitrate-reducing bacteria, these observations suggest that the link between oral health issues such as chronic periodontitis and cardiovascular disease may be due in part to decreased abundance of nitrate reducers and concurrent increase of pathogenic bacterial species in the oral cavity. Interestingly, the strictly anaerobic bacteria associated with chronic periodontitis convert nitrate to ammonia rather than nitrite. Importantly, this relationship occurs independently from (though sometimes concurrently with) changes in blood pressure, indicating that sodium nitrate supplementation could improve cardiovascular health through the action of oral nitrate-reducing bacteria in multiple ways.

Disruption of nitrite and NO production in the oral cavity may contribute to the oral-systemic link between oral hygiene and cardiovascular risk and disease. The identification of new biomarkers for NO insufficiency and the exploitation of the oral microbiota to increase cardiovascular health will be enabled by further characterization of the enzymatic activities of native oral bacterial communities from larger healthy cohorts and specific patient populations. These cohorts should consist not only of specific U.S. population, but also of other around-the-world (European, Asian) populations. It is likely that the oral microbiomes of different ethnic groups, even those within different regions of the U.S., vary widely. It will be important to determine whether different nitrate-reducing communities are more prevalent in geographically dispersed healthy populations; likewise, it will also be important to determine whether different nitrate-reducing communities are lacking in specific patient populations from around the world. If certain patient populations lack specific nitrate-reducing bacteria, personalized treatments to enrich for nitrate reducers may be warranted. Is it tempting to speculate whether the use of mouthwash may be discouraged as part of such treatments? Indeed, studies have shown that chlorhexidine-based bactericidal mouthwashes raise blood pressure in animal models and in humans [24, 27]; however, it has yet to be determined whether other mouthwashes, such as alcohol-based mouthwashes, have similar negative effects. Additionally, while antibiotics are sometimes used to target specific bacterial species, it is possible that potential deleterious effects of antibiotic usage on nitrate-reducing communities may preclude the use of antibiotics in specific patient populations.

Clearly, the potential for the entero-salivary nitrate–nitrite–NO pathway to serve as a NO bioavailability maintenance system by harnessing the nitrate reductase activity of specific commensal bacteria calls for studies that may be profound and truly transformative. Future studies are likely to unveil new paradigms on the regulation and production of endogenous NO that are likely to be new targets for specialized, multifaceted, and potentially personalized therapeutic interventions. These studies will have the potential to: (1) redefine the meaning of "healthy oral microbiome" to include microbes associated with NO production; (2) provide a new target for NO-based therapies and open a new direction in cardiovascular research; and (3) allow development of new diagnostics targeted at specific oral microbial communities or select bacteria, the absence of which may reflect a state of NO insufficiency and change the treatment strategies for NO restoration in a number of different diseases. With the loss of NO signaling and homeostasis being one of the earliest events in the onset and progression of cardiovascular disease, targeting microbial communities early in the process may lead to better preventative interventions in cardiovascular medicine. This may also affect the way oral health professionals recommend oral hygienic practices.

References

1. Robles Alonso V, Guarner F. Linking the gut microbiota to human health. Br J Nutr. 2013;109 Suppl 2:S21–6.
2. Lundberg JO, Weitzberg E, Gladwin MT. The nitrate-nitrite-nitric oxide pathway in physiology and therapeutics. Nat Rev Drug Discov. 2008;7(8):156–67.
3. Lundberg JO, Weitzberg E, Cole JA, Benjamin N. Nitrate, bacteria and human health. Nat Rev Microbiol. 2004;2(7):593–602.
4. Bryan NS, editor. Food, nutrition and the nitric oxide pathway: biochemistry and bioactivity. Lancaster: DesTech Publishing; 2009.
5. Bryan NS, Loscalzo J, editors. Nitrite and nitrate in human health and disease. New York: Humana Press; 2011.
6. Bryan NS, Fernandez BO, Bauer SM, Garcia-Saura MF, Milsom AB, Rassaf T, et al. Nitrite is a signaling molecule and regulator of gene expression in mammalian tissues. Nat Chem Biol. 2005;1(5):290–7.
7. Zweier JL, Wang P, Samouilov A, Kuppusamy P. Enzyme-independent formation of nitric oxide in biological tissues. Nat Med. 1995;1(8):804–9.
8. Lundberg JO, Weitzberg E, Lundberg JM, Alving K. Intragastric nitric oxide production in humans: measurements in expelled air. Gut. 1994;35(11):1543–6.
9. Benjamin N, O'Driscoll F, Dougall H, Duncan C, Smith L, Golden M, et al. Stomach NO synthesis. Nature. 1994;368(6471):502.

10. Kleinbongard P, Dejam A, Lauer T, Rassaf T, Schindler A, Picker O, et al. Plasma nitrite reflects constitutive nitric oxide synthase activity in mammals. Free Radic Biol Med. 2003;35(7):790–6.
11. Bryan NS. Nitrite in nitric oxide biology: cause or consequence? A systems-based review. Free Radic Biol Med. 2006;41(5):691–701.
12. Angelo M, Singel DJ, Stamler JS. An S-nitrosothiol (SNO) synthase function of hemoglobin that utilizes nitrite as a substrate. Proc Natl Acad Sci U S A. 2006;103(22):8366–71.
13. Tannenbaum SR, Sinskey AJ, Weisman M, Bishop W. Nitrite in human saliva. Its possible relationship to nitrosamine formation. J Natl Cancer Inst. 1974;53(1):79–84.
14. van Maanen JM, van Geel AA, Kleinjans JC. Modulation of nitrate-nitrite conversion in the oral cavity. Cancer Detect Prev. 1996;20(6):590–6.
15. Duncan C, Dougall H, Johnston P, Green S, Brogan R, Leifert C, et al. Chemical generation of nitric oxide in the mouth from the enterosalivary circulation of dietary nitrate. Nat Med. 1995;1(6):546–51.
16. Lundberg JO, Govoni M. Inorganic nitrate is a possible source for systemic generation of nitric oxide. Free Radic Biol Med. 2004;37(3):395–400.
17. Spiegelhalder B, Eisenbrand G, Preussmann R. Influence of dietary nitrate on nitrite content of human saliva: possible relevance to in vivo formation of N-nitroso compounds. Food Cosmet Toxicol. 1976;14(6):545–8.
18. Hunault CC, van Velzen AG, Sips AJ, Schothorst RC, Meulenbelt J. Bioavailability of sodium nitrite from an aqueous solution in healthy adults. Toxicol Lett. 2009;190(1):48–53.
19. Bryan NS, Calvert JW, Elrod JW, Gundewar S, Ji SY, Lefer DJ. Dietary nitrite supplementation protects against myocardial ischemia-reperfusion injury. Proc Natl Acad Sci U S A. 2007;104(48):19144–9.
20. Bryan NS, Calvert JW, Gundewar S, Lefer DJ. Dietary nitrite restores NO homeostasis and is cardioprotective in endothelial nitric oxide synthase-deficient mice. Free Radic Biol Med. 2008;45(4):468–74.
21. Webb AJ, Patel N, Loukogeorgakis S, Okorie M, Aboud Z, Misra S, et al. Acute blood pressure lowering, vasoprotective, and antiplatelet properties of dietary nitrate via bioconversion to nitrite. Hypertension. 2008;51(3):784–90.
22. Carlstrom M, Larsen FJ, Nystrom T, Hezel M, Borniquel S, Weitzberg E, et al. Dietary inorganic nitrate reverses features of metabolic syndrome in endothelial nitric oxide synthase-deficient mice. Proc Natl Acad Sci U S A. 2010;107(41):17716–20.
23. Carlstrom M, Persson AE, Larsson E, Hezel M, Scheffer PG, Teerlink T, et al. Dietary nitrate attenuates oxidative stress, prevents cardiac and renal injuries, and reduces blood pressure in salt-induced hypertension. Cardiovasc Res. [Research Support, Non-U.S. Gov't]. 2011;89(3):574–85.
24. Petersson J, Carlstrom M, Schreiber O, Phillipson M, Christoffersson G, Jagare A, et al. Gastroprotective and blood pressure lowering effects of dietary nitrate are abolished by an antiseptic mouthwash. Free Radic Biol Med. 2009;46(8):1068–75.
25. Webb A, Bond R, McLean P, Uppal R, Benjamin N, Ahluwalia A. Reduction of nitrite to nitric oxide during ischemia protects against myocardial ischemia-reperfusion damage. Proc Natl Acad Sci U S A. 2004;101:13683–8.
26. Hendgen-Cotta UB, Luedike P, Totzeck M, Kropp M, Schicho A, Stock P, et al. Dietary nitrate supplementation improves revascularization in chronic ischemia. Circulation. 2012;126(16):1983–92.
27. Kapil V, Haydar SM, Pearl V, Lundberg JO, Weitzberg E, Ahluwalia A. Physiological role for nitrate-reducing oral bacteria in blood pressure control. Free Radic Biol Med. 2013;55:93–100.
28. Bryan NS, Rassaf T, Maloney RE, Rodriguez CM, Saijo F, Rodriguez JR, et al. Cellular targets and mechanisms of nitros(yl)ation: an insight into their nature and kinetics in vivo. Proc Natl Acad Sci U S A. 2004;101(12):4308–13.
29. Walters CL, Casselden RJ, Taylor AM. Nitrite metabolism by skeletal muscle mitochondria in relation to haem pigments. Biochim Biophys Acta. 1967;143(2):310–8.
30. Kozlov AV, Staniek K, Nohl H. Nitrite reductase activity is a novel function of mammalian mitochondria. FEBS Lett. 1999;454(1–2):127–30.
31. Cosby K, Partovi KS, Crawford JH, Patel RP, Reiter CD, Martyr S, et al. Nitrite reduction to nitric oxide by deoxyhemoglobin vasodilates the human circulation. Nat Med. 2003;9(12):1498–505.
32. Li H, Samouilov A, Liu X, Zweier JL. Characterization of the effects of oxygen on xanthine oxidase-mediated nitric oxide formation. J Biol Chem. 2004;279(17):16939–46.
33. Duranski MR, Greer JJ, Dejam A, Jaganmohan S, Hogg N, Langston W, et al. Cytoprotective effects of nitrite during in vivo ischemia-reperfusion of the heart and liver. J Clin Invest. 2005;115(5):1232–40.
34. Pluta RM, Dejam A, Grimes G, Gladwin MT, Oldfield EH. Nitrite infusions to prevent delayed cerebral vasospasm in a primate model of subarachnoid hemorrhage. JAMA. 2005;293(12):1477–84.
35. Hunter CJ, Dejam A, Blood AB, Shields H, Kim-Shapiro DB, Machado R, et al. Inhaled nebulized nitrite is a hypoxia-sensitive NO-dependent selective pulmonary vasodilator. Nat Med. 2004;10:1122–7.
36. Hardwick JB, Tucker AT, Wilks M, Johnston A, Benjamin N. A novel method for the delivery of nitric oxide therapy to the skin of human subjects using a semi-permeable membrane. Clin Sci (Lond). 2001;100(4):395–400.
37. Bjorne HH, Petersson J, Phillipson M, Weitzberg E, Holm L, Lundberg JO. Nitrite in saliva increases gastric mucosal blood flow and mucus thickness. J Clin Invest. 2004;113(1):106–14.

38. Tsuchiya K, Kanematsu Y, Yoshizumi M, Ohnishi H, Kirima K, Izawa Y, et al. Nitrite is an alternative source of NO in vivo. Am J Physiol Heart Circ Physiol. 2005;288(5):H2163–70.
39. Kleinbongard P, Dejam A, Lauer T, Jax T, Kerber S, Gharini P, et al. Plasma nitrite concentrations reflect the degree of endothelial dysfunction in humans. Free Radic Biol Med. 2006;40(2):295–302.
40. Larsen FJ, Ekblom B, Sahlin K, Lundberg JO, Weitzberg E. Effects of dietary nitrate on blood pressure in healthy volunteers. N Engl J Med. 2006;355(26):2792–3.
41. Kapil V, Haydar SM, Pearl V, Lundberg JO, Weitzberg E, Ahluwalia A. Physiological role for nitrate-reducing oral bacteria in blood pressure control. Free Radic Biol Med. 2012;55C:93–100.
42. Bryan NS, Bian K, Murad F. Discovery of the nitric oxide signaling pathway and targets for drug development. Front Biosci. 2009;14:1–18.
43. Bum EN, Schmutz M, Meyer C, Rakotonirina A, Bopelet M, Portet C, et al. Anticonvulsant properties of the methanolic extract of Cyperus articulatus (Cyperaceae). J Ethnopharmacol. 2001;76(2):145–50.
44. Doel J, Benjamin N, Hector M, Rogers M, Allaker R. Evaluation of bacterial nitrate reduction in the human oral cavity. Eur J Oral Sci. 2005;113:14–9.
45. Hyde ER, Andrade F, Vaksman Z, Parthasarathy K, Jiang H, Parthasarathy DK, et al. Metagenomic analysis of nitrate-reducing bacteria in the oral cavity: implications for nitric oxide homeostasis. PLoS One. 2014;9(3), e88645.
46. Qin L, Liu X, Sun Q, Fan Z, Xia D, Ding G, et al. Sialin (SLC17A5) functions as a nitrate transporter in the plasma membrane. Proc Natl Acad Sci U S A. 2012;109(33):13434–9.
47. Pfaffe T, Cooper-White J, Beyerlein P, Kostner K, Punyadeera C. Diagnostic potential of saliva: current state and future applications. Clin Chem. 2011;57(5):675–87.
48. Kapil V, Milsom AB, Okorie M, Maleki-Toyserkani S, Akram F, Rehman F, et al. Inorganic nitrate supplementation lowers blood pressure in humans: role for nitrite-derived NO. Hypertension. 2010;56(2):274–81.
49. Govoni M, Jansson EA, Weitzberg E, Lundberg JO. The increase in plasma nitrite after a dietary nitrate load is markedly attenuated by an antibacterial mouthwash. Nitric Oxide. 2008;19(4):333–7.
50. Bryan NS, Torregrossa AC, Mian AI, Berkson DL, Westby CM, Moncrief JW. Acute effects of hemodialysis on nitrite and nitrate: potential cardiovascular implications in dialysis patients. Free Radic Biol Med. 2013;58:46–51.
51. Go AS, Mozaffarian D, Roger VL, Benjamin EJ, Berry JD, Blaha MJ, et al. Heart disease and stroke statistics—2014 update: a report from the American Heart Association. Circulation. 2014;129(3):e28–292.
52. Hyde ER, Luk B, Cron S, Kusic L, McCue T, Bauch T, et al. Characterization of the rat oral microbiome and the effects of dietary nitrate. Free Radic Biol Med. 2014;77:249–57.
53. Takeshita T, Suzuki N, Nakano Y, Shimazaki Y, Yoneda M, Hirofuji T, et al. Relationship between oral malodor and the global composition of indigenous bacterial populations in saliva. Appl Environ Microbiol. 2010;76(9):2806–14.
54. Chen C, Wang T, Chen W. Occurrence of Aggregatibacter actinomycetemcomitans serotypes in subgingival plaque from United States subjects. Mol Oral Microbiol. 2010;25(3):207–14.
55. Reddy D, Lancaster Jr JR, Cornforth DP. Nitrite inhibition of Clostridium botulinum: electron spin resonance detection of iron-nitric oxide complexes. Science. 1983;221(4612):769–70.
56. Allaker RP, Silva Mendez LS, Hardie JM, Benjamin N. Antimicrobial effect of acidified nitrite on periodontal bacteria. Oral Microbiol Immunol. 2001;16(4):253–6.
57. Torregrossa AC, Aranke M, Bryan NS. Nitric oxide and geriatrics: implications in diagnostics and treatment of the elderly. J Geriatr Cardiol. 2011;8:230–42.
58. Joshipura K, Ritchie C, Douglass C. Strength of evidence linking oral conditions and systemic disease. Compend Contin Educ Dent Suppl. 2000;30:12–23; quiz 65.

Chapter 4
Epigenetics and the Regulation of Nitric Oxide

Diane E. Handy and Joseph Loscalzo

Key Points

- Epigenetics involves chromatin-based modifications to DNA or histone proteins that influence patterns of gene expression.
- Nitric oxide synthase genes and other genes involved in nitric oxide production are targets for epigenetic modification.
- Histone deacetylase (HDAC) inhibitors alter histone acetylation patterns.
- HDAC inhibitors have differential effects on expression of the various nitric oxide genes.
- Crosstalk between HDAC and nitric oxide pathways can modulate gene expression and NOS enzyme function.

Keywords Epigenetics • Promoter • Histone • Histone methyltransferases • Histone demethylases • Histone acetyltransferases • Histone deacetylases

Abbreviations

5hmC	5-Hydroxymethylcytosine
5mC	5-Methylcytosine
CBP	CREB-binding protein
DMD	Duchenne muscular dystrophy
DNMT	DNA methyltransferase
eNOS	Endothelial NOS enzyme
GNAT	Gcn5-related *N*-acetyl transferases
H3	Histone 3

D.E. Handy, Ph.D.
Cardiovascular Division, Department of Medicine, Brigham and Women's Hospital
and Harvard Medical School, NRB 0630L, 77 Avenue Louis Pasteur, Boston 02115, MA, USA

J. Loscalzo, M.D., Ph.D. (✉)
Department of Medicine, Brigham and Women's Hospital,
75 Francis Street, Boston, MA 02115, USA

Department of Medicine, Harvard Medical School, Boston, MA, USA
e-mail: jloscalzo@partners.org

N.S. Bryan, J. Loscalzo (eds.), *Nitrite and Nitrate in Human Health and Disease*, Nutrition and Health,
DOI 10.1007/978-3-319-46189-2_4, © Springer International Publishing AG 2017

H4	Histone 4
HAT	Histone acetyltransferase
HDAC	Histone deacetylase
iNOS	Inducible NOS enzyme
Jmj	Jumonji
KDM	Lysine demethylase
KLF2	Krüppel-like factor 2
KMT	Lysine methyltransferase
me1	Mono-methylation
me2	Di-methylation
me3	Tri-methylation
MeCp	Methylcytosine binding protein
miR	microRNA
MKP1	MAPK phosphatase 1
nNOS	Neuronal NOS enzyme
NOS	Nitric oxide synthase
NOS1	Neuronal NOS gene
NOS2	Inducible NOS gene
NOS3	Endothelial NOS gene
PP2A	Protein phosphatase 2A
S-NO	S-nitrosylation
Tet	Ten–eleven translocation
Tyr-NO$_2$	Tyrosine nitration
Ub	Ubiquitination

Introduction

Epigenetics involves heritable changes to chromatin that influence gene expression without changes to the underlying DNA sequence. Chromatin consists of genomic DNA and the tightly associated histone proteins that serve to organize and package the DNA in nucleosomal units. At its core, the nucleosome contains an octameric histone complex formed by two subunits each of the histones 2A, 2B, 3, and 4. DNA (146 bp) wraps around each histone core to form the nucleosome. In between the nucleosomes, linker DNA associates with histone 1. Epigenetic modifications to DNA and the N-terminal tails of the core histone proteins can be permissive for transcription, allowing an open conformation of chromatin (loosely packed nucleosomes), or limit the accessibility of transcription factors by condensing chromatin (tightly packed nucleosomes) (Fig. 4.1). Importantly, epigenetic tags regulate development and differentiation, contribute to transgenerational inheritance, and may be influenced by diet and the environment.

Our understanding of epigenetics and its role in biological systems remain incomplete; nonetheless, the continued study of these mechanisms may provide a powerful means to control essential cellular and physiological processes. Several studies have suggested that prenatal exposure to pharmacological agents or special diets can promote epigenetic changes that affect gene expression and physiology in offspring [1–4]. In one recent study, the use of pentaerythritol tetranitrate, an antioxidant and NO donor, decreased blood pressure in the female offspring of spontaneous hypertensive rats, an inbred strain used as an experimental model of essential hypertension [1]. The protective effects of this treatment on offspring correlated with an altered epigenetic profile consistent with reduced blood pressure and augmented vascular function, as these epigenetic changes are permissive for increased expression of the nitric oxide synthase 3 (*NOS3*) gene, as well as genes encoding other antioxidant enzymes. Other studies have similarly shown that the maternal diet can affect epigenetic

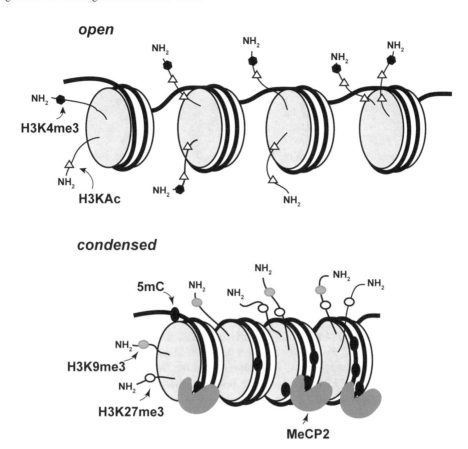

Fig. 4.1 Epigenetic regulation of chromatin structure. Major epigenetic tags that are discussed throughout this chapter are included in this illustration of nucleosomes. DNA (*thick black line*) is wrapped around histone octamers (represented by the *disks*). The open conformation of chromatin permissive for transcription is associated with particular modifications to histone tails, including histone acetylation by histone transferase enzymes. Several lysine residues in histones 3 and 4 are susceptible to acetylation. Histone 3 lysine 4 mono-, di-, and tri-methylation (H3K4me3) have all been associated with transcriptionally active chromatin. These methylation tags are regulated by Set 1 family methyltransferases and the Jard1a, Jard1b, and LSD2 demethylases. The closed conformation of chromatin is promoted by DNA methylation at the 5 position of cytosines in CpG dinucleotides (5mC). This epigenetic tag recruits DNA methyl-binding proteins, such as MeCP2. Deacetylation of the histone tails by histone deacetylase enzymes and establishment of H3K9me3 and H3K27me3 marks by the G9a/GLP and EZH2 methyltransferases, respectively, contribute to the repressive, condensed state of the chromatin

regulation of gene expression [2, 3] and modulate disease outcomes [4, 5]. In this chapter, we provide an overview of epigenetics, discuss how these processes may influence nitric oxide production, and provide an update on the crosstalk between nitric oxide and epigenetic mechanisms.

Epigenetic Tags

DNA Methylation

Overall, methylation at the five position of cytosine residues in CpG dinucleotides is associated with gene suppression, especially when it occurs at regions with high concentrations of CpG dinucleotides (CpG islands) or at nuclear factor binding sites [6]. The 5-methylcytosine (5mC) modification is an

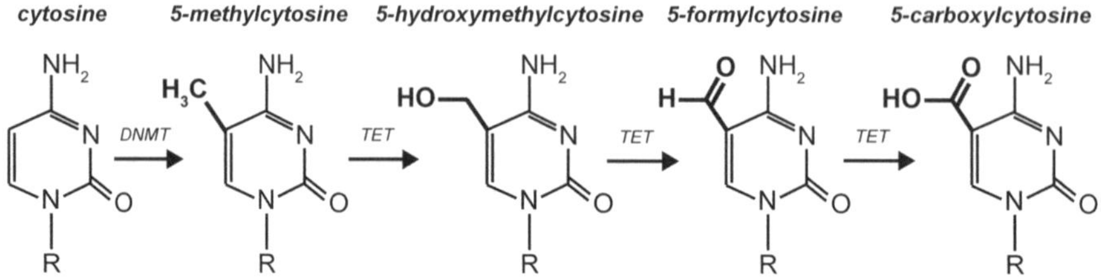

Fig. 4.2 Epigenetic modifications to cytosine. DNA methyltransferase (DNMT) 3a and 3b are responsible for the formation of 5-methylcytosine. The ten-eleven translocation (Tet) dioxygenase converts 5-methylcytosine to 5-hydroxymethylcytosine and additional oxidized forms, 5-formylcytosine and 5-carboxylcytosine. DNMT1 retains the 5-methylcytosine modification during DNA replication; however, the other modified forms are not recognized by this enzyme and revert to cytosine during replication

essential mechanism that regulates gene expression during differentiation and development. 5mC represses transcription of transposons and repeat elements, maintains the silencing of imprinted genes, and is essential for X-chromosome inactivation [7]. The 5mC mark leads to chromatin condensation, as these DNA modifications foster binding of specific proteins (MeCp binding proteins) that subsequently recruit histone-modifying enzymes to remodel further chromatin (reviewed in [8] and later).

A family of DNA methyltransferases (DNMTs) is involved in the *de novo* methylation of cytosines (DNMT3a and DNMT3b) and the maintenance methylation of daughter strands during DNA replication (DNMT1) [9–11]. Until recently, it was thought that the 5mC mark was irreversible and only removed following the loss of DNMT1 function; however, there is growing evidence for additional mechanisms that result in removal of this repressive epigenetic mark. Specifically, the ten–eleven translocation (Tet) dioxygenase family of enzymes can convert 5mC sites to hydroxymethylation sites (5hmC) in a 2-oxoglutarate, oxygen, and iron-dependent manner [12, 13]. Although the 5hmC mark may regulate chromatin structure [14], this modification is notable because it can lead to the effective demethylation of DNA [14, 15]. Essentially the epigenetic mark is lost during DNA replication as the 5hmC mark is not recognized by the maintenance DNA methyltransferase, DNMT1 [16]. Additionally, Tet enzymes can oxidize further 5hmC to 5-formylcytosine and 5-carboxylcytosine, which facilitates DNA excision-repair mechanisms that restore the unmodified cytosine (Fig. 4.2) [15, 17].

Histone Modification

The N-terminal histone tails are accessible sites in the histone octamer that can be tagged by a variety of post-translational modifications, such as phosphorylation, ubiquitination, sumoylation, acetylation, and methylation [18–20]. Of these, the most widely studied are the reversible acetylation and methylation at the ε-amino group of lysine residues in the H3 and H4 histone tails. Arginine residues may also be targets for methylation; however, the effects of this modification on gene expression are not well studied [21].

Histone Acetylation

Histone lysine acetylation is associated with an open conformation of nucleosomes leading to greater accessibility of nuclear factors, which, in general, leads to a more permissive environment for transcription. It should be mentioned, however, that in some cases acetylation may also lead to the recruitment of repressor complexes [22]. Histone acetyltransferase (HAT) cofactors can be grouped into three main families of enzymes: the E1A-associated proteins, which includes CREB-binding protein (CBP) and p300; the MYST family, and the Gcn5-related *N*-acetyl transferases or GNAT family [23]. These HAT enzymes do not directly bind to DNA; rather, they are directed to target genes by their interactions with sequence-specific transcription factors. In addition, HAT enzymes can target other proteins for acetylation.

Opposing the action of the HAT enzymes are the histone deacetylases (HDAC). The targeting of these enzymes may be due, in part, to their recruitment by specific repressor complexes [24, 25]. HDAC may also be recruited to histones by MeCp binding proteins, providing a mechanism by which DNA methylation leads to histone remodeling and repressive remodeling of chromatin [26]. There are at least 18 known mammalian HDACs, including the class I (HDAC1, 2, 3, and 8); class II (HDAC 4, 5, 6, 7, 9, and 10); class III (NAD+-dependent sirtuins, SIRT1-7); and class IV (HDAC11) [27]. Some class I and class II HDACs can shuttle in and out of the nucleus. Additionally, many HDACs can deacetylate proteins other than histones, such as nuclear factors, or enzymes, such as eNOS.

Histone Methylation

Histone Lysine Methylation

Histone lysine methylation can occur at multiple sites in histones H3 and H4. Depending on the site and the extent of modification (mono-, di-, or tri-methylation), methylation can contribute to the active or inactive conformation of chromatin. Overall, triple methylation of histone 3 at lysine 4 (H3K4me3) is permissive for transcription, as is H3K36me3, whereas H3K9me3 and H3K27me3 are repressive modifications [28].

Many different lysine methyltransferases and demethylases contribute to the complex pattern of histone methylation [28]. Most of the lysine methyltransferases (KMT) are members of the large Set domain family, which has several subfamilies, whereas the Dot1-like proteins comprise a small group of KMT that lack the Set domain. Each methyltransferase has limited specificity for particular lysine residues. For example, Set1-family enzymes, such as Set7 and MLL1, target H3K4 methylation, whereas EZH2, which is the active subunit of the polycomb repressor complex (PRC), targets di- and tri-methylation of H3K27me3. Similar to EZH2, Set-1 methyltransferases associate with other proteins in multisubunit complexes. Accumulating evidence suggests that the recruitment and activity of Set-family methyltransferases relies on the composition of these multisubunit complexes. For example, the extent of H3K4 methylation depends on the presence of WRAD in these complexes, although various Set1 methyltransferases have different propensities for di- and tri-methylation even in the presence of WRAD [29, 30]. Similarly, the EZH2-containing PRC may be recruited to particular genes, in part, via the association of accessory proteins including G9a and GLP methyltransferases [31]. G9a and GLP are methyltransferases that target H3K9 for mono- and di-methylation [32] and may also mono-methylate H3K27 [31]. The association between G9a/Glp and EZH2 may also explain the coexistence of H3K9me and H3K27me3 in the same regions of some target genes.

The state of lysine methylation is achieved by a balance between the action of targeted methylases and demethylases [33, 34]. Lysine demethylases (KDM) belong to either the FAD-dependent amine oxidases or the jumonji domain (Jmj)-containing dioxygenases that are dependent on iron, oxygen, and alpha-ketoglutarate. Similar to the methyltransferases, the demethylases are specific for certain histone modifications. Thus, the FAD-dependent LSD2 (KDM1b) demethylates H3K4me1 and H3K4me2 [35], whereas the Jmj and PHD finger domain-containing proteins Jard1a (KDM5A) and Jarid1b (KDM5B) silence genes by removing H3K4me marks [36]. The repressive H3K27me3 is targeted by the JmjD demethylases UTX (KDM6a) and JmjD3 (KDM6b) [37].

Histone Arginine Methylation

Arginine residues can be monomethylated or dimethylated in a symmetrical or asymmetrical manner. These common protein modifications are regulated by members of the PRMT family of enzymes, some of which target nonhistone proteins [21]. Overall, the function of many of the histone arginine methylation marks is not well understood, although some arginine methylation sites have been shown to affect the frequency of specific lysine modifications. Thus, histone 3 arginine 2 asymmetric dimethylation was found to attenuate H3K4me3 formation in some promoters [21, 38], thereby repressing gene transcription. In contrast, the H3R2 symmetrical methylation has been shown to colocalize with the transcriptionally permissive H3K4me3 in active promoters, suggesting the symmetrical modification may have the opposite effect on transcription [39]. The role of many of these sites in transcription is not clear, and some of the sites for arginine methylation occur in the core region rather than the tail region of histones (reviewed in [21]).

Epigenetic Regulation of NOS Genes

The three *NOS* genes, *NOS1*, *NOS2A*, *NOS3*, that encode the neuronal (nNOS), inducible (iNOS), and endothelial (eNOS) enzymes, respectively, can all be regulated by epigenetic mechanisms. In this section, we discuss the direct and indirect epigenetic mechanisms that regulate *NOS* gene expression and NO production. We also address how crosstalk between HDAC and NO pathways can modulate histone modification and NOS activity.

NOS1 *Regulation and Epigenetic Crosstalk*

nNOS is the major source of neuronal NO, although it also contributes to NO production in other cells, such as pulmonary epithelium, skeletal muscle, and keratinocytes. The *NOS1* gene structure is complex with several alternative promoters, at least 12 first exons, and its transcript has multiple splicing isoforms that are differentially expressed in various cell types [40]. Compared to the other *NOS* genes, relatively few studies have examined the epigenetic regulation of *NOS1*; however, evidence exists to suggest that DNA methylation represses *NOS1* gene expression as downregulation of DNMT1 increased *NOS1* expression in a subset of neuroblastoma lines [41]. Although many genes had changes in DNA methylation following retinoic acid-stimulated suppression of DNMT1, fewer genes had both a decrease in DNA methylation and significant increases in their expression. *NOS1* was upregulated more than 2-fold by DNMT1 suppression and its promoter was also demethylated. In particular, the differentially methylated *NOS1* promoter identified in this screen was located upstream of exon1f in the *NOS1* gene. [Exons 1c, 1d, 1f, and 1g containing-mRNA are the predominant forms of *NOS1* transcripts in a variety of tissues, including brain, testes, skin, brain, skeletal muscle, kidney, and gastrointestinal cells [42, 43]]. Furthermore, in some cells the promoter upstream of exon 1F may also control the upregulation of *NOS1* transcripts in response to certain stimuli, such as dibutyryl cAMP [43], and this promoter has an NFkB site that may contribute to its expression in neuroblastoma cells (Fig. 4.3) [44, 45].

The ubiquitously expressed HAT p300 was found to regulate *NOS1* expression via its effects on NFkB-mediated transcription [44, 45]. Interestingly, p300 enhanced the expression of *NOS1* through multiple mechanisms. p300 directly acetylated p50 and p65 subunits of the transcription factor NFkB and enhanced its binding to the *NOS1* promoter. Furthermore, p300 was found to stimulate *NOS1* expression in the absence of its HAT activity, most likely due to its ability to form a complex with NFkB at its binding site in the *NOS1* promoter. Although it is well known that p300 can promote

NOS1

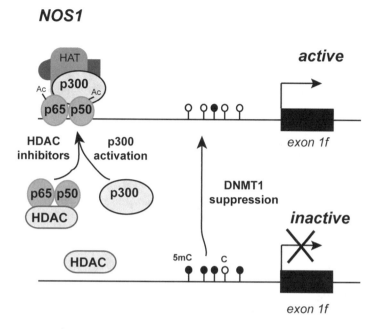

Fig. 4.3 Epigenetic regulation of *NOS1* gene transcription. Studies have implicated regions of the *NOS1* promoter upstream of exon 1f in epigenetic regulation in neuronal cells. Suppression of DNMT1 significantly decreases DNA methylation (5mC, illustrated by the *black circles*) in this region and upregulates *NOS1* gene expression. By increasing acetylation of histones and other proteins, HDAC inhibitors promote *NOS1* gene and protein expression in some cells, resulting in an increase in cellular NO. The histone acetyltransferase, p300, regulates *NOS1* gene expression via the acetylation of the p65 and p50 subunits of the transcription factor NFkB. NFkB recruits p300 to the promoter, where it may attract additional transcriptional cofactors, including other HAT enzymes

transcription through its coactivating domain and may recruit additional HAT enzymes, it remains unclear whether p300 directly or indirectly regulates histone acetylation at the *NOS1* promoter, as histone acetylation was not assessed in this study. Nonetheless in these cells, trichostatin A, a HDAC inhibitor, augmented *NOS1* gene and protein expression resulting in an increase in nNOS-derived NO production in these cells, presumably due, in part, to the increase in p65 and p50 acetylation and their subsequent enhanced binding to the *NOS1* promoter [45].

NO production is important for neuronal differentiation. It has been shown that nNOS-mediated production of NO regulates HDAC2 via S-nitrosylation of Cys262 and Cys272 [46]. Nitrosylation of HDAC2 induces its release from chromatin to promote histone acetylation, thereby providing an epigenetic mechanism by which NO regulates patterns of gene expression associated with neuronal development [46, 47]. NO can also regulate HDAC2-mediated effects in other cell types, such as skeletal muscle. Thus, recent studies have implicated nNOS pathways in Duchenne muscular dystrophy (DMD) pathogenesis, in part, via NO's effects on HDAC2 [48, 49]. In DMD, the lack of dystrophin in muscle sarcolemma decreases the levels of muscle nNOS and its production of NO. In mouse models of this disease, dystrophin deficiency was found to decrease HDAC2 nitrosylation and, thereby, decreased histone acetylation at key regulatory genes, such as specific microRNAs (miR) [48]. Consequently, decreased histone acetylation correlated with a suppression of miR-1 and miR-29. It was suggested that loss of these miRNAs contribute to oxidative stress and fibrosis in the murine DMD model. In the context of dystrophin deficiency in mice, class I HDAC inhibitors or targeted HDAC2 knockdown promoted muscle fiber maturation, in part, by restoring targeted gene acetylation lost with dystrophin deficiency [49, 50]. In contrast, during hindlimb ischemia, a class I HDAC inhibitor was found to be detrimental: it prevented muscle regeneration and delayed the normal upregulation of dystrophin, eNOS, and nNOS during the regeneration process [51].

NOS2A Regulation and Epigenetic Crosstalk

In most cells, the expression of *NOS2A* is inducible by exposure to proinflammatory stimuli, such as endotoxin or cytokines. Studies have shown that the degree of CpG methylation in the *NOS2A* promoter inversely correlates with the ability of immune stimuli to upregulate its expression in a variety of cell types [52, 53]. Demethylation by the knockdown of DNMT3b or exposure to 5-azacytidine, a DNA methylation inhibitor, was reported to increase significantly the response to cytokines in some cell types owing to the subsequent decrease of 5mC within the *NOS2A* promoter. Interestingly, in human umbilical vein endothelial cells (HUVEC), which had an almost complete CpG methylation in the *NOS2A* promoter at baseline, the *NOS2A* promoter was relatively resistant to 5-azacytidine-induced demethylation. Consistent with the continued presence of 5mC in the promoter, the induction of *NOS2A* gene expression remained low in these cells even after prolonged 5-azacytidine treatment. In endothelial cells, the presence of 5mC also correlated with increased DNA binding of MeCp2 and an increase in the inhibitory H3K9me2 and H3K9me3 marks at the *NOS2A* promoter. The repressive histone marks were not found in *VCAM1*, a gene readily induced by cytokine stimulation in these cells. Furthermore, the repressive histone marks were not found in the *NOS2A* promoter in other human cell lines in which *NOS2A* gene expression was readily induced by cytokines or by 5-azacytidine exposure. Unexpectedly, a recent study found that human macrophages generate little to no NO in response to inflammatory stimuli compared to murine macrophages [54]. Similar to the findings in human endothelial cells, human macrophages had a high degree of 5mC marks, along with a decreased levels of the permissive H3K4me3 and increased levels of the repressive H3K27me3 marks at the *NOS2A* promoter. These findings suggest that dense 5mC modification of DNA together with repressive histone modifications maintain the repressive state of the *NOS2A* gene in some human cells.

Knockdown of DNMTs or treatment with 5-azacytidine is not gene-specific treatment, and, therefore, they will alter the expression of many genes simultaneously. Thus, to regulate *NOS2A* expression selectively, Gregory and colleagues [55] constructed artificial zinc finger constructs to direct thymidine DNA glycosylase, an enzyme thought to be involved in DNA demethylation by excision repair, to specific target regions in the proximal promoter and distal CpG island of *NOS2A* in NIH-3T3 cells. The simultaneous use of multiple constructs to target various 5mC sites resulted in a decrease of DNA methylation and an increase in inducible *NOS2A* expression in response to LPS and IFNγ. Although many questions remain regarding the specificity of these 'targeting' zinc fingers and the process of DNA demethylation, these data support the notion that specific 5mC sites silence the *NOS2A* promoter, and that modification of these sites may enhance *NOS2A* induction by inflammatory stimuli.

Several epidemiological studies have examined the epigenetic patterns of *NOS2A* gene methylation in human populations. In a study of elderly men, aging was associated with a decrease in *NOS2A* methylation in blood lymphocytes [56]. Although the relationship of these methylation changes to inflammation, *NOS2A* expression, and NO production were not assessed in this study, the effects of aging on this and other genes suggest that alterations in gene-specific DNA methylation may reflect underlying biological processes that change with age. Given other experimental findings discussed earlier, however, the degree of promoter methylation may inversely correlate with the potential inducibility of the *NOS2A* gene. In other studies, associations have been made between decreased *NOS2A* promoter methylation and increased fractional exhaled NO in children with asthma or increased exposure to particulate pollution. These findings suggest that alterations in *NOS2A* promoter methylation may reflect airway inflammation and increased expression of *NOS2A*, although these changes in methylation were identified in nasal and buccal DNA samples, respectively, rather than in lung tissue [57, 58].

Activation of NFkB enhances *NOS2A* expression (Fig. 4.4). Induction of *NOS2A* gene expression in response to inflammatory stimuli involves NFkB activation and its subsequent binding to the promoter. NFkB then recruits cofactors, such as HAT enzymes, to promote histone acetylation. Multiple studies

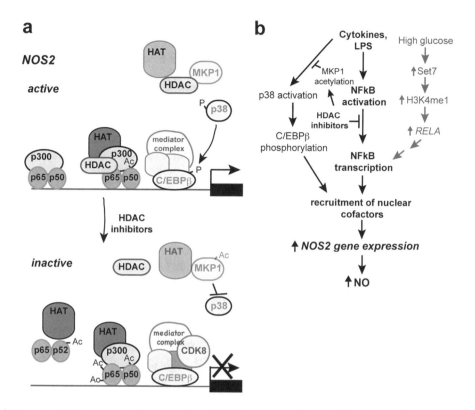

Fig. 4.4 Direct and indirect epigenetic modulation of *NOS2* gene transcription. (**a**) Protein acetylation plays an important role in *NOS2* gene transcription. *Upper panel.* Acetylation of NFkB subunits p65 and p50 by the HAT, p300, promotes *NOS2* gene transcription. Additionally, LPS stimulation can promote *NOS2* gene transcription by activating the MAPK, p38. p38-dependent phosphorylation of C/EBPβ promotes transcription, possibly via the recruitment of the Mediator complex. *Lower panel.* Paradoxically, HDAC inhibitors suppress induced *NOS2* gene transcription. Multiple mechanisms may contribute to this suppression: illustrated are two potential mechanisms that may regulate NFkB function, although their role in *NOS2* regulation is speculative. NFkB subunits may be acetylated at inhibitory sites, or additional NFkB subunits, such as p52, may be acetylated to form a complex with p65 that does not bind to DNA. Alternatively, HDAC inhibition may result in the acetylation of other proteins. Thus, acetylation of the MAPK phosphatase, MKP1, activates it, to inhibit p38 kinase. As a result, an unphosphorylated C/EBPβ recruits CDK8 to repress transcription. (**b**) Pathways of *NOS2* gene regulation. Shown is an overview of the pathways illustrated in (**a**), along with an additional pathway (*light text*) that involves the indirect epigenetic regulation of *NOS2* gene transcription under conditions of high glucose. High glucose has been shown to enhance the expression of the Set7 methyltransferase to increase the accumulation of H3K4me1 at the *RELA* gene (encoding the NFkB p65 subunit). Excess production of p65 leads to increased expression of *NOS2*

show that p300 recruitment by NFkB promotes *NOS2A* gene expression. Furthermore, these studies show that in mouse macrophage cells (RAW264.7) overexpression of p300 with an intact HAT domain increases expression of *NOS2A* reporter gene constructs and the endogenous gene, that p300 acetylates NFkB p50 and p65, and that p300 as well as acetylated p50 and p65 NFkB subunits bind to the *NOS2A* NFkB site [59–61]. Paradoxically, in many cell types HDAC inhibitors decrease NFkB-dependent upregulation of *NOS2A* gene expression, lessening cellular NO. Thus, although HDAC class I and II inhibitors, such as trichostatin A and sodium butyrate, promote histone acetylation, the cytokine-stimulated expression of the *NOS2A* gene is suppressed by these treatments, thereby decreasing cytokine-induced NO production [62–64]. The suppressive effects of HDAC inhibitors on *NOS2A* expression was also shown to coincide with increased histone acetylation at the distal NFkB binding sites of the *NOS2A* promoter [65]. Consistent with the suppressive effect of HDAC inhibition on

NOS2A gene expression, overexpression of HDACs enhanced *NOS2A* promoter activity in response to cytokines by augmenting the ability of NFkB to promote transcription [62]. These seemingly contradictory effect of HDAC inhibitors on *NOS2A* gene expression holds true for other NFkB-induced genes, as well, in many cell systems. HDAC inhibition may suppress NFkB transcriptional activity by altering NFkB subunit acetylation without affecting its ability to bind DNA [62, 66]; however, some changes in NFkB acetylation may alter its DNA binding. Furthermore, the effects of HDAC inhibitors on NFkB function may depend on the biological context, such as the complement of HDACs and HATs present in a particular cell type. In one study in rat beta cell lines, the differential effects of individual HDACs on regulating cytokine-induced *nos2* gene expression and NFkB function were analyzed by individual knockdown of each expressed HDAC [67]. Thus, although cytokine-induced *nos2* gene expression was suppressed by a deficiency of HDAC1 or HDAC3, only the HDAC3 knockdown resulted in decreased p65 binding to the proximal and distal promoter region of *nos2*. Thus, different HDAC may promote different acetylation reactions: there are multiple sites for p65 acetylation, some that promote transcription [44] and others that are inhibitory [68]. In addition, other NFkB subunits are also subject to acetylation. Thus, in an epidermal mouse cell line, HDAC inhibition alleviated cellular activation by TNF-α by altering the expression and acetylation of p52, a noncanonical NFkB protein [69]. Excess acetylated p52 increased the formation of nuclear p52/p65 heterodimers with reduced DNA binding, effectively suppressing NFkB-mediated gene transcription in these cells. In addition to the effects of HDAC inhibitors on histone and transcription factor acetylation, these agents may also alter acetylation of other proteins. This mechanism was analyzed in murine macrophages where the HDAC inhibitor TSA was found to suppress LPS-induced *nos2* [70]. In these cells, TSA had no effect on LPS-stimulated binding of the transcription factors NFkB or C/EBPβ; rather, it promoted recruitment of the corepressor CDK8 to the Mediator complex at the murine *nos2* promoter. [The Mediator complex is a multi-subunit complex that coordinates positive and negative transcriptional signals; association of the CDK8 kinase with the Mediator complex leads to a repression of transcriptional initiation.] Acetylation was suggested to play a key role in CDK8 recruitment, by regulating cell signaling: thus, increased acetylation of MAPK phosphatase 1 decreases p38 MAPK activation and phosphorylation of C/EBPβ. Apparently, the C/EBPβ phosphorylation state alters the recruitment of activator or repressor complexes to gene promoters. Although this latter study was performed in murine cells, C/EBPβ sites are also in the human *NOS2A* promoter [71] and pharmacological inhibition of p38 MAPK was found to decrease cytokine-mediated upregulation of *NOS2A* gene expression and nitrite production in cultured human chondrocytes [72], suggesting that p38/C/EBPβ pathways may also regulate human *NOS2A* expression.

As discussed earlier, regulation of epigenetic processes may indirectly affect *NOS2A* expression by modulating NFkB activity. Furthermore, epigenetic processes that affect transcription of the p65 gene (*RELA*) have also been shown to regulate *NOS2A* expression [73]. Typically, NFkB activity is held in check by protein–protein interactions between the p65/p50 heterodimer and the inhibitor of kappa B (IkB) family of proteins that sequester it in the cytoplasm. Inflammatory stimulation of NFkB signaling leads to phosphorylation and degradation of IkB proteins, and the subsequent translocation of p65/p50 dimers to the nucleus. In endothelial cells, exposure to high glucose can promote persistent activation of *RELA* gene transcription that remains following a return to normal glucose exposure [74]. The mechanism for this transcriptional upregulation involves the Set7 methyltransferase. Increased expression and nuclear localization of this methyltransferase causes an increase in H3K4me1 at the *RELA* promoter that contributes to excess p65 expression and the subsequent transcriptional activation of many oxidant/inflammatory genes including *NOS2A* [73–75]. In endothelial cells, the high glucose-mediated upregulation of *NOS2A* and the other inflammatory genes did not involve a direct increase in H3K4me1 at the inflammatory gene promoters; rather, their upregulation was found to be dependent on Set7-mediated *RELA* expression [73]. In human lens epithelial cells, high glucose was similarly found to upregulate *NOS2A* gene expression through the accumulation of nuclear p65 and an increase of NFkB-dependent transcription; the expression of *RELA* and epigenetic changes were not examined in this study [76]. In TNF-α-mediated activation, Set7 recruitment to NFkB-regulated genes promoted an increase of H3K4me3 that correlated with an upregulation of inflammatory genes. Other studies have found that Set7 and NFkB directly associate, and in addition to its methylation of histones, Set 7 may directly methylate NFkB to enhance its transcriptional activity [75, 77].

NOS3 Regulation and Epigenetic Crosstalk

eNOS is constitutively expressed in endothelial cells, although it may be found in many other cells such as pulmonary epithelium, macrophages, osteoclasts, and neurons [78]. The *NOS3* gene promoter has three alternative transcription start sites, but the major start site is in exon 2, immediately downstream of the proximal promoter element. *NOS3* transcription is regulated by many ubiquitously expressed transcription factors, and proximal promoter reporter gene constructs have been shown to have similar patterns of expression in endothelial and nonendothelial cell lines, notwithstanding the differential expression of the endogenous *NOS3* gene in these cells [79]. Thus, it has been proposed that epigenetic mechanisms provide a means to regulate differentially *NOS3* expression in a cell-specific manner [79, 80]. Analysis of the proximal promoter and upstream enhancer of the *NOS3* gene provides some evidence to support this theory: in multiple cell lines that do not express eNOS, the proximal promoter region of the *NOS3* gene (−361/+3) was found to be heavily methylated, whereas in endothelial cells this region of the gene was unmethylated [79]. Distal regions (−4912/−4587) were less methylated in all cell types. Consequently, chromatin immunoprecipitation assays showed a greater recruitment of the transcription factors Sp1, Sp3, and Ets to their sites in the *NOS3* proximal promoter in endothelial cells compared to other cell types, indicating that a heavily methylated promoter may block association of key transcriptional factors to limit *NOS3* gene expression in nonendothelial cells. Furthermore, CpG methylation of recombinant promoter reporter constructs significantly repressed promoter activity in endothelial cells, providing additional proof that DNA methylation inhibits *NOS3* gene transcription. Subsequent studies have supported these findings. In vascular smooth muscle cells, DNA methylation at the *NOS3* promoter enhanced the binding of MeCP2 [81], a protein involved in transcriptional repression that may recruit histone-modifying enzymes. A comparison of human placental arterial and venous cells also reported that the extent of *NOS3* proximal promoter methylation inversely correlated with the levels of its expression [82]. Similarly, in proangiogenic cells, such as early EPCs, mesoangioblasts, and bone marrow-derived-CD34$^+$ cells, the *NOS3* promoter was silenced by a combination of DNA methylation and repressive histone marks in the proximal promoter region [83]. In comparison, the *NOS3* promoter in eNOS-expressing microvascular and umbilical vein endothelial cells was unmethylated [83]. In this study of proangiogenic cells, *NOS3* gene expression was induced in the nonexpressing cells by pharmacologically suppressing repressive histone marks. Furthermore, following this induction, the *NOS3* promoter DNA remained methylated. Other studies, however, have upregulated the expression of *NOS3* in nonendothelial cells, such as HeLa and vascular smooth muscle cells, by inhibiting DNA methylation with 5-azacytidine [79] or by using a combination of HDAC inhibitors combined with 5-azacytidine [84]. In this latter study, 5-azacytidine alone had minimal effect on upregulating *NOS3* gene expression in nonendothelial cells; however, it synergistically enhanced trichostatin A-induced upregulation of *NOS3* expression.

The role of HDAC inhibitors on *NOS3* expression is complex: in the context of nonexpressing cells, trichostatin A augmented *NOS3* expression; however, in endothelial cells HDAC inhibitors were found to suppress *NOS3* expression [81, 84]. The basis for the suppression of *NOS3* expression by HDAC inhibitors is unclear, but as discussed earlier, HDAC and HAT enzymes may target nonhistone proteins, including transcription factors or enzymes affecting signaling pathways, to modify their activities. The eNOS protein is also a target for acetylation and HDAC1 and HDAC3 may directly suppress eNOS activity by lysine deacetylation [85, 86].

Several studies have reported that permissive histone marks correlate with *NOS3* gene expression. Thus, in endothelial cells, the *NOS3* proximal promoter was enriched in acetylated H3K9 and H4K12, as well as the permissive methylation marks H3K4me2 and H3K4me3, compared to nonendothelial cell types [81]. Similarly, in a comparison of eNOS expressing and nonexpressing proangiogenic cells, permissive histone marks were found at regulatory domains of the *NOS3* promoter in endothelial cells, whereas repressive marks predominated in the nonexpressing cells. In this study, the positive regulatory domains I and II from −169/−99, which overlaps Sp1 and Ets binding sites, the AP-1 binding site (−695/−623), the Smad2 binding sites (−934/−871), and the upstream hypoxia responsive element (HRE) (−5347/5290) were assessed. Each of these regions had increased H3 acetylation and H3K4me3 marks in eNOS-expressing endothelial cells. Repressive histone marks, especially H3K27me3, were enriched in proangiogenic early EPCs and CD34+ cells that do not express eNOS [83]. Treatment of these nonexpressing cells with a combination of an HDAC inhibitor, trichostatin A, together with 3-deazaneplanocin A, an inhibitor of EZH2, increased the ratio of H3K4me3/H3K27me3 and depressed *NOS3* gene expression in the proangiogenic cells.

In addition to its function as a constitutively expressed protein in endothelial cells, eNOS is induced during neurogenesis, where its transient expression plays an essential role in regulating neural cell proliferation [87]. The role of epigenetics in this process was studied in an in vitro system involving retinoic acid (RA)-induced differentiation of human teratocarcinoma NT2 cells. In this study, the transient increase in *NOS3* gene expression was inversely correlated with a gradual reduction (over days) in DNA methylation of the *NOS3* promoter region from −734 to −989. Following peak expression of eNOS, *NOS3* gene expression gradually diminished over time, as the DNA methylation was restored in the *NOS3* promoter. Similarly, H3 acetylation patterns transiently increased in the −734 to −989 region, as well as a more proximal promoter region, in concert with the upregulation of *NOS3* gene expression. In this study, eNOS protein localized to the nucleus of these differentiating neuronal cells, suggesting a source for nuclear NO that may *S*-nitrosylate HDAC2 or other nuclear factors.

Studies have also implicated epigenetic mechanisms in modulating the transcription of *NOS3* in response to various physiological stimuli (Fig. 4.5). Thus, laminar shear stress has been shown to lead to increased acetylation of H3 and H4 histones in the *NOS3* promoter that corresponds with an increase in *NOS3* gene expression [88]. In particular, p300 was implicated in increasing histone acetylation following shear stress. Subsequent studies have shown that Krüppel-like factor 2 (KLF2), a transcription factor known to regulate *NOS3* expression, is also upregulated by fluid shear stress through mechanisms that lead to increased histone acetylation of the *KLF2* gene [89]. *KLF2* transcription is regulated by the transcription factor MEF2C, which in the absence of flow associates with HDAC5. Shear stress enhanced MEF2C-mediated *KLF2* transcription by increasing the phosphorylation and nuclear export of HDAC5 via activation of a calcium/calmodulin-dependent kinase [90]. Mechanistically, the phosphorylation-induced dissociation of HDAC5 and MEF2C may allow for chromatin acetylation by HAT enzymes. Although histone acetylation was not assessed in this study, increased expression of KLF2 by shear stress resulted in increased *NOS3* transcription. In other studies, KLF2 was shown also to directly bind with HDAC5 to repress its ability to promote *NOS3* transcription; knockdown of HDAC5 in cultured endothelial cells or in a mouse model increased eNOS expression [91]. Interestingly, in endothelial cells, NO production may cause the activation of protein phosphatase 2A to shuttle the class II HDACs 4 and 5 from the cytoplasm to the nucleus where they can decrease histone acetylation and modulate gene expression [92]. Taken together, these data suggest a possible feedback loop that may regulate *NOS3* transcription via HDAC translocation, depending on the levels of cellular NO.

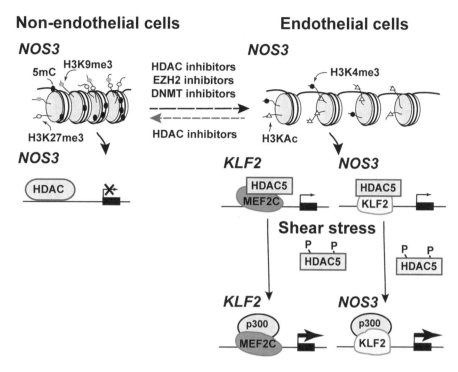

Fig. 4.5 Epigenetic regulation of *NOS3* gene expression. In nonendothelial cells, epigenetic tags, such as DNA methylation (5mC) and histone methylation (H3K9me3 and H3K27me3), may contribute to the lack of eNOS in these cells. In nonendothelial cells, removal of 5mC and/or repressive histone tags promotes *NOS3* gene expression. These changes can be induced by targeting the inhibition of DNMT, HDAC, or EZH2 (see text for details). Interestingly, in a model of neuronal cell differentiation, retinoic acid was shown to mediate chromatin remodeling to transiently upregu- late *NOS3* gene expression; differentiation in these cells leads to a subsequent suppression of *NOS3* expression and a reversion of the repressive epigenetic tags. In endothelial cells, *NOS3* gene expression is paradoxically suppressed by HDAC inhibition. *NOS3* gene expression can be upregulated by fluid shear stress in endothelial cells due to direct and indirect epigenetic mechanisms. Briefly, shear stress enhances the calcium/calmodulin-dependent phosphorylation of HDAC5, to promote its translocation from the nucleus. Unphosphorylated HDAC5 binds to MEF2C thereby repressing *KLF2* gene transcription. Similarly, HDAC5 binds to KLF2, to suppress directly KLF2-driven *NOS3* gene transcription. Loss of nuclear HDAC5 under shear stress allows for the recruitment of p300 by KLF2 or MEF2C nuclear factors at the *NOS3* and *KLF2* genes, respectively, to promote histone acetylation and gene expression

Overview of NO Regulation of HDAC

Nitric oxide modulates HDAC activity via S-nitrosylation and NO-induced activation of phosphory- lation pathways. Thus, HDAC2 is a target for direct S-nitrosylation, which attenuates its chromatin binding and promotes its cytoplasmic localization, leading to increased histone acetylation. This mechanism is crucial to neuronal and skeletal muscle differentiation. In lung disease, such as COPD, severe asthma, or smoking, loss of HDAC2 (a class I HDAC) function owing to its S-nitrosylation or possibly tyrosine nitration is thought to contribute to the pathobiology of these diseases [93–95]. In contrast, NO-mediated activation of protein phosphatase 2A fosters the dephosphorylation and subse- quent nuclear translocation of class II HDAC, decreasing histone acetylation. The consequences of these opposing actions of cellular NO on histone acetylation may depend on subcellular localization of NOS enzymes, cell-specific expression of various HDACs, and the association of various HDACs

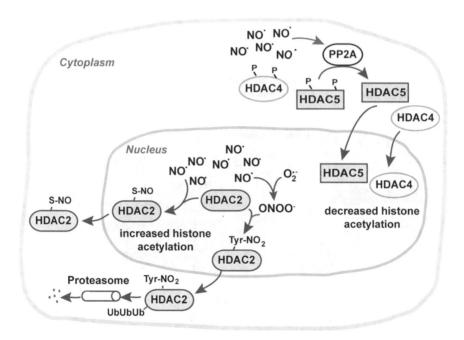

Fig. 4.6 NO modulation of HDAC function. NO has been shown to increase histone acetylation by causing the cyto-plasmic translocation of HDAC2 by S-nitrosylation (S-NO) or tyrosine nitration (Tyr-NO$_2$). Tyr-NO$_2$ leads to the subsequent ubiquitination (Ub) and proteasomal degradation of HDAC2. NO may also decrease histone acetylation by indirect pathways involving the activation of protein phosphatase 2A (PP2A). Dephosphorylation of HDACII members, including HDAC5 and HDAC4, promotes their migration to the nucleus to reduce histone acetylation. It is unclear whether these pathways coexist in the same cells, however, these mechanisms provide additional means by which nitric oxide may modulate gene expression

with specific gene targets. It has been suggested that protein–protein trans-*S*-nitrosylation may be one mechanism by which to transmit S-nitrosylation between cytoplasmic and nuclear compartments in the absence of nuclear NOS localization [96]. Regardless of the precise mechanism, the influence of NO on HDAC function provides a mechanism by which cellular NO influences gene expression by regulating epigenetic processes (Fig. 4.6).

Nitric oxide has been used to promote mesodermic differentiation of mouse embryonic stem cells, in part, owing to its ability to activate class II HDAC [97]. Thus, NO exposure of stem cells grown in the absence of LIF induced global H3 deacetylation and suppressed the expression of stem cell genes, while increasing the expression of mesodermal markers. Injection of these cells in a model of hindlimb ischemia increased revascularization, whereas an HDAC class IIa-specific inhibitor lessened this effect. Many questions remain, however, regarding the beneficial effects of NO in this process that await further investigation. New HDAC inhibitors with an NO functional group have been synthesized and screened for their abilities to promote myogenic differentiation [98]. These compounds have been shown both to stimulate cGMP pathways and inactivate HDACs by multiple mechanisms, including S-nitrosylation in cell culture systems, but the in vivo effects of these dual-acting agents have not yet been tested.

Although the full extent by which NO affects gene expression via epigenetic mechanisms is unclear, it seems likely that dietary and pharmacological interventions that alter NO production or stores may similarly affect gene expression via epigenetic mechanism. Thus, agents such as pentaerythritol tetranitrate or dietary nitrite and nitrate may have beneficial effects that extend beyond their regulation of cGMP-mediated events. Nitrate and nitrite are formed endogenously via oxidation of NO. They are also consumed in normal diets as natural components of vegetables or food additives. In the oral cavity, nitrate is converted to nitrite by nitrate reducing bacteria. The role of nitrate/nitrite in human health has been controversial because of the cancer risks associated with nitrosamines, especially in the context of nitrate/nitrite use as food preservatives in meat processing. The formation of nitrosamines, however, is favored by the coadministration of other carcinogenic precursors [100] and recent findings on nitrite/ nitrate supplements support their beneficial effects on the cardiovascular system [101–104]. Thus, administration of these supplements promotes an improvement in endothelium-dependent vasodilation and a suppression of platelet aggregation. These NO-like properties may be due, in part, to the conversion of nitrite to NO. Furthermore, dietary nitrite supplements have been shown to increase S-nitrosylation of proteins in the gastrointestinal tract (added in part by the acidic gastric environment [105]). Similarly, increased consumption of nitrite increases the levels of circulating plasma nitrite and, theoretically, may contribute to tissue-specific protein S-nitrosylation, as well [106, 107], and perhaps even the modulation of HDAC nitrosylation. Further study, however, is needed to understand better how NO-modulates epigenetic events and the contributions of dietary nitrite and nitrate to these processes.

Overview of the Epigenetic Regulation of NO Production

The three *NOS* genes have unique promoters that are differentially transcribed during differentiation, tissue remodeling and repair, or in response to stimuli, such as shear stress and cytokines. Epigenetic mechanisms contribute to many of these processes. In general, DNA methylation, possibly in combination with repressive histone marks, silences these genes in nonexpressing tissues. Although the signals that regulate the establishment and removal of these marks are not well understood, evidence exists that a reversal of these inhibitory epigenetic marks can increase expression of these genes, at least in some cell types. In neuronal differentiation, the chromatin remodeling at the *NOS3* gene promoter correlates with the transient expression of this gene. In many cells, stimulation of cytokine receptors or Toll-like receptors of the innate immune system promotes chromatin remodeling and *NOS2A* gene expression. Furthermore, enhancement of *NOS3* gene expression by shear stress is associated with increases in its promoter-associated histone acetylation.

Although reversal of DNA methylation may occur under some physiological conditions, histone modifications are more readily altered to modulate gene transcription. In the setting of inflammation, HDAC inhibitors suppress the induction of *NOS2A* gene expression by interfering with NFkB function or via disruption of other pathways. These inhibitors may be useful as therapies in autoimmune diseases such as rheumatoid arthritis or systemic lupus erythematosus and possibly even during the development of type 1 diabetes, where they may limit cytotoxicity [99]. Surprisingly, these agents can also decrease normal endothelial cell production of NO by lessening *NOS3* gene expression by mechanisms that have not been fully elucidated. Similarly, HDAC inhibitors are useful in promoting muscle regeneration in the context of the DMD mouse, where they may compensate for the loss of NO-mediated HDAC2 regulation, but they have shown no utility in promoting muscle recovery following hindlimb ischemia, possibly owing to the suppression of the normal induction of *NOS1* and *NOS3* expression. The means by which HDAC inhibitors suppress *NOS1* and *NOS3* expression in the context of hindlimb ischemia are not clear; in neuroblastoma cells *NOS1* gene expression was upregulated by HDAC inhibition that enhanced NFkB acetylation.

Conclusion

Thus, although HDAC inhibitors may prove useful in some disease settings, it is difficult to predict their effects on *NOS* genes (or other genes) under all physiological conditions. Nonetheless, epigenetics-based research may provide additional insights into *NOS* gene regulation and the development of novel therapies that regulate NO production.

References

1. Wu Z, Siuda D, Xia N, Reifenberg G, Daiber A, Munzel T, Forstermann U, Li H. Maternal treatment of spontaneously hypertensive rats with pentaerythritol tetranitrate reduces blood pressure in female offspring. Hypertension. 2015;65(1):232–7.
2. Burdge GC, Slater-Jefferies J, Torrens C, Phillips ES, Hanson MA, Lillycrop KA. Dietary protein restriction of pregnant rats in the F0 generation induces altered methylation of hepatic gene promoters in the adult male offspring in the F1 and F2 generations. Br J Nutr. 2007;97(3):435–9.
3. Lillycrop KA, Slater-Jefferies JL, Hanson MA, Godfrey KM, Jackson AA, Burdge GC. Induction of altered epigenetic regulation of the hepatic glucocorticoid receptor in the offspring of rats fed a protein-restricted diet during pregnancy suggests that reduced DNA methyltransferase-1 expression is involved in impaired DNA methylation and changes in histone modifications. Br J Nutr. 2007;97(6):1064–73.
4. Bogdarina I, Welham S, King PJ, Burns SP, Clark AJ. Epigenetic modification of the renin-angiotensin system in the fetal programming of hypertension. Circ Res. 2007;100(4):520–6.
5. Turunen MP, Aavik E, Yla-Herttuala S. Epigenetics and atherosclerosis. Biochim Biophys Acta. 2009;1790(9):886–91.
6. Bird AP. CpG-rich islands and the function of DNA methylation. Nature. 1986;321(6067):209–13.
7. Reik W. Stability and flexibility of epigenetic gene regulation in mammalian development. Nature. 2007;447(7143):425–32.
8. Handy DE, Castro R, Loscalzo J. Epigenetic modifications: basic mechanisms and role in cardiovascular disease. Circulation. 2011;123(19):2145–56.
9. Chen T, Ueda Y, Dodge JE, Wang Z, Li E. Establishment and maintenance of genomic methylation patterns in mouse embryonic stem cells by Dnmt3a and Dnmt3b. Mol Cell Biol. 2003;23(16):5594–605.
10. Liang G, Chan MF, Tomigahara Y, Tsai YC, Gonzales FA, Li E, Laird PW, Jones PA. Cooperativity between DNA methyltransferases in the maintenance methylation of repetitive elements. Mol Cell Biol. 2002;22(2):480–91.
11. Okano M, Bell DW, Haber DA, Li E. DNA methyltransferases Dnmt3a and Dnmt3b are essential for de novo methylation and mammalian development. Cell. 1999;99(3):247–57.
12. Ito S, D'Alessio AC, Taranova OV, Hong K, Sowers LC, Zhang Y. Role of Tet proteins in 5mC to 5hmC conversion, ES-cell self-renewal and inner cell mass specification. Nature. 2010;466(7310):1129–33.
13. Tahiliani M, Koh KP, Shen Y, Pastor WA, Bandukwala H, Brudno Y, Agarwal S, Iyer LM, Liu DR, Aravind L, Rao A. Conversion of 5-methylcytosine to 5-hydroxymethylcytosine in mammalian DNA by MLL partner TET1. Science. 2009;324(5929):930–5.
14. Scourzic L, Mouly E, Bernard OA. TET proteins and the control of cytosine demethylation in cancer. Genome Med. 2015;7(1):9.
15. Hahn MA, Szabo PE, Pfeifer GP. 5-Hydroxymethylcytosine: a stable or transient DNA modification? Genomics. 2014;104(5):314–23.
16. Hashimoto H, Liu Y, Upadhyay AK, Chang Y, Howerton SB, Vertino PM, Zhang X, Cheng X. Recognition and potential mechanisms for replication and erasure of cytosine hydroxymethylation. Nucleic Acids Res. 2012;40(11):4841–9.
17. He YF, Li BZ, Li Z, Liu P, Wang Y, Tang Q, Ding J, Jia Y, Chen Z, Li L, Sun Y, Li X, Dai Q, Song CX, Zhang K, He C, Xu GL. Tet-mediated formation of 5-carboxylcytosine and its excision by TDG in mammalian DNA. Science. 2011;333(6047):1303–7.
18. Strahl BD, Allis CD. The language of covalent histone modifications. Nature. 2000;403(6765):41–5.
19. Kouzarides T. Chromatin modifications and their function. Cell. 2007;128(4):693–705.
20. Latham JA, Dent SY. Cross-regulation of histone modifications. Nat Struct Mol Biol. 2007;14(11):1017–24.
21. Cheng X. Structural and functional coordination of DNA and histone methylation. Cold Spring Harb Perspect Biol. 2014;6(8).pii:a018747.
22. Ooi L, Belyaev ND, Miyake K, Wood IC, Buckley NJ. BRG1 chromatin remodeling activity is required for efficient chromatin binding by repressor element 1-silencing transcription factor (REST) and facilitates REST-mediated repression. J Biol Chem. 2006;281(51):38974–80.
23. Berndsen CE, Denu JM. Catalysis and substrate selection by histone/protein lysine acetyltransferases. Curr Opin Struct Biol. 2008;18(6):682–9.
24. Li J, Wang J, Wang J, Nawaz Z, Liu JM, Qin J, Wong J. Both corepressor proteins SMRT and N-CoR exist in large protein complexes containing HDAC3. EMBO J. 2000;19(16):4342–50.
25. Kao HY, Downes M, Ordentlich P, Evans RM. Isolation of a novel histone deacetylase reveals that class I and class II deacetylases promote SMRT-mediated repression. Genes Dev. 2000;14(1):55–66.
26. Nan X, Ng HH, Johnson CA, Laherty CD, Turner BM, Eisenman RN, Bird A. Transcriptional repression by the methyl-CpG-binding protein MeCP2 involves a histone deacetylase complex. Nature. 1998;393(6683):386–9.

27. Pham TX, Lee J. Dietary regulation of histone acetylases and deacetylases for the prevention of metabolic diseases. Nutrients. 2012;4(12):1868–86.
28. Greer EL, Shi Y. Histone methylation: a dynamic mark in health, disease and inheritance. Nat Rev Genet. 2012;13(5):343–57.
29. Shinsky SA, Monteith KE, Viggiano S, Cosgrove MS. Biochemical reconstitution and phylogenetic comparison of human SET1 family core complexes involved in histone methylation. J Biol Chem. 2015;290(10):6361–75.
30. Zhang P, Chaturvedi CP, Tremblay V, Cramet M, Brunzelle JS, Skiniotis G, Brand M, Shilatifard A, Couture JF. A phosphorylation switch on RbBP5 regulates histone H3 Lys4 methylation. Genes Dev. 2015;29(2):123–8.
31. Mozzetta C, Pontis J, Fritsch L, Robin P, Portoso M, Proux C, Margueron R, Ait-Si-Ali S. The histone H3 lysine 9 methyltransferases G9a and GLP regulate polycomb repressive complex 2-mediated gene silencing. Mol Cell. 2014;53(2):277–89.
32. Tachibana M, Ueda J, Fukuda M, Takeda N, Ohta T, Iwanari H, Sakihama T, Kodama T, Hamakubo T, Shinkai Y. Histone methyltransferases G9a and GLP form heteromeric complexes and are both crucial for methylation of euchromatin at H3-K9. Genes Dev. 2005;19(7):815–26.
33. Martin C, Zhang Y. The diverse functions of histone lysine methylation. Nat Rev Mol Cell Biol. 2005;6(11):838–49.
34. Mosammaparast N, Shi Y. Reversal of histone methylation: biochemical and molecular mechanisms of histone demethylases. Annu Rev Biochem. 2010;79:155–79.
35. Fang R, Chen F, Dong Z, Hu D, Barbera AJ, Clark EA, Fang J, Yang Y, Mei P, Rutenberg M, Li Z, Zhang Y, Xu Y, Yang H, Wang P, Simon MD, Zhou Q, Li J, Marynick MP, Li X, Lu H, Kaiser UB, Kingston RE, Xu Y, Shi YG. LSD2/KDM1B and its cofactor NPAC/GLYR1 endow a structural and molecular model for regulation of H3K4 demethylation. Mol Cell. 2013;49(3):558–70.
36. Chicas A, Kapoor A, Wang X, Aksoy O, Evertts AG, Zhang MQ, Garcia BA, Bernstein E, Lowe SW. H3K4 demethylation by Jarid1a and Jarid1b contributes to retinoblastoma-mediated gene silencing during cellular senescence. Proc Natl Acad Sci U S A. 2012;109(23):8971–6.
37. Hubner MR, Spector DL. Role of H3K27 demethylases Jmjd3 and UTX in transcriptional regulation. Cold Spring Harb Symp Quant Biol. 2010;75:43–9.
38. Guccione E, Bassi C, Casadio F, Martinato F, Cesaroni M, Schuchlautz H, Luscher B, Amati B. Methylation of histone H3R2 by PRMT6 and H3K4 by an MLL complex are mutually exclusive. Nature. 2007;449(7164):933–7.
39. Yuan CC, Matthews AG, Jin Y, Chen CF, Chapman BA, Ohsumi TK, Glass KC, Kutateladze TG, Borowsky ML, Struhl K, Oettinger MA. Histone H3R2 symmetric dimethylation and histone H3K4 trimethylation are tightly correlated in eukaryotic genomes. Cell Rep. 2012;1(2):83–90.
40. Bros M, Boissel JP, Godtel-Armbrust U, Forstermann U. Transcription of human neuronal nitric oxide synthase mRNAs derived from different first exons is partly controlled by exon 1-specific promoter sequences. Genomics. 2006;87(4):463–73.
41. Das S, Foley N, Bryan K, Watters KM, Bray I, Murphy DM, Buckley PG, Stallings RL. MicroRNA mediates DNA demethylation events triggered by retinoic acid during neuroblastoma cell differentiation. Cancer Res. 2010;70(20):7874–81.
42. Saur D, Seidler B, Paehge H, Schusdziarra V, Allescher HD. Complex regulation of human neuronal nitric-oxide synthase exon 1c gene transcription. Essential role of Sp and ZNF family members of transcription factors. J Biol Chem. 2002;277(28):25798–814.
43. Boissel JP, Zelenka M, Godtel-Armbrust U, Feuerstein TJ, Forstermann U. Transcription of different exons 1 of the human neuronal nitric oxide synthase gene is dynamically regulated in a cell- and stimulus-specific manner. Biol Chem. 2003;384(3):351–62.
44. Li Y, Li C, Sun L, Chu G, Li J, Chen F, Li G, Zhao Y. Role of p300 in regulating neuronal nitric oxide synthase gene expression through nuclear factor-kappaB-mediated way in neuronal cells. Neuroscience. 2013;248:681–9.
45. Li Y, Zhao Y, Li G, Wang J, Li T, Li W, Lu J. Regulation of neuronal nitric oxide synthase exon 1f gene expression by nuclear factor-kappaB acetylation in human neuroblastoma cells. J Neurochem. 2007;101(5):1194–204.
46. Nott A, Watson PM, Robinson JD, Crepaldi L, Riccio A. S-Nitrosylation of histone deacetylase 2 induces chromatin remodelling in neurons. Nature. 2008;455(7211):411–5.
47. Nott A, Nitarska J, Veenvliet JV, Schacke S, Derijck AA, Sirko P, Muchardt C, Pasterkamp RJ, Smidt MP, Riccio A. S-nitrosylation of HDAC2 regulates the expression of the chromatin-remodeling factor Brm during radial neuron migration. Proc Natl Acad Sci U S A. 2013;110(8):3113–8.
48. Cacchiarelli D, Martone J, Girardi E, Cesana M, Incitti T, Morlando M, Nicoletti C, Santini T, Sthandier O, Barberi L, Auricchio A, Musaro A, Bozzoni I. MicroRNAs involved in molecular circuitries relevant for the Duchenne muscular dystrophy pathogenesis are controlled by the dystrophin/nNOS pathway. Cell Metab. 2010;12(4):341–51.
49. Colussi C, Mozzetta C, Gurtner A, Illi B, Rosati J, Straino S, Ragone G, Pescatori M, Zaccagnini G, Antonini A, Minetti G, Martelli F, Piaggio G, Gallinari P, Steinkuhler C, Clementi E, Dell'Aversana C, Altucci L, Mai A, Capogrossi MC, Puri PL, Gaetano C. HDAC2 blockade by nitric oxide and histone deacetylase inhibitors reveals a common target in Duchenne muscular dystrophy treatment. Proc Natl Acad Sci U S A. 2008;105(49):19183–7.

50. Minetti GC, Colussi C, Adami R, Serra C, Mozzetta C, Parente V, Fortuni S, Straino S, Sampaolesi M, Di Padova M, Illi B, Gallinari P, Steinkuhler C, Capogrossi MC, Sartorelli V, Bottinelli R, Gaetano C, Puri PL. Functional and morphological recovery of dystrophic muscles in mice treated with deacetylase inhibitors. Nat Med. 2006;12(10):1147–50.

51. Spallotta F, Tardivo S, Nanni S, Rosati JD, Straino S, Mai A, Vecellio M, Valente S, Capogrossi MC, Farsetti A, Martone J, Bozzoni I, Pontecorvi A, Gaetano C, Colussi C. Detrimental effect of class-selective histone deacetylase inhibitors during tissue regeneration following hindlimb ischemia. J Biol Chem. 2013;288(32):22915–29.

52. Chan GC, Fish JE, Mawji IA, Leung DD, Rachlis AC, Marsden PA. Epigenetic basis for the transcriptional hypo-responsiveness of the human inducible nitric oxide synthase gene in vascular endothelial cells. J Immunol. 2005;175(6):3846–61.

53. Yu Z, Kone BC. Hypermethylation of the inducible nitric-oxide synthase gene promoter inhibits its transcription. J Biol Chem. 2004;279(45):46954–61.

54. Gross TJ, Kremens K, Powers LS, Brink B, Knutson T, Domann FE, Philibert RA, Milhem MM, Monick MM. Epigenetic silencing of the human NOS2 gene: rethinking the role of nitric oxide in human macrophage inflammatory responses. J Immunol. 2014;192(5):2326–38.

55. Gregory DJ, Zhang Y, Kobzik L, Fedulov AV. Specific transcriptional enhancement of inducible nitric oxide synthase by targeted promoter demethylation. Epigenetics. 2013;8(11):1205–12.

56. Madrigano J, Baccarelli A, Mittleman MA, Sparrow D, Vokonas PS, Tarantini L, Schwartz J. Aging and epigenetics: longitudinal changes in gene-specific DNA methylation. Epigenetics. 2012;7(1):63–70.

57. Baccarelli A, Rusconi F, Bollati V, Catelan D, Accetta G, Hou L, Barbone F, Bertazzi PA, Biggeri A. Nasal cell DNA methylation, inflammation, lung function and wheezing in children with asthma. Epigenomics. 2012;4(1):91–100.

58. Salam MT, Byun HM, Lurmann F, Breton CV, Wang X, Eckel SP, Gilliland FD. Genetic and epigenetic variations in inducible nitric oxide synthase promoter, particulate pollution, and exhaled nitric oxide levels in children. J Allergy Clin Immunol. 2012;129(1):232–9.e1-7.

59. Deng WG, Wu KK. Regulation of inducible nitric oxide synthase expression by p300 and p50 acetylation. J Immunol. 2003;171(12):6581–8.

60. Granja AG, Sabina P, Salas ML, Fresno M, Revilla Y. Regulation of inducible nitric oxide synthase expression by viral A238L-mediated inhibition of p65/RelA acetylation and p300 transactivation. J Virol. 2006;80(21):10487–96.

61. Hwang SY, Hwang JS, Kim SY, Han IO. O-GlcNAc transferase inhibits LPS-mediated expression of inducible nitric oxide synthase through an increased interaction with mSin3A in RAW264.7 cells. Am J Physiol Cell Physiol. 2013;305(6):C601–8.

62. Yu Z, Zhang W, Kone BC. Histone deacetylases augment cytokine induction of the iNOS gene. J Am Soc Nephrol. 2002;13(8):2009–17.

63. Susick L, Veluthakal R, Suresh MV, Hadden T, Kowluru A. Regulatory roles for histone deacetylation in IL-1beta-induced nitric oxide release in pancreatic beta-cells. J Cell Mol Med. 2008;12(5A):1571–83.

64. Larsen L, Tonnesen M, Ronn SG, Storling J, Jorgensen S, Mascagni P, Dinarello CA, Billestrup N, Mandrup-Poulsen T. Inhibition of histone deacetylases prevents cytokine-induced toxicity in beta cells. Diabetologia. 2007;50(4):779–89.

65. Yu Z, Kone BC. Targeted histone H4 acetylation via phosphoinositide 3-kinase- and p70s6-kinase-dependent pathways inhibits iNOS induction in mesangial cells. Am J Physiol Renal Physiol. 2006;290(2):F496–502.

66. Furumai R, Ito A, Ogawa K, Maeda S, Saito A, Nishino N, Horinouchi S, Yoshida M. Histone deacetylase inhibitors block nuclear factor-kappaB-dependent transcription by interfering with RNA polymerase II recruitment. Cancer Sci. 2011;102(5):1081–7.

67. Lundh M, Christensen DP, Damgaard Nielsen M, Richardson SJ, Dahllof MS, Skovgaard T, Berthelsen J, Dinarello CA, Stevenazzi A, Mascagni P, Grunnet LG, Morgan NG, Mandrup-Poulsen T. Histone deacetylases 1 and 3 but not 2 mediate cytokine-induced beta cell apoptosis in INS-1 cells and dispersed primary islets from rats and are differentially regulated in the islets of type 1 diabetic children. Diabetologia. 2012;55(9):2421–31.

68. Dekker FJ, van den Bosch T, Martin NI. Small molecule inhibitors of histone acetyltransferases and deacetylases are potential drugs for inflammatory diseases. Drug Discov Today. 2014;19(5):654–60.

69. Hu J, Colburn NH. Histone deacetylase inhibition down-regulates cyclin D1 transcription by inhibiting nuclear factor-kappaB/p65 DNA binding. Mol Cancer Res. 2005;3(2):100–9.

70. Serrat N, Sebastian C, Pereira-Lopes S, Valverde-Estrella L, Lloberas J, Celada A. The response of secondary genes to lipopolysaccharides in macrophages depends on histone deacetylase and phosphorylation of C/EBPbeta. J Immunol. 2014;192(1):418–26.

71. Kolyada AY, Madias NE. Transcriptional regulation of the human iNOS gene by IL-1beta in endothelial cells. Mol Med. 2001;7(5):329–43.

72. Schmidt N, Pautz A, Art J, Rauschkolb P, Jung M, Erkel G, Goldring MB, Kleinert H. Transcriptional and post-transcriptional regulation of iNOS expression in human chondrocytes. Biochem Pharmacol. 2010;79(5):722–32.

73. Paneni F, Costantino S, Battista R, Castello L, Capretti G, Chiandotto S, Scavone G, Villano A, Pitocco D, Lanza G, Volpe M, Luscher TF, Cosentino F. Adverse epigenetic signatures by histone methyltransferase Set7 contribute to vascular dysfunction in patients with type 2 diabetes mellitus. Circ Cardiovasc Genet. 2015;8(1):150–8.

74. El-Osta A, Brasacchio D, Yao D, Pocai A, Jones PL, Roeder RG, Cooper ME, Brownlee M. Transient high glucose causes persistent epigenetic changes and altered gene expression during subsequent normoglycemia. J Exp Med. 2008;205(10):2409–17.

75. Keating ST, El-Osta A. Chromatin modifications associated with diabetes. J Cardiovasc Transl Res. 2012;5(4):399–412.

76. Jia J, Liu Y, Zhang X, Liu X, Qi J. Regulation of iNOS expression by NF-kappaB in human lens epithelial cells treated with high levels of glucose. Invest Ophthalmol Vis Sci. 2013;54(7):5070–7.

77. Li Y, Reddy MA, Miao F, Shanmugam N, Yee JK, Hawkins D, Ren B, Natarajan R. Role of the histone H3 lysine 4 methyltransferase, SET7/9, in the regulation of NF-kappaB-dependent inflammatory genes. Relevance to diabetes and inflammation. J Biol Chem. 2008;283(39):26771–81.

78. Mattila JT, Thomas AC. Nitric oxide synthase: non-canonical expression patterns. Front Immunol. 2014;5:478.

79. Chan Y, Fish JE, D'Abreo C, Lin S, Robb GB, Teichert AM, Karantzoulis-Fegaras F, Keightley A, Steer BM, Marsden PA. The cell-specific expression of endothelial nitric-oxide synthase: a role for DNA methylation. J Biol Chem. 2004;279(33):35087–100.

80. Fish JE, Marsden PA. Endothelial nitric oxide synthase: insight into cell-specific gene regulation in the vascular endothelium. Cell Mol Life Sci. 2006;63(2):144–62.

81. Fish JE, Matouk CC, Rachlis A, Lin S, Tai SC, D'Abreo C, Marsden PA. The expression of endothelial nitric-oxide synthase is controlled by a cell-specific histone code. J Biol Chem. 2005;280(26):24824–38.

82. Joo JE, Hiden U, Lassance L, Gordon L, Martino DJ, Desoye G, Saffery R. Variable promoter methylation contributes to differential expression of key genes in human placenta-derived venous and arterial endothelial cells. BMC Genomics. 2013;14:475.

83. Ohtani K, Vlachojannis GJ, Koyanagi M, Boeckel JN, Urbich C, Farcas R, Bonig H, Marquez VE, Zeiher AM, Dimmeler S. Epigenetic regulation of endothelial lineage committed genes in pro-angiogenic hematopoietic and endothelial progenitor cells. Circ Res. 2011;109(11):1219–29.

84. Gan Y, Shen YH, Wang J, Wang X, Utama B, Wang J, Wang XL. Role of histone deacetylation in cell-specific expression of endothelial nitric-oxide synthase. J Biol Chem. 2005;280(16):16467–75.

85. Hyndman KA, Ho DH, Sega MF, Pollock JS. Histone deacetylase 1 reduces NO production in endothelial cells via lysine deacetylation of NO synthase 3. Am J Physiol Heart Circ Physiol. 2014;307(5):H803–9.

86. Jung SB, Kim CS, Naqvi A, Yamamori T, Mattagajasingh I, Hoffman TA, Cole MP, Kumar A, Dericco JS, Jeon BH, Irani K. Histone deacetylase 3 antagonizes aspirin-stimulated endothelial nitric oxide production by reversing aspirin-induced lysine acetylation of endothelial nitric oxide synthase. Circ Res. 2010;107(7):877–87.

87. Jezierski A, Deb-Rinker P, Sodja C, Walker PR, Ly D, Haukenfrers J, Sandhu JK, Bani-Yaghoub M, Sikorska M. Involvement of NOS3 in RA-Induced neural differentiation of human NT2/D1 cells. J Neurosci Res. 2012;90(12):2362–77.

88. Chen W, Bacanamwo M, Harrison DG. Activation of p300 histone acetyltransferase activity is an early endothelial response to laminar shear stress and is essential for stimulation of endothelial nitric-oxide synthase mRNA transcription. J Biol Chem. 2008;283(24):16293–8.

89. Huddleson JP, Ahmad N, Srinivasan S, Lingrel JB. Induction of KLF2 by fluid shear stress requires a novel promoter element activated by a phosphatidylinositol 3-kinase-dependent chromatin-remodeling pathway. J Biol Chem. 2005;280(24):23371–9.

90. Wang W, Ha CH, Jhun BS, Wong C, Jain MK, Jin ZG. Fluid shear stress stimulates phosphorylation-dependent nuclear export of HDAC5 and mediates expression of KLF2 and eNOS. Blood. 2010;115(14):2971–9.

91. Kwon IS, Wang W, Xu S, Jin ZG. Histone deacetylase 5 interacts with Kruppel-like factor 2 and inhibits its transcriptional activity in endothelium. Cardiovasc Res. 2014;104(1):127–37.

92. Illi B, Dello Russo C, Colussi C, Rosati J, Pallaoro M, Spallotta F, Rotili D, Valente S, Ragone G, Martelli F, Biglioli P, Steinkuhler C, Gallinari P, Mai A, Capogrossi MC, Gaetano C. Nitric oxide modulates chromatin folding in human endothelial cells via protein phosphatase 2A activation and class II histone deacetylases nuclear shuttling. Circ Res. 2008;102(1):51–8.

93. Barnes PJ. Histone deacetylase-2 and airway disease. Ther Adv Respir Dis. 2009;3(5):235–43.

94. Osoata GO, Yamamura S, Ito M, Vuppusetty C, Adcock IM, Barnes PJ, Ito K. Nitration of distinct tyrosine residues causes inactivation of histone deacetylase 2. Biochem Biophys Res Commun. 2009;384(3):366–71.

95. Yao H, Rahman I. Role of histone deacetylase 2 in epigenetics and cellular senescence: implications in lung inflammaging and COPD. Am J Physiol Lung Cell Mol Physiol. 2012;303(7):L557–66.

96. Kornberg MD, Sen N, Hara MR, Juluri KR, Nguyen JV, Snowman AM, Law L, Hester LD, Snyder SH. GAPDH mediates nitrosylation of nuclear proteins. Nat Cell Biol. 2010;12(11):1094–100.

97. Spallotta F, Rosati J, Straino S, Nanni S, Grasselli A, Ambrosino V, Rotili D, Valente S, Farsetti A, Mai A, Capogrossi MC, Gaetano C, Illi B. Nitric oxide determines mesodermic differentiation of mouse embryonic

stem cells by activating class IIa histone deacetylases: potential therapeutic implications in a mouse model of hindlimb ischemia. Stem Cells. 2010;28(3):431–42.

98. Borretto E, Lazzarato L, Spallotta F, Cencioni C, D'Alessandra Y, Gaetano C, Fruttero R, Gasco A. Synthesis and biological evaluation of the first example of NO-donor histone deacetylase inhibitor. ACS Med Chem Lett. 2013;4(10):994–9.

99. Christensen DP, Gysemans C, Lundh M, Dahllof MS, Noesgaard D, Schmidt SF, Mandrup S, Birkbak N, Workman CT, Piemonti L, Blaabjerg L, Monzani V, Fossati G, Mascagni P, Paraskevas S, Aikin RA, Billestrup N, Grunnet LG, Dinarello CA, Mathieu C, Mandrup-Poulsen T. Lysine deacetylase inhibition prevents diabetes by chromatin-independent immunoregulation and beta-cell protection. Proc Natl Acad Sci U S A. 2014;111(3):1055–9.

100. Bryan NS, Alexander DD, Coughlin JR, Milkowski AL, Boffetta P. Ingested nitrate and nitrite and stomach cancer risk: an updated review. Food Chem Toxicol. 2012;50(10):3646–65.

101. Garg HK, Bryan NS. Dietary sources of nitrite as a modulator of ischemia/reperfusion injury. Kidney Int. 2009;75(11):1140–4.

102. Omar SA, Webb AJ. Nitrite reduction and cardiovascular protection. J Mol Cell Cardiol. 2014;73:57–69.

103. Weitzberg E, Lundberg JO. Novel aspects of dietary nitrate and human health. Annu Rev Nutr. 2013;33:129–59.

104. Machha A, Schechter AN. Dietary nitrite and nitrate: a review of potential mechanisms of cardiovascular benefits. Eur J Nutr. 2011;50(5):293–303.

105. Pereira C, Barbosa RM, Laranjinha J. Dietary nitrite induces nitrosation of the gastric mucosa: the protective action of the mucus and the modulatory effect of red wine. J Nutr Biochem. 2015;26(5):476–83.

106. Jiang H, Torregrossa AC, Potts A, Pierini D, Aranke M, Garg HK, Bryan NS. Dietary nitrite improves insulin signaling through GLUT4 translocation. Free Radic Biol Med. 2014;67:51–7.

107. Milsom AB, Fernandez BO, Garcia-Saura MF, Rodriguez J, Feelisch M. Contributions of nitric oxide synthases, dietary nitrite/nitrate, and other sources to the formation of NO signaling products. Antioxid Redox Signal. 2012;17(3):422–32.

Chapter 5
The Mitochondrion: A Physiological Target of Nitrite

Danielle A. Guimaraes, Chris Reyes, and Sruti Shiva

Key Points

- The mitochondrial respiratory chain can metabolize nitrite by reducing nitrite to nitric oxide in hypoxia.
- Nitrite-mediated protection against ischemia/reperfusion injury is dependent on the S-nitrosation of mitochondrial complex I, which attenuates mitochondrial reactive oxygen species generation at reperfusion.
- Heme proteins such as myoglobin form a metabolome with nitrite and mitochondria to catalyze the reduction of nitrite to NO, which can reversibly inhibit mitochondrial complex IV to conserve tissue oxygen during hypoxia.
- Nitrite modulates mitochondrial protein expression to increase exercise efficiency.
- Nitrite stimulates specific cellular signaling pathways to increase hypoxic mitochondrial biogenesis and propagate mitochondrial fusion.

Keywords Mitochondria • Metabolism • Bioenergetics • Nitric oxide • Nitrite • Ischemia/reperfusion • Exercise • Hypoxia

Introduction

The mitochondrion is a highly dynamic organelle present in most eukaryotic cells in significant numbers. Once known solely as the "powerhouse" of the cell, the homeostatic role of the mitochondrion is now recognized to span far beyond ATP production and encompass a number of functions including

D.A. Guimaraes, Ph.D.
Vascular Medicine Institute, University of Pittsburg, 1240E Biomedical Science Tower, 200 Lothrop St., Pittsburgh, PA 15261, USA

C. Reyes, B.S.
Department of Bioengineering, University of Pittsburgh, 1240E Biomedical Science Tower, 200 Lothrop St., Pittsburgh, PA 15261, USA

S. Shiva, Ph.D. (✉)
Department of Pharmacology and Chemical Biology, Vascular Medicine Institute, Center for Metabolism and Mitochondrial Medicine (C3M), University of Pittsburgh School of Medicine, 1240E BST, 200 Lothrop St., Pittsburgh, PA 15261, USA
e-mail: sss43@pitt.edu

N.S. Bryan, J. Loscalzo (eds.), *Nitrite and Nitrate in Human Health and Disease*, Nutrition and Health, DOI 10.1007/978-3-319-46189-2_5, © Springer International Publishing AG 2017

redox signal transduction, lipid metabolism, Ca^{2+} homeostasis, and heme formation. Additionally, the central role of mitochondrial cytochrome c release in the apoptotic cascade demonstrates that the organelle not only maintains life but also regulates cell death. Interactions between nitrite and mito-chondria were first observed in the 1960s [1], long before nitrite was postulated to have physiological bioactivity. These early studies focused on mitochondrial-dependent reduction of nitrite to nitric oxide (NO). However, with the recognition that nitrite represents an endocrine reserve of NO activity, which mediates a number of physiological responses [2, 3], interest in nitrite-dependent regulation of mitochondria has greatly expanded. It is now evident that the mitochondrion is a significant subcel-lular target for nitrite and that specific modulation of mitochondrial function, structure, and number underlies many of the physiological effects of nitrite [4–11]. This chapter will first provide a brief overview of mitochondrial function, basic nitrite biology, and mitochondrial sites of nitrite reduction; however, the majority of this chapter will focus on the known mechanisms by which nitrite regulates mitochondrial structure and function to mediate its physiological effects.

Mitochondrial Function: ATP Generation and Beyond

Mitochondrial function is tightly linked to the double membrane structure of the organelle, which is comprised of a thin outer membrane (OM) and much thicker inner membrane (IM), separated by the intermembrane space. Buried within the inner mitochondrial membrane is the electron transport chain (ETC), consisting of five protein complexes, each comprised of multiple subunits, encoded by both nuclear and mitochondrial DNA (excluding Complex II which is entirely encoded in nuclear DNA). These five complexes catalyze ATP production via oxidative phosphorylation. In this coordinated process, substrates donate electrons to either complex I or complex II, from which they are shuttled through complex III and then by the small electron transport protein cytochrome c to complex IV ultimately to reduce oxygen to water. The movement of electrons through these complexes is coupled to the pumping of protons by complexes I, III, and IV from the matrix, across the mitochondrial inner membrane into the intermembrane space, where the buildup of protons maintains a membrane poten-tial ($\Delta\psi$) across the inner membrane. This proton motive force drives the generation of ATP at com-plex V (the F_1F_0 ATPase) and couples oxygen consumption to ATP production. On a molecular level, ATP synthase acts as a molecular motor whereby the passing of a proton through its transmembrane channel back into the mitochondrial matrix results in a conformational rotation of its γ subunit facili-tating phosphorylation of ADP to generate ATP [8].

While the majority of electrons entering the electron transport chain are utilized at complex IV to reduce oxygen, a small proportion escape the chain prior to complex IV, predominantly at complexes I and II, and facilitate a one electron reduction of oxygen to generate superoxide ($O_2^{\bullet}$) [12, 13]. While it is approximated that 1–3 % of electrons entering the electron transport chain undergo this fate, it is established that the level of superoxide production can be dynamically modulated depending on the membrane potential and electron transport flux of the respiratory chain [14, 15]. Although excessive ROS production results in detrimental oxidation of proteins and lipids [16, 17], low levels of mitochon-drial ROS have been linked to protective physiological signaling including induction of transcription factors such as hypoxia-inducible factor-1 (HIF-1α) and inducing ischemic preconditioning [18–20].

Beyond bioenergetics, mitochondria are key regulators of the apoptotic pathway [21, 22]. Release of the electron transport protein cytochrome c from the mitochondrion is an irreversible initiating signal for apoptosis. When released into the cytosol, the interaction of cytochrome c with apoptosis protease activating factor-1 (APAF1) forms an apoptosome, which then activates caspase 9 to trigger a downstream cascade resulting in apoptosis [21, 22]. The precise mechanisms by which cytochrome c is released from the mitochondrion is unclear; however, its release is closely associated with the opening of a voltage and calcium sensitive channel that spans the inner and outer mitochondrial membrane, known as the permeability transition pore (PTP). Assembly and opening of this pore causes eventual depolarization of the mitochondrion, rupture of the mitochondrial membranes, and cytochrome c release [23–25].

Nitrite Formation and Reduction

Nitric oxide (NO) is a diatomic endogenous signaling molecule with a biological half-life in the millisecond range. NO is generated enzymatically in vivo by nitric oxide synthase (NOS) and, once produced, reacts rapidly with metal centers or other free radicals, and is oxidized to nitrite and nitrate [26]. While NO oxidation is the predominant source of nitrite in the body, dietary nitrate consumption also contributes to nitrite levels in vivo. Particularly, leafy green vegetables are high in nitrate content, and ingested nitrate is reduced to nitrite through a complex entero-salivary pathway (reviewed in [27, 28]). Briefly, through this pathway, nitrate is absorbed by the gastrointestinal tract and enters the circulation, from which it is taken up by the salivary gland. Once in the saliva, nitrate is actively reduced to nitrite by the commensal bacteria in the oral cavity [27, 28]. Total nitrite levels are maintained in the nanomolar range (100–400 nM) in blood [29] and in the micromolar range (1–10 μM) in tissue [30], with levels varying depending on diet, exercise, and levels of inflammation [28, 31–35].

While nitrite was once considered physiologically inert, it is now recognized as an endocrine reservoir of NO bioactivity that mediates a number of physiological effects, including, but not limited to, hypoxic vasodilation [34, 36, 37], modulation of gene/protein expression [6, 11, 30, 38], induction of angiogenesis [39, 40], prevention of ischemia/reperfusion injury (for a review see [41, 42], and regulation of exercise capacity [11]). The majority of nitrite's signaling effects are dependent on its reduction to bioavailable NO, a process that is optimized in conditions of hypoxia and acidosis. The reduction of nitrite is catalyzed by a number of proteins, including deoxygenated heme globins (hemoglobin [34, 43], myoglobin [8, 44], neuroglobin [45]), xanthine oxidoreductase [46], nitric oxide synthase [47], and cytochrome P450 enzymes [48], and these proteins appear to contribute differentially depending on tissue type and level of hypoxia [49]. Notably, this reduction of nitrite that is operative when oxygen is limited is viewed physiologically as an alternative source of bioavailable NO that works in concert with NOS, which requires oxygen as a substrate to generate NO [2, 3].

Nitrite Reduction by Mitochondria

The first reported interaction between nitrite and mitochondria was published by Taylor and colleagues in 1965. This study demonstrated that isolated porcine mitochondria metabolize sodium nitrite to nitrosylated cytochrome c, indicative of NO formation [1]. Notably, no speculation was made in this study about the physiological implications of this nitrite reduction. Several decades later, Nohl and colleagues confirmed that isolated mitochondria, indeed, reduce nitrite to NO and further localized this activity to the electron transport chain utilizing submitochondrial particles. More specifically, inhibitor studies identified the bc_1 site of complex III to be responsible for nitrite reduction [50, 51].

Poyton and colleagues identified cytochrome c oxidase (complex IV) as another site of hypoxic nitrite reduction [52, 53]. While the exact mechanism by which complex IV reduces nitrite is unclear, it has been shown that enzymatic turnover is required for this reduction, and inhibition of cytochrome c oxidase with carbon monoxide completely abolished the NO production. Additionally, nitrite reduction at this site requires complete anoxia, and NO production is potentiated as pH decreases below 7. Notably, nitrite reduction by cytochrome c oxidase requires the presence of its substrate cytochrome c [52]. However, studies by Kim-Shapiro and colleagues have identified cytochrome c as another putative site of nitrite reduction, independent of cytochrome c oxidase [54]. Cytochrome c is a 12 kDa heme protein that normally exists in a hexacoordinate state, with the heme iron bound by histidine-18 and methionine-80. However, upon oxidation or nitration of the protein or when anionic inner membrane lipids are associated with the protein, the iron-methionine bond may weaken and break, resulting in a pentacoordinated heme. This pentacoordinated cytochrome c is able to effectively reduce nitrite in anoxic and acidic conditions to generate bioavailable NO. Furthermore, in vitro experiments utilizing submitochondrial particles reconstituted with cytochrome c and anionic lipid demonstrated that the NO generated by nitrite reduction could bind the active site of cytochrome c oxidase, resulting in respiratory inhibition [54]. While these data suggest that cytochrome c-dependent nitrite reduction has potential physiological consequences, many questions remain about the relevance of this reaction in vivo, including whether a large enough proportion of pentacoordinate cytochrome c, which cannot mediate electron transfer, exists to generate enough NO to inhibit respiration.

Studies by Feelisch and colleagues demonstrate a positive correlation in rat tissue between the tissue rate of nitrite reduction and mitochondrial content, supporting the physiological role of mitochondria-dependent nitrite reduction [48]. However, while many mitochondrial reduction sites have been identified, the relative role of each of these proteins and whether they work together remain unclear. It has been suggested that nitrite concentration governs the site of nitrite reduction, with complex III-mediated reduction occurring at lower (micromolar) concentrations while cytochrome c oxidase-mediated reduction occurring at higher (millimolar) concentrations [55]. Further work is required to determine whether this hypothesis is true in vivo and in vitro, as well as to compare mitochondrial nitrite reduction efficiency to that of other nonmitochondrial enzymes that possess this activity.

Mitochondria as a Physiological Target of Nitrite

While the physiological role of mitochondrial-dependent nitrite reduction remains in question, accumulated evidence clearly demonstrates that the mitochondrion is a physiological subcellular target through which nitrite mediates signaling and cytoprotection [7, 11, 41]. Several sites within the mitochondrion have been described to be modified by nitrite specifically, leading to regulation of mitochondrial function [4, 6, 8–10, 35, 56]. This regulation of respiration, ROS production, and apoptotic signaling underlies many of the physiologic and cytoprotective effects of nitrite, particularly in the context of physiological hypoxia and pathological ischemic disease. The remainder of this chapter will focus on the known mitochondrial targets of nitrite, the mechanisms by which modification of these targets regulate organelle function, and the physiological implications of this nitrite-dependent control.

S-Nitrosation of Complex I to Attenuate Ischemia/Reperfusion Injury

Ischemia/reperfusion (I/R) injury is a major component of a number of pathologies (e.g., stroke, myocardial infarction, organ transplant, etc.) and occurs when blood supply to an organ is disrupted for minutes to hours (ischemia) and then rapidly restored (reperfusion) [57]. While restoration of blood flow to ischemic tissue is desirable in order to generate ATP, the rapid reentry of oxygen on top of existing ischemic damage exacerbates cell and tissue injury. On a tissue level, reperfusion induces a dysregulated inflammatory response characterized by cytokine production, increased leukocyte adhesion, and upregulation of endothelial adhesion markers [58, 59]. Cellular swelling occurs due to the disruption of the sodium/potassium ATPase, oxidative enzymes such as NADPH oxidase are released, and NO levels are pathologically decreased [60–63]. Additionally, mitochondrial dysfunction is central to the progression of I/R injury and is characterized in ischemia by the inability to synthesize ATP, the buildup of electrons within the electron transport chain, and an increase in mitochondrial calcium concentration [64–67].

Upon reperfusion, electrons built up within the electron transport chain leak from the chain onto the high concentration of oxygen and generate ROS, which then oxidize critical proteins and lipids. Furthermore, a large influx of calcium along with oxidation of the inner membrane leads to the opening of the permeability transition pore, cytochrome c release, and eventual cell death [23, 68–70].

One of the most well characterized and reproducible physiological effects of nitrite is its ability to mediate potent cytoprotection against I/R injury (reviewed in [41, 42]). Nitrite-mediated cytoprotection has been confirmed in a number of animal models of I/R and in all major organ systems, including heart [9, 44, 71–75], liver [9, 71, 76–78], brain [79, 80], and kidney [81], as well as limb ischemia [39, 40], and in systemic cardiac arrest and resuscitation models [82, 83]. While each individual study is too extensive to discuss in this forum (see [41, 42] for a review), it is notable that nitrite-dependent cytoprotection is versatile in both its temporal and dose range and potent in both in vitro and in vivo models. For example, nitrite mediates the same degree of protection when administered 15–30 min before the ischemic episode as when administered in the middle of the ischemic period or just 5 min prior to reperfusion. Dose–response studies are remarkably reproducible across different laboratories and show that nitrite has a biphasic concentration-dependent protective effect. In vivo, 0.1–100 μM/ kg provides cytoprotection in most animal models, with peak protection occurring at concentrations that increase plasma nitrite levels by 10 μM [9, 71, 72, 79, 83, 84]. Notably, depletion of endogenous nitrite renders the heart and liver susceptible to exacerbated I/R injury while dietary nitrite supplementation attenuates injury [38, 78, 85, 86]. These studies are consistent with nitrite being a physiological modulator of the ischemic response.

On a mechanistic level, nitrite-mediated posttranslational modification of mitochondrial complex I underlies the cytoprotective effects of nitrite observed after I/R [9, 75]. Complex I (NADH ubiquinone oxidoreductase) is a large (~980 kDa) 46 subunit complex that catalyzes the transfer of electrons from NADH to ubiquinone through its nine iron sulfur centers and flavin mononucleotide active site. Additionally, the complex contains several reactive cysteine residues that are integral to its catalytic function. Importantly, complex I not only represents the entry point of electrons into the electron transport chain, but is also a significant site of damage during I/R, which leads to its production of high concentrations of ROS at reperfusion [87, 88]. The role of complex I ROS production in the pathogenesis of I/R injury is confirmed by studies in which complex I inhibitors such as rotenone and amobarbital decrease reperfusion ROS production and attenuate I/R injury [87, 89]. Nitrite recapitulates these studies by inhibiting complex I activity, preventing cell death, and attenuating I/R injury [9, 75, 78, 83].

On a molecular level, nitrite inhibits complex I through the specific S-nitrosation (addition of an NO^+ group to a cysteine) of complex I. In 2008, Shiva and colleagues demonstrated that nitrite-mediated protection after hepatic I/R was strongly associated with complex I S-nitrosation and the subsequent attenuation of mitochondrial reperfusion ROS production [9]. Chouchani and colleagues followed up on this study and later identified cysteine 39 of the ND3 subunit as the critical residue that is S-nitrosated to confer cytoprotection [75]. Furthermore, they demonstrated that removal of this modification abolishes cytoprotection. It is postulated that Cys39 serves as an "ischemic switch" for complex I. In conditions of high enzymatic activity, Cys39 is buried within the enzyme complex and is thereby protected from modification. However, in conditions of low enzymatic activity, particularly in the absence of NADH supply, as occurs in ischemic conditions, the ND3 subunit undergoes a conformational change such that Cys39 is exposed. Posttranslational modification of the exposed Cys39 is thought to "lock" the complex in its low activity conformation during ischemia, which decreases electron transfer but also attenuates reperfusion ROS production, inhibiting oxidation of critical proteins and lipids [75]. Importantly, S-nitrosation is reversible. Thus, at reperfusion, the initial burst of mitochondrial ROS is attenuated, but S-nitrosation is gradually reversed leading to the ultimate restoration of electron transport function. Brookes and colleagues have termed this phenomenon "gradual wake up" [69].

Several studies have now confirmed the role of complex I S-nitrosation in cytoprotection since its original discovery [78, 83]. Dezfulian and colleagues demonstrated in a murine model of cardiopulmonary arrest and resuscitation that mitochondrial complex I in the heart was S-nitrosated in response to nitrite, an effect that was associated with improved survival and cardiac function. This study also

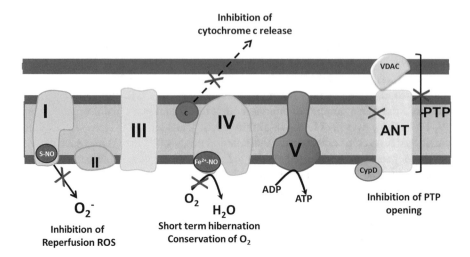

Fig. 5.1 Mitochondrial mechanisms of nitrite-mediated protection during I/R. During ischemia, nitrite is reduced to NO which binds the heme center on complex IV, resulting in the inhibition of respiration to conserve oxygen and facilitate short-term hibernation. Nitrite also S-nitrosates complex I which inhibits reperfusion ROS generation and cytochrome c release. Nitrite inhibits permeability transition pore (PTP) opening. While the exact mechanism is unknown, nitrite is known to downregulate the ANT, a major component of the PTP

defined the reversibility of S-nitrosation as time course experiments showed that S-nitrosation, while present 5 min after the initiation of reperfusion, was completely reversed by 60 min into reperfusion. This effect was accompanied by a 50 % inhibition of complex I activity at 5 min that returned to uninhibited basal levels at 60 min [83]. In a separate study, Raat and colleagues showed that dietary nitrite supplementation inhibited liver complex I and mediated protection against hepatic I/R in mice. Importantly, the extent to which complex I was inhibited was directly correlated with the level of protection after I/R, as well as the concentration of nitrite in the diet [78]. These data demonstrate that mitochondrial complex I represents an important physiological and therapeutic target for nitrite-dependent ischemic tolerance.

It is important to note that beyond inhibition of mitochondrial ROS production, nitrite has been shown to attenuate both mitochondrial permeability pore opening and cytochrome c release after I/R. This has been demonstrated in isolated mitochondria subjected to anoxia/reoxygenation as well as in cultured myoblasts, and in vivo this ultimately results in the prevention of I/R-induced apoptosis [4, 9, 73, 78, 79, 83, 90]. Oxidation of cardiolipin and other mitochondrial inner membrane phospholipids has been shown to propagate cytochrome c release [91]. Thus, it is likely that nitrite-dependent complex I S-nitrosation at least partially prevents cytochrome c release through the attenuation of lipid oxidation. However, more work is required to determine whether nitrite directly modulates other targets, such as components of the permeability transition pore, to further prevent mitochondrial membrane potential collapse and initiation of the apoptotic cascade (Fig. 5.1).

Heme Nitrosylation of Complex IV: Conservation of Oxygen and Short-Term Hibernation

Cytochrome c oxidase (complex IV), the tersminal complex of the respiratory chain, catalyzes the reduction of oxygen to water. The active site of the enzyme consists of a $heme_{a3}$-$copper_B$ binuclear center to which oxygen binds and is reduced by electrons transported into the complex by cytochrome c. Notably, the binuclear center binds not only oxygen but also other small gaseous signaling molecules, including NO and CO. Specifically, NO binds to the heme of the binuclear center, precluding the binding of oxygen and thus inhibiting mitochondrial oxygen consumption (and oxidative phosphorylation) [92–94]. Importantly, this heme-nitrosylation and resultant inhibition of respiration is completely reversible [93, 95]. Thus, several groups have hypothesized that NO-mediated complex IV inhibition represents a mechanism to conserve oxygen in hypoxic tissues [96–98]. In the myocardium, this concept is known as "short-term hibernation" and is thought to preserve cardiac function in the ischemic heart.

While physiological concentrations of nitrite do not directly bind complex IV, NO generated by the reduction of nitrite has been shown to inhibit complex IV and mitochondrial respiration [8, 43–45, 54]. This phenomenon has been studied most extensively in the heart where myoglobin is the predominant nitrite reductase. The monomeric heme protein myoglobin has traditionally been considered solely as an oxygen storage and transport protein which supports mitochondrial respiration through facilitated diffusion in the hypoxic heart and skeletal muscle [99]. However, it is now evident that myoglobin possesses nitrite reductase activity [8, 44]. Deoxygenated ferrous myoglobin reacts with nitrite to yield bioactive NO, oxidizing myoglobin in the process [8].

$$deoxyMb\left(Fe^{2+}\right) + nitrite + H^+ \rightarrow deoxyMb\left(Fe^{3+}\right) + NO + OH \qquad (5.1)$$

A number of studies have now tested the capability of nitrite-derived NO to inhibit mitochondrial respiration. Early studies by Shiva, et al., using purified myoglobin and isolated mitochondria, showed that nitrite inhibited respiration only in the presence of deoxygenated myoglobin [8]. In isolated cardiomyocytes, nitrite inhibits mitochondrial respiration when this effect is lost when myoglobin is oxidized such that it cannot reduce nitrite [8]. Follow-up studies by Rassaf and colleagues recapitulated these results in a Langendorff murine isolated perfused heart model in which nitrite treatment simulated "short-term hibernation" by downregulating cardiac energy status in hearts from wild-type mice but had no effect in mice lacking myoglobin [44, 97].

As mentioned earlier, heme-nitrosylation and, thus, respiratory inhibition is completely reversible and dependent on oxygen concentration [95]. Hence, it is possible to imagine a scenario in the ischemic heart in which coordinated nitrite-dependent inhibition of complexes I and IV mediate protection. During ischemia, myoglobin-mediated complex IV inhibition serves to downregulate metabolism to conserve oxygen. However, once oxygen is restored, this inhibition is reversed to reestablish ATP production, while complex I inhibition persists for a longer duration in order to attenuate ROS generation (Fig. 5.1). Consistent with a role for myoglobin-dependent nitrite reduction in ischemic tolerance, mice genetically lacking myoglobin are not protected by nitrite after I/R [44]. While a functional metabolome consisting of nitrite, mitochondria, and myoglobin appears to be required for myocardial protection, other heme globins, including neuroglobin [45, 100] and cytoglobin [101], have also been shown to catalyze nitrite reduction; it remains to be explored whether these proteins modulate ischemic tolerance in other organs.

Nitrite-Dependent Regulation of Mitochondrial Ion Channels: Putative Targets in I/R

Recent studies suggest that nitrite modulates mitochondrial ion channels in the heart [56, 74]. While the modulation of these channels has not been extensively investigated, these targets may potentially contribute to nitrite-mediated cardioprotection after myocardial I/R. The mitochondrial K_{ATP} channel has been closely associated with cardioprotection. While the exact mechanism by which this channel operates and mediates protection remains unclear, opening of the channel has been linked to prevention of permeability pore opening during reperfusion [102–104]. In 2007, Baker and colleagues reported in a Langendorff perfused rat heart model that pharmacological inhibition of the mitochondrial K_{ATP} channel significantly attenuated the protective effect of nitrite after I/R [74]. This study has not been reproduced to date, and further investigation is required to determine if this is an operative mechanism in vivo. However, separate studies have shown that NO stimulates opening of the K_{ATP} channel through the S-nitrosation of critical cysteines on the channel raising the possibility that nitrite mediates a similar response [68, 105].

More recently, Shulz and colleagues demonstrated that nitrite mediates S-nitrosation of mitochondrial connexin 43 [56]. Connexin 43, the predominant plasma membrane gap junction protein, has been shown to be expressed in the mitochondrial inner membrane [106]. Formation of a nonspecific pore by the association of six connexin proteins regulates potassium gradients and ROS production by the mitochondrion, as well as protein movement into the organelle. Notably, formation of this channel has been linked to the effects of cardioprotective agents, including NO [107]. Shulz et al. showed that administration of nitrite into the ventricle of wild-type mice at physiological oxygen tension resulted in S-nitrosation of mitochondrial connexin-43 [56]. However, the physiological implication of this modification remains unclear in terms of whether it regulates mitochondrial function and whether it is present during I/R. Further work is required to test the role of nitrite-mediated modification of both the K_{ATP} and connexin-43 channels in I/R and other physiological models.

Downregulation of Adenine Nucleotide Translocase and Uncoupling Protein 3: Enhanced Exercise Efficiency

In contrast to pathological I/R, exercising muscle represents a physiological situation of hypoxia and acidosis, an environment that is conducive to nitrite reduction and signaling. To that end, several studies have demonstrated the effects of dietary nitrate (which increases circulating nitrite levels through entero-salivary reduction) on exercise in humans (for a review see [11, 108]). In 2007, Larsen and colleagues were the first to show in healthy subjects that dietary supplementation with sodium nitrate decreased oxygen cost without increasing lactate production during submaximal exercise. These data demonstrate that nitrate significantly increased exercise efficiency compared to placebo treatment [109]. While this initial study consisted of well-trained subjects exercising on a cycle ergometer, this beneficial effect of nitrite was quickly confirmed by studies by a number of different laboratories utilizing different forms of exercise and subjects ranging from athletes to patients with peripheral artery disease. In addition, separate studies have shown that nitrate supplementation confers resistance to exercise fatigue and this is associated with the phenomenon of enhanced efficiency [10, 35, 110–112].

Mechanistic investigation into this phenomenon revealed that regulation of exercise efficiency was at the level of the mitochondrion [10, 11]. Respirometry of skeletal muscle biopsies from healthy subjects after dietary placebo or nitrate supplementation showed a significant increase in mitochondrial efficiency with nitrate supplementation versus placebo as evidenced by an approximately 19 % increase in the ratio of ATP production to oxygen consumption (P:O ratio) [10]. Consistent with this, increased mitochondrial coupling of oxygen consumption to ATP production was observed. On a molecular level, nitrate decreased the expression of two key proteins that regulate mitochondrial coupling. Nitrite significantly downregulated the expression of the adenine nucleotide transferase (ANT), which not only catalyzes the export and import of ATP and ADP from the mitochondrion respectively, but also facilitates proton leak from the intermembrane space into the mitochondrial matrix [10]. Nitrite also decreased expression of uncoupling protein 3, another protein that mediate proton leak, although this did not reach statistical significance. Downregulation of these proteins significantly decreases the leak of protons across the inner membrane, raising mitochondrial membrane potential and ensuring that a greater number of protons pumped into the intermembrane space by the electron transport chain are utilized for ATP production. This leads to increased mitochondrial coupling and the enhanced efficiency observed in these studies.

The precise mechanism by which nitrite modulates ANT and UCP3 (uncoupling protein 3) protein expression remains under unclear. Skeletal muscle contains micromolar concentrations of myoglobin, so it is interesting to consider the role of myoglobin-dependent nitrite reduction in the modulation of ANT and UCP3 expression. Furthermore, although it has been shown that nitrite regulates ANT protein expression [10], it is possible that nitrite also regulates the protein's activity. This is a particularly interesting prospect given that ANT contains a number of surface thiols that are susceptible to S-nitrosation and dictate the protein's activity [113]. Further investigation is required to elucidate these mechanistic details as well as determine whether other mitochondrial targets play a role in the enhancement of exercise efficiency.

Nitrite Regulation of Mitochondrial Dynamics and Number

Nitrite Stimulates Mitochondrial Biogenesis

Mitochondrial function is directly dependent on the number of distinct mitochondria within the cell. The absolute number of mitochondria can increase to meet cellular bioenergetic and signaling requirements through the process of mitochondrial biogenesis in which mitochondrial and nuclear gene expression is coordinated to increase mitochondrial number [114, 115]. The master regulator of mitochondrial biogenesis is nuclear-localized PPARγ coactivator 1α (PGC1α), which acts as a transcriptional coactivator with nuclear respiratory factor 1 (Nrf1) to stimulate mitochondrial transcription factor A (TFAM)-mediated mtDNA transcription and mitochondrial protein expression. This biogenesis pathway is stimulated by a number of physiologic conditions including caloric restriction [116, 117] and is associated with cellular adaptation and stress resistance. Conversely, decreased biogenesis has been linked to attenuated OXPHOS gene transcription in several diseases [118–121], suggesting biogenesis is critical in maintaining proper mitochondrial function. Notably, NO has been characterized as a potent stimulator of biogenesis as supplementation with NO donors stimulates biogenesis through a mechanism dependent on the activation of soluble guanylyl cyclase and resultant cGMP production to activate PGC1α [122]. Consistent with NO being a physiological regulator of biogenesis, endothelial NOS knockout mice have markedly lowered mitochondrial content than wild-type mice [122].

In 2012 Mo and colleagues demonstrated that nitrite stimulates hypoxic mitochondrial biogenesis in vascular smooth muscle cells [6]. This effect is initiated by nitrite-dependent activation of adenylate kinase which, through the alteration of the AMP/ATP ratio, activates the central metabolic sensor AMP Kinase (AMPK). AMPK in turn activates sirtuin 1 (SIRT1), which de-acetylates PGC1α, a vital step in its activation. Nitrite-dependent PGC1α activation initiates the biogenesis cascade through which expression of nuclear respiratory factor-2 and TFAM generate functional mitochondria (Fig. 5.2) [6]. Importantly, unlike NO-mediated biogenesis, nitrite-dependent activation and upregulation of PGC1α is independent of soluble guanylyl cyclase activation and cGMP production. Genetic deletion of the β-subunit of soluble guanylate cyclase in smooth muscle cells significantly attenuated NO-dependent but not nitrite-mediated increases in PGC1α mRNA levels. Moreover, inhibition of nitrite reduction to NO did not significantly attenuate nitrite-mediated biogenesis [6]. These data, taken together, suggest that nitrite-mediated biogenesis occurs by a mechanism completely independent of NO formation.

The physiological implications of this phenomenon are vast given that mitochondrial biogenesis represents a mechanism of cellular adaptation and repair. It has been most clearly demonstrated in a rat carotid vascular injury model that nitrite treatment increases PGC1α mRNA levels and mitochondrial number in vivo. This effect was associated with a protection against injury-induced vascular intimal hyperplasia [6]. Notably, nitrite-mediated biogenesis is not an operative mechanism in all disease models as a number of studies utilizing long-term nitrite or nitrate supplementation have found no indication of nitrite-stimulated biogenesis [9, 11, 35]. Speculatively, the lack of biogenesis in these models is potentially due to the lack of sufficient hypoxia and/or acidosis as a catalyst. However, further investigation is required to determine the physiological conditions that optimize nitrite-mediated biogenesis.

Nitrite Modulates Mitochondrial Dynamics to Mediate Preconditioning

Mitochondrial function is continually regulated and altered to meet the demands of the cell. One major mechanism by which mitochondrial function can be modulated on a minute to hour timescale is by the dynamic remodeling of cellular mitochondrial networks through fusion and fission events. The elongation of mitochondrial networks through fusion generally enhances mitochondrial activity and confers cellular resilience. This process is countered by fission, the fragmentation of mitochondrial networks, which is often observed in cellular injury and can be an initial step in the degradation of dysfunctional mitochondria. Fission and fusion processes are mediated predominantly by a family of small GTPase proteins that include dynamin related protein-1 (drp-1), the main catalyst of fission, and mitofusins 1 and 2, whose expression propagate fusion [123].

Studies from our laboratory have recently shown that nitrite is a potent stimulator of mitochondrial fusion in cultured myocytes, adipocytes, and in the heart [4, 5]. This modulation of mitochondrial dynamics occurs predominantly through the inhibition of drp-1 activity which, through the attenuation of fission, propagates mitochondrial fusion. Mechanistically, nitrite mediates this effect by directly activating protein kinase A (PKA). A major target of PKA is drp-1, which possesses an inhibitory phosphorylation site at serine 656. PKA-dependent phosphorylation of this site decreases translocation of drp-1 from the cytosol to the mitochondrial membrane and also inhibits the mitochondrial dividing activity of the GTPase. This results in the inhibition of fission and the elongation of mitochondrial networks in the cell (fusion) (Fig. 5.2) [4, 5].

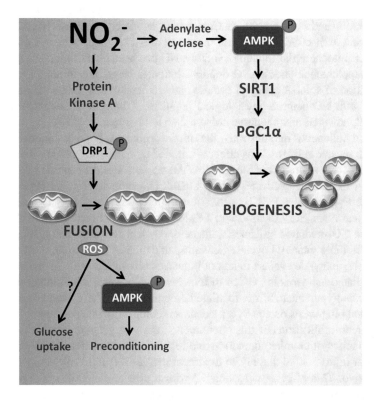

Fig. 5.2 Nitrite modulation mitochondrial number and dynamics. In normoxia, nitrite activates Protein kinase A to phosphorylate dynamin related protein 1 (drp1) leading to mitochondrial fusion. This results in mitochondrial ROS production which has been shown to oxidize and activate AMPK in myocytes leading to preconditioning. Fusion also stimulates glucose uptake in adipocytes by an unknown mechanism. In hypoxia nitrite activates adenylate cyclase to activate AMPK, which activates sirtuin-1 (SIRT1) leading to the de-acetylation and activation of PGC1α and subsequent biogenesis

The nitrite-dependent propagation of mitochondrial fusion has known physiological implications. Nitrite-mediated fusion appears to underlie preconditioning conferred by nitrite. As mentioned in previous sections, nitrite confers protection against I/R injury in a wide temporal window. Not only does nitrite confer acute protection (when administered immediately prior to I/R or during ischemia) as detailed earlier but the molecule also confers delayed preconditioning. In this phenomenon, one administration of nitrite protects against an ischemic episode that occurs 24–72 h later [4, 9, 85, 124]. Importantly, in this condition, nitrite is metabolized minutes after it is administered, so delayed protection is due to downstream signaling events and not to S-nitrosation events that occur when nitrite is present during the ischemic period as in acute protection [4, 9]. Kamga and colleagues demonstrated that nitrite-induced mitochondrial fusion resulted in the increased production of mitochondrial superoxide in cultured myocytes. This increase in ROS was responsible for the mild oxidation of the metabolic sensor AMPK, which led to its auto-phosphorylation and activation. Activation of AMPK has been shown to mediate protection after I/R by a number of mechanisms and silencing of AMPK abolishes nitrite-mediated preconditioning in cultured cells as well as the Langendorff heart [4].

A second physiological effect of nitrite-induced fusion is the modulation of glucose uptake by the adipocyte [5]. Khoo and colleagues reported in cultured murine adipocytes that nitrite treatment induced mitochondrial fusion through not only the PKA-dependent inhibition of drp-1 but also the concomitant increase in the pro-fusion proteins mitofusin 1 and 2. This nitrite-dependent fusion enhanced mitochondrial respiration, fatty acid oxidation, and significantly increased glucose uptake by the cells [5]. While the exact mechanism by which fusion increases glucose uptake remains under investigation, fusion-dependent mitochondrial ROS production is potentially responsible. This increased glucose uptake by adipocytes is significant in a physiological context given that other groups have reported that nitrite (and dietary nitrate) attenuate many symptoms of the metabolic syndrome including dysregulated glucose homeostasis [125]. Further study is required to determine whether improved glucose uptake by adipocytes mediate these protective effects of nitrite in vivo.

It is important to note that nitrite-mediated fusion, unlike all other nitrite-dependent mitochondrial modulatory mechanisms described earlier, does not require hypoxia. While the exact mechanism by which nitrite activates PKA in normoxia is under investigation, these data suggest that nitrite may mediate its effects through biochemical mechanisms independent of hypoxic reduction to NO [4]. This may include nitrite-dependent nitration or formation of nitrogen dioxide as a signaling molecule. The mechanisms by which nitrite mediates signaling in nonhypoxic environments as well as the physiological and potential therapeutic applications of nitrite as a mitochondrial fusion stimulator are currently under investigation.

Conclusions

There has been a long-standing interest in the interactions of nitrite with the mitochondrion. While very early studies focused on the role of mitochondria as nitrite metabolizers [1, 50, 51], the landscape drastically shifted with the recognition that nitrite is an intrinsic signaling molecule, mediator of physiological responses and potential therapeutic agent. It is now clear that the mitochondrion represents a physiological subcellular target for nitrite and that nitrite can regulate essential parameters of mitochondrial function including respiration [8, 43, 44], ROS generation [9, 75, 83] and apoptotic signaling [4, 9, 73]. Furthermore, through specific cellular signaling, nitrite also exerts control over mitochondrial number [6] and dynamics [4, 5]. This coordinated modulation of mitochondrial function is now known to underlie many of the protective effects of nitrite (and dietary nitrate).

The role of nitrite-dependent modulation of mitochondria has been most extensively studied in the context of I/R, and it is now known that S-nitrosation of complex I is essential for nitrite-mediated cytoprotection after I/R [9, 75]. Several clinical trials have been initiated to test the therapeutic effect of nitrite after I/R [126, 127]. As these trials move forward, ongoing studies in the field are determining whether complex I S-nitrosation could potentially be utilized as a physiological biomarker for cardiovascular disease risk as well as whether dietary modulation of complex I activity can attenuate this risk. Beyond I/R, mitochondrial regulation and signaling has been implicated in a number of other physiological and therapeutic effects of nitrite. Moving forward, clarification of the role of mitochondria in these effects, particularly enhancement of exercise efficiency and attenuation of metabolic syndrome, will likely lead to development of mitochondrial based therapies and protective dietary strategies.

Since the early 2000s several mechanisms of nitrite-dependent mitochondrial modulation have been discovered. Importantly, these mechanisms are diverse as each is operative in different conditions (hypoxia versus physiological oxygen tension) and involves varied biochemistry. Future study will, no doubt, not only reveal the physiological context for these existing mechanisms, but also uncover additional ways in which nitrite controls the mitochondrion.

References

1. Walters CL, Taylor AM. The reduction of nitrite by skeletal-muscle mitochondria. Biochim Biophys Acta. 1965;96:522–4.
2. Lundberg JO, Gladwin MT, Ahluwalia A, Benjamin N, Bryan NS, Butler A, et al. Nitrate and nitrite in biology, nutrition and therapeutics. Nat Chem Biol. 2009;5(12):865–9.
3. Lundberg JO, Weitzberg E, Gladwin MT. The nitrate-nitrite-nitric oxide pathway in physiology and therapeutics. Nat Rev Drug Discov. 2008;7(2):156–67.
4. Kamga Pride C, Mo L, Quesnelle K, Dagda RK, Murillo D, Geary L, et al. Nitrite activates protein kinase A in normoxia to mediate mitochondrial fusion and tolerance to ischaemia/reperfusion. Cardiovasc Res. 2014;101(1):57–68.
5. Khoo NK, Mo L, Zharikov S, Kamga-Pride C, Quesnelle K, Golin-Bisello F, et al. Nitrite augments glucose uptake in adipocytes through the protein kinase A-dependent stimulation of mitochondrial fusion. Free Radic Biol Med. 2014;70:45–53.
6. Mo L, Wang Y, Geary L, Corey C, Alef MJ, Beer-Stolz D, et al. Nitrite activates AMP kinase to stimulate mitochondrial biogenesis independent of soluble guanylate cyclase. Free Radic Biol Med. 2012;53(7):1440–50.
7. Shiva S. Nitrite: a physiological store of nitric oxide and modulator of mitochondrial function. Redox Biol. 2013;1(1):40–4.
8. Shiva S, Huang Z, Grubina R, Sun J, Ringwood LA, MacArthur PH, et al. Deoxymyoglobin is a nitrite reductase that generates nitric oxide and regulates mitochondrial respiration. Circ Res. 2007;100(5):654–61.
9. Shiva S, Sack MN, Greer JJ, Duranski M, Ringwood LA, Burwell L, et al. Nitrite augments tolerance to ischemia/reperfusion injury via the modulation of mitochondrial electron transfer. J Exp Med. 2007;204(9):2089–102.
10. Larsen FJ, Schiffer TA, Borniquel S, Sahlin K, Ekblom B, Lundberg JO, et al. Dietary inorganic nitrate improves mitochondrial efficiency in humans. Cell Metab. 2011;13(2):149–59.
11. Larsen FJ, Schiffer TA, Weitzberg E, Lundberg JO. Regulation of mitochondrial function and energetics by reactive nitrogen oxides. Free Radic Biol Med. 2012;53(10):1919–28.
12. Turrens JF. Mitochondrial formation of reactive oxygen species. J Physiol. 2003;552(Pt 2):335–44.
13. Boveris A, Cadenas E, Stoppani AO. Role of ubiquinone in the mitochondrial generation of hydrogen peroxide. Biochem J. 1976;156(2):435–44.
14. Guzy RD, Schumacker PT. Oxygen sensing by mitochondria at complex III: the paradox of increased reactive oxygen species during hypoxia. Exp Physiol. 2006;91(5):807–19.
15. Hoffman DL, Brookes PS. Oxygen sensitivity of mitochondrial reactive oxygen species generation depends on metabolic conditions. J Biol Chem. 2009;284(24):16236–45.
16. Chen Q, Camara AK, Stowe DF, Hoppel CL, Lesnefsky EJ. Modulation of electron transport protects cardiac mitochondria and decreases myocardial injury during ischemia and reperfusion. Am J Physiol Cell Physiol. 2007;292(1):C137–47.
17. Paradies G, Petrosillo G, Pistolese M, Di Venosa N, Federici A, Ruggiero FM. Decrease in mitochondrial complex I activity in ischemic/reperfused rat heart: involvement of reactive oxygen species and cardiolipin. Circ Res. 2004;94(1):53–9.
18. Sena LA, Chandel NS. Physiological roles of mitochondrial reactive oxygen species. Mol Cell. 2012;48(2):158–67.
19. Stowe DF, Kevin LG. Cardiac preconditioning by volatile anesthetic agents: a defining role for altered mitochondrial bioenergetics. Antioxid Redox Signal. 2004;6(2):439–48.

20. Vanden Hoek TL, Becker LB, Shao Z, Li C, Schumacker PT. Reactive oxygen species released from mitochondria during brief hypoxia induce preconditioning in cardiomyocytes. J Biol Chem. 1998;273(29):18092–8.
21. Ow YP, Green DR, Hao Z, Mak TW. Cytochrome c: functions beyond respiration. Nat Rev Mol Cell Biol. 2008;9(7):532–42.
22. Oberst A, Bender C, Green DR. Living with death: the evolution of the mitochondrial pathway of apoptosis in animals. Cell Death Differ. 2008;15(7):1139–46.
23. Di Lisa F, Canton M, Menabo R, Dodoni G, Bernardi P. Mitochondria and reperfusion injury. The role of permeability transition. Basic Res Cardiol. 2003;98(4):235–41.
24. Halestrap AP, Richardson AP. The mitochondrial permeability transition: a current perspective on its identity and role in ischaemia/reperfusion injury. J Mol Cell Cardiol. 2015;78C:129–41.
25. Hausenloy DJ, Yellon DM. The mitochondrial permeability transition pore: its fundamental role in mediating cell death during ischaemia and reperfusion. J Mol Cell Cardiol. 2003;35(4):339–41.
26. Lancaster Jr JR. A tutorial on the diffusibility and reactivity of free nitric oxide. Nitric Oxide. 1997;1(1):18–30.
27. Lundberg JO, Weitzberg E. Biology of nitrogen oxides in the gastrointestinal tract. Gut. 2013;62(4):616–29.
28. Weitzberg E, Lundberg JO. Novel aspects of dietary nitrate and human health. Annu Rev Nutr. 2013;33:129–59.
29. Dejam A, Hunter CJ, Pelletier MM, Hsu LL, Machado RF, Shiva S, et al. Erythrocytes are the major intravascular storage sites of nitrite in human blood. Blood. 2005;106(2):734–9.
30. Bryan NS, Fernandez BO, Bauer SM, Garcia-Saura MF, Milsom AB, Rassaf T, et al. Nitrite is a signaling molecule and regulator of gene expression in mammalian tissues. Nat Chem Biol. 2005;1(5):290–7.
31. Crawford JH, Chacko BK, Pruitt HM, Piknova B, Hogg N, Patel RP. Transduction of NO-bioactivity by the red blood cell in sepsis: novel mechanisms of vasodilation during acute inflammatory disease. Blood. 2004;104(5):1375–82.
32. Woessner M, Smoliga JM, Tarzia B, Stabler T, Van Bruggen M, Allen JD. A stepwise reduction in plasma and salivary nitrite with increasing strengths of mouthwash following a dietary nitrate load. Nitric Oxide. 2016;54:1–7.
33. Hon YY, Lin EE, Tian X, Yang Y, Sun H, Swenson ER, et al. Increased consumption and vasodilatory effect of nitrite during exercise. Am J Physiol Lung Cell Mol Physiol. 2016;310(4):L354–64.
34. Cosby K, Partovi KS, Crawford JH, Patel RP, Reiter CD, Martyr S, et al. Nitrite reduction to nitric oxide by deoxyhemoglobin vasodilates the human circulation. Nat Med. 2003;9(12):1498–505.
35. Larsen FJ, Weitzberg E, Lundberg JO, Ekblom B. Dietary nitrate reduces maximal oxygen consumption while maintaining work performance in maximal exercise. Free Radic Biol Med. 2010;48(2):342–7.
36. Crawford JH, Isbell TS, Huang Z, Shiva S, Chacko BK, Schechter AN, et al. Hypoxia, red blood cells, and nitrite regulate NO-dependent hypoxic vasodilation. Blood. 2006;107(2):566–74.
37. Totzeck M, Hendgen-Cotta UB, Luedike P, Berenbrink M, Klare JP, Steinhoff HJ, et al. Nitrite regulates hypoxic vasodilation via myoglobin-dependent nitric oxide generation. Circulation. 2012;126(3):325–34.
38. Bryan NS, Calvert JW, Gundewar S, Lefer DJ. Dietary nitrite restores NO homeostasis and is cardioprotective in endothelial nitric oxide synthase-deficient mice. Free Radic Biol Med. 2008;45(4):468–74.
39. Kumar D, Branch BG, Pattillo CB, Hood J, Thoma S, Simpson S, et al. Chronic sodium nitrite therapy augments ischemia-induced angiogenesis and arteriogenesis. Proc Natl Acad Sci U S A. 2008;105(21):7540–5.
40. Hendgen-Cotta UB, Luedike P, Totzeck M, Kropp M, Schicho A, Stock P, et al. Dietary nitrate supplementation improves revascularization in chronic ischemia. Circulation. 2012;126(16):1983–92.
41. de Lima Portella R, Lynn Bickta J, Shiva S. Nitrite confers preconditioning and cytoprotection after ischemia/reperfusion injury through the modulation of mitochondrial function. Antioxid Redox Signal. 2015;23(4):307–27.
42. Dezfulian C, Raat N, Shiva S, Gladwin MT. Role of the anion nitrite in ischemia-reperfusion cytoprotection and therapeutics. Cardiovasc Res. 2007;75(2):327–38.
43. Shiva S, Rassaf T, Patel RP, Gladwin MT. The detection of the nitrite reductase and NO-generating properties of haemoglobin by mitochondrial inhibition. Cardiovasc Res. 2011;89(3):566–73.
44. Hendgen-Cotta UB, Merx MW, Shiva S, Schmitz J, Becher S, Klare JP, et al. Nitrite reductase activity of myoglobin regulates respiration and cellular viability in myocardial ischemia-reperfusion injury. Proc Natl Acad Sci U S A. 2008;105(29):10256–61.
45. Tiso M, Tejero J, Basu S, Azarov I, Wang X, Simplaceanu V, et al. Human neuroglobin functions as a redox-regulated nitrite reductase. J Biol Chem. 2011;286(20):18277–89.
46. Zhang Z, Naughton DP, Blake DR, Benjamin N, Stevens CR, Winyard PG, et al. Human xanthine oxidase converts nitrite ions into nitric oxide (NO). Biochem Soc Trans. 1997;25(3):524S.
47. Vanin AF, Bevers LM, Slama-Schwok A, van Faassen EE. Nitric oxide synthase reduces nitrite to NO under anoxia. Cell Mol Life Sci. 2007;64(1):96–103.
48. Feelisch M, Fernandez BO, Bryan NS, Garcia-Saura MF, Bauer S, Whitlock DR, et al. Tissue processing of nitrite in hypoxia: an intricate interplay of nitric oxide-generating and -scavenging systems. J Biol Chem. 2008;283(49):33927–34.
49. Curtis E, Hsu LL, Noguchi AC, Geary L, Shiva S. Oxygen regulates tissue nitrite metabolism. Antioxid Redox Signal. 2012;17(7):951–61.

50. Kozlov AV, Staniek K, Nohl H. Nitrite reductase activity is a novel function of mammalian mitochondria. FEBS Lett. 1999;454(1–2):127–30.
51. Nohl H, Staniek K, Sobhian B, Bahrami S, Redl H, Kozlov AV. Mitochondria recycle nitrite back to the bioregulator nitric monoxide. Acta Biochim Pol. 2000;47(4):913–21.
52. Castello PR, David PS, McClure T, Crook Z, Poyton RO. Mitochondrial cytochrome oxidase produces nitric oxide under hypoxic conditions: implications for oxygen sensing and hypoxic signaling in eukaryotes. Cell Metab. 2006;3(4):277–87.
53. Castello PR, Woo DK, Ball K, Wojcik J, Liu L, Poyton RO. Oxygen-regulated isoforms of cytochrome c oxidase have differential effects on its nitric oxide production and on hypoxic signaling. Proc Natl Acad Sci U S A. 2008;105(24):8203–8.
54. Basu S, Azarova NA, Font MD, King SB, Hogg N, Gladwin MT, et al. Nitrite reductase activity of cytochrome c. J Biol Chem. 2008;283(47):32590–7.
55. Nohl H, Staniek K, Kozlov AV. The existence and significance of a mitochondrial nitrite reductase. Redox Rep. 2005;10(6):281–6.
56. Soetkamp D, Nguyen TT, Menazza S, Hirschhauser C, Hendgen-Cotta UB, Rassaf T, et al. S-nitrosation of mitochondrial connexin 43 regulates mitochondrial function. Basic Res Cardiol. 2014;109(5):433.
57. Eltzschig HK, Eckle T. Ischemia and reperfusion—from mechanism to translation. Nat Med. 2011;17(11):1391–401.
58. Frangogiannis NG, Smith CW, Entman ML. The inflammatory response in myocardial infarction. Cardiovasc Res. 2002;53(1):31–47.
59. Vilahur G, Badimon L. Ischemia/reperfusion activates myocardial innate immune response: the key role of the toll-like receptor. Front Physiol. 2014;5:496.
60. Hearse DJ, Manning AS, Downey JM, Yellon DM. Xanthine oxidase: a critical mediator of myocardial injury during ischemia and reperfusion? Acta Physiol Scand Suppl. 1986;548:65–78.
61. Lefer AM, Lefer DJ. The role of nitric oxide and cell adhesion molecules on the microcirculation in ischaemia-reperfusion. Cardiovasc Res. 1996;32(4):743–51.
62. McCord JM. Oxygen-derived free radicals in postischemic tissue injury. N Engl J Med. 1985;312(3):159–63.
63. McCord JM, Roy RS, Schaffer SW. Free radicals and myocardial ischemia. The role of xanthine oxidase. Adv Myocardiol. 1985;5:183–9.
64. Rouslin W, Broge CW, Grupp IL. ATP depletion and mitochondrial functional loss during ischemia in slow and fast heart-rate hearts. Am J Physiol. 1990;259(6 Pt 2):H1759–66.
65. King LM, Opie LH. Glucose and glycogen utilisation in myocardial ischemia—changes in metabolism and consequences for the myocyte. Mol Cell Biochem. 1998;180(1–2):3–26.
66. Murphy E, Steenbergen C. Mechanisms underlying acute protection from cardiac ischemia-reperfusion injury. Physiol Rev. 2008;88(2):581–609.
67. Namekata I, Shimada H, Kawanishi T, Tanaka H, Shigenobu K. Reduction by SEA0400 of myocardial ischemia-induced cytoplasmic and mitochondrial Ca2+ overload. Eur J Pharmacol. 2006;543(1–3):108–15.
68. Burwell LS, Brookes PS. Mitochondria as a target for the cardioprotective effects of nitric oxide in ischemia-reperfusion injury. Antioxid Redox Signal. 2008;10(3):579–99.
69. Burwell LS, Nadtochiy SM, Brookes PS. Cardioprotection by metabolic shut-down and gradual wake-up. J Mol Cell Cardiol. 2009;46(6):804–10.
70. Garcia-Dorado D, Ruiz-Meana M, Inserte J, Rodriguez-Sinovas A, Piper HM. Calcium-mediated cell death during myocardial reperfusion. Cardiovasc Res. 2012;94(2):168–80.
71. Duranski MR, Greer JJ, Dejam A, Jaganmohan S, Hogg N, Langston W, et al. Cytoprotective effects of nitrite during in vivo ischemia-reperfusion of the heart and liver. J Clin Invest. 2005;115(5):1232–40.
72. Webb A, Bond R, McLean P, Uppal R, Benjamin N, Ahluwalia A. Reduction of nitrite to nitric oxide during ischemia protects against myocardial ischemia-reperfusion damage. Proc Natl Acad Sci U S A. 2004;101(37):13683–8.
73. Gonzalez FM, Shiva S, Vincent PS, Ringwood LA, Hsu LY, Hon YY, et al. Nitrite anion provides potent cytoprotective and antiapoptotic effects as adjunctive therapy to reperfusion for acute myocardial infarction. Circulation. 2008;117(23):2986–94.
74. Baker JE, Su J, Fu X, Hsu A, Gross GJ, Tweddell JS, et al. Nitrite confers protection against myocardial infarction: role of xanthine oxidoreductase, NADPH oxidase and K(ATP) channels. J Mol Cell Cardiol. 2007;43(4):437–44.
75. Chouchani ET, Methner C, Nadtochiy SM, Logan A, Pell VR, Ding S, et al. Cardioprotection by S-nitrosation of a cysteine switch on mitochondrial complex I. Nat Med. 2013;19(6):753–9.
76. Lu P, Liu F, Yao Z, Wang CY, Chen DD, Tian Y, et al. Nitrite-derived nitric oxide by xanthine oxidoreductase protects the liver against ischemia-reperfusion injury. Hepatobiliary Pancreat Dis Int. 2005;4(3):350–5.
77. Elrod JW, Calvert JW, Gundewar S, Bryan NS, Lefer DJ. Nitric oxide promotes distant organ protection: evidence for an endocrine role of nitric oxide. Proc Natl Acad Sci U S A. 2008;105(32):11430–5.
78. Raat NJ, Noguchi AC, Liu VB, Raghavachari N, Liu D, Xu X, et al. Dietary nitrate and nitrite modulate blood and organ nitrite and the cellular ischemic stress response. Free Radic Biol Med. 2009;47(5):510–7.

79. Jung KH, Chu K, Ko SY, Lee ST, Sinn DI, Park DK, et al. Early intravenous infusion of sodium nitrite protects brain against in vivo ischemia-reperfusion injury. Stroke. 2006;37(11):2744–50.

80. Jung KH, Chu K, Lee ST, Sunwoo JS, Park DK, Kim JH, et al. Effects of long term nitrite therapy on functional recovery in experimental ischemia model. Biochem Biophys Res Commun. 2010;403(1):66–72.

81. Tripatara P, Patel NS, Webb A, Rathod K, Lecomte FM, Mazzon E, et al. Nitrite-derived nitric oxide protects the rat kidney against ischemia/reperfusion injury in vivo: role for xanthine oxidoreductase. J Am Soc Nephrol. 2007;18(2):570–80.

82. Dezfulian C, Alekseyenko A, Dave KR, Raval AP, Do R, Kim F, et al. Nitrite therapy is neuroprotective and safe in cardiac arrest survivors. Nitric Oxide. 2012;26(4):241–50.

83. Dezfulian C, Shiva S, Alekseyenko A, Pendyal A, Beiser DG, Munasinghe JP, et al. Nitrite therapy after cardiac arrest reduces reactive oxygen species generation, improves cardiac and neurological function, and enhances survival via reversible inhibition of mitochondrial complex I. Circulation. 2009;120(10):897–905.

84. Jung KH, Chu K, Lee ST, Park HK, Kim JH, Kang KM, et al. Augmentation of nitrite therapy in cerebral ischemia by NMDA receptor inhibition. Biochem Biophys Res Commun. 2009;378(3):507–12.

85. Bryan NS, Calvert JW, Elrod JW, Gundewar S, Ji SY, Lefer DJ. Dietary nitrite supplementation protects against myocardial ischemia-reperfusion injury. Proc Natl Acad Sci U S A. 2007;104(48):19144–9.

86. Shiva S, Wang X, Ringwood LA, Xu X, Yuditskaya S, Annavajjhala V, et al. Ceruloplasmin is a NO oxidase and nitrite synthase that determines endocrine NO homeostasis. Nat Chem Biol. 2006;2(9):486–93.

87. Lesnefsky EJ, Chen Q, Moghaddas S, Hassan MO, Tandler B, Hoppel CL. Blockade of electron transport during ischemia protects cardiac mitochondria. J Biol Chem. 2004;279(46):47961–7.

88. Lesnefsky EJ, Moghaddas S, Tandler B, Kerner J, Hoppel CL. Mitochondrial dysfunction in cardiac disease: ischemia—reperfusion, aging, and heart failure. J Mol Cell Cardiol. 2001;33(6):1065–89.

89. Chen Q, Moghaddas S, Hoppel CL, Lesnefsky EJ. Reversible blockade of electron transport during ischemia protects mitochondria and decreases myocardial injury following reperfusion. J Pharmacol Exp Ther. 2006;319(3):1405–12.

90. Li W, Meng Z, Liu Y, Patel RP, Lang JD. The hepatoprotective effect of sodium nitrite on cold ischemia-reperfusion injury. J Transplant. 2012;2012:635179.

91. Kagan VE, Chu CT, Tyurina YY, Cheikhi A, Bayir H. Cardiolipin asymmetry, oxidation and signaling. Chem Phys Lipids. 2014;179:64–9.

92. Brunori M, Giuffre A, Forte E, Mastronicola D, Barone MC, Sarti P. Control of cytochrome c oxidase activity by nitric oxide. Biochim Biophys Acta. 2004;1655(1–3):365–71.

93. Cleeter MW, Cooper JM, Darley-Usmar VM, Moncada S, Schapira AH. Reversible inhibition of cytochrome c oxidase, the terminal enzyme of the mitochondrial respiratory chain, by nitric oxide. Implications for neurodegenerative diseases. FEBS Lett. 1994;345(1):50–4.

94. Cooper CE, Brown GC. The inhibition of mitochondrial cytochrome oxidase by the gases carbon monoxide, nitric oxide, hydrogen cyanide and hydrogen sulfide: chemical mechanism and physiological significance. J Bioenerg Biomembr. 2008;40(5):533–9.

95. Shiva S, Brookes PS, Patel RP, Anderson PG, Darley-Usmar VM. Nitric oxide partitioning into mitochondrial membranes and the control of respiration at cytochrome c oxidase. Proc Natl Acad Sci U S A. 2001;98(13):7212–7.

96. Thomas DD, Liu X, Kantrow SP, Lancaster Jr JR. The biological lifetime of nitric oxide: implications for the perivascular dynamics of NO and O2. Proc Natl Acad Sci U S A. 2001;98(1):355–60.

97. Rassaf T, Flogel U, Drexhage C, Hendgen-Cotta U, Kelm M, Schrader J. Nitrite reductase function of deoxymyoglobin: oxygen sensor and regulator of cardiac energetics and function. Circ Res. 2007;100(12):1749–54.

98. Brookes PS, Shiva S, Patel RP, Darley-Usmar VM. Measurement of mitochondrial respiratory thresholds and the control of respiration by nitric oxide. Methods Enzymol. 2002;359:305–19.

99. Wittenberg BA, Wittenberg JB. Myoglobin-mediated oxygen delivery to mitochondria of isolated cardiac myocytes. Proc Natl Acad Sci U S A. 1987;84(21):7503–7.

100. Jayaraman T, Tejero J, Chen BB, Blood AB, Frizzell S, Shapiro C, et al. 14-3-3 binding and phosphorylation of neuroglobin during hypoxia modulate six-to-five heme pocket coordination and rate of nitrite reduction to nitric oxide. J Biol Chem. 2011;286(49):42679–89.

101. Petersen MG, Dewilde S, Fago A. Reactions of ferrous neuroglobin and cytoglobin with nitrite under anaerobic conditions. J Inorg Biochem. 2008;102(9):1777–82.

102. Ertracht O, Malka A, Atar S, Binah O. The mitochondria as a target for cardioprotection in acute myocardial ischemia. Pharmacol Ther. 2014;142(1):33–40.

103. Liu Y, Sato T, Seharaseyon J, Szewczyk A, O'Rourke B, Marban E. Mitochondrial ATP-dependent potassium channels. Viable candidate effectors of ischemic preconditioning. Ann N Y Acad Sci. 1999;874:27–37.

104. Ardehali H, O'Rourke B. Mitochondrial K(ATP) channels in cell survival and death. J Mol Cell Cardiol. 2005;39(1):7–16.

105. Sasaki N, Sato T, Ohler A, O'Rourke B, Marban E. Activation of mitochondrial ATP-dependent potassium channels by nitric oxide. Circulation. 2000;101(4):439–45.
106. Boengler K, Dodoni G, Rodriguez-Sinovas A, Cabestrero A, Ruiz-Meana M, Gres P, et al. Connexin 43 in cardiomyocyte mitochondria and its increase by ischemic preconditioning. Cardiovasc Res. 2005;67(2):234–44.
107. Heinzel FR, Luo Y, Li X, Boengler K, Buechert A, Garcia-Dorado D, et al. Impairment of diazoxide-induced formation of reactive oxygen species and loss of cardioprotection in connexin 43 deficient mice. Circ Res. 2005;97(6):583–6.
108. Lundberg JO, Carlstrom M, Larsen FJ, Weitzberg E. Roles of dietary inorganic nitrate in cardiovascular health and disease. Cardiovasc Res. 2011;89(3):525–32.
109. Larsen FJ, Weitzberg E, Lundberg JO, Ekblom B. Effects of dietary nitrate on oxygen cost during exercise. Acta Physiol (Oxf). 2007;191(1):59–66.
110. Bailey SJ, Winyard P, Vanhatalo A, Blackwell JR, Dimenna FJ, Wilkerson DP, et al. Dietary nitrate supplementation reduces the O2 cost of low-intensity exercise and enhances tolerance to high-intensity exercise in humans. J Appl Physiol. 2009;107(4):1144–55.
111. Lansley KE, Winyard PG, Fulford J, Vanhatalo A, Bailey SJ, Blackwell JR, et al. Dietary nitrate supplementation reduces the O2 cost of walking and running: a placebo-controlled study. J Appl Physiol. 2011;110(3):591–600.
112. Vanhatalo A, Bailey SJ, Blackwell JR, DiMenna FJ, Pavey TG, Wilkerson DP, et al. Acute and chronic effects of dietary nitrate supplementation on blood pressure and the physiological responses to moderate-intensity and incremental exercise. Am J Physiol Regul Integr Comp Physiol. 2010;299(4):R1121–31.
113. Chang AH, Sancheti H, Garcia J, Kaplowitz N, Cadenas E, Han D. Respiratory substrates regulate S-nitrosylation of mitochondrial proteins through a thiol-dependent pathway. Chem Res Toxicol. 2014;27(5):794–804.
114. McLeod CJ, Pagel I, Sack MN. The mitochondrial biogenesis regulatory program in cardiac adaptation to ischemia—a putative target for therapeutic intervention. Trends Cardiovasc Med. 2005;15(3):118–23.
115. Spiegelman BM. Transcriptional control of mitochondrial energy metabolism through the PGC1 coactivators. Novartis Found Symp. 2007;287:60–3; discussion 63–9.
116. Ruetenik A, Barrientos A. Dietary restriction, mitochondrial function and aging: from yeast to humans. Biochim Biophys Acta. 2015;1847(11):1434–47.
117. Cerqueira FM, Cunha FM, Laurindo FR, Kowaltowski AJ. Calorie restriction increases cerebral mitochondrial respiratory capacity in a NO*-mediated mechanism: impact on neuronal survival. Free Radic Biol Med. 2012;52(7):1236–41.
118. Zamora M, Pardo R, Villena JA. Pharmacological induction of mitochondrial biogenesis as a therapeutic strategy for the treatment of type 2 diabetes. Biochem Pharmacol. 2015;98(1):16–28.
119. Liang Q, Kobayashi S. Mitochondrial quality control in the diabetic heart. J Mol Cell Cardiol. 2016;95:57–69.
120. Ren J, Pulakat L, Whaley-Connell A, Sowers JR. Mitochondrial biogenesis in the metabolic syndrome and cardiovascular disease. J Mol Med. 2010;88(10):993–1001.
121. Jornayvaz FR, Shulman GI. Regulation of mitochondrial biogenesis. Essays Biochem. 2010;47:69–84.
122. Nisoli E, Clementi E, Paolucci C, Cozzi V, Tonello C, Sciorati C, et al. Mitochondrial biogenesis in mammals: the role of endogenous nitric oxide. Science. 2003;299(5608):896–9.
123. Youle RJ, van der Bliek AM. Mitochondrial fission, fusion, and stress. Science. 2012;337(6098):1062–5.
124. Rassaf T, Totzeck M, Hendgen-Cotta UB, Shiva S, Heusch G, Kelm M. Circulating nitrite contributes to cardioprotection by remote ischemic preconditioning. Circ Res. 2014;114(10):1601–10.
125. Carlstrom M, Larsen FJ, Nystrom T, Hezel M, Borniquel S, Weitzberg E, et al. Dietary inorganic nitrate reverses features of metabolic syndrome in endothelial nitric oxide synthase-deficient mice. Proc Natl Acad Sci U S A. 2010;107(41):17716–20.
126. Siddiqi N, Neil C, Bruce M, MacLennan G, Cotton S, Papadopoulou S, et al. Intravenous sodium nitrite in acute ST-elevation myocardial infarction: a randomized controlled trial (NIAMI). Eur Heart J. 2014;35(19):1255–62.
127. Jones DA, Pellaton C, Velmurugan S, Andiapen M, Antoniou S, van Eijl S, et al. Randomized phase 2 trial of intra-coronary nitrite during acute myocardial infarction. Circ Res. 2015;116:437–47.

Chapter 6
Sources of Exposure to Nitrogen Oxides

Andrew L. Milkowski

Key Points

- There are five primary sources of exposure to nitrite and nitrate:

    - Endogenous production of nitric oxide
    - Dietary exposure to nitrite and nitrates in food
    - Drinking water
    - Swallowing saliva
    - Environmental/atmospheric exposure to nitric oxide and nitrogen dioxide

- The majority of exposure to nitrite and nitrate in most people is through dietary intake of vegetables.
- Cured and processed meats contribute very little to the overall burden of exposure.
- Individual levels of exposure are dependent upon where and how one lives and works, what types of food one eats, and lifestyle habits.
- Low to moderate exposure to these sources poses little risk with known benefits while overexposure can lead to health risks.

Keywords Dietary nitrate/nitrite • Drinking water nitrite/nitrate • Environmental nitrate/nitrite • Salivary nitrate/nitrite • Nitric oxide • Cured meats • Nitrite safety

Introduction

Humans are exposed to nitrogen oxides that include the toxicologically significant gases nitric oxide (NO) and nitrogen dioxide (NO_2) and the nonvolatile salts of nitrate and nitrite ions. Nitric oxide is a sharp, sweet-smelling gas at room temperature, whereas nitrogen dioxide (NO_2) has a strong, harsh odor and is a liquid at refrigerated and cool room temperatures, becoming a reddish-brown gas above 70 °F. Both molecules are free radicals. Before it was realized that NO is produced in the body, NO

A.L. Milkowski, Ph.D. (✉)
Department of Animal Sciences, Meat Science and Muscle Biology, University of Wisconsin—Madison, 1805 Linden Drive, Madison, WI 53706, USA
e-mail: milkowski@wisc.edu

N.S. Bryan, J. Loscalzo (eds.), *Nitrite and Nitrate in Human Health and Disease*, Nutrition and Health, DOI 10.1007/978-3-319-46189-2_6, © Springer International Publishing AG 2017

was considered a poisonous toxic air pollutant. The majority of the gaseous nitrogen oxides are metabolized to nitrite and nitrate in the body. Many chapters in this monograph refer to multiple sources of nitrite, nitrate, and nitric oxide in human physiology. In an attempt to summarize the discussion relative to human exposure, they can be grouped into some key areas, and this chapter will focus primarily on exposure to NO, nitrite, and nitrate. In 2006, the International Agency for Research on Cancer (IARC) developed an extensive exposure estimate of nitrite and nitrate during the preparation of their monograph on the carcinogenic potential of nitrate and nitrite under conditions of endogenous nitrosation [1]. This monograph provides an excellent source of exposure information that has been supplemented with additional reports [1]. Owing to the chemical reactivity of nitric oxide and nitrogen dioxide, estimates for human exposure have largely defaulted to those of nitrate and nitrite.

Endogenous Production

The demonstration of NO formation by an enzyme in vascular endothelial cells in the 1980s has since had profound implications in research and medicine. NO was shown to be a potent vasodilator, inhibitor of platelet aggregation, and active species of nitroglycerin before the discovery of endothelium-derived relaxing factor (EDRF) in 1980 [2]. Subsequent studies revealed that EDRF is NO and is synthesized by mammalian cells from L-arginine through a complex oxidation reaction catalyzed by the flavo-hemoprotein NO synthase (NOS) [3, 4]. NOS catalyzes the NADPH- and oxygen-dependent oxidation of L-arginine to NO plus L-citrulline in a reaction that requires several cofactors in the presence of oxygen, including NADPH, FAD, FMN, tetrahydrobiopterin, heme, reduced glutathione, and calmodulin. Enzyme-bound calmodulin facilitates the transfer of electrons from NADPH to the flavoprotein domain of NOS and also from the flavins to the heme domain of NOS [3, 4]. These electrons are used to reduce iron to the ferrous state so that it can bind oxygen, which is incorporated into the substrate, arginine, to generate NO plus citrulline. The endogenous production of NO by NOS has been established as playing an important role in vascular homeostasis, neurotransmission, and host defense mechanisms [5]. The major removal pathway for NO is the stepwise oxidation to nitrite and nitrate [6]. Although some NO can be oxidized to nitrogen dioxide, it will quickly nitrosate water to form nitrite. In plasma, NO is oxidized to nitrite, with a half-life of 20–30 min [7, 8]. In contrast, NO and nitrite are rapidly oxidized to nitrate in whole blood. NO can also be enzymatically oxidized to nitrite by ceruloplasmin or other metal-containing proteins [9]. The half-life of nitrite in human blood is about 110 s [10]. Nitrate, by contrast, has a circulating half-life of 5–8 h [11, 12], while tissue nitrite and nitrate both have in vivo half-lives of tens of minutes [13]. During fasting conditions with a low previous intake of nitrite/nitrate, enzymatic NO formation accounts for the majority of nitrite [14]. On the basis of these studies, it was believed that NO metabolism acutely terminates by oxidation to nitrite and nitrate. Early studies on nitrogen balance in humans and analyses of fecal and ileostomy samples indicated that nitrite and nitrate are formed de novo in the intestine. It was these findings by Tannenbaum et al. [15] that significantly altered our conceptions of human exposure to exogenous nitrite and nitrate and represented the original observations that would eventually lead to the discovery of the L-arginine:NO pathway. Prior to these studies, it was thought that steady-state levels of nitrite and nitrate originated solely from the diet and from nitrogen-fixing enteric bacteria. The steady-state concentrations of nitrite in the body are tightly regulated and vary depending on each tissue or compartment and relative NOS activity [16, 17]. However, there is usually more nitrite concentrated in tissue than in the circulatory system [16]. Being limited primarily by blood levels of NO in humans, blood nitrite and nitrate levels may be an underestimation of what occurs in specific tissues [18].

The endogenous production of NO plays an important role in vascular homeostasis, neurotransmission, and host defense mechanisms [5]. It has been estimated that a healthy 70 kg human body produces on average 1.68 mmole NO per day (based on 1 μmole/kg/h NO production). The majority of this NO will end up as nitrite and nitrate. In plasma, steady-state concentrations of nitrite are conserved across various mammalian species, including humans, in the range of 150–600 nM [19]. Apart from plasma, nitrite can also be transported within red blood cells [20]. The net concentration in plasma is a result of its formation and consumption. It has been shown that up to 70–90 % of plasma nitrite is derived from eNOS activity in fasted humans and other mammals [19]. Humans, unlike prokaryotes, are thought to lack the enzymatic machinery to reduce nitrate back to nitrite; however, the commensal bacteria that reside within the human body can reduce nitrate, thereby supplying an alternative source of nitrite [1, 21–23]. Plasma nitrite increases after ingestion of large amounts of nitrate. This increase is entirely due to enterosalivary circulation of nitrate (as much as 25 % is actively taken up by the salivary glands) and reduction to nitrite by commensal bacteria in the mouth [24, 25]. Therefore, dietary and enzymatic sources of nitrate are actually a potentially large source of nitrite in the human body. Nitrate is rapidly absorbed in the small intestine and readily distributed throughout the body [26]. This nitrate pathway to NO has been shown to help reduce gastrointestinal tract infection and increase mucous barrier thickness and gastric blood flow [26–31]. Nitrite and nitrate are excreted through the kidneys. Nitrate is excreted in the urine as such or after conversion to urea [32]. Clearance of nitrate from blood to urine approximates 40 μmole/h in adults [33], indicating considerable renal tubular reabsorption of this anion. There is little detectable nitrite or nitrate in feces [34]. There is some loss of nitrate and nitrite in sweat, but this is not a major route of excretion [35].

Diet

Diet has been suggested to be the largest source of nitrogen oxides, primarily via nitrite and nitrate intake. Much of the quantified exposure of nitrite and nitrate has been based on the focus of dietary ingestion and endogenous formation of nitrosamines as a cause of human cancers [36–38]. What is often downplayed in this arena is the fact that nitrogen is essential for growth and reproduction of all plant and animal life. An essential form of that nitrogen is nitrate. Nitrogen is a basic constituent of proteins. The form of nitrogen within plants when consumed by animals has important effects on growth and reproduction. Several different groups of nitrogen-containing compounds may be found in plants. The amount of each form depends on the specific plant species, maturity, and environmental conditions during growth. These nitrogen compounds may be broadly classified as either protein or nonprotein compounds. Under normal growing conditions, plants use nitrogen to form plant proteins. When normal growth is altered, protein formation may be slowed and the nitrogen remains in the plant as nonprotein nitrogen. Nitrate, nitrite, amides, free amino acids, and small peptides make up most of the nonprotein nitrogen fraction. We will focus specifically on nitrite and nitrate occurring naturally in plants and that added exogenously in meat processing. First, it is worth a cursory review of how plants accumulate nitrate from the soil in order to understand how they become a significant source of nitrite and nitrate.

Nitrate is a natural molecule in soils. Adequate supply of nitrate is necessary for good plant growth. Probably more than 90 % of the nitrogen absorbed by plants is in the nitrate form. Soils can be supplemented with nitrogen often in the form of ammonium nitrogen (NH_4^+), which is rapidly converted to nitrate (NO_3^-) in the soil. The amount of crop growth is essentially the same whether nitrogen fertilizer is applied as ammonia (NH_3), ammonium, or nitrate (NO_3^-). Chemical fertilizers may be composed of ammonium nitrate, ammonium phosphates, ammonium sulfate, alkaline metal nitrate salts, urea, and other organic forms of nitrogen. Microorganisms must change organic nitrogen to ammonium or nitrate ionic species before plants can use it. Usual release of available nitrogen from soil organic matter is 1–4 % annually, depending on soil texture and weather conditions. *Animal manure* is an excellent source of nitrogen and can contribute significantly to soil improvement. Animal manure contains about 10 lb of nitrogen per ton, poultry manure about 20 lb per ton, and legume residues

20–80 lb per ton. About half of this organic nitrogen may be converted to nitrate–nitrogen and become available for plant use the year it is added to the soil; however, it is low in phosphorus content. Excessive manure applications can result in toxic levels of nitrate in forage crops the same as excessive use of chemical nitrogen fertilizer. Adding phosphate fertilizer to manure can reduce the nitrate content in the crop produced.

Nitrate content of plants is determined by their inherited metabolic pattern (genetics) and the available nitrate in the soil. Applying fertilizer in amounts beyond the ability of the vegetable crop to use them may result in an accumulation of nitrate, particularly if some other essential nutrient is inadequate. Leafy green vegetables and some root crops naturally contain nitrates, with wide variations between species. Vegetables produced from high organic soils, and even where no fertilizer nitrogen is applied, frequently have higher nitrate content than the same species grown in fertilized soil. Other organically grown vegetables contain less nitrate than conventionally grown [39]. There are also seasonal differences: it is reported that nitrate content in spinach was 1.5-fold higher in the autumn–winter harvest than in spring (2580 vs. 1622 mg/kg) [40]. Disruption of normal plant growth increases the probability for nitrate accumulation in leaves, stems, and stalks. Factors favoring accumulation of nitrate in plants include drought, high temperatures (rapid transpiration of water), shading and cloudiness (reduced protein synthesis), deficiency of other plant growth nutrients (phosphorus, potassium, calcium), excessive soil nitrogen (manures, legume, residues, effluents and solid wastes, and nitrogen fertilizer), damage from insects and certain weed control chemicals, and immaturity at harvest [41, 42]. As a result of all of these factors, nitrate content of vegetables is highly variable but still represents the main dietary source of nitrate.

The National Research Council report *The Health Effects of Nitrate, Nitrite, and N-Nitroso Compounds* (NRC 1981) reported estimates of nitrite and nitrate intake based on food consumption tables and calculated that the average total nitrite and nitrate intake in the U.S. was 0.77 and 76 mg, respectively, per day. Meah et al. estimated 1985 dietary intake in the UK at 4.2 mg/day for nitrite and 54 mg/day for nitrate with the largest contribution to both being potatoes [43]. In 1975, White [44] estimated cured meats were a source of 10 % of dietary nitrate and 21 % of dietary nitrite exposure. The White calculations were based on the relatively higher levels of nitrate (208 ppm residual measured) and nitrite (52.5 ppm residual) in cured meats that were used prior to concerns about nitrosamine formation. Although the amount of nitrite and nitrate present in cured and processed meats is currently much less than what is found naturally in vegetables, there are a number of epidemiological reports erroneously implicating the nitrite and nitrate in meats with increased risk of some cancers [45, 46]. Past intense investigations into nitrosation chemistry with respect to the use of nitrite and nitrate in meat preservation have resulted in reduction and, in many cases, the elimination of nitrate as a curing salt, reduction in levels of nitrite used for curing, as well as the widespread use of ascorbate or erythorbate salts to control better curing chemistry and greatly reduce the likelihood of nitrosamine formation [47–49]. Thus, changes to cured meat manufacturing procedures over the past 35 years have made earlier estimates of the contribution of cured meats to dietary nitrate and nitrite intake obsolete. A survey of products by Cassens et al. in 1996, representing about 1/3 of ham, bacon, bologna, and frankfurters manufactured in the US, found residual nitrate at <10 ppm and average nitrite at 10 ppm [50, 51]. This represented more than an 80 % reduction in nitrite in cured meats versus earlier reports. Keeton et al. repeated the survey in 2008–2009 and found similar values [4, 52, 53]. Temme et al. also summarized updated dietary exposure data for Belgians in 2011. In 2009, Hord et al. summarized dietary intake estimates of nitrite and nitrate for the "DASH" (dietary approaches to stop hypertension) diet that included two scenarios for vegetables and fruit consumption [54]. At a recommended level of vegetable and fruit intake that encompassed high nitrate sources, nitrate intake could be over 1200 mg/day and nitrite would be approximately 0.35 mg/day from food sources exclusive of processed meats. A diet with low nitrate containing vegetable and fruit intake could have as little as 174 mg/day nitrate but still contain 0.41 mg/day nitrite. This latter estimate was also close to values reported by the IARC committee in their estimates of human diet intake based on data from the

United Kingdom [51, 55–57]. Adding 75 g/day consumption of cured meats (as done by White in 1975) to the Hord estimates would add 1.5–6 mg/day of nitrate and 0.05–0.54 mg/day of nitrite using the mean residual levels of nitrate and nitrite reported in a wide range of cured meats collected at American retail outlets [52]. Thus, cured meats contribute negligible nitrate compared to other dietary sources and amounts of nitrite equivalent to other dietary sources such as vegetables and fruits. Other foods such as dairy items, eggs, fish, and grains have been reported to supply relatively minor fractions of the total intake of nitrite and nitrate [43, 54, 58–60].

The indirect exposure to nitrite from dietary nitrate, however, must also be considered because of the enterosalivary recirculation of nitrate and biochemical reduction of salivary nitrate to nitrite by commensal bacteria in the oral cavity. Hord [54] estimated the nitrite exposure from this source to be approximately 5.2 mg/day, about half of which was a result of recycling of ingested nitrate, and the other half a result of oxidation and recycling of endogenously produced nitric oxide. Thus, oral nitrite ingestion can be summarized as 6–7 % directly from fruits and vegetables, 0.5–6.5 % directly from cured meats, approximately 50 % indirectly from vegetables and fruits, and 40–50 % from endogenous production.

Water

Even though drinking water usually contains very low levels of nitrate and nitrite, when large volumes are consumed, this can be a significant source for ingestion. The U.S. Environmental Protection Agency regulates nitrate and nitrite contamination levels in approximately 160,000 different water supplies and does periodic monitoring [61–64]. The regulatory maximum contaminant level (MCL) for nitrate is 10 mg/L nitrogen from nitrate which equates to 44 mg nitrate ion/L. For nitrite, the MCL is 1 mg/L as nitrogen or 3.3 mg/L nitrite ion. The WHO has published similar recommendations of 50 mg/L as nitrate ion and 3 mg/L as nitrite ion in drinking water guidelines [65]. Using these maximum values as an estimate for intake may provide a crude estimate of an upper limit for the drinking water contribution to ingestion of nitrite and nitrate in the US. Recommended dietary consumption of water and beverages is listed at 2.7 L/day for adult males and 2.2 L/day for females [66]. However, IARC has used an average water intake of 1.4 L/day on which to base their estimates [58]. Thus, as much as 60–119 mg/day nitrate ingestion from drinking water may occur in drinking water meeting legal standards. Similarly, as much as 4.6–8.9 mg/day of nitrite may also be ingested from drinking water.

The natural background level of nitrate and nitrite is very low in water, generally <10 mg/L for the former and rarely over 3 mg/L for the latter. The increased use of nitrogen fertilizers in agriculture during the second half of the twentieth century contributed to potential contamination of water supplies by nitrate containing runoff of from farms into watersheds that ultimately became sources of drinking water. Due to their very high solubility, nitrates can enter groundwater. Levels of nitrate can be significantly higher in shallow wells that are connected to agricultural runoff. Environmental Protection Agency monitoring in 2009 revealed violations of federal standards affecting 4.5 million people with over 70 % of these violations for population centers of less than 100,000 [67]. Much less is known about private wells, which in the U.S. are usually required to be tested only when the well is constructed or when the property is sold. Tentative estimates for Western Europe and the U.S. are that 2–3 % of the population is potentially exposed to drinking water nitrate exceeding 50 mg/L [68]. Thus, higher exposures from drinking water are not an uncommon event. Excessive levels of nitrate and nitrate in drinking water have been associated with illness and death in newborns and young infants. Thus, today, the regulations that exist for nitrate and nitrite in drinking water in many parts of the world are based on this related toxic effect. Newborn infants are particularly sensitive to oxidation of hemoglobin in erythrocytes and, if the level approaches 20 % of total hemoglobin, methemoglobinemia or cyanosis can develop which interferes with blood transport of oxygen [69]. Methemoglobinemia became better understood in the mid-twentieth century through medical and epidemiological investigations which resulted in special cautions being applied to diet and water sources for this segment of the human population. It is noteworthy that the few human nitrate and nitrite exposure studies, including those in both children and adults, have not reported methemoglobinemia. Infants exposed to 175–700 mg nitrate per day did not experience methemoglobin levels above 7.5%, suggesting that nitrate alone is not causative for methemoglobinemia [70]. A more recent randomized 3-way crossover study exposed healthy volunteer adults to single doses of sodium nitrite that ranged from 150 to 190 mg per volunteer up to 290–380 mg per volunteer [71]. Observed methemoglobin concentrations were 12.2% for volunteers receiving the higher dose of nitrite ion and 4.5% for those receiving the lower dose. Recent nitrite infusion studies of up to 110 μg/kg/min for 5 min induced methemoglobin concentrations of only 3.2% in adults [72]. These data have led to alternative explanations for the observed methemoglobinemia in infants, including gastroenteritis and associated iNOS-mediated production of NO induced by bacteria-contaminated water [73, 74] as a result of immune activation against the infection. Experts have questioned the veracity of the evidence supporting the hypothesis that nitrates and nitrites are toxic for healthy postinfant populations [75–77]. It appears that the earlier biologically plausible hypothesis of nitrite toxicity (e.g., methemoglobinemia) has essentially been transformed into sacrosanct dogma [75], despite lack of proof [73, 74]. These studies call into question the mechanistic basis for exposure regulations for nitrate and nitrite. At best, these findings highlight a serious, but context-specific, risks associated with nitrite overexposure in infants.

Saliva

Similar to the environmental nitrogen cycle, there is now a recognized human nitrogen cycle whereby nitrate is reduced to nitrite and to NO [78]. Nitrate, after its absorption in the upper gastrointestinal tract, reaches the salivary glands via the blood circulation where it is secreted into the oral cavity and partially reduced to nitrite by the oral microflora. The protein sialin has been proposed as a transporter for the secretion of nitrate into saliva [79]. There is a linear relationship between the amounts of nitrate ingested and amounts of nitrate and nitrite found in saliva. The ability of the oral microflora to reduce nitrate to nitrite depends on the individual's age. Mean salivary nitrite was found to increase from well below 1 ppm in infants of up to 6 months of age to about 7 ppm or higher in adults [80], but levels over 500 mg/L have been reported to occur subsequent to consumption of nitrate-rich foods [24].

Early work by Tannenbaum et al. reported that nitrite [22, 81] in saliva of healthy individuals is 6–10 mg/L and higher levels in the hundreds of mg/L could form under conditions of very high nitrate intake and active bacterial reduction in saliva. More recently, the importance of the microbiome in the oral cavity on the reduction of nitrate to nitrite in saliva has been described [82, 83]. Additionally, the effect of antimicrobial mouthwashes is to reduce the formation of nitrite in saliva [84]. Values of 97 mg/L nitrate and 10 mg/L nitrite were reported in a Chinese study after an overnight fasting period [85]. Bjorne et al. [30] also reported 1.9 mg/L fasting levels of salivary nitrite, which were increased to near 100 mg/L after a standardized nitrate load of 4.6 mg/kg body weight was administered. Certainly the high variation in many reports reflects biological variability and dietary interactions. Thus, it is extremely difficult to quantify accurately the contribution of saliva to nitrate and nitrite ingestion, but it must be considered to be a major source. The current consensus is that as much as 25% of the combined dietary, water solubilized, and endogenously generated nitrate is recycled in the body via the enterosalivary route and that up to 25% of that nitrate pool is reduced in the saliva by commensal bacteria [25, 86]. The earlier discussed Hord values for salivary nitrite contribution to ingestion were based on a 2.0 mg/L concentration of nitrite in saliva and secretion of 1.5 L/day. Thus, saliva can commonly be a source of as much 15 mg ingested nitrite per day and under conditions of sustained nitrate intake from vegetables, this could go even higher.

Nondietary Sources of Nitrogen Oxide Exposure: Occupational, Smoking, Toothpaste

On a global scale, quantities of nitric oxide and nitrogen dioxide produced naturally by bacterial growth, volcanic action, and lightning by far outweigh those generated by man's activities; however, as they are distributed over the entire earth's surface, the resulting background atmospheric concentrations are very small. The major source of man-made emissions of nitrogen oxides into the atmosphere is the combustion of fossil fuels in stationary sources (heating, power generation) and in motor vehicles (internal combustion engines). Other contributions to the atmosphere come from specific noncombustion industrial processes, such as the manufacture of nitric acid and explosives. Indoor sources include smoking, gas-fired appliances, and oil stoves. Differences in the nitrogen dioxide emission of various countries are mainly due to differences in fossil fuel consumption. Worldwide emissions of oxides of nitrogen were estimated in 1994 at approximately 50 million metric tons [87]. Nitrogen dioxide is a respiratory pollutant and the US EPA first set standards for NO_2 in 1971 at 0.053 parts per million (53 ppb) averaged annually. While reviewed periodically, they have not been changed since then, and all areas of the US currently meet these standards [88]. Nitrogen oxides are released to the air from the exhaust of motor vehicles; the burning of coal, oil, or natural gas; and during processes such as arc welding, electroplating, engraving, and dynamite blasting. When it is NO that is released, it quickly reacts with oxygen to form NO_2, which is the brown gas formed over cities with high pollution [89]. Photocatalysis in the troposphere results in reformation of nitric oxide, and because of this cycle, they are both typically referred to as NOx in air pollution literature (not to be confused as nitrite + nitrate which is also commonly referred to as NO_x in biological literature).

Nitrogen oxides are used in the production of nitric acid, lacquers, dyes, and other chemicals and also used in rocket fuels, nitration of organic chemicals, and the manufacture of explosives. As a result, humans are chronically exposed to these nitrogen oxides, including nitric oxide at rates dependent upon many factors, but primarily where they live and the industry in which they work. Everybody is exposed to small amounts of nitrogen oxides in ambient air. Exposure to high levels of nitrogen oxides, particularly nitrogen dioxide, can damage the respiratory airways. Contact with the skin or eyes can cause burns. People who live near combustion sources such as coal burning power plants or areas with heavy motor vehicle use may be exposed to higher levels of nitrogen oxides. Households that burn wood or use kerosene heaters and gas stoves tend to have higher levels of nitrogen oxides in air when compared to houses without these appliances. Exposure from indoor sources such as home appliances and smoking should not be underestimated. In the immediate proximity of domestic gas-fired appliances, nitrogen dioxide concentrations of up to 2000 µg/m³ (1.1 ppm) have been measured. Workers employed in facilities that produce nitric acid or certain explosives, like dynamite and trinitrotoluene (TNT), as well as workers involved in the welding of metals may breathe in nitrogen oxides during their work. Some of the human toxicological studies of nitrate have relied on workers in explosives and fertilizer manufacturing and usage occupations. While they are not representative of the general human population, they can be a significant source for some people. Ammonium and potassium nitrate are important fertilizers. There have been estimates of nitrate exposure as high as >10 mg/m³ from dust [58, 90]. Much literature in this area examines nitrate-containing dust without actual measurements of the dust composition; hence, extensive data is lacking [91, 92]. This is understandable given the complexity of crystalline forms of nitrate salts in fertilizers, changes with storage, and mixtures with other fertilizer components [93, 94]. Farmers who were exposed to silo gases from the fermentation of harvested crops were acutely affected by nitrogen oxides, some of them fatally. It has been estimated that exposure to nitrogen dioxide levels of 560–940 mg/m³ (300–500 ppm) may result in fatal pulmonary edema or asphyxia, and that levels of 47–140 mg/m³ (25–75 ppm) can cause bronchitis or pneumonia. Miners who used explosives repeatedly in their work were reported to develop chronic respiratory diseases. Analysis of the products of explosion showed the presence of oxides of nitrogen at concentrations of 88–167 ppm [95].

Nitric oxide and nitrogen dioxide are found in tobacco smoke, exposing people who smoke or breathe in second-hand smoke to nitrogen oxides. Nitrogen oxides in cigarette smoke have been implicated in the genesis of chronic obstructive pulmonary disease (COPD) [96] and as precursors to carcinogenic N-nitrosamines [97]. Jenkins et al. reported a range of 3–47 μmoles of NO_x and that was highly variable depending on type and brand of cigarette [98]. Total NO_x produced from smoking one cigarette is of the order of 0.5–2 mg [99], which can be a significant exposure, being potentially equal to that from endogenous production for heavy smokers. Tobacco smoke has been reported to contain nitric oxide levels of about 98–135 mg/m^3 (80–110 ppm) with up to 1000 ppm reported [100, 101], and nitrogen dioxide levels of about 150–226 mg/m^3 (80–120 ppm), but these levels may fluctuate considerably with the conditions of combustion. Normal, metabolic production of small amounts of nitric oxide causes airways to expand. The large amount of nitric oxide in tobacco smoke acutely impacts individuals in two ways: [1] when smokers are smoking, it expands their airways even further, making it easier for their lungs to absorb nicotine and other smoke constituents; [2] when they are not smoking, it shuts off endogenous nitric oxide production, causing their airways to constrict. This may be one reason why regular smokers often have difficulty with normal breathing.

Although it is not widely realized, potassium nitrate is used in many toothpaste formulations to reduce sensitivity. Potassium nitrate is the active antisensitivity ingredient that penetrates the exposed tubules (tiny holes) of the teeth to relieve the pain from sensitive nerves inside teeth. Additionally, it is one of the ADA-approved ingredients to treat tooth hypersensitivity. Toothpaste designed for sensitive teeth is formulated with up to 5 % (50,000 ppm) potassium nitrate, which is higher than the concentration found in vegetables [99, 102–105]. Since toothpaste is not commonly swallowed, the net contribution to nitrate and nitrite exposure cannot be accurately estimated. But, nitrate and nitrite as water soluble ions are likely to be extracted from the toothpaste and mixed with saliva. The total exposure could therefore be equivalent to or greater than that contributed by water or cured processed meats.

Nitrosative Chemistry of Nitrite and Nitrate

Regardless of the source of exposure, the primary cause for concern of exposure to nitrite, nitrate, or other nitrogen oxides is the nitrosative chemistry that can occur. The risk–benefit spectrum from nitrate may very well depend upon the specific metabolism and the presence of other components that may be concomitantly ingested. The stepwise reduction to nitrite and NO may account for the benefits, while pathways leading to nitrosation of low-molecular-weight amines or amides may account for the health risks. Understanding and affecting those pathways will certainly help mitigate the risks. The discussion later will describe these two pathways.

Nitrate itself is generally considered to be harmless at low concentrations. Nitrite, on the other hand, is reactive especially in the acidic environment of the stomach where it can nitrosate other molecules, including proteins, amines, and amides. Nitrite is occasionally found in the environment but most human exposure occurs through ingested nitrate that can be chemically reduced to nitrite by commensal bacteria often found in saliva [22, 81, 82]. The specter of a cancer risk posed by nitrite and nitrate is invariably accompanied by concern about exposure to preformed N-nitrosamines or N-nitrosamines formed in the stomach from ingesting foods enriched in nitrite and nitrate. This concern has been raised because some low-molecular-weight amines can be converted (nitrosated) to their carcinogenic N-nitroso derivatives by reaction with nitrite (see reaction later) [106]. Nitrosamines are a class of chemical compounds that were first described in the chemical literature over 100 years ago, but not until 1956 did they receive much attention. In that year, 2 British scientists, John Barnes and Peter Magee, reported that dimethylnitrosamine produced liver tumors in rats [107]. This discovery was made during a routine screening of chemicals that were being proposed for use as solvents in the dry cleaning industry. Magee and Barnes's landmark discovery caused scientists around the world to investigate the carcinogenic properties of other nitrosamines. Approximately 300 of these compounds have been tested, and 90 % of them have been found to be carcinogenic in a wide variety of experimental animals. Most nitrosamines are mutagens and a number are transplacental carcinogens. Most are organ specific. For instance, dimethylnitrosamine causes liver cancer in experimental animals, whereas some of the tobacco-specific nitrosamines cause lung cancer.

Nitrosamines can occur because their chemical precursors—amines and nitrosating agents—occur commonly, and the chemical reaction for nitrosamine formation is quite plausible.

Nitrite ion will be in equilibrium with nitrous acid (HNO_2), an extremely unstable compound, which is likewise in equilibrium with nitrous anhydride (N_2O_3), the required nitrosating species. N_2O_3 is produced from two molecules of HONO. The pKa value for the nitrous acid-to-nitrite anion equilibrium is 3.34, indicating that at a pH of 3.34, only 50 % HONO (the nitrosating agent) can exist in solution, such as in the stomach contents, while at a higher pH there is much less nitrosating potential because the nitrite anion will predominate. The following reactions summarize the nitrosation reactions:

$$NaNO_2 \left(\text{sodium nitrite} \right) + H^+ \rightarrow HNO_2 + Na^+$$

$$2HNO_2 \leftrightarrow N_2O_3 + H_2O$$

$$R_2NH \left(\text{amines} \right) + N_2O_3 \rightarrow R_2N - N = O \left(\text{nitrosamine} \right) + HNO_2$$

In the presence of acid (such as in the stomach) or high temperature (such as via frying), nitrosamines are converted to diazonium ions.

$$R_2N - N = O \left(\text{nitrosamine} \right) + \left(\text{acid or heat} \right) \rightarrow R - N^+ - N = O \left(\text{diazonium ion} \right)$$

Certain nitrosamines, such as dimethylnitrosamine and N-nitrosopyrrolidine, form carbocations that react with biological nucleophiles (such as DNA or an enzyme) in the cell.

$$R - N^+ - N = O \left(\text{diazonium ion} \right) \rightarrow R^+ \left(\text{carbocation} \right)$$
$$+ N_2 \left(\text{leaving group} \right) + : Nu \left(\text{biological nucleophiles} \right) \rightarrow R - Nu$$

If this reaction occurs at a crucial site in a biomolecule, it can disrupt normal cell function.

In 1973 it was reported that ascorbic acid inhibits nitrosamine formation [108–110]. Another antioxidant, alpha-tocopherol (vitamin E), has also been shown to inhibit nitrosamine formation [111]. Ascorbic acid, erythorbic acid, and alpha-tocopherol inhibit nitrosamine formation due to their oxidation–reduction properties. For example, when ascorbic acid is oxidized to dehydroascorbic acid, nitrous anhydride, a potent nitrosating agent formed from sodium nitrite, is reduced to nitric oxide, which is not a nitrosating agent. Most vegetables that are enriched in nitrate are also rich in antioxidants, such as vitamins C and E that can act to prevent the unwanted nitrosation chemistry. These compounds are also now almost universally added to cured, processed meats. Nitrate in drinking water, on the other hand, has no such protective nitrosation inhibitor present and may be cause for concern. Controlling the metabolic fate of nitrate and nitrite away from nitrosation and toward reduction to NO may provide a strategy to promote health benefits while mitigating the health risks. Adverse health effects may be the result of a complex interaction of the amount of nitrite and nitrate ingested, the concomitant ingestion of nitrosation cofactors and precursors, and specific medical conditions such as chronic inflammation that increase nitrosation. Controlling for such factors is essential for defining the safety of nitrite and nitrate.

Table 6.1 Ranges of nitrate, nitrite, and nitric oxide exposure from diet, endogenous synthesis, and recycling for adult humans expressed as mg/day

Source	Nitrate, NO_3^-	Nitrite, NO_2^-	Nitric oxide, NO
Diet (excluding cured processed meat)[a]	50–220	0–0.7	–
From 75 g/day cured processed meat intake[b]	1.5–6	0.05–0.6	–
Water[c]	0–132	0–10	–
Saliva[d]	>30–1000	5.2–8.6	–
Endogenous synthesis[e]	–	–	70

[a]Based on IARC Table 1.8 ([58], Table 1.8)
[b]Based on Keeton et al. [52] average values and intake described in White [44, 49]
[c]Based on none present to US EPA MCL for water and 2.7 L consumption/day
[d]Based on White [44, 49] and Hord [54] and includes both recycling of diet-derived nitrate via the enterosalivary route and that from endogenous NO
[e]Based on 1 mg/kg/day endogenous synthesis for 70 kg adults

Conclusion

Human exposure to nitrate and nitrite can vary widely based on both dietary habits and geographical location of their water supply, as well as individual exposure to atmospheric nitrogen oxides. A summary of estimates is presented in Table 6.1. The largest source of nitrogen oxides in the body is derived from plant sources in the diet. While drinking water can be the next highest source, endogenous production is likely higher. Salivary nitrate and nitrite reflect a recycling of the large dietary nitrate consumption and endogenous nitric oxide synthesis, and in some cases may be the largest source of nitrate and nitrite that is swallowed. Processed meats, dairy products, and poultry products provide only minor contributions to the human exposure. Overall human exposure to nitrate is primarily from exogenous sources and nitrite is primarily from endogenous sources. Nitric oxide exposure is exclusively from endogenous synthesis or smoking. Nitrogen oxide exposure may also occur from administered pharmaceuticals, which have not been considered in this analysis. Safety and health concerns can be addressed by inhibiting unwanted nitrosation reactions.

References

1. Ishiwata H, Tanimura A, Ishidate M. Nitrite and nitrate concentrations in human saliva collected from salivary ducts. J Food Hyg Soc Jpn. 1975;16:89–92.
2. Moncada S, Palmer RMJ, Higgs A. Nitric oxide: physiology, pathophysiology and pharmacology. Pharmacol Rev. 1991;43(2):109–42.
3. Abu-Soud HM, Yoho LL, Stuehr DJ. Calmodulin controls neuronal nitric-oxide synthase by a dual mechanism. J Biol Chem. 1994;269(51):32047–50.
4. Palmer RMJ, Ashton DS, Moncada S. Vascular endothelial cells synthesize nitric oxide from L-arginine. Nature. 1988;333(6174):664–6.
5. Maryuri T, Nuñez De González MT, Osburn WN, Hardin MD, Longnecker M, Garg HK, Bryan NS, Keeton JT. A survey of nitrate and nitrite concentrations in conventional and organic-labeled raw vegetables at retail. J Food Sci. 2015;80(5):C942–9.
6. Yoshida K, Kasama K, Kitabatake M, Imai M. Biotransformation of nitric oxide, nitrite and nitrate. Int Arch Occup Environ Health. 1983;52:103–15.
7. Grube R, Kelm M, Motz W, Strauer BE. The biology of nitric oxide. In: Moncada S, Feelisch M, Busse R, Higgs EA, editors. Enzymology, biochemistry, and immunology, vol. 4. London: Portland Press; 1994. p. 201–4.
8. Kelm M, Feelisch M, Grube R, Motz W, Strauer BE. The biology of nitric oxide. In: Moncada S, Marletta MA, Hibbs JB, Higgs EA, editors. Physiological and clinical aspects, vol. 1. London: Portland Press; 1992. p. 319–22.

9. Shiva S, Wang X, Ringwood LA, et al. Ceruloplasmin is a NO oxidase and nitrite synthase that determines endocrine NO homeostasis. Nat Chem Biol. 2006;2(9):486–93.

10. Kelm M. Nitric oxide metabolism and breakdown. Biochim Biophys Acta. 1999;1411:273–89.

11. Tannenbaum SR. Nitrate and nitrite: origin in humans. Science. 1994;205:1333–5.

12. Kelm M, Yoshida K. Metabolic fate of nitric oxide and related N-oxides. In: Feelisch M, Stamler JS, editors. Methods in nitric oxide research. Chichester: Wiley; 1996. p. 47–58.

13. Bryan NS, Fernandez BO, Bauer SM, et al. Nitrite is a signaling molecule and regulator of gene expression in mammalian tissues. Nat Chem Biol. 2005;1(5):290–7.

14. Rhodes P, Leone AM, Francis PL, Struthers AD, Moncada S. The L-arginine:nitric oxide pathway is the major source of plasma nitrite in fasted humans. Biochem Biophys Res Commun. 1995;209:590–6.

15. Tannenbaum SR, Fett D, Young VR, Land PD, Bruce WR. Nitrite and nitrate are formed by endogenous synthesis in the human intestine. Science. 1978;200:1487–8.

16. Bryan NS, Rassaf T, Maloney RE, et al. Cellular targets and mechanisms of nitros(yl)ation: an insight into their nature and kinetics in vivo. Proc Natl Acad Sci U S A. 2004;101(12):4308–13.

17. Rodriguez J, Maloney RE, Rassaf T, Bryan NS, Feelisch M. Chemical nature of nitric oxide storage forms in rat vascular tissue. Proc Natl Acad Sci U S A. 2003;100:336–41.

18. Bryan NS. Nitrite in nitric oxide biology: cause or consequence? A systems-based review. Free Radic Biol Med. 2006;41(5):691–701.

19. Kleinbongard P, Dejam A, Lauer T, et al. Plasma nitrite reflects constitutive nitric oxide synthase activity in mammals. Free Radic Biol Med. 2003;35(7):790–6.

20. Dejam A, Hunter CJ, Pelletier MM, et al. Erythrocytes are the major intravascular storage sites of nitrite in human blood. Blood. 2005;106(2):734–9.

21. Goaz PW, Biswell HA. Nitrite reduction in whole saliva. J Dent Res. 1961;40:355–65.

22. Tannenbaum SR, Sinskey AJ, Weisman M, Bishop W. Nitrite in human saliva. Its possible relationship to nitrosamine formation. J Natl Cancer Inst. 1974;53:79–84.

23. van Maanen JM, van Geel AA, Kleinjans JC. Modulation of nitrate-nitrite conversion in the oral cavity. Cancer Detect Prev. 1996;20(6):590–6.

24. Lundberg JO, Govoni M. Inorganic nitrate is a possible source for systemic generation of nitric oxide. Free Radic Biol Med. 2004;37(3):395–400.

25. Weitzberg E, Lundberg JO. Novel aspects of dietary nitrate and human health. Annu Rev Nutr. 2013;13:129–59.

26. Walker R. The metabolism of dietary nitrites and nitrates. Biochem Soc Trans. 1996;24(3):780–5.

27. McKnight GM, Smith LM, Drummond RS, Duncan CW, Golden M, Benjamin N. Chemical synthesis of nitric oxide in the stomach from dietary nitrate in humans. Gut. 1994;40(2):211–4.

28. Pique JM, Whittle BJ, Esplugues JV. The vasodilator role of endogenous nitric oxide in the rat gastric microcirculation. Eur J Pharmacol. 1989;174(2–3):293–6.

29. Brown JF, Hanson PJ, Whittle BJ. Nitric oxide donors increase mucus gel thickness in rat stomach. Eur J Pharmacol. 1992;223(1):103–4.

30. Bjorne HH, Petersson J, Phillipson M, Weitzberg E, Holm L, Lundberg JO. Nitrite in saliva increases gastric mucosal blood flow and mucus thickness. J Clin Invest. 2004;113(1):106–14.

31. Lundberg JO, Weitzberg E, Lundberg JM, Alving K. Intragastric nitric oxide production in humans: measurements in expelled air. Gut. 1994;35(11):1543–6.

32. Green LC, Ruiz de Luzuriaga K, Wagner DA, et al. Nitrate biosynthesis in man. Proc Natl Acad Sci U S A. 1981;78(12):7764–8.

33. Wennmalm A, Benthin G, Edlund A, et al. Metabolism and excretion of nitric oxide in humans. An experimental and clinical study. Circ Res. 1993;73(6):1121–7.

34. Bednar C, Kies C. Nitrate and vitamin C from fruits and vegetables: impact of intake variations on nitrate and nitrite excretions in humans. Plant Foods Hum Nutr. 1994;45(1):71–80.

35. Weller R, Pattullo S, Smith L, Golden M, Ormerod A, Benjamin N. Nitric oxide is generated on the skin surface by reduction of sweat nitrate. J Invest Dermatol. 1996;107(3):327–31.

36. National Academy of Sciences. The health effects of nitrate, nitrite and N-nitroso compounds. Washington, DC: National Academy Press; 1981.

37. National Academy of Sciences. Alternatives to the current use of nitrite in foods. Washington, DC: National Academy Press; 1982.

38. Program NT. Toxicology and carcinogenesis studies of sodium nitrite (CAS NO. 7632-00-0) in F344/N rats and B6C3F1 mice (drinking water studies). NIH publication no. 01-3954: National Institutes of Health; 2001. p. 7–273.

39. Nunez de Gonzalez MT, Osburn WN, Hardin MD, Longnecker M, Garg HK, Bryan NS, Keeton JT. A survey of nitrate and nitrite concentrations in conventional and organic-labeled raw vegetable at retail. J Food Sci. 2015;80(5):C942–9.

40. Santamaria P, Elia A, Serio F, Todaro E. A survey of nitrate and oxalate content in fresh vegetables. J Sci Food Agric. 1999;79:1882–8.

41. Salomez J, Hofman G. Nitrogen nutrition effects on nitrate accumulation of soil-grown Greenhouse Butterhead Lettuce. Soil Sci Plant Anal. 2009;40(1):620–32.
42. Anjana SU, Iqbal M. Nitrate accumulation in plants, factors affecting the process, and human health implications. A review. Agronomy. 2007;27(1):45–57.
43. Meah MN, Harrison N, Davies A. Nitrate and nitrite in foods and the diet. Food Addit Contam. 1994;11(4):519–32.
44. White Jr JW. Relative significance of dietary sources of nitrate and nitrite. J Agric Food Chem. 1975;23(5):886–91.
45. Cross AJ, Ferrucci LM, Risch A, et al. A large prospective study of meat consumption and colorectal cancer risk: an investigation of potential mechanisms underlying this association. Cancer Res. 2010;70(6):2406–14.
46. Sinha R, Cross AJ, Graubard BI, Leitzmann MF, Schatzkin A. Meat intake and mortality: a prospective study of over half a million people. Arch Intern Med. 2009;169(6):562–71.
47. Honikel KO. The use and control of nitrate and nitrite for the processing of meat products. Meat Sci. 2008;78:68–76.
48. Sebranek JG, Fox JB. A review of nitrite and chloride chemistry: interactions and implications for cured meats. J Sci Food Agric. 1985;36:1169–82.
49. Cassens RG. Use of sodium nitrite in cured meats today. Food Technol. 1995;49:72–81.
50. Cassens RG. Residual nitrite in cured meat. Food Technol. 1997;51:53–5.
51. Cassens RG. Composition and safety of cured meats in the USA. Food Chem. 1997;59:561–6.
52. Nuñez De González MT, Osburn WN, Hardin MD, Longnecker M, Garg HK, Bryan NS, Keeton JT. Survey of residual nitrite and nitrate in conventional and organic/natural/uncured/indirectly cured meats available at retail in the United States. J Agric Food Chem. 2012;60(15):3981–90.
53. Temme EH, Vandevijvere S, Vinkx C, Huybrechts I, Goeyens L, Van Oyen H. Average daily nitrate and nitrite intake in the Belgian population older than 15 years. Food Addit Contam Part A Chem Anal Control Expo Risk Assess. 2011;28(9):1193–204.
54. Hord NG, Tang Y, Bryan NS. Food sources of nitrates and nitrites: the physiologic context for potential health benefits. Am J Clin Nutr. 2009;90(1):1–10.
55. FSA. UK Monitoring programme for nitrate in lettuce and spinach. Food Survey Information Sheet 74/05; 2004.
56. Sebranek JG, Bacus JN. Cured meat products without direct addition of nitrate or nitrite: what are the issues? Meat Sci. 2007;77(1):136–47.
57. FSA. Survey of nitrite and nitrate in bacon and cured meat products. Food Surveillance Information Sheet No. 142, London; 1998.
58. IARC, editor. Ingested nitrate and nitrite and cyanobacterial peptide toxins. Lyon: IARC Press; 2010.
59. FSA. Total diet study, nitrate and nitrite. Food Surveillance Information Sheet No. 163, London; 1998.
60. World Health Organization. Nitrate and nitrite. Geneva: World Health Organization; 2007.
61. Jakszyn PG, Ibanez R, Pera G, Agudo A, García-Closas R, Amiano P, González CA. Food content of potential carcinogens. European prospective investigation on cancer report. Barcelona; 2004.
62. EPA US EPA. Technical fact sheet on nitrate/nitrite.
63. EPA US EPA. EPA community water regulations handbook; 2010. www.epa.gov/sites/production/files/documents/EnvRegSC_Hndbk.pdf.
64. EPA US EPA. Listing of local drinking water contaminant results. http://www.epa.gov/ccr, http://www3.epa.gov/storet/dbtop.html.
65. World Health Organization. Guidelines for drinking water quality. 4th ed; 2011. p. 398–403. http://www.who.int/water_sanitation_health/publications/dwq_guidelines/en/.
66. Food Nutrition Board, National Academy of Sciences. Institute of Medicine dietary reference intakes: water, potassium, sodium, chloride, and sulfate. Washington, DC: National Academies Press; 2004.
67. EPA US EPA. Water quality monitoring report; 2009. http://www.epa.gov/waterdata/2004-national-water-quality-inventory-report-congress.
68. van Grinsven HJ, Ward MH, Benjamin N, de Kok TM. Does the evidence about health risks associated with nitrate ingestion warrant an increase of the nitrate standard for drinking water? Environ Health. 2006;5:26.
69. Wright RO, Lewander WJ, Woolf AD. Methemoglobinemia: etiology, pharmacology, and clinical management. Ann Emerg Med. 1999;34(5):646–56.
70. Cornblath M, Hartmann AF. Methhemoglobinaemia in young infants. J Pediatr. 1948;33:421–5.
71. Kortboyer J, Olling M, Zeilmaker MJ. The oral bioavailability of sodium nitrite investigated in healthy adult volunteers. Bilthoven: National Institute of Public Health and the Environment; 1997.
72. Dejam A, Hunter CJ, Tremonti C, et al. Nitrite infusion in humans and nonhuman primates: endocrine effects, pharmacokinetics, and tolerance formation. Circulation. 2007;116(16):1821–31.
73. L'Hirondel JL, Avery AA, Addiscott T. Dietary nitrate: where is the risk? Environ Health Perspect. 2006;114(8):A458–9; author reply A459–61.
74. Powlson DS, Addiscott TM, Benjamin N, et al. When does nitrate become a risk for humans? J Environ Qual. 2008;37(2):291–5.

75. McKnight GM, Duncan CW, Leifert C, Golden MH. Dietary nitrate in man: friend or foe? Br J Nutr. 1999;81(5):349–58.
76. L'Hirondel JL. Nitrate and man: toxic, harmless or beneficial? Wallingford: CABI Publishing; 2001.
77. EFS Authority. Nitrate in vegetables: scientific opinion of the panel on contaminants in the food chain. EFSA J. 2008;289:1–79.
78. Lundberg JO, Weitzberg E, Gladwin MT. The nitrate-nitrite-nitric oxide pathway in physiology and therapeutics. Nat Rev Drug Discov. 2008;7(8):156–67.
79. Qin L, Liu X, Sun Q, Fan Z, et al. Sialin (SLC17A5) functions as a nitrate transporter in the plasma membrane. Proc Natl Acad Sci U S A. 2012;109(33):13434–9.
80. Eisenbrand G, Spiegelhalder B, Preussmann R. Nitrate and nitrite in saliva. Oncology. 1980;37(4):227–31.
81. Tannenbaum SR, Weisman M, Fett D. The effect of nitrate intake on nitrite formation in human saliva. Food Cosmet Toxicol. 1976;14(6):549–52.
82. Hyde ER, Andrade F, Vaksman Z, Parthasarathy K, et al. Metagenomic analysis of nitrate-reducing bacteria in the oral cavity: implications for nitric oxide homeostasis. PLoS One. 2014;9(3), e88645.
83. Hezel MP, Weitzberg E. The oral microbiome and nitric oxide homoeostasis. Oral Dis. 2015;21(1):7–16.
84. Bondonno CP, Liu AH, Croft KD, Considine MJ, et al. Antibacterial mouthwash blunts oral nitrate reduction and increases blood pressure in treated hypertensive men and women. Am J Hypertens. 2015;28(5):572–5.
85. Xia D, Deng D, Wang S. Alterations of nitrate and nitrite content in saliva, serum, and urine in patients with salivary dysfunction. J Oral Pathol Med. 2003;32(2):95–9.
86. Lundberg JO, Weitzberg E, Cole JA, Benjamin N. Nitrate, bacteria and human health. Nat Rev Microbiol. 2004;2(7):593–602.
87. World Resources Institute. World resources 1994–95: a guide to the global environment. New York: Oxford University Press; 1994.
88. US EPA National trends in nitrogen dioxide levels 2014. http://www3.epa.gov/airtrends/nitrogen.html. Accessed 28 Dec 2015.
89. Lightning and atmospheric chemistry: the rate of atmospheric NO production. In: Volland H, editor. Handbook of atmospheric dynamics, No. 1. Boca Raton: CRC Press; 1995.
90. Al-Dabbagh S, Forman D, Bryson D, Stratton I, Doll R. Mortality of nitrate fertiliser workers. Br J Ind Med. 1986;43(8):507–15.
91. Fandrem SI, Kjuus H, Andersen A, Amlie E. Incidence of cancer among workers in a Norwegian nitrate fertiliser plant. Br J Ind Med. 1993;50(7):647–52.
92. Addiscott TM. Fertilizers and nitrate leaching. Issues Environ Sci Technol. 1996;5:1–26.
93. Miller P, Saeman WC. Properties of monocrystalline ammonium nitrate. Fertilizer Ind Eng Chem. 1948;40:154–9.
94. European Fertilizer Manufacturers Association. Production of ammonium nitrate and calcium ammonium nitrate. Brussels: EFMA; 2000.
95. World Health Organization. Environmental health criteria 4: oxides of nitrogen. Geneva: WHO; 1977.
96. Anderson HR, Spix C, Medina S, et al. Air pollution and daily admissions for chronic obstructive pulmonary disease in 6 European cities: results from the APHEA project. Eur Respir J. 1997;10(5):1064–71.
97. Hoffmann D, Rivenson A, Hecht SS. The biological significance of tobacco-specific N-nitrosamines: smoking and adenocarcinoma of the lung. Crit Rev Toxicol. 1996;26(2):199–211.
98. Jenkins RA, Gill BE. Determination of oxides of nitrogen (NOx) in cigarette smoke by chemiluminescent analysis. Anal Chem. 1980;52:925–8.
99. Lofroth G, Burton RM, Forehand L, Hammond SK, Seila RL, Zweidinger RB, Lewtas J. Characterization of environmental tobacco smoke. Environ Sci Technol. 1989;23:610–4.
100. Norman V, Keith CG. Nitrogen oxides in tobacco smoke. Nature. 1965;205:915–6.
101. Cueto R, Pryor W. A cigarette smoke chemistry: conversion of nitric oxide to nitrogen dioxide and reactions of nitrogen oxides with other smoke components as studied by Fourier transform infrared spectroscopy. Vibrat Spectrosc. 1994;7:97–111.
102. Manochehr-Pour M, Bhat M, Bissada N. Clinical evaluation of two potassium nitrate toothpastes for the treatment of dental hypersensitivity. Periodontal Case Rep. 1984;6(1):25–30.
103. Orchardson R, Gillam DG. The efficacy of potassium salts as agents for treating dentin hypersensitivity. J Orofac Pain. 2000;14(1):9–19.
104. Poulsen S, Errboe M, Hovgaard O, Worthington HW. Potassium nitrate toothpaste for dentine hypersensitivity. Cochrane Database Syst Rev. 2001;(2):CD001476.
105. Wara-Aswapati N, Krongnawakul D, Jiraviboon D, Adulyanon S, Karimbux N, Pitiphat W. The effect of a new toothpaste containing potassium nitrate and triclosan on gingival health, plaque formation and dentine hypersensitivity. J Clin Periodontol. 2005;32(1):53–8.
106. Spiegelhalder B, Eisenbrand G, Preussmann R. Influence of dietary nitrate on nitrite content of human saliva: possible relevance to in vivo formation of N-nitroso compounds. Food Cosmet Toxicol. 1976;14(6):545–8.

107. Magee PN, Barnes JM. The production of malignant primary hepatic tumours in the rat by feeding dimethylnitrosamine. Br J Cancer. 1956;10(1):114–22.
108. Fan TY, Tannenbaum SR. Natural inhibitors of nitrosation reactions: the concept of available nitrite. J Food Sci. 1973;38(6):1067–9.
109. Fiddler W, Pensabene JW, Piotrowski EG, Doerr RC, Wasserman AE. Use of sodium ascorbate or erythorbate to inhibit formation of N-nitrosodimethlyamine in frankfurters. J Food Sci. 1973;38(6):1084.
110. Mirvish SS. Blocking the formation of N-nitroso compounds with ascorbic acid in vitro and in vivo. Ann N Y Acad Sci. 1975;258:175–80.
111. Mirvish SS. Inhibition by vitamins C and E of in vivo nitrosation and vitamin C occurrence in the stomach. Eur J Cancer Prev. 1996;5 Suppl 1:131–6.

Part II
Food and Environmental Exposures to Nitrite and Nitrate

Chapter 7
History of Nitrite and Nitrate in Food

Jimmy T. Keeton

Key Points

- Saltpeter or "niter" (potassium nitrate), a natural contaminant of salt, contributed historically to the pinkish-red color in salted meats.
- Nitric oxide, derived from nitrate/nitrite reduction, when combined with the intracellular protein myoglobin produces the pink colored pigment in cured meat products.
- Nitrite prevents sporulation of *Clostridium botulinum* in cured meats.
- In 1925, the United States Department of Agriculture (USDA) approved nitrite's use in curing brines and formulas at a maximum ingoing level of 200 ppm.
- In the U.S. residual nitrite content declined in cured meats from the 1930s to 1970s and has remained low since the 1980s, thus mitigating the formation of *N*-nitrosamines.
- USDA and EU regulations restrict ingoing levels of nitrite and nitrate (if allowed) to specific levels in various meat product categories.
- Residual nitrite levels in conventional and "organic" cured meats are <10 ppm in the U.S.
- Naturally occurring nitrate in raw vegetables at retail in the U.S. ranges from 200 to 3000 ppm.
- Nitrates in the municipal water supply of 25 major metropolitan U.S. cities are <10 mg/L, but some cities report levels near the 10 mg/L upper limit.
- Nitrite contributes to the safety of cured meats and currently no suitable alternative is available.

Keywords Nitrate • Nitrite • Cured meats • *Clostridium botulinum* • Dietary load • USDA regulations • Vegetables • Water

Introduction

Salting as means of preserving meat, poultry, fish, seafood, and vegetables predates written history and was essential in ancient times for providing nutrient-dense foods during scarcity or population migration. The application of rock salt to animal tissues led to the curing practices of today that

J.T. Keeton, B.S., M.S., Ph.D. (✉)
Department of Nutrition and Food Science, Texas A&M University,
1090 Pace Rd., College Station, TX 38059, USA
e-mail: jkeeton@tamu.edu

N.S. Bryan, J. Loscalzo (eds.), *Nitrite and Nitrate in Human Health and Disease*, Nutrition and Health,
DOI 10.1007/978-3-319-46189-2_7, © Springer International Publishing AG 2017

include brine injection, marination, dry rubs, or a combination of each that preserve, protect, and flavor present-day cured meat products. Meat curing is historically defined as the addition of salt (sodium chloride) to fresh meat cuts to remove moisture and reduce the water activity of tissues to prevent spoilage. In ancient times, salt was obtained from crystalline deposits by mining directly from the earth, or evaporating water from brine pools or seawater. As a consequence, it often contained natural contaminants such as sodium or potassium nitrate (saltpeter or "niter") that contributed directly to the curing reaction and the preservation process. These contaminants, it was later learned, were the primary components in curing reactions that allowed reduction of nitrate salts to gaseous nitric oxide and its subsequent reaction with myoglobin, an intracellular transport protein.

The fundamental utility of nitrite has been its ability to inhibit the growth of a number of aerobic and anaerobic microorganisms and especially suppress the outgrowth of *Clostridium botulinum* spores. Nitrite in combination with salt and other curing factors may also control the growth of other pathogens such as *Bacillus cereus*, *Staphylococcus aureus*, and *Clostridium perfringens*. However, under conditions of prolonged temperature fluctuations in a range that promotes bacterial growth, it does not prevent pathogen outgrowth and may allow toxin production or spoilage. Other properties of nitrite include its reduction and reaction with myoglobin to produce the characteristic reddish-pink cured meat color, its ability to inhibit lipid oxidation (and reduce oxidative rancidity or warmed-over flavor), and its contribution to the cured flavor of meat products.

Meat curing today is understood to mean the incorporation of salt, sodium nitrite, and occasionally sodium or potassium nitrate into meat products. Additional ingredients in a curing mixture may include water, potassium chloride, sodium ascorbate (or its isomer erythorbate), sweeteners (sucrose, glucose, honey), alkaline phosphates, lactic acid, citric acid, acetic acid, sodium or potassium lactate, sodium citrate, glucono delta lactone, lactic acid starter cultures, spices and seasonings, antioxidants, nonmeat binders and extenders, hydrolyzed proteins, gelatin, modified starch, hydrocolloids, and liquid or natural smoke. Some of these ingredients (saltpeter, vinegar, honey, sugar, spices, natural smoke, lactic acid bacteria) were used historically to impart flavor, functional properties, and textural characteristics to cured meats but were self-limiting in most cases and unregulated in the United States prior to the turn of the twentieth century. These ingredients, and especially nitrates and nitrites, have been studied to identify their specific functions in the curing process and their role in the potential formation of *N*-nitrosamines, some of which are known carcinogens in several animal species. The purpose of this chapter is to provide a brief historical perspective on the use of nitrates and nitrites in preserving meat products, the regulatory limits of cure ingredients as established by the United States Department of Agriculture—Food Safety Inspection Service (USDA-FSIS) or Food and Drug Administration (FDA), and to give a synopsis of the residual levels of nitrates and nitrites in present-day cured meat products. A brief overview of the concentrations of naturally occurring nitrates in selected vegetables and municipal water supplies in the United States is also presented since these contribute to dietary nitrate and nitrite load.

History of Nitrate and Nitrite in Meat Curing

In a review of the history of nitrate and nitrite in meat curing, Binkerd and Kolari [1] concluded that salting as a means of meat preservation was first practiced in the deserts of Asia. Saline salts from this area contained impurities such as nitrates that contributed to the characteristic red color of cured meats. As early as 3000 BC in Mesopotamia, cooked meats and fish were preserved in sesame oil and dried, salted meat and fish were part of the Sumerian diet. Salt from the Dead Sea was in common use by Jewish inhabitants around 1600 BC, and by 1200 BC, the Phoenicians were trading salted fish in the Eastern Mediterranean region. By 900 BC, salt was being produced in "salt gardens" in Greece and dry salt curing and smoking of meat were well established. The Romans (200

BC) acquired curing procedures from the Greeks and further developed techniques to "pickle" various kinds of meats in a brine marinade. It was during this time that the reddening effect of salting was noted. Saltpeter (potassium nitrate) is mentioned as being gathered in China and India prior to the Christian era for use in meat curing along with "wall" saltpeter (calcium nitrate), which is formed by nitrifying bacteria and deposited on the walls of caves and stables. In Medieval times, the application of salt and saltpeter as curing ingredients was commonplace and the reddening effect on meat was attributed to saltpeter.

As early as 1835, saltpeter was cited as imparting juiciness and flavor to bacon when applied at 0.5 lb per 100 lb of meat (5000 ppm or mg/kg). In 1873, Edward Smith described salt as "the oldest and best known of preserving agents…its chief action appears to be due to its power of attracting moisture, and thus extracting fluid to harden the tissues." He further described the development of a "reddish color throughout" in meat preserved with saltpeter as compared to preservation by salt alone, which allowed the meat color to fade. Meat curing was more of an art than a science in the early nineteenth century, but as a greater understanding of the curing process evolved in the late 1800s and early 1900s, the role of nitrate and nitrite in the formation of cured meat color and flavor became apparent. In 1891, Polenske first reported finding nitrite in cured meat and reused curing pickle [1]. He concluded that nitrite was the result of the bacterial reduction of nitrate added to the pickle [2]. In 1899, Lehman and Kisskalt (cited in [1]) demonstrated that the color development in cured meats was actually due the presence of nitrite and not nitrate. Later, Haldane [3] and Hoagland [4] explained the chemical reactions involved in the development of red color in cooked, cured meats in a series of studies with hemoglobin. Haldane demonstrated the formation of nitrosylhemoglobin when nitrite was combined with hemoglobin, and the subsequent conversion of nitrosylhemoglobin to its red pigment form, nitrosylhemochromogen, with heating. Hoagland, also using "hemoglobin," later described the reduction of nitrate to nitrite and nitrite to nitrous acid with subsequent formation of nitric oxide by bacterial and/or enzymatic reactions [5]. Conversion of nitrous acid to nitric oxide allowed reaction with "hemoglobin" resulting in the formation of nitrosylhemoglobin and nitrosohemochromogen. Additional studies by Hoagland in 1914 [6] with cooked salted meats showed that nitrous acid, or its metabolite nitric oxide, reacted with myoglobin to form nitrosylmyoglobin and nitrosylprotoheme, the characteristic pink color compounds observed in uncooked and cooked cured meat products, respectively [7, 8].

Regulatory Restrictions of Nitrate and Nitrite

Following the work of Haldane and Hoagland, the German government in 1909 recommended the use of partially reduced nitrate in curing mixes and marketed these across Europe. Up until this time, U.S. processors had limited control, if any, over the production of nitrite in a nitrate brine or pickle. Curing establishments, prior to the early 1900s, frequently used the same lot of pickle more than once, adding more salt and nitrate to maintain the original concentration in the pickle, failing to realize that the pickle now contained more nitrite derived from the reduction of the original nitrate in the brine. The effect was a pickle relatively rich in nitrite, as well as highly contaminated with bacteria. Products produced using these brines were susceptible to spoilage, quite variable in nitrite concentration (often ranging from 2 to 960 ppm), and in some cases had exceptionally high levels of residual nitrite [2]. To resolve this problem, the Bureau of Animal Industry of the USDA authorized a series of experiments in 1923 to study the direct use of nitrite as a curing agent rather than nitrate. Kerr et al. [9], under the supervision of Bureau of Animal Industry inspectors, cured hams with a closely controlled amount of nitrite (~2000 ppm) in a pickle, analyzed the hams for residual nitrite and compared these results to the residual nitrite in nitrate-only cured products. They established that the addition of approximately 2000 ppm of sodium nitrite in the curing pickle resulted in an average nitrite content of 42–150 ppm in hams with no more than 200 ppm in any part of the product. Nitrate-cured hams, on the other hand, had a maximum residual nitrite content of 45 ppm. Subsequent evaluation of the flavor and quality of the nitrite-cured hams indicated that they were equivalent to traditionally cured (nitrate) hams. Based on these results, a tentative limit of 200 ppm residual nitrite was established as the maximum nitrite level allowed in finished cured meat products. Additional commercial-scale experiments in 17 meat curing plants with shoulders, loins, tongues, hams, bacon, corned beef, dried beef, and sausages enabled Kerr et al. to conclude that "from 1/4th to 1 oz. of sodium nitrite is sufficient to fix the color in 100 lb (of meat), the exact quantity depending on the meat to be cured and the process employed." Based on these experiments, sodium nitrite was allowed as a curing ingredient in federally inspected establishments in 1925 [10, 11]. The Bureau authorization later stated that "Extended experiments have demonstrated that successful curing may be accomplished by the addition of as small a quantity as one-fourth of an ounce of sodium nitrite to each 100 lb of meat; therefore, pending further ruling by the Bureau, the finished product shall not contain sodium nitrite in excess of 200 parts per million." U.S. meat packers found that by using nitrite in place of nitrate in curing pickles, they gained more control over the process with more uniform results. The direct addition of nitrite also shortened pickling times and resulted in a curing process that did not require the presence of reducing microorganisms for nitrate, thereby allowing for a process that limited the presence of spoilage bacteria.

In 1931, the USDA issued a rule stating that curing solutions containing both nitrite and nitrate be limited to ingredient levels that deliver no more than 156 ppm nitrite and 1716 ppm nitrate per 100 lb of pumped, brine-cured meats. Additional regulations for cured meat were issued in 1970 that allowed the use of sodium or potassium nitrate at 7 lb per 100 gal pickle, 3.5 oz per 100 lb meat in dry cure, or 2.75 oz per 100 lb of chopped meat. Sodium nitrite was permitted at 2 lb per 100 gal pickle (assuming a 10 % pump level), while 1 oz could be applied directly to 100 lb meat as a dry cure or 0.25 oz incorporated into 100 lb of chopped meat or meat product that would result in not more than 200 ppm nitrite in the finished product.

Current U.S. regulations allow the use of nitrite and selective use of nitrate in meat products based on product category and method of curing. Immersion cured, massaged, or pumped products, such as hams or pastrami, are limited to a maximum ingoing level of 200 ppm of sodium or potassium nitrite and/or 700 ppm of nitrate, respectively, based on the raw product weight [12]. Dry-cured products, however, are allowed a maximum ingoing level of 625 ppm nitrite and/or 2187 ppm nitrate since these products have longer curing times that allow for immediate nitrite reaction with myoglobin and longer term conversion of nitrate to nitrite. If a combination of nitrite and nitrate is used, the combination must not result in more than 200 ppm sodium nitrite in the finished product. Comminuted products, such as frankfurters, bologna, and other cured sausages, are limited to a maximum ingoing level of 156 ppm of sodium or potassium nitrite based on the raw weight of the meat block. Nitrate may be added to these products at 1718 ppm regardless of the type of salt used. In 1978, USDA regulations lowered the ingoing sodium nitrite in bacon to 120 ppm (148 ppm potassium nitrite), required addition of 547 ppm sodium ascorbate or its isomer erythorbate, and eliminated the use of nitrates in bacon [13]. Bacon regulations were again changed in 1986 to give processors three alternatives that would allow lower levels of nitrite in combination with other processing procedures. Skinless bacon was required to have 120 ppm of sodium nitrite (148 ppm potassium nitrite) in combination with 547 ppm of sodium ascorbate or erythorbate to reduce the ingoing nitrite level and the potential for N-nitrosamine formation. These regulations also allowed for a ±20 % (96–144 ppm) variance from the specified concentration of nitrite at the time of injection or massaging. Other exceptions to these regulations include reducing sodium nitrite to 100 ppm (123 ppm potassium nitrite) with an "appropriate partial quality control program" [14], or 40–80 ppm of sodium nitrite (49–99 ppm potassium nitrite) if sugar and lactic acid starter culture are included in the curing brine. Dry-cured bacon was limited to 200 ppm sodium nitrite or 246 ppm potassium nitrite.

In other countries, the levels of nitrite and nitrate in cured meats are regulated, but vary depending upon the maximum input levels allowed by each regulatory authority and the specific processing procedures followed by manufacturers. The European Union rules, as specified by Directive No. 95/2/EC [15] and modified in Directive in 2006/52/EC, fix the input nitrite level in bacon at 150 ppm and residual amounts between 50 and 175 ppm. In Denmark, a lesser amount is allowed (60–150 ppm) in semipreserved products and special cured hams. Directive No. 95/2/EC allows only 250 ppm residual nitrate (sodium) in salted meats, but for unheated products 150 ppm nitrite +150 ppm nitrate are allowed. Wiltshire or dry-cured bacon may have 175 ppm residual nitrite +250 ppm residual nitrite [7]. In Canada, the maximum allowable limit of nitrite in meat products, such as hams, loins, shoulders, cooked sausages, and corned beef, is 200 ppm, but further reductions to 120–180 ppm have been taken by the Canadian meat industry [8]. Bacon may not contain more than 120 ppm nitrite (input) to reduce the risk of *N*-nitrosamine formation.

Residual Nitrate and Nitrite Levels in Meat Products

From 1925 until 1970, many U.S. meat processors continued to use nitrate in their curing formulations. In 1930, Mohler (cited in [1]) reported that 54 % of meat processors were still using sodium nitrate as a curing ingredient, 17 % used sodium nitrite, and 30 % used a combination of nitrate and nitrite. By 1970, approximately 50 % of meat processors were reported using nitrate in canned, shelf-stable products, but by 1974, all manufacturers surveyed had discontinued the use of nitrate in these products as well as cured bacon, hams, canned sterile meats, and frankfurters. Reasons for this change have been attributed to a technological shift to more rapid curing and processing procedures, careful control of nitrite quantities that in turn eliminated the need for nitrate, and concern over nitrate being a precursor to the formation of volatile *N*-nitrosamines during processing or after consumption.

Early surveys of nitrate and nitrite levels in U.S. cured meats by Mighton [16] and Lewis [17] showed lower levels of nitrite and nitrate than were previously observed by Kerr et al. [9] just 10 years earlier. By 1972, no nitrates were reported in a survey of cured products such as hams, bacon, corned beef, bologna, or frankfurters (Table 7.1) [18]. As noted in Table 7.1, residual nitrite levels were lower in bologna (36 ppm) and frankfurters (38 ppm) in the 1972 survey compared to mean nitrite levels of 61–72 ppm and 84–102 ppm observed in two studies in the mid-1930s [16, 17]. In the early 1970s, the use of nitrite to cure meat was questioned due to the potential formation of *N*-nitrosamines based upon the level of residual nitrite contributed to the total dietary load [19]. Recommendations derived from this concern and the subsequent publication of additional studies on the use of nitrite in curing led to a change in USDA regulations in 1978 and 1986 that reduced the allowable levels of nitrite and nitrate in meat products and made provision for the use of reductants such as sodium ascorbate to decrease the level of residual nitrite and reduce the potential for *N*-nitroso compound formation [20]. The use of nitrate was restricted to only certain long-term cured products (dry-cured hams, prosciutto) and stricter handling controls of nitrites were instituted in the manufacturing process. In 1980, The National Academy of Sciences (NAS) entered into a contract with USDA and the FDA and established the *Committee on Nitrate and Alternative Curing Agents in Food*. The committee was charged with the task of assessing the health risks associated with overall exposure to nitrate, nitrate, and *N*-nitroso compounds derived from both natural sources and the nitrate and nitrite added to foods. They also were to review the status of research and future prospects for developing feasible alternatives to the use of nitrite as a preservative. In 1981, a comprehensive report entitled, "The Health Effects of Nitrate, Nitrite and *N*-Nitroso Compounds," was published by the NAS to assess the human health risk of these compounds. Eleven recommendations, which are shortened as follows [21, 22], were made to reduce the risk associated with consumption of nitrites, nitrates, and *N*-nitroso compounds in cured meats.

1. Nitrate is neither carcinogenic nor mutagenic. Some human population studies have indicated an association of exposure to high nitrate levels with certain cancers. Future studies were therefore recommended.
2. Nitrite does not act directly as a carcinogen in animal studies. Further testing may be warranted.

Table 7.1 Residual nitrite and nitrate levels (ppm or mg/kg) in cured meat products surveyed at retail in 1936 and 1937

Product	Nitrite (ppm)		Nitrate (ppm)	
	Average	Range	Average	Range
Hams	52[M,B]	7–145	360	120–610
Smoked	80[L,B]	34–184	600	100–1200
Boiled	59[M,B]	11–87	370	30–910
	49[L,B]	31–63	700	300–1100
Bacon (raw)	13[M,B]	4–22	1400	700–2100
	16[L,B]	11–29	1200	300–1900
	96[K]	24–170	–	–
Corned beef	3[M,B]	3–5	200	169–270
	75[L,B]	1–216	1700	500–3000
Bologna	61[M,B]	44–86	380	30–790
	72[L,B]	60–114	650	100–1020
	36[K]	0–76	–	–
Frankfurters	84[M,B]	55–146	710	570–880
	102[L,B]	13–195	513	100–1090
	38[K]	15–80	–	–

M Mighton [16]; *L* Lewis [17]; *K* Kolari and Aunan (1972) [18]; *B* Binkerd and Kolari (1975) [1]

3. Most *N*-nitroso compounds are carcinogenic in laboratory animals, mutagenic in microbial and mammalian test systems, and some are teratogenic in laboratory animals. Future work should emphasize quantitative assessment of potency and outcome.

4. Because nitrate and nitrite can exert acute toxic effects and contribute to the total body burden of *N*-nitroso compounds, it was recommended that exposure to these agents be reduced. Reduction in nitrate use should not compromise protection against botulism. The use of nitrate salts in curing should be eliminated, with the exception of a few special products. Attention should be given to reducing nitrate content of vegetables and drinking water.

5. Sources of exposure to *N*-nitroso compounds in various environmental media should be determined so that it can be reduced. Analytical procedures should be improved, especially for nonvolatile compounds and for both free and bound nitrite.

6. The exposure of humans to amines and nitrosamines can be reduced in some instances by modifying manufacturing practices, such as with certain pesticides and drugs.

7. Additional studies are needed to increase the understanding of the metabolism and pharmacokinetics of nitrate in humans.

8. Further study of inhibitors of nitrosation is needed.

9. Further studies should be made to determine the mechanism(s) whereby nitrite controls the outgrowth of *C. botulinum* spores, and also its effect against other spoilage and pathogenic microorganisms.

10. Although it is not possible to estimate the potential morbidity or mortality from *C. botulinum* in the absence of nitrite as a curing agent in certain products, the prudent approach to protecting public health requires consideration of the possibility that certain preserved food items may be contaminated and may be abused.

A subsequent report entitled, "Alternatives to the Current Use of Nitrite in Foods" [23], made recommendations for reducing nitrite and nitrate levels in cured meats and recommended alternatives that could serve as partial or complete replacements for nitrate, and agents that block the formation of nitrosamines in products containing conventional concentrations of nitrite. The following alternatives were found by the committee to be the most promising: a combination of ascorbate, α-tocopherol, and nitrite; irradiation (with or without nitrate); lactic acid-producing organisms (with or without nitrite); potassium sorbates with low concentrations of nitrite; sodium hypophosphite (with or without nitrite);

Table 7.2 Mean residual nitrite and nitrate concentrations (ppm or mg/kg) of cured meat products in the United States from the 1970s to present

Product category	Nunez et al. [26][a] Nitrite (ppm)	Nunez et al. [26][a] Nitrate (ppm)	Cassens [19] Nitrite (ppm)	NAS [21, Table 5.1] Nitrite (ppm)	NAS [21, Table 5.1][b] Nitrate (ppm)
Cured, dried sausages (uncooked)	0.7	79	–	6–17	78–89
German air-dry sausage, chorizo, Italian dry sausage					
Cured sausages (cooked)	7	28	8–15	7–31	32–110
Bologna, frankfurters, polish sausage					
Fermented/acidified sausages (cooked)	0.6	36	–	6–17	78–89
Pepperoni, summer sausage, snack sticks					
Whole-muscle brine cured (uncooked)	7	26	3–5	12–42	33–96
Bacon					
Whole-muscle brine cured (cooked)	7	15	4–7	16–99	140–150
Hams, bacon (precooked), cured poultry, pastrami					
Corned beef					
Whole muscle dry-cured (uncooked)	2	67	–	280	150
Dry-cure country style hams, dry-cure bacon, prosciutto					

[21] With permission © National Academies Press
[26] With permission © North American Meat Institute
[a]Combined means of the nitrite and nitrate concentrations in conventional and organic cured meat products at retail
[b]National Academy of Sciences (1981)

and several fumarate esters. Combined, these reports alleviated public concern about cured meat as a human health risk. However, an epidemiological report that processed meats (specifically hot dogs) caused cancer in children [24] led to additional concerns about the potential risk of N-nitroso compounds and the levels of nitrite in these products. Following the publication of the NAS reports and other studies linking nitrite as a curing ingredient to cancer, a significant amount of research was initiated. However, a suitable alternative to nitrite with the same protective effect against *Clostridium botulinum* has not been identified.

To substantiate that levels of nitrite were lower in cured meat products since USDA regulatory changes were instituted in 1978 and 1986, a multicity survey of U.S. cured meat products (bacon, ham bologna, and wieners from major manufacturers) was conducted by Cassens [19] to assess the levels of residual nitrite and ascorbate in these products. The overall mean residual nitrite and ascorbate levels in 164 samples were 10 and 209 ppm, respectively. In comparison, the average nitrite value was lower than those reported by White [25] and those given in the NAS [21] report. White's mean values for nitrite and nitrate were 52.5 and 235 ppm (361 ppm for canned items), respectively, for a broad range of U.S. cured meat products taken from different databases in the early 1970s. Overall mean residual nitrite levels in Cassens' survey indicated that the residual nitrite in cured meats had been reduced significantly from the levels reported in the 1970s.

A national U.S. retail survey of the residual nitrate and nitrite concentrations in cured meat products was conducted by Keeton in 2009 to verify that levels contributed by cured meats have remained low and are minor contributors to the total human dietary intake of nitrate and nitrite [26, 27]. A comparison of the mean nitrite and nitrate concentrations in the major cured meat categories is shown in Table 7.2. Nitrite means of cured products selected from the NAS [21] study and derived from

Table 7.3 Mean residual nitrite and nitrate concentrations (ppm or mg/kg) of retail cured meat categories classified as conventional or organic

Product category		Conventional				Organic			
		N	Mean	Min	Max	N	Mean	Min	Max
Cured dried uncooked sausage	Nitrite	40	0.8	0.03	10	24	0.6	0.01	8
	Nitrate	40	113	0.1	2289	23	19	0.1	137
Cured cooked sausage	Nitrite	59	8	0.1	29	30	5	0.05	35
	Nitrate	59	32	0.8	541	30	18	0.4	74
Fermented cooked sausage	Nitrite	52	0.8	0	27	19	0.1	0	0.5
	Nitrate	52	46	2	320	19	7	0	57
Whole-muscle brine cured uncooked	Nitrite	20	7	1	36	19	8	0.1	36
	Nitrate	20	14	3	32	18	38	2	462
Whole-muscle brine cured cooked	Nitrite	97	7	0.03	28	44	7	0.1	22
	Nitrate	97	16	0.2	108	44	13	0	143
Whole-muscle dry-cured uncooked	Nitrite	39	1	0.02	16	24	3	0.02	29
	Nitrate	38	106	0.4	1367	24	7	0.2	50

See Table 7.2 for examples of actual products in each product category. Nunez et al. [26] with permission © North American Meat Institute

databases amassed in the 1970s ranged from 6 to 42 ppm except for dry-cure products, which had nitrite levels of 280 ppm. Compared to mean nitrite values (3–15 ppm) in three product categories reported by Cassens [19], residual nitrite levels have decreased significantly since the NAS study was conducted. Keeton [26, 27] likewise evaluated 470 cured meat products and found comparable mean nitrite values (0.6–7 ppm) to those of Cassens [19]. Keeton found much lower nitrate levels (15–79 ppm) within each product category compared to those in the NAS survey (24–640 ppm). The nitrite and nitrate concentrations of cured meat products classified as conventional or organic are given in Table 7.3. In general, all residual nitrite levels were not different between conventional and organic categories. However, the nitrate concentrations in organic labeled cured meat products tended to be lower in dried sausage, cooked sausage, fermented sausage, and dry-cured products when compared to conventional products. Thus, it appears that implementation of regulatory restrictions by the USDA-FSIS in 1978 and 1986, the use of reductants such as ascorbate, and controlled manufacturing procedures have dramatically reduced the levels of residual nitrite, and nitrate, in all cured meat product categories.

Nitrates and Nitrites Occurring Naturally in Food

Vegetables

Nitrates are intrinsic, naturally occurring constituents of plants and as a consequence are consumed with vegetables and fruits. Nitrates are produced in the soil by microorganisms acting on manure, urine, and vegetable waste, thus making them available to growing plants [28]. Several factors can affect the nitrate content of vegetables [21, 29–31]. These include:

- Related plant strains (cultivars) that systematically differ in nitrate content.
- Different levels and timing of nitrogen fertilizer application. Nitrate accumulation increases as the amount of nitrogen fertilizer used increases and if the fertilizer is applied shortly before harvest.

Table 7.4 Mean nitrite and nitrate concentrations (ppm or mg/kg) of raw vegetables at retail classified as conventional or organic

Product category		Conventional				Organic			
		N	Mean	Min	Max	N	Mean	Min	Max
Broccoli	Nitrite	20	0.6	0.01	9	20	0.1	0.01	0.7
	Nitrate	20	394	29	1140	19	204	3	683
Cabbage	Nitrite	20	0.1	0.01	0.4	20	1	0.01	11
	Nitrate	20	418	37	1831	20	552	2	2114
Celery	Nitrite	18	0.1	0.02	0.5	20	1	0.02	9
	Nitrate	19	1495	20	4269	20	912	0.7	3589
Lettuce	Nitrite	19	0.6	0.01	10	20	0.1	0	0.3
	Nitrate	19	850	79	2171	20	844	58	2013
Spinach	Nitrite	19	8	0	137	20	1	0	17
	Nitrate	19	2797	65	8000	20	1318	16	4089

Nunez et al. [26] with permission © North American Meat Institute

- Nitrate levels tend to increase as daytime temperatures drop below an optimal temperature; thus, geographical region and season of harvest affect nitrate content.
- Greenhouse plants tend to accumulate higher levels of nitrate than do plants grown outdoors perhaps because nitrogen fertilizers are used more heavily indoors.
- Plants grown in shade, at high altitudes with limited sunlight, and during drought accumulate higher levels of nitrate than do plants grown under optimal conditions of water and light supply.
- Leafy plants harvested on a sunny afternoon often contain less nitrate than those harvested in the morning or during inclement weather.
- Some plant diseases, insect damage, or exposure to herbicides, such as those used in weed control, often increase nitrate accumulation.
- Soil deficiencies, such as insufficient molybdenum or potassium, or acidic, organically rich (peat) soils lead to elevated nitrate content.

These factors may alter nitrate levels in vegetables by affecting one or more plant processes such as nitrogen uptake, nitrogen transport, or nitrate reduction and assimilation. Processes such as canning and blanching can reduce the nitrate levels by 20–50 % in some vegetables, while storage, particularly at higher temperatures, can result in an increase of nitrite due to the reduction of nitrate to nitrite by reductase enzymes present in raw plant tissues and by contaminating bacteria. While the nitrate and nitrite content of vegetables varies greatly, the initial nitrate content can be modified by certain modifications in growing conditions including water source; soil conditions; time of harvest; plant-specific factors; and by the amount, kind, and timing of nitrogen fertilization [21].

In general, the nitrate concentration in raw vegetables is extremely variable and also varies from country to country and region to region due to the factors previously mentioned. Results from a recent national U.S. survey of nitrite and nitrate concentrations of raw vegetables at retail classified as conventional or organic are given in Table 7.4. In general, the concentrations of nitrate and nitrite were not different between conventional or organic labeled broccoli, cabbage, celery, and lettuce. One exception was the nitrate level in organic spinach (1318 ppm), which was lower than its conventional counterpart (2797 ppm). The concentrations (ppm) of nitrate from highest to lowest in raw vegetables presented in Table 7.4 were spinach (1318–2797), celery (912–1495), lettuce (844–850), and broccoli (204–394). Other vegetables such as beets, radishes, eggplant, celery, lettuce, collards, and turnip greens also contain similar high concentrations of nitrates.

Drinking Water

The nitrate ion (NO_3^-), the oxidized form of dissolved nitrogen, is the most common groundwater contaminant worldwide and is highly soluble making it easily leached from soils into the available groundwater. It is derived from both natural and human sources and is a relatively stable form of nitrogen in oxygen-rich soils and aquifers. Nitrogen is an inert gas that makes up 78 % of the atmosphere and through natural processes is converted into a variety of common bioavailable compounds, the most common being ammonia (NH_3), ammonium (NH_4^+), nitrate nitrogen (NO_3–N), nitrite (NO_2^-), N-nitrosamines, or organic nitrogen (R-N_2H) [32]. Natural sources of nitrate from nitrogen include fixation by lightning, bacterial conversion in plants and to a lesser extent igneous rocks, deep geothermal fluids, and dissolution of some evaporite minerals. During the decomposition of plants, stored nitrogen can be released to the soil where it is then converted to nitrate and incorporated into aquifers by precipitation.

The primary sources of nitrates that contaminate groundwater are derived from human activity and also include waste from farm animals, fertilizers, manure applied to soils, human waste from septic tanks, and waste water treatment systems (discharge) [33]. In the Southwestern United States and other agricultural areas, inorganic fertilizer and animal manure are the most common nitrate sources while urban areas without proper sewer containment contribute to the nitrate levels in groundwater. Ammonia used in fertilizers may volatilize, be used by plants, or may be denitrified by microbial action, thus releasing gaseous nitrogen.

The U.S. Environmental Protection Agency (EPA) established the maximum contaminant level (MCL) for nitrate–nitrogen (NO_3^-N) at 10 mg/L in 1975 to regulate nitrate levels in drinking water and protect human health [32]. It has been known for more than 50 years that high concentrations of nitrate (>20 mg/L as nitrogen) could cause methemoglobinemia ("blue baby disease") in infants less than 6 months of age and was part of the impetus for the EPA regulation. Other conditions such as hypertension, central nervous system birth defects, certain cancers, non-Hodgkins lymphoma, and diabetes mellitus have been linked to nitrate in drinking water. Having similar concerns, the World Health Organization in 1993 established the Maximum Concentration Limit (MCL) for nitrate at 11.3 mg/L nitrate–nitrogen which has been the limit adopted by many other countries throughout the world. However, evidence shows that nitrate can actually mitigate the above-mentioned conditions [34, 35]. In the United States, a survey conducted by Nolan and Stoner [36] of 33 regional aquifers found that more than 15 % of the wells drawing from the aquifers had nitrate concentrations above the EPA maximum limit of 10 mg/L nitrate-N. In other studies, nitrate was the most frequently reported groundwater contaminant in over 40 states [37, 38].

The permissible concentration of nitrate in drinking water is 50 mg nitrate/L in the European Union and 44 mg/L in the United States [39]. The U.S. EPA MCL or exposure level for drinking water is 10 mg/L nitrate and 1 mg/L nitrite. The Joint Food and Agricultural Organization/World Health Organization set the Acceptable Daily Intake (ADI) for nitrate ion at 3.7 mg/kg body weight and for the nitrite ion at 0.6 mg/kg body weight. The EPA has set a Reference Dose for nitrate at 1.6 mg nitrate ion/kg-day body weight and for the nitrite ion a level of 0.1 mg/kg-day body weight. Lower doses are set for infants and children due to the potential for methemoglobinemia.

A survey of EPA reports giving the nitrate and nitrite concentrations in the municipal water supply of 25 major cities across the United States indicated that the water from each city was compliant for nitrate and nitrite content [26]. Philadelphia, PA; Atlantic City, NJ; and Los Angeles, CA reported the highest levels of nitrate of the 25 cities at 4.9, 4.6, and 2.2–9.2 ppm (mg/L), respectively. It is interesting to note that the highest nitrate concentrations in Los Angeles came from groundwater that was taken from wells. All drinking water sources were below the allowable limits for nitrate (10 mg/L) and nitrite (1 mg/L) established by the EPA.

Current and future treatment options to lower nitrate levels in drinking water include (1) blending high-nitrate water with low-nitrate water; (2) ion exchange of potable water (most widely used nitrate removal method, but may contribute very low levels of nitrosamines or their precursors to water from the membrane resins); (3) membrane separation (reverse osmosis and electrodialysis are used in small communities, but require high energy inputs); (4) biological denitrification with selective microorganisms that convert NO_3^- to N_2; and (5) chemical denitrification (under development) that reduces nitrate. Both denitrification systems require low-dissolved oxygen levels.

Conclusion

Nitrite, and in some cases nitrate, are functional food ingredients that serve as effective antimicrobials to inhibit pathogens such as *Clostridium botulinum*, impart a distinctive reddish-pink cured color to meat products, provide antioxidant properties to retard lipid oxidation, and extend the shelf-life of cured meat products. Some concern has been raised about the use of these ingredients and whether their level of use truly poses a sufficient health risk to warrant their restriction or removal from cured meats.

Historically, meat curing consisted of the addition of salt to fresh meat cuts to remove moisture and reduce the water activity of the tissues to prevent spoilage. In ancient times, salt was obtained from crystalline deposits left by evaporating water from brine pools, seawater, or mining directly from the earth. As a consequence, it often contained natural contaminants such as sodium or potassium nitrate (saltpeter) that contributed directly to the curing reaction and the preservation process. Written records of meat curing are available as early as 3000 BC and dry salt curing and smoking of meat were well established by 900 BC in Greece. The Romans (200 BC) acquired curing procedures from the Greeks, developed "pickling" techniques of meats with a brine marinade, and noted a reddening effect when salt extracted from natural sources was used. Meat curing was more of an art than a science in the early nineteenth century, but as a better understanding of the curing process evolved in the late 1800s and early 1900s, the role of nitrate, and specifically nitrite, in the formation of a distinct cured meat color and its suppression of the outgrowth of *Clostridium botulinum* spores became apparent.

In 1925, for the first time the U.S. Department of Agriculture, Bureau of Animal Industry allowed the use of nitrite salts in meat curing formulas to reduce product spoilage and high levels of residual nitrite in cured products. They also established a maximum limit of 200 ppm nitrite in finished cured meat products. From the 1930s to the mid-1970s, the use of nitrate declined in cured meat products, and today, nitrate use is limited to only dry-cured and specialty meat products. With the implementation of additional regulatory restrictions in 1978 and 1986, the use of nitrite in cured meat products is now allowed at no more than 120–200 ppm depending upon the meat product category. These restrictions have dramatically reduced the levels of residual nitrite, and nitrate, in all meat product categories and their levels have remained low since the implementation regulatory restrictions. The most recent survey indicates an average of 0.6–7 ppm residual nitrite and 15–67 ppm nitrate across the major categories of cured meat products consumed in the United States. This could be significant when evaluating cured meat's contribution to the total dietary load of nitrite and nitrate and their contribution to increased risk of gastrointestinal cancer or their potential benefits to cardiovascular health and gastrointestinal immune function.

The nitrate content of vegetables varies greatly due to cultivar; variety; growing conditions; water source; soil conditions; time of harvest; plant-specific factors; the amount, kind, and timing of nitrogen fertilization; storage period and conditions; and processing procedures. Although vegetables are considered a major source of nitrate in the diet, the fact that the nitrate content of vegetables is variable poses a potential dilemma in determining nitrate's actual contribution to the nitrate/nitrite load in the diet. This variation might be of sufficient magnitude to alter epidemiological predictions if not considered appropriately by region and vegetable category.

EPA reports of the nitrate and nitrite levels in 25 of the largest municipal water supplies across the United States indicate that water from these sources meets acceptable nitrate and nitrite levels based on EPA limits. However, some municipalities were close to the EPA maximum indicating that water in certain regions of the United States should be given consideration when evaluating dietary nitrate and nitrite intake.

Although nitrite and nitrate have only recently received public scrutiny since the 1960s, these two molecules have a rich history dating back thousands of years. They are a natural part of our diet through the consumption of vegetables and are essential for food safety when used as an additive. Understanding the rich history of the use of nitrite and nitrate will hopefully help in educating consumers and scientists alike on the safety and potential risks associated with their ingestion.

References

1. Binkerd EF, Kolari OE. The history and use of nitrate and nitrite in the curing of meat. Food Cosmet Toxicol. 1975;13(6):655–61.
2. Jones O. Nitrite in cured meats. Analyst. 1933;58(684):140.
3. Haldane J. The red colour of salted meat. J Hyg (Lond). 1901;1(1):115–22.
4. Hoagland R. The action of saltpeter upon the color of meat. In: The 25th annual report of the Bureau of Animal Industry, U.S. Department of Agriculture; 1908. p. 301–14.
5. Pegg RB, Shahidi F. History of the curing process. Nitrite curing of meat, the N-nitrosamine problem and nitrite alternatives. Trumbull: Food & Nutrition Press; 2000. p. 7–21.
6. Hoagland R. Coloring matter of raw and cooked salted meats. J Agric Res. 1914;3:211–25.
7. Honikel K-O. The use and control of nitrate and nitrite for the processing of meat products. Meat Sci. 2008;78:68–76.
8. Shahidi F, Samaranayaka AGP. Brine. In: Jensen WK, Devine C, Dikeman M, Jensen WK, Devine C, Dikeman M, editors. Encyclopedia of meat sciences. Oxford: Elsevier; 2004. p. 366–74.
9. Kerr RH, Marsh CTN, Schroeder WF, Boyer EA. The use of sodium nitrite in curing meat. J Agric Res. 1926;33:541–51.
10. USDA. Service and regulatory announcements. Washington, DC: Bureau of Animal Industry; 1925. p. 102.
11. USDA. Service and regulatory announcements. Washington, DC: Bureau of Animal Industry; 1926. p. 2.
12. USDA. Processing inspectors' calculations handbook. FSIS Directive 7620.3; 1995.
13. IFT. Nitrate, nitrite and nitroso compounds in foods. Food Technol. 1987;41(4):127–36.
14. Sebranek JG, Bacus JN. Cured meat products without direct addition of nitrate or nitrite: what are the issues? Meat Sci. 2007;77(1):136–47.
15. Leth T, Fagt S, Nielsen S, Andersen R. Nitrite and nitrate content in meat products and estimated intake in Denmark from 1998 to 2006. Food Addit Contam Part A Chem Anal Control Expo Risk Assess. 2008;25(10):1237–45.
16. Mighton CJ. An analytical survey of cured meats. Chicago: Institute of American Meat Packers; 1936.
17. Lewis WL. The use of nitrite of soda in curing meat. Chicago: Institute of American Meat Packers; 1937.
18. Kolari OE, Aunan WJ. The residual levels of nitrite in cured meat products. Meat Sci Rev. 1972;6:27.
19. Cassens RG. Residual nitrite in cured meat. Food Technol. 1997;51:53–5.
20. Milkowski AL. The controversy about sodium nitrite. In: Savoy: meat industry research conference, American Meat Science Association; 2006.
21. NAS. The health effects of nitrate, nitrite, and N-nitroso compounds. Washington, DC: National Academy Press; 1981.
22. Cassens RG. Use of sodium nitrite in cured meats today. Food Technol. 1995;49:72–81.
23. NAS. Alternatives to the current use of nitrite in foods. Washington, DC: National Academy Press; 1982.
24. Peters JM, Preston-Martin S, London SJ, Bowman JD, Buckley JD, Thomas DC. Processed meats and risk of childhood leukemia (California, USA). Cancer Causes Control. 1994;5(2):195–202.
25. White Jr JW. Relative significance of dietary sources of nitrate and nitrite. J Agric Food Chem. 1975;23(5):886–91.
26. Nuñez MTG, Osburn WN, Hardin MD, Longnecker M, Garg H, Bryan NS, Keeton JT. Survey of nitrate and nitrite contents of conventional and organic labeled raw vegetables available at retail in the United States. J Food Sci. 2015;80(5):C942–9.
27. Nunez De Gonzalez MT, Osburn WN, Hardin, MD, Longnecker M, Garg, HK, Bryan, NS, Keeton JT. Survey of residual nitrite and nitrate in conventional and organic/natural/uncured/indirectly cured meats available at retail in the United States. J Agric Food Chem. 2012;60:3981–90.
28. Barnum DW. Some history of nitrates. J Chem Educ. 2003;80(12):1393–6.

29. Corre WJ, Breimer, T. Nitrate and nitrite in vegetables. In: Centre for Agricultural Publishing and Documentation. Wageningen: Pudoc; 1979.
30. Maynard DN, Barker AV. Regulation of nitrate accumulation in vegetables. Acta Hortic. 1979;93:153–62.
31. Maynard DN. Nitrate accumulation in vegetables. Adv Agron. 1976;28:71–118.
32. Rosen MR, Kropf C. Nitrates in Southwest groundwater. Southwest Hydrol. 2009;8(4):20–1, 34.
33. Harder T. Agricultural impacts on groundwater nitrate. Hydrology. 2009;8(4):22–3, 35.
34. Carlström M, Larsen FJ, Nyström T, Hezel M, Borniquel S, Weitzberg E, Lundberg JO. Dietary inorganic nitrate reverses features of metabolic syndrome in endothelial nitric oxide synthase-deficient mice. Proc Natl Acad Sci U S A. 2010;107(41):17716–20.
35. Kapil V, Milsom AB, Okorie M, Maleki-Toyserkani S, Akram F, Rehman F, Arghandawi S, Pearl V, Benjamin N, Loukogeorgakis S, Macallister R, Hobbs AJ, Webb AJ, Ahluwalia A. Inorganic nitrate supplementation lowers blood pressure in humans: role for nitrite-derived NO. Hypertension. 2010;56(2):274–81.
36. Nolan BT, Stoner JD. Nutrients in ground waters of the conterminous United States, 1992–1995. Environ Sci Technol. 2000;34(7):1156–65.
37. Fetter CS. Contaminant hydrology. New York: MacMillan; 1993.
38. UEPA. Nitrates and nitrites. TEACH chemical summary. US Environmental Protection Agency; 2007.
39. Hord NG, Tang Y, Bryan NS. Food sources of nitrates and nitrites: the physiologic context for potential health benefits. Am J Clin Nutr. 2009;90(1):1–10.

Chapter 8
Nutritional Epidemiology of Nitrogen Oxides: What Do the Numbers Mean?

Martin Lajous and Walter C. Willett

Key Points

- When evaluating epidemiologic data on nutrition, careful consideration of the study design, implementation, and accuracy of dietary assessment is required.
- Prospective cohort studies on diet may avoid biases commonly observed in case–control studies and may be less problematic than large randomized experiments.
- Food-frequency questionnaires provide useful information about intake of major nutrients and are a practical option for dietary assessment in large studies.
- Biochemical measurements of nutrients are particularly important when a specific nutrient is poorly measured by other methods but it is essential to evaluate its sensitivity and the validity of this measure as an indicator of long-term intake.
- The great variability in nitrate and nitrite contents of foods and the short half-lives of their biochemical indicators complicate exposure measurement; however, some studies suggest that nitrate and nitrite intake in epidemiologic studies is feasible.
- The epidemiologic evidence for the relation between dietary nitrate and nitrite and cancer is limited and weak. Higher intake of nitrate has been associated with lower risk of primary open-angle glaucoma in one prospective study.
- There is no direct epidemiologic evidence relating nitrite and nitrate intake and cardiovascular disease, and this deserves investigation.

Keywords Nutrition • Diet • Epidemiology • Food frequency questionnaire • Cancer • Cardiovascular disease

M. Lajous, M.D.
Centro de Investigación en Salud Poblacional; Instituto Nacional de Salud Pública,
Av. Universidad # 655, Col. Santa Maria Ahuacatitlán, Cuernavaca, Morelos 62508, Mexico

Department of Global Health and Population, Harvard School of Public Health, Boston, MA, USA

W.C. Willett, M.D., Dr.P.H. (✉)
Channing Division of Network Medicine, Department of Medicine, Brigham and Women's Hospital, Boston, MA, USA

Department of Nutrition, Harvard T.H. Chan School of Public Health, Boston, MA, USA

Department of Epidemiology, Harvard School of Public Health, Brigham and Women's Hospital and Harvard Medical School, Boston, MA, USA
e-mail: wwillett@hsph.harvard.edu

N.S. Bryan, J. Loscalzo (eds.), *Nitrite and Nitrate in Human Health and Disease*, Nutrition and Health, DOI 10.1007/978-3-319-46189-2_8, © Springer International Publishing AG 2017

Abbreviations

CHD   Coronary heart disease
CI   Confidence interval
FFQ   Food-frequency questionnaire
g   Grams
HDL   High-density lipoprotein
POAG   Primary open-angle glaucoma
RR   Relative risk
WCFR   World Cancer Research Fund

Introduction

Understanding the role of diet in health is important because dietary variables are potentially modifiable risk factors on which preventive efforts may focus. New insights into nitrogen oxide metabolism and beneficial effects of nitrite intake on cardiovascular health in experimental models have opened new areas of investigation for prevention of cardiovascular and metabolic disease [1]. When evaluating epidemiologic data on nutrition, careful consideration of the study design and the implementation and accuracy of dietary assessment is required. Case–control studies of nutrition may afford important insights, but as compared to prospective cohort studies these are more often susceptible to bias. Affected individuals may associate their disease with foods perceived to be poor in nutritional value and overreport them relative to unaffected controls. Thus, prospective studies are strongly preferred.

Dietary exposures can be assessed using questionnaires or biochemical indicators. These methods have different strengths and limitations, which need to be considered for a particular application. The current review is a summary of the available epidemiologic evidence relating dietary nitrates and nitrites and cancer and cardiovascular disease that will focus on methods and study designs commonly used in nutritional epidemiology.

Nutritional Epidemiology and Dietary Assessment

Study Designs

The complex nature of human diets has posed difficult challenges for the evaluation of diet–disease relations [2]. For other exposures, such as cigarette smoking or aspirin use, individuals can easily recall the presence or absence, level, and pattern of lifetime exposure. In contrast, diet comprises a set of strongly correlated variables where all individuals usually have some level of intake; everyone eats fat, carbohydrates, and vitamin C. In addition, individuals rarely make abrupt changes in their diet and are unaware of the nutrient content of the foods they eat; thus nutrient intake is determined indirectly based on reports of foods. Hence, the major limitation in research on nutrition and health has been the accurate measurement of dietary factors. However, well-conducted nutritional epidemiology studies have provided solid evidence to advance population health [3].

Initially, nutritional epidemiology relied on ecologic or correlation studies to investigate dietary factors and health. This strategy relies on comparisons of disease rates and per capita consumption of specific dietary factors across populations. Ecologic studies in nutritional epidemiology have been profoundly important in developing hypotheses on diet and health, but they are considered to provide insufficient evidence for decision making by individuals or for public health policy. The primary reason is the potential for confounding by factors that are difficult to measure. There may be determinants of the disease other than the dietary factor under study that vary between areas with high and low disease incidence and could explain the observed relation. In addition to this important limitation, ecologic studies rely on national dietary data that may vary in quality.

Two study designs commonly employed in epidemiology have been used to avoid the limitations of ecologic studies. In case–control studies, information about previous diet is obtained from diseased individuals and compared with that from individuals free of disease. In contrast, prospective cohort studies obtain data on diet from healthy individuals who are then followed to determine disease rates according to their level of intake of the dietary factors under study. Both study types allow for control of confounding effects of other factors when these have been measured. Investigators may limit the potential effect of these factors by matching individuals to be compared on the basis of known risk factors or by restriction to a particular level of that factor. Alternatively, the effect of these factors may be taken into account in the analysis stage of the study through the use of multivariate modeling. However, case–control and cohort studies have limitations that also need to be considered when evaluating the evidence for diet–disease relations. Case–control studies in nutritional epidemiology may be at a particular risk for recall bias. Diseased individuals associate their disease to an unhealthy diet and overreport consumption of foods considered unhealthy, whereas disease-free individuals may not have this selective recall of certain foods. Thus, recall bias may result in illusory associations between disease and dietary factors considered to be unhealthy [4]. In addition, like in other nondiet applications, case–control studies are subject to selection bias which results from inappropriate enrollment of control individuals. Typically control individuals should represent the population from which the diseased individuals arose and their participation in the study should be unrelated to their level of exposure to the dietary factor considered. However, in practice investigators often include as controls individuals who have a different disease and make strong assumption that this disease is unrelated to the dietary factor of interest. Even when individuals free of disease from the general population are invited to participate, problems may arise. Participation rates for epidemiologic studies are often low, and because diet is particularly associated to health consciousness, the diet of those who decide to participate may be very different from those who do not. There is mounting evidence that case–control studies, even when consistent throughout different populations, may be biased.

Prospective cohort studies may avoid both recall and selection bias because diet is assessed before the diagnosis of disease and even if few individuals decide to participate from the eligibility pool, this low participation rate will not distort observed associations as long as the enrolled persons are not lost to follow-up. For this reason, evidence from prospective cohort studies is considered to be stronger than that from case–control studies. However, prospective cohorts are limited for one strong logistical reason: a large number of people need to be enrolled over a relatively long period of follow-up in order to accrue sufficient cases. Thus, in instances where a rare disease is under study there will never be sufficient cases to conduct analyses of diet when the true effect is relatively small.

The strongest evidence for establishing causal relations in epidemiology comes from double-blind randomized trials; the primary advantage being that randomization and assignment to an exposure by the investigator results in equal distribution of factors that may distort the observed relation. With the exception of studies on dietary supplements, randomized experiments of diet are problematic. Individuals who enroll must change their diet substantially and follow a dietary regime for several years. Only recently a large randomized experiment in women aimed at reducing fat consumption [5] showed the difficulty in maintaining the intervention in women who had consumed a high fat diet most of their life, making results difficult to interpret [6]. Specifically, in this large trial, there was no difference in plasma HDL cholesterol levels between the groups assigned to low fat or usual diets, despite abundant information from controlled feeding trials and other studies showing that reducing the percentage of calories from fat decreases HDL levels.

Diet Assessment Methods

In addition to the methodological concerns related to study design, the main hurdle in nutritional epidemiology is the accurate measurement of diet. In epidemiologic settings, measurement of diet has taken two general paths: assessment by structured questionnaire or interview or measurement by biochemical methods. Today, the food-frequency questionnaire (FFQ) is the primary method for measuring diet in nutritional epidemiology research. As compared to other methods like short-term recall (i.e., 24-h recalls) and diet records, the FFQ focuses on intake over a long period of time, which is the relevant exposure period for chronic disease etiology, rather than in a few specific days. The FFQ has two components: a food list, typically comprising 50–200 food items, and a frequency response section that requests information on the average frequency of intake over the last year (typically in nine or ten categories, ranging from never to six or more times a day). Food intake is then transformed to nutrients using a food composition database. The FFQ captures food preferences and average frequencies of consumption reasonably well; however, because it has a restricted list of food items it may miss important foods that were consumed, resulting in error in nutrient calculations. Further error in classification of nutrient intakes may occur if the nutrient content of a particular nutrient in food varies substantially. Thus, it is essential to evaluate the performance of a questionnaire in terms of reproducibility and validity to determine the impact of these sources of error.

The reproducibility of FFQs is often assessed by repeating the same questionnaire at two time points within a realistic period of time, typically 1–4 years. To assess the validity of FFQs, most studies compare the FFQ with a series of 24-h recalls or dietary records, or with biomarkers for specific nutrients. One early study evaluating the reproducibility and validity of a FFQ administered the questionnaire twice to 173 women 1 year apart [7]. Within that year, participants also completed four 1-week dietary records. When evaluating the reproducibility, the correlation coefficients for nutrients assessed 1 year apart using the FFQs ranged between 0.49 and 0.71. After adjustment for energy, a comparison between the second FFQ and the average of the dietary records showed correlation coefficients ranging from 0.36 for vitamin A without supplements to 0.75 for vitamin C. Correlation coefficients for carbohydrate, protein, and fat intake were within this range. Plasma levels of folate and an N-3 fatty acid in subcutaneous adipose tissue have been used to validate dietary folate and N-3 fatty acid intake from the FFQ. Adjusted correlation coefficients were 0.63 for folate and 0.47 for the N-3 fatty acid [8, 9]. An extensive literature has developed on the validation of FFQs, which have become more detailed and extensive [10]. Using the average of three such questionnaires over a period of years, the correlations with dietary intakes can be as high as 0.8 or 0.9, as found for total and saturated fat. Taken together, these results show that the questionnaire provides useful information about intake of major nutrients and is a practical option for dietary assessment that is easy for individuals to complete, as well as being simple and inexpensive to process.

Biochemical indicators of dietary intake in blood or urine seem at first glance particularly attractive for their objectiveness as compared to dietary questionnaires. However, these measures are subject to the same problems of measurement error bias due to confounding factors as questionnaires, and careful consideration needs to be given regarding their use for particular applications. First, it is important to know the degree to which the marker is sensitive to dietary intake. Certain measurable markers in blood or urine may be strongly affected by factors other than diet like genetics or homeostatic mechanisms, so that large changes in diet are necessary to produce any effect. Serum sodium provides an extreme example; the levels can be measured precisely but are not informative regarding dietary intake. Serum cholesterol levels are also largely determined by factors other than dietary intake. The second consideration is whether the measure provides an indication of time-integrated intake. Similar to the FFQ, the conceptually important exposure is long term rather than a few hours, the previous meal, or a couple of days. The third consideration to be made is that blood levels of a biochemical indicator may be affected by other factors and that adjustment for these may improve the estimation of dietary intake. The last, but important consideration is the laboratory error in measuring the biochemical parameter. However, as compared to other applications, in epidemiological research consistency in the laboratory is more important than accuracy on an absolute scale.

Biochemical measurements of nutrients are particularly important when a specific nutrient is poorly measured by other methods due to large within-food variation in nutrient content. However, similar to FFQs, it is essential to evaluate the sensitivity of a particular nutrient biomarker to intake of that nutrient and the validity of this measure as an indicator of long-term intake. One important strategy to evaluate biomarkers is repeating measurements of the biomarker among the same group of people on more than one occasion to assess the indicator's stability over time, usually over an interval of several months to several years. If the correlation is high, and the biomarker has been shown to be responsive to intake, the biomarker may be a well-integrated measure of long-term intake.

Assessment of Nitrate and Nitrite Intake

Nitrite and nitrite oxide are biologically active compounds involved in vasodilatation, inhibition of endothelial inflammation, and platelet aggregation. Nitrates and nitrites in diet can serve as substrate for a stepwise reduction of nitrate to nitrite to nitric oxide and are the relevant compounds for nutritional epidemiology. However, as compared to other nutrients, quantifying nitrate and nitrite intake is challenging. The major sources of nitrate exposure are foods and drinking water. Large within-food variation of nitrates and nitrites and large geographical and seasonal differences in nitrate content in drinking water complicate accurate estimation of exposure [11]. A recent survey of vegetables conducted in five U.S. cities found variation in nitrate concentration when comparing conventional and organically produced vegetables as well as regional differences in nitrate [12]. Assessment methods for the two major sources of intake are different. Assessment of exposure to nitrates and nitrites from food relies for the most part on 24-h recalls and FFQs; while drinking water nitrate exposure assessment is based on environmental epidemiologic methods that often rely on drinking water quality registries.

The validity of nutrient intake assessment using 24-h recalls, dietary records, or FFQs relies on the availability of an accurate nutrient content database. Most epidemiologic studies rely on food composition values from the U.S. Department of Agriculture's National Nutrient Database. Currently, there is no standard database for the nitrate and nitrite content of foods. There have been some efforts to develop food databases [13–15]. The most recent effort was undertaken by the National Cancer Institute and demonstrated that FFQs could assess adequately dietary nitrate and nitrite intake [15]. However, most investigators use ad hoc databases based on publications judged to be most relevant for a particular population and on drinking water quality registries [16–21]. In general, the main sources of nitrate intake are green leafy vegetables, while for nitrite intake the major sources are processed meats to which nitrite is added as a preservative [11]. Relying on literature-based nutrient content database, in the Nurses' Health Study a prospective cohort of 78,191 women and in similar study, the Health Professional's Health Study, among 31,537 men the main contributors of nitrate intake were lettuce, spinach, celery, and broccoli (Table 8.1). Nitrate and nitrite values can vary substantially within foods. The level of nitrate intake from vegetables depends importantly on soil quality, nitrate content of the fertilizer used, nitrate level on the water supply, and storage and transport practices [11]. For meat, variation in nitrite depends on meat processing practices. Nitrate concentrations in nonorganic spinach collected in five cities in the U.S. varied from 564 to 4923 mg/kg [12]. Comparing estimated nitrate and nitrite content in foods from a convenience sample in the United States [11] to a recent literature-based nutrient content database showed dramatic differences [13]. Measured nitrate in spinach was 741 mg/100 g while the nutrient database estimated 210 mg/100 g. For broccoli, the convenience sample estimated 39.5 mg/100 g of nitrate and the database reported 34.2 mg/100 g. Laboratory analysis of nitrites in bacon showed 0.38 mg/100 g, while the database reported 2.91 mg/100 g. Similarly, for hot dog the nitrite estimates varied substantially: measured nitrite was 0.05 mg/100 g and literature estimated nitrite 2.7 mg/100 g.

Table 8.1 Percent contribution of different foods to nitrate intake from a food-frequency questionnaire in two large epidemiologic studies

Nurses' health study		Health professionals' health study	
Food	% Contribution	Food	% Contribution
Iceberg lettuce	28.2	Iceberg lettuce	27.8
Romaine lettuce	26.1	Romaine lettuce	22.4
Cooked spinach	6.1	Cooked spinach	6.5
Celery	5.5	Broccoli	5.2
Broccoli	5.1	Celery	4.6
Potatoes	3.1	Kale	3.2
Raw spinach	2.1	Potatoes	3.0
Tomato sauce	2.0	Raw spinach	2.5
Kale	1.9	Tomato sauce	2.3
String beans	1.7	Cabbage	1.6

Based on 2002 nitrate estimates for both cohorts

The great variability in nitrate and nitrite contents of foods highlights the value of identifying valid and practical blood or urine biomarkers and of validating questionnaire-based assessments of nitrate and nitrite intake. There are several factors that make the estimation of nitrate and nitrite exposure in vivo challenging. Serum and urinary nitrate levels are affected by dietary nitrate intake and endogenous nitric oxide production [22]; however, nitrate and nitrite have short half-lives: 5–8 h for nitrate and 1–5 min for nitrite [23]. In addition, other factors like exercise have been shown to affect circulating nitrite and nitric oxide levels [24]. Thus, plasma and urinary nitrate and nitrite may not accurately reflect long-term exposure, and measured levels may be affected by other factors that need to be taken into account in any analysis. One strategy is assessing overnight nitrate and nitrite metabolite excretion in urine. However, this measurement may still not be reflective of long-term exposure and its logistical difficulty limits its applicability in large-scale population studies. A recent pilot study that evaluated the feasibility of using plasma nitrate and nitrite in epidemiologic studies found this composite biomarker to be stable and the within-person variability was modest but comparable to commonly used biomarkers (personal communication Tianying Wu).

Repeated measurements of urine nitrate may be useful to account for within-person variation and have been used to evaluate the validity of nitrate intake assessed by FFQ [25]. This type of validation study was conducted among 59 individuals who responded to a FFQ; the correlation between estimated intake and nitrate excretion was estimated. The crude correlation was 0.20, but after correction for within-person variation and adjustment for sex, gender, and body mass intake the correlation coefficient was 0.59. Thus, FFQ assessment of nitrate intake may provide useful information on usual nitrate intake [25]. This study also suggests that the within-food variation in nitrate content is not so large as to preclude useful assessment of intake by dietary questionnaires. This is possible if the average nitrate content of a food does not vary substantially among individuals. For example, even though the nitrate content of spinach may vary from sample to sample, over a year the average nitrate content of spinach consumed by one person may not be substantially different from the average nitrate content of spinach consumed by other persons. Because spinach consumption does vary greatly among individuals, it can be possible to distinguish among persons according to their nitrate intake.

Nitrate and nitrite are widely distributed in fruits, vegetables, and processed meats and drinking water, however, evaluating exposure to these substances is challenging. Their wide distribution in the food and water supply makes clinical studies of endogenous nitrogen oxides difficult as designing nitrate-free diets is complex. Also, estimates of dietary and drinking water nitrate and nitrites are subject to important measurement error that may complicate the interpretation of epidemiological studies. However, the report by van den Brandt et al. [25] does suggest that assessment of nitrate intake in epidemiologic studies is feasible. In the context of extensive measurement error that is independent of the outcome, the absence of an association should be interpreted with caution.

Nitrate and Nitrite Intake and Cancer

Nitrate and nitrite intake and their relation to cancer, and more specifically gastric and other gastrointestinal cancers, have been extensively studied [26]. Interest in nitrates and nitrites as risk factors for cancer stems from the generation of N-nitroso compounds, potential carcinogens, from nitrites (inorganic nitrates are not subject to nitrosation reactions) in the stomach [27]. For gastric cancer, data are still inconclusive as very few prospective cohorts have evaluated this relation and results from these studies show no association [28]. More recently, although preliminary and inconclusive, two prospective studies suggested a role of nitrate from drinking water and from meat products in bladder cancer incidence [21, 29]. Observations from prospective studies for glioma and thyroid are even less compelling [20, 30–32]. The World Cancer Research Fund's (WCFR) report on Nutrition, Physical Activity and Prevention of Cancer did evaluate nitrate and nitrite intake; however, data were too limited to allow conclusions. Nevertheless, the report suggests that they may be potential carcinogens "under conditions that promote nitrosation" [33]. A more recent International Agency for Research in Cancer monograph concluded that, based on the association with gastric cancer, there was limited evidence of carcinogenicity of nitrites in food and inadequate evidence for dietary nitrates [34]. Nevertheless, as mentioned earlier, the epidemiologic evidence supporting these reports is based on nitrate and nitrite assessments with potential important measurement error.

An additional approach to evaluate these relations is to determine associations between the main food sources of nitrates and nitrites. The WCFR report concluded that the evidence from epidemiologic studies that vegetables or foods protect against cancer was not as impressive as originally thought [33]; the initial suggestions of benefit came mainly from case–control studies, and these had generally not been supported by prospective cohort studies. Green leafy vegetables, the major source of dietary nitrates, were independently evaluated for several cancers. Evidence from cohort studies showed a slightly reduced risk of lung cancer when comparing high to low intake groups (meta-analysis RR=0.91 (95 % CI 0.89–0.93) per serving/day). One important consideration when analyzing dietary data is that, as compared to other exposures, nutrients are not consumed independently because many different nutrients are found in the same food. Thus, it may be difficult to separate the effects of nutrients that are highly correlated. Table 8.2 presents the correlation coefficients of nitrates with other nutrients in two large prospective studies of women and men. Nitrate intake is highly correlated with intake of folate, which is considered to have anticarcinogenic effects [35], so it is possible that this and other nutrients may explain the observed inverse relation with lung cancer. The WCRF report concluded there is convincing evidence that processed meat, to which nitrite is added, increases the risk of colorectal cancer, as well as red meat where no nitrite is added (red meat meta-analysis RR=1.29 (95 % CI 1.04–1.60) per 100 g/day; processed meat meta-analysis RR=1.21 (95 % CI 1.04–1.42) per 50 g/day) [33]. There are several proposed mechanisms for these robust associations: formation of N-nitroso compounds through exposure to nitrites, which are added as preservatives; accumulation of iron free iron that can lead to production of free radicals; and exposure to heterocyclic amines and polycyclic aromatic hydrocarbons produced during cooking.

Table 8.2 Correlation coefficient between dietary nitrate and other nutrients from two large epidemiologic studies

Nutrient	Nurses' health study	Health professional's health study
Carbohydrates	−0.01	0.04
Saturated fat	−0.20	−0.20
Protein	0.18	0.20
Animal protein	0.12	0.08
Fiber	0.45	0.43
Vitamin C	0.20	0.19
Beta-carotene	0.66	0.60
Folate	0.64	0.63

Based on the 2002 diet from both cohorts. Nutrients were log transformed and energy adjusted

Nitrate and Nitrite Intake and Cardiovascular Disease

Organic nitrates at pharmacologic doses have been used as vasodilators for a long time. Inorganic salts of nitrate have been used for centuries to preserve food. More recently, data supporting the vasoprotective, blood-pressure lowering and antiplatelet aggregation effects of low-dose dietary nitrate intake have emerged [36–38]. To date there is no epidemiologic evidence relating nitrate and nitrite intake and cardiovascular outcomes such as coronary heart disease (CHD) and stroke. Nevertheless, intake of fruits and vegetables, green leafy vegetables, red and processed meats, and cardiovascular disease have been extensively studied.

A meta-analysis of 23 cohort studies found a significant inverse relation between fruits and vegetables intake and CHD [39]. A pooled analysis of two studies included in the meta-analysis, the NHS and HPFS with 24 and 22 years of follow-up, respectively, and repeated dietary measures reported 16 % lower risk of CHD for 4–4.9 daily servings fruit and vegetable compared to less than 3 servings per day [40]. For stroke, a meta-analysis of 20 prospective cohorts observed a consistent inverse relation [41] and the NHS reported a significant 21 % lower risk of stroke for an increase in one daily serving of green leafy vegetables [42].

Processed meats are the main contributors of nitrite intake (fresh red meat has little or no residual nitrite). However, a recent meta-analysis found a significant direct association with CHD for processed meat (RR = 1.42 (95 % CI 1.07–1.89)) but not for red (unprocessed) meat (RR = 1.00 (95 % CI 0.81–1.23)), although the data for unprocessed meats were extremely limited [43]. No significant association for processed and unprocessed meats and stroke was observed. In the largest prospective study included in the meta-analysis, with 322,263 men and 223,390 women, a direct association for processed meats was observed [44]. There was a 9 % increase in CHD risk comparing extreme quintiles in men (RR = 1.09 (95 % CI, 1.03–1.15)), and a 38 % increase in women (RR = 1.38 (95 % CI, 1.26–1.51)). The increased risks with processed meats may be due in part to the content of preservatives in these foods; however, a significant association was also observed for red meat to which no preservatives are added (RR = 1.27 (95 % CI, 1.20–1.35) in men and RR − 1.50 (95 % CI, 1.37–1.65) in women). A pooled analysis of the NHS and HPFS, processed and unprocessed meats were associated to incident stroke. The risk of stroke increased by 32 % (RR = 1.32 (95 % CI, 1.11–1.57)) and 16 % (RR = 1.16 (95 % CI, 1.00–1.33)) for a daily serving of processed and unprocessed meats [45].

Nitrate Intake and Other Outcomes

Because alterations in nitric oxide signaling have been implicated in the etiology of primary open-angle glaucoma (POAG), Kang et al. [46] investigated the relation between dietary nitrate intake and risk of this disease in the Nurses' Health Study and Health Professional's Health Study. Diet was assessed by a semiquantitative food frequency questionnaire every 4 years and the cumulative average intake was used to evaluate long-term diet. Among over 100,000 men and women followed for up to 28 years, 1483 cases of POAG were documented. After adjustment for potentially confounding variables, the relative risk of POAG was 0.79 (95 % CI 0.66–0.93; P, trend = 0.02) for those in the highest compared to the lowest quintile. Fifty-seven percent of the nitrate consumption was from green leafy vegetables, and a similar inverse association was seen for these foods. The inverse association with nitrate intake was particularly strong (a 40–50 % lower risk) for POAG with early paracentral visual field loss, which is specifically associated with vascular degeneration. These findings need to be confirmed in other studies but support the feasibility of investigating nitrate intake in relation to health outcomes in epidemiologic studies.

Conclusion

Elucidating the relationship between nitrate and nitrite intake and different health outcomes is of great importance because intakes of these molecules may be amenable to intervention. When evaluating the evidence to establish these relationships it is important to take into account both the strengths and limitations of individual studies with respect to the design, validity of the dietary assessment of nitrates and nitrites, adjustment for other dietary and nondietary factors that may account for the observed relation, and the distinction of different sources of nitrates and nitrites. In one validation study using urinary nitrate excretion as the gold standard, there did appear to be sufficient validity of nitrate intake assessed by FFQ to be useful for evaluating hypotheses in epidemiologic studies. Evaluating foods and food groups that are main contributors to intakes of nitrate and nitrite in relation to health outcomes may be helpful in the absence of a standard nitrate and nitrite content database and wide within-food variability. However, the intake of a particular food of food group may be closely correlated to intake of other foods or health-related behaviors. Therefore, care should be taken in the interpretation of results.

To date, the epidemiologic evidence for the relation between dietary nitrate and nitrite and cancer is limited and weak. There is no direct epidemiologic evidence relating dietary nitrite and nitrate to risk of cardiovascular disease. However, the primary source of dietary nitrate, green leafy vegetables, has been associated with lower risk of cardiovascular disease. Processed meats are important contributors of nitrite intake. However, the increased risks for cardiovascular disease are also observed for red meat to which no preservatives are added. Future studies, should more thoroughly investigate nitrate and nitrite intake from both vegetable and meat sources, in relation to vascular diseases.

References

1. Bryan NS, Ivy JL. Inorganic nitrite and nitrate: evidence to support consideration as dietary nutrients. Nutr Res. 2015;35(8):643–54.
2. Willett W. Nutritional epidemiology: issues and challenges. Int J Epidemiol. 1987;16(2):312–7.
3. Willett WC, Stampfer MJ. Current evidence on healthy eating. Annu Rev Public Health. 2013;34:77–95.
4. Giovannucci E, Stampfer MJ, Colditz GA, Manson JE, Rosner BA, Longnecker M, et al. A comparison of prospective and retrospective assessments of diet in the study of breast cancer. Am J Epidemiol. 1993;137(5):502–11.
5. Prentice RL, Caan B, Chlebowski RT, Patterson R, Kuller LH, Ockene JK, et al. Low-fat dietary pattern and risk of invasive breast cancer: the Women's Health Initiative Randomized Controlled Dietary Modification Trial. JAMA. 2006;295(6):629–42.
6. Willett WC. The WHI, joins MRFIT: a revealing look beneath the covers. Am J Clin Nutr. 2010;91(4):829–30.
7. Willett WC, Sampson L, Stampfer MJ, Rosner B, Bain C, Witschi J, et al. Reproducibility and validity of a semi-quantitative food frequency questionnaire. Am J Epidemiol. 1985;122(1):51–65.
8. Hunter DJ, Rimm EB, Sacks FM, Stampfer MJ, Colditz GA, Litin LB, et al. Comparison of measures of fatty acid intake by subcutaneous fat aspirate, food frequency questionnaire, and diet records in a free-living population of US men. Am J Epidemiol. 1992;135(4):418–27.
9. Jacques PF, Sulsky SI, Sadowski JA, Phillips JC, Rush D, Willett WC. Comparison of micronutrient intake measured by a dietary questionnaire and biochemical indicators of micronutrient status. Am J Clin Nutr. 1993;57(2):182–9.
10. Willett WC. Nutritional epidemiology. 3rd ed. New York: Oxford University Press; 2012.
11. Hord NG, Tang Y, Bryan NS. Food sources of nitrates and nitrites: the physiologic context for potential health benefits. Am J Clin Nutr. 2009;90(1):1–10.
12. Nunez de Gonzalez MT, Osburn WN, Hardin MD, Longnecker M, Garg HK, Bryan NS, et al. A survey of nitrate and nitrite concentrations in conventional and organic-labeled raw vegetables at retail. J Food Sci. 2015;80(5):C942–9.
13. Griesenbeck JS, Steck MD, Huber Jr JC, Sharkey JR, Rene AA, Brender JD. Development of estimates of dietary nitrates, nitrites, and nitrosamines for use with the Short Willet Food Frequency Questionnaire. Nutr J. 2009;8:16.

14. Jakszyn P, Agudo A, Ibanez R, Garcia-Closas R, Pera G, Amiano P, et al. Development of a food database of nitrosamines, heterocyclic amines, and polycyclic aromatic hydrocarbons. J Nutr. 2004;134(8):2011–4.

15. Inoue-Choi M, Virk-Baker MK, Aschebrook-Kilfoy B, Cross AJ, Subar AF, Thompson FE, et al. Development and calibration of a dietary nitrate and nitrite database in the NIH-AARP Diet and Health Study. Public Health Nutr. 2016;19:1934–43.

16. Huber Jr JC, Brender JD, Zheng Q, Sharkey JR, Vuong AM, Shinde MU, et al. Maternal dietary intake of nitrates, nitrites and nitrosamines and selected birth defects in offspring: a case-control study. Nutr J. 2013;12:34.

17. Keszei AP, Goldbohm RA, Schouten LJ, Jakszyn P, van den Brandt PA. Dietary N-nitroso compounds, endogenous nitrosation, and the risk of esophageal and gastric cancer subtypes in the Netherlands Cohort Study. Am J Clin Nutr. 2013;97(1):135–46.

18. Dellavalle CT, Xiao Q, Yang G, Shu XO, Aschebrook-Kilfoy B, Zheng W, et al. Dietary nitrate and nitrite intake and risk of colorectal cancer in the Shanghai Women's Health Study. Int J Cancer. 2014;134(12):2917–26.

19. Aschebrook-Kilfoy B, Ward MH, Gierach GL, Schatzkin A, Hollenbeck AR, Sinha R, et al. Epithelial ovarian cancer and exposure to dietary nitrate and nitrite in the NIH-AARP Diet and Health Study. Eur J Cancer Prev. 2012;21(1):65–72.

20. Michaud DS, Holick CN, Batchelor TT, Giovannucci E, Hunter DJ. Prospective study of meat intake and dietary nitrates, nitrites, and nitrosamines and risk of adult glioma. Am J Clin Nutr. 2009;90(3):570–7.

21. Jones RR, Weyer PJ, Dellavalle CT, Inoue-Choi M, Anderson KE, Cantor KP, et al. Nitrate from drinking water and diet and bladder cancer among postmenopausal women in Iowa. Environ Health Perspect. 2016;124(11):1751–8.

22. Pannala AS, Mani AR, Spencer JP, Skinner V, Bruckdorfer KR, Moore KP, et al. The effect of dietary nitrate on salivary, plasma, and urinary nitrate metabolism in humans. Free Radic Biol Med. 2003;34(5):576–84.

23. Lundberg JO, Weitzberg E. NO generation from nitrite and its role in vascular control. Arterioscler Thromb Vasc Biol. 2005;25(5):915–22.

24. Rassaf T, Lauer T, Heiss C, Balzer J, Mangold S, Leyendecker T, et al. Nitric oxide synthase-derived plasma nitrite predicts exercise capacity. Br J Sports Med. 2007;41(10):669–73, discussion 673.

25. Van den Brandt PA, Willett WC, Tannenbaum SR. Assessment of dietary nitrate intake by a self-administered questionnaire and by overnight urinary measurement. Int J Epidemiol. 1989;18(4):852–7.

26. Mirvish SS. Role of N-nitroso compounds (NOC) and N-nitrosation in etiology of gastric, esophageal, nasopharyngeal and bladder cancer and contribution to cancer of known exposures to NOC. Cancer Lett. 1995;93(1):17–48.

27. International Agency for Research on Cancer. Supplement 7. Overall evaluations of carcinogenicity: an updating of IARC monographs volumes 1 to 42. Lyon: International Agency for Research on Cancer; 1987.

28. Song P, Wu L, Guan W. Dietary nitrates, nitrites, and nitrosamines intake and the risk of gastric cancer: a meta-analysis. Nutrients. 2015;7(12):9872–95.

29. Ferrucci LM, Sinha R, Ward MH, Graubard BI, Hollenbeck AR, Kilfoy BA, et al. Meat and components of meat and the risk of bladder cancer in the NIH-AARP Diet and Health Study. Cancer. 2010;116(18):4345–53.

30. Dubrow R, Darefsky AS, Park Y, Mayne ST, Moore SC, Kilfoy B, et al. Dietary components related to N-nitroso compound formation: a prospective study of adult glioma. Cancer Epidemiol Biomarkers Prev. 2010;19: 1709–22.

31. Aschebrook-Kilfoy B, Shu XO, Gao YT, Ji BT, Yang G, Li HL, et al. Thyroid cancer risk and dietary nitrate and nitrite intake in the Shanghai women's health study. Int J Cancer. 2013;132(4):897–904.

32. Kilfoy BA, Zhang Y, Park Y, Holford TR, Schatzkin A, Hollenbeck A, et al. Dietary nitrate and nitrite and the risk of thyroid cancer in the NIH-AARP Diet and Health Study. Int J Cancer. 2011;129(1):160–72.

33. World Cancer Research Fund/American Institute for Cancer Research. Food, nutrition, physical activity, and the prevention of cancer: a global perspective. Washington, DC: AICR; 2007.

34. International Agency for Research on Cancer. Ingested nitrate and nitrite, and cyanobacterial peptide toxins, Vol. 94. Lyon: IARC; 2010.

35. Mason JB. Folate, cancer risk, and the Greek god, Proteus: a tale of two chameleons. Nutr Rev. 2009;67(4): 206–12.

36. Larsen FJ, Ekblom B, Sahlin K, Lundberg JO, Weitzberg E. Effects of dietary nitrate on blood pressure in healthy volunteers. N Engl J Med. 2006;355(26):2792–3.

37. Webb AJ, Patel N, Loukogeorgakis S, Okorie M, Aboud Z, Misra S, et al. Acute blood pressure lowering, vasoprotective, and antiplatelet properties of dietary nitrate via bioconversion to nitrite. Hypertension. 2008;51(3):784–90.

38. Bryan NS, Calvert JW, Elrod JW, Gundewar S, Ji SY, Lefer DJ. Dietary nitrite supplementation protects against myocardial ischemia-reperfusion injury. Proc Natl Acad Sci U S A. 2007;104(48):19144–9.

39. Gan Y, Tong X, Li L, Cao S, Yin X, Gao C, et al. Consumption of fruit and vegetable and risk of coronary heart disease: a meta-analysis of prospective cohort studies. Int J Cardiol. 2015;183:129–37.

40. Bhupathiraju SN, Wedick NM, Pan A, Manson JE, Rexrode KM, Willett WC, et al. Quantity and variety in fruit and vegetable intake and risk of coronary heart disease. Am J Clin Nutr. 2013;98(6):1514–23.

41. Hu D, Huang J, Wang Y, Zhang D, Qu Y. Fruits and vegetables consumption and risk of stroke: a meta-analysis of prospective cohort studies. Stroke. 2014;45(6):1613–9.

42. Joshipura KJ, Ascherio A, Manson JE, Stampfer MJ, Rimm EB, Speizer FE, et al. Fruit and vegetable intake in relation to risk of ischemic stroke. JAMA. 1999;282(13):1233–9.

43. Micha R, Wallace SK, Mozaffarian D. Red and processed meat consumption and risk of incident coronary heart disease, stroke, and diabetes mellitus: a systematic review and meta-analysis. Circulation. 2010;121(21):2271–83.

44. Sinha R, Cross AJ, Graubard BI, Leitzmann MF, Schatzkin A. Meat intake and mortality: a prospective study of over half a million people. Arch Intern Med. 2009;169(6):562–71.

45. Bernstein AM, Pan A, Rexrode KM, Stampfer M, Hu FB, Mozaffarian D, et al. Dietary protein sources and the risk of stroke in men and women. Stroke. 2012;43(3):637–44.

46. Kang JH, Willett WC, Rosner BA, Buys E, Wiggs JL, Pasquale LR. Association of dietary nitrate intake with primary open-angle glaucoma: a prospective analysis From the Nurses' Health Study and Health Professionals follow-up study. JAMA Ophthalmol. 2016;134(3):294–303.

Chapter 9
Nutritional Impact on the Nitric Oxide Pathway

Wing Tak Wong and John P. Cooke

Key Points

- The endothelial nitric oxide synthase (eNOS) pathway is highly modulated by nutrition.
- Dietary choices and interventions may modulate endothelial function.
- The Mediterranean diet enhances endothelial vasodilator function and reduces major adverse cardiovascular events (MACE).
- Increased dietary consumption of fish and other sources of omega-3 fatty acids improve NO activity, and also reduce MACE in those with cardiovascular disease.
- A number of nutritional supplements may improve endothelial function by reducing oxidative stress, by restoring elements of the NOS pathway, or by ameliorating insulin resistance, hypertension, or hyperlipidemia; however, their long-term effects on MACE are unknown.
- Some dietary supplements known to enhance the NOS pathway and improve endothelial vasodilator function in humans (Vitamin E, the B vitamins, and L-arginine) were ineffective at reducing MACE in large randomized clinical trials.
- Thus, dietary interventions that improve endothelial vasodilator function in the short term might not necessarily have long-term benefit.

Keywords Diet • Nutrition • Mediterranean diet • Vitamins • Cardiovascular disease • Functional foods

Vascular Homeostasis and Its Modulation by the Diet

Diet has a profound and immediate impact upon the bioactivity of nitric oxide (NO) generated by the endothelium. Chronic nutritional interventions, by modulating the expression and activity of endothelial nitric oxide synthase (eNOS), can dramatically affect vascular homeostasis. Endothelium-derived

W.T. Wong, Ph.D.
Department of Cardiovascular Sciences, Houston Methodist Hospital, 6670 Bertner Ave, Houston, TX 77030, USA
e-mail: wong.jwt@gmail.com

J.P. Cooke, M.D., Ph.D. (✉)
Department of Cardiovascular Sciences, Houston Methodist Hospital, 6670 Bertner Ave, Houston, TX 77030, USA

Center for Cardiovascular Regeneration, Houston Methodist Hospital, Houston, TX 77030, USA
e-mail: jpcooke@houstonmethodist.org

N.S. Bryan, J. Loscalzo (eds.), *Nitrite and Nitrate in Human Health and Disease*, Nutrition and Health,
DOI 10.1007/978-3-319-46189-2_9, © Springer International Publishing AG 2017

nitric oxide (NO) is a powerful regulator of vascular homeostasis. By virtue of its ability to activate soluble guanylyl cyclase and increase intracellular cyclic GMP, NO relaxes the underlying vascular smooth muscle to improve vascular compliance and to reduce vascular resistance. In addition, endothelium-derived NO inhibits platelet adhesion and aggregation, suppresses leukocyte adhesion and vascular inflammation, and limits the proliferation of the underlying vascular smooth muscle cells. Furthermore, NO is mitogenic for endothelial cells and increases the regeneration of the endothelial monolayer. In large conduit vessels such as the coronary artery, NO plays a critical role in defending against vascular inflammation and lesion formation. This knowledge may be useful to the clinician in management of patients with, or at risk of, cardiovascular disease. We begin by considering the points along the NOS pathway that are most susceptible to modification by diet.

Regulation of the NOS Pathway

L-arginine is converted into NO and L-citrulline by the enzymatic action of eNOS (Fig. 9.1). Endothelial shear stress, as well as a variety of humoral or paracrine factors such as acetylcholine, adenosine diphosphate, thrombin, and vasopressin, is known to induce vasodilation, secondary to phosphorylation and activation of eNOS [1, 2]. Impairment of the NOS pathway may lead to endothelial dysfunction [3]. The ability of the endothelium to respond to shear stress or other stimuli, and to induce relaxation of the underlying vascular smooth muscle, is impaired in older individuals [4–6] and those with diabetes, hypertension, hypercholesterolemia, or tobacco exposure [7, 8].

An impairment of endothelial NOS not only reduces the ability of a blood vessel to relax but also broadly disrupts vascular homeostasis. In addition to relaxing vascular smooth muscles, NO is a potent inhibitor of platelet adhesion and aggregation [9]. In addition, NO suppresses vascular inflammation by reducing the expression of leukocyte adhesion molecules and inflammatory cytokines [10–13]. Consistent with these observations, in animal models, enhancement of NO synthesis (as with L-arginine administration or overexpression of eNOS protein) reduces the progression of atherosclerosis and myointimal hyperplasia [14–16]. The importance of NO in vascular homeostasis is supported by a large number of preclinical and clinical studies revealing that an impairment of endothelial vasodilator function is an independent risk factor for cardiovascular morbidity and mortality [17–19].

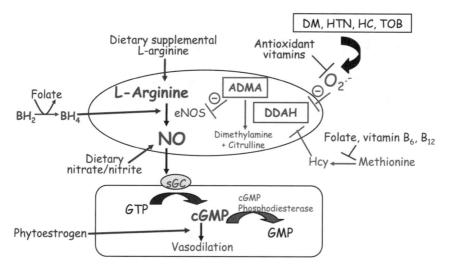

Fig. 9.1 The potential dietary interventions that may enhance the function of the nitric oxide synthase (NOS) pathway

A number of risk factors associated with cardiovascular diseases are also known to impair the NOS pathway. For example, diabetes mellitus is associated with mitochondrial dysfunction and oxidative stress that can accelerate the degradation of NO [20, 21]. Furthermore, diabetes mellitus favors the production of advanced glycosylation end products (AGEs) which can also disrupt eNOS activation, as with formation of N-acetylglucosamine (OGlcNAc) adducts with serine phosphorylation sites on eNOS [22, 23]. Aging also impairs the NOS pathway and increases the risk for major adverse cardiovascular events (MACE). Aging alters the phosphorylation and activation of eNOS in experimental animals [24]. Dyslipidemia is another major cause of endothelial dysfunction. Hypercholesterolemia enhances the inhibitory interaction of caveolin-1 with eNOS; an effect that can be reversed by diet and exercise [25].

Components of Nutrition and Endothelial Function

As can be seen in Fig. 9.1, there are multiple dietary interventions that theoretically may enhance the function of the NOS pathway. One might utilize dietary interventions to treat the cardiovascular risk factors that are known to impair the NOS pathway. Evidence indicates that optimal management of diabetes mellitus, hypertension, hypercholesterolemia, and hyperhomocysteinemia may improve endothelial vasodilator function [26–28]. Furthermore, it is possible that one might supplement portions of the pathway that are impaired. For example, one might provide exogenous dietary arginine to reverse the effects of asymmetric dimethylarginine (ADMA) or provide antioxidant vitamins to reduce the oxidative degradation of NO and/or the impairment of dimethylarginine dimethylaminohydrolase (DDAH). In what follows, we review the effect of components of the diet on endothelial function, and the evidence that dietary interventions might favorably influence endothelial vasodilator function.

Lipids

Saturated and trans-fatty acids: Evidence suggests that increased intake of saturated or trans-fatty acids impairs endothelial vasodilator function within hours of the meal. For example, Plotnick et al. [29] showed that a single high-fat meal could impair endothelium-dependent vasodilation in human subjects, an effect that could be prevented by simultaneous consumption of antioxidant vitamins. Antioxidant vitamins may act to keep NOS cofactors such as tetrahydrobiopterin (BH4) from becoming oxidized, thereby increasing L-arginine conversion to NO; they may protect NO from oxidative degradation once produced and/or prevent oxidative impairment of DDAH, which is exquisitely oxidant sensitive [8]. Evidence for the latter comes from Fard et al. [30] who discovered that ADMA increases in the blood after eating a high-fat meal.

Evidence suggests that consumption of saturated fat may reduce cell membrane fluidity, vascular compliance, and the ability of the endothelium to respond appropriately to shear stress [31]. In addition, dietary saturated fat contributes to dyslipidemia [32], which can impair endothelial function [7]. Trans-fatty acids (e.g., partially hydrogenated vegetable oil) are found in processed foods, such as pretzels and potato chips. Trans-fatty acids also raise LDL cholesterol and impair endothelial vasodilator function [33, 34]. In a study conducted in the Netherlands, researchers found that men who consumed the most trans-fatty acids (6.4 % of total calories) had twice the risk of developing heart disease than those who consumed the least amounts (2.4 %) [35].

Mono and polyunsaturated fats: As a general rule, the amount of saturated fat in a plant-based diet tends to be lower than that in a diet rich in animal proteins. Furthermore, vegetables, nuts, and plant oils, such as olive or canola oil, are a source of monounsaturated fats (e.g., omega-9 fatty acid or oleic acid).

Monounsaturated fats reduce LDL cholesterol in the blood and have no effect on HDL cholesterol [33, 34]. In this regard, the Adventist Health study found that greater consumption of nuts reduces the risks of cardiovascular morbidity and mortality [36], possibly related to the monounsaturated fats in nuts.

Polyunsaturated fats include the omega-3 and omega-6 fatty acids. Both omega-3 and omega-6 fatty acids are considered essential fatty acids. Omega-6 fatty acids (linoleic acid) are polyunsaturated fats that are found in vegetable oils and favorably influence the lipid profile. The marine omega-3 fatty acids, found in the oils of fish, seem to be vasoprotective. In 1980, Bang and Dyerberg reported that Greenland Eskimos have lower cardiovascular morbidity despite a high fat diet that includes whale blubber [37]. It is likely that the resistance of Eskimos to cardiovascular events is due to the high intake of marine lipids. About 40 % of the fatty acids in fish are omega-3 polyunsaturated fatty acids (specifically Eicosapentaenoic acid (EPA) and Docosahexaenoic acid (DHA)). These fatty acids enhance membrane fluidity and vascular compliance [38]. Furthermore, omega-3s also improve endo-thelial function and enhance the production of NO [39]. In addition, they substitute for arachidonic acid in the synthesis of an active form of prostacyclin (PGI3) and an inactive form of thromboxane (TxA3), thus improving blood vessel relaxation [40]. Accordingly, fish oil supplementation improves endothelium-dependent relaxation in patients with atherosclerosis [41]. Fish with the highest content of omega-3 tend to be fatty, cold-water fish such as mackerel, salmon, and trout (Table 9.1). Although the primary dietary source of omega-3 is from fish, some plants are also rich in omega-3 including flaxseed (about 50 % of its oil is linolenic acid), some nuts, and dark leafy vegetables.

Table 9.1 Omega-3 content of foods (g/100 g of food)

Foods	g/100 g
Mackerel	5.3
Herring	3.1
Salmon	2.0
Trout	1.6
Sardine, canned in oil	1.3
Halibut	0.9
Swordfish	0.8
Surimi	0.6
Shrimp	0.5
Catfish	0.4[a]
Fast food fish sandwich	0.4[a]
Fish sticks	0.4
Crab	0.3
Cod	0.3
Clams	0.3
Plaice	0.2
Haddock	0.2
Flounder	0.2
Tuna, canned in oil	0.2
Pike, northern	0.1
Flaxseed	22.8
Walnuts, English	6.8
Soybeans, roasted	1.5
Wheat germ	0.7
Beans, dry	0.6
Almonds	0.4
Avocado	0.1

Adapted from: http://www.nal.usda.gov/fnic/foodcomp/Data/index.html
[a]However, in these products, the fat is oxidized by frying so that there is minimal or no "good" fat

Protein

In general, vegetable protein (e.g., soy and beans) has greater cardiovascular benefit than animal protein, partly because vegetable protein contains little saturated fat. Furthermore, vegetable protein contains less methionine, which may be converted to homocysteine. Finally, vegetable proteins also have more phytonutrients and fiber than animal sources. These differences in the compositions of animal versus vegetable protein may explain the observation that oxidative stress in the blood goes up after eating a high-fat meal containing animal protein, while it goes down after a plant-based meal [42].

A complete protein source is a food that contains all nine essential amino acids in quantities that are similar to those needed by the body. Animal protein is generally complete. Proteins from plants are usually incomplete, with the exception of soy protein [43]. By combining them in a meal, vegetable proteins can become a complete source of the essential amino acids, e.g., beans and rice. Nonessential amino acids are those that can be synthesized endogenously. However in times of physiological stress, intense exercise, or trauma, some of these may become "conditionally" essential. We have suggested that L-arginine may be conditionally essential in patients with elevated levels of plasma ADMA (see section "Arginine"). In these subjects, we have shown that L-arginine supplementation may improve endothelium-dependent vasodilation, at least transiently [44].

Carbohydrates

Fruits and vegetables are the best source of carbohydrates because, in addition to fiber, they are also loaded with micronutrients (such as Vitamins C and E, and β-carotene), enzyme cofactors (such as B_6, B_{12}, and folate), and minerals (such as potassium, zinc, and magnesium). By consuming at least seven servings of fruit and vegetables daily, most people can obtain adequate amounts of the majority of these micronutrients and vitamins.

The nutrient value of fresh fruits, nuts, vegetables, beans, and whole grains is generally higher than refined or commercially prepared food. This is because most of the micronutrients and fiber are lost in the refining process. For example, when wheat is refined, the bran (or outer shell) of the grain is removed. In an epidemiological study of about 34,000 Norwegians, those who ate the highest amounts of whole grains had a 23 % reduction in cardiovascular mortality compared to those who ate less or no whole grains [45].

Fiber is found only in plants. There are two different types of fiber: soluble and insoluble. Certain grains and plants, such as beans, legumes, and fruits, contain soluble fiber such as pectin and psyllium. Soluble fiber lowers blood cholesterol levels [32, 46], thereby improving endothelial vasodilator function. Insoluble fiber is found in whole wheat goods, seeds, nuts, wheat bran, and the skin of fruits and vegetables. Insoluble fiber does not affect cholesterol levels but may help in weight management as it increases intestinal transit time.

Phytonutrients have antioxidant capacity, including carotenes (Vitamin A or β-carotene), tocopherols (Vitamin E compounds), and ascorbate (Vitamin C and polyphenols). Fruits and vegetables contain over 4000 polyphenols which may have cardiovascular benefit. Polyphenols are often lost in processed foods. For example, the peel of many fruits may be enriched in polyphenols in comparison to the pulp (the peel also contains more fiber).

The most common groups of plant antioxidants are the flavonoids and the carotenoids. Plant flavonoids (e.g., from green tea, cranberries, or grapes) have been shown to reduce oxidation of LDL cholesterol, inhibit platelet aggregation, suppress inflammation, and enhance endothelium-dependent vasodilation [47–49]. Chocolate contains polyphenols similar to those in tea, fruits, and vegetables. These antioxidants reduce the oxidation of cholesterol and protect and/or increase the production of NO [50]. However, because most sources of chocolate are high in calories and saturated fat, as well as the caffeine analog theobromine, moderation is advisable.

The French Paradox

How is it that the French eat large amounts of fatty foods, yet purportedly have fewer heart attacks than other Europeans? A possible factor could be the moderate consumption of red wine, which contains the antioxidant flavonoids, resveratrol and quercetin. One glass of red wine has the same antioxidant effect as 500 mg of Vitamin C and can blunt the adverse effect of a high-fat meal on the endothelium [51]. Fitzpatrick et al. [52] found that red wine (as opposed to white) relaxes vascular rings in vitro, an effect related to antioxidant activity in the skin of the grape (red wine is incubated with the skin of the grape, as opposed to most white wines). Teetotalers may take comfort in the observation that purple grape juice also improves the ability of the blood vessels to relax in people with impaired endothelial function [47].

That being said, several large population studies have suggested that any form of alcohol, when consumed in moderation, reduces the risk of cardiovascular events [53]. Moderate alcohol consumption may increase HDL cholesterol and have antiplatelet and anti-inflammatory effects [54, 55]. Epidemiological studies suggest that people with diabetes who consume one drink daily may have as much as an 80 % reduction in risk of MACE by comparison to those who do not drink [56]. However, more than two or three drinks daily predisposes to hypertension, atrial arrhythmias, and cardiomyopathy.

The Western Diet and Endothelial Dysfunction

A single meal high in saturated fat transiently impairs endothelial vasodilator function [57]. The typical western diet is high in saturated fat and is associated with dyslipidemia and insulin resistance, each of which conditions are associated with impaired endothelial vasodilator function and elevated levels of ADMA [7, 30, 58, 59]. Even in young children with hypercholesterolemia, elevated ADMA and endothelial vasodilator dysfunction is observed [60]. In these metabolic disorders, endothelial vasodilator dysfunction is due in part to increased vascular oxidative stress, as an oral dose of antioxidant Vitamin C can improve flow-mediated vasodilation [61, 62].

A diet high in animal protein and low in fruits and vegetables may provide insufficient dietary levels of B vitamins, namely, folate, B_6, and B_{12}. This relative vitamin B deficiency may cause hyperhomocysteinemia, which is further exacerbated by the higher levels of methionine in animal protein. Within hours, oral administration of methionine increases plasma homocysteine, with a temporally related increase in plasma ADMA and a decline in endothelial vasodilator function [63]. Homocysteine impairs the activity of DDAH which may account for the correlation of plasma ADMA and homocysteine levels [63, 64].

The Mediterranean Diet

The Lyon Diet Heart Study compared a Mediterranean-style diet to the standard low fat diet recommended by the American Heart Association (AHA) at that time. The Lyon study was terminated after 27 months because of a marked (70 %) reduction in MACE by comparison to the AHA diet. In a follow-up study, those assigned to the Mediterranean diet enjoyed a 50 % reduction in risk of myocardial infarction. The Mediterranean diet of the Lyon Diet Heart Study consisted of less saturated fats, cholesterol, and linoleic acid, and more oleic and α-linolenic acid [65]. The diet included more whole grain bread, more root and green vegetables, more fish and less red meat, more fruit, and replacement of butter and cream with canola or olive oil. The most notable distinction between the diets was the higher amount of polyunsaturated oils, namely, omega-3 fatty acid. This is significant in view of the prior Diet and Reinfarction Trial (DART). The DART study enrolled men who had sustained a myocardial infarction [66]. Those subjects that were instructed to increase their fish consumption enjoyed a 29 % decline in all-cause mortality. The benefit of the Mediterranean diet may be due in part to its effect to enhance endothelial function [67–69]. Emerging data also seem to suggest the benefits of a Mediterranean diet may be due to higher levels of nitrite and nitrate [70, 71].

The DASH Diet

The DASH (Dietary Approaches to Stop Hypertension) study indicated that diet may be as effective as any single antihypertensive agent in reducing blood pressure [72, 73]. This effect may be partly due to the high levels of nitrate consumed through such a diet [71, 74]. The DASH diet was enriched with vegetables, fruits, and low-fat dairy products and was low in both fat and saturated fat. In a follow-up study, the diet also included techniques to reduce sodium intake (the use of spices, herbs, and fruit juices to season food; rinsing canned vegetables to remove excess salt; limiting cured meats and vegetables in brine solutions; and avoidance of canned ready-to-eat foods like soup). The findings of the DASH diet support the current AHA recommendations for at least seven servings of vegetables and fruit daily.

High Protein Diets

These diets (e.g., Atkins, Zone, and Stillman diets) are characterized by increased percentage of the diet as protein (30–60 %) and fat (30–65 %) with reductions in the percentage of diet as carbohydrate (5–40 %). Some data suggest that high protein diets are somewhat more efficacious in reducing weight [75], but there is no evidence that they improve endothelial vasodilator function or reduce MACE. Because these diets are high in animal protein, they may theoretically impair endothelial function. Notably, epidemiological studies indicate that those subjects who consume more red meat are twice as likely to have a heart attack or stroke [76].

Ultra-Low Fat Diets

Diets such as the Ornish or Kurzweil diets demand a reduction in fat to 10 % of total calories. Ornish et al. [77] have investigated the effects of a regimen that includes a vegetarian diet with less than 10 % of calories from fat and minimal amounts of saturated fat (the "Reversal Diet"), together with moderate exercise, daily use of stress management techniques, group support and counseling, and a smoking cessation program. This program has been documented to reduce the progression of coronary artery disease [77]. However, potentially adverse effects of an ultra-low fat diet include a reduction in HDL cholesterol, as well as an increase in triglycerides and insulin resistance [78]. Nevertheless, for individuals with coronary artery disease and high LDL cholesterol, particularly those that do not have low HDL cholesterol and high triglycerides, ultra-low fat vegetarian diets may be a reasonable alternative to the Mediterranean diet.

Dietary Supplements

About 40 % of all Americans use some form of dietary supplement, such as vitamins and botanicals [79]. Dietary supplements are regulated by the Dietary Supplement Health and Education Act (DSHEA). However, evidence-based medicine is not required for the marketing of dietary supplements. Furthermore, quality control for dietary supplements is not as rigorous as that for drugs. Although some supplements described later have strong scientific evidence to support their utility in some indications, there may be heterogeneity in products within a particular class.

Omega-3 Fatty Acids

The omega-3 fatty acids DHA and EPA improve endothelial vasodilator function, perhaps due to their favorable effects on the lipid profile [80–82]. These agents also reduce oxidative stress and have anti-inflammatory effects [83–86]. Finally, there are data indicating that omega-3 fatty acids reduce myocardial dysrhythmia [87–89]. In the GISSI study, patients with cardiovascular disease that were given supplements of fish oil (1 g twice daily) were less likely to experience sudden death [90].

Supplemental DHA and EPA enhance the flow-induced vasodilation of forearm in healthy individuals [91], as well as hypercholesterolemic subjects [38, 92]. In diabetic patients, blood flow response to acetylcholine and NO release is enhanced after postprandial omega-3 fatty acid supplement [93]. Fish oil supplement increases endothelium-dependent vasodilation in patients with coronary artery disease [41, 94, 95]. Long-term consumption of fish or omega-3 fatty acids also lowers the blood pressure, lowers the incidence of coronary heart disease, preserves renal function, and reduces mortality [95, 96].

Unfortunately, unprocessed fish oil may contain traces of heavy metals (cadmium, copper, iron, magnesium, and lead) and/or pesticides [97]. Commercial processing of fish oil removes contaminants but also removes natural antioxidants. Most commercial suppliers add antioxidants back to the fish oil, but this does not ensure stability. To avoid the problems with commercial processing of fish oils, some manufacturers are now obtaining EPA and DHA from algae [98, 99].

B Vitamins

Vitamin B3 (Niacin): Food sources of vitamin B_3 include peanuts, brewer's yeast, fish, and meat, and (to a lesser extent) whole grains. Supplemental niacin in high doses (100–2000 mg/day) can favorably alter the lipid profile, and thereby improve endothelium-dependent vasodilation [100]. Furthermore, evidence suggests that high-dose niacin can reduce the progression of atherosclerosis and reduce MACE [101, 102]. Niacin improves endothelial dysfunction in patients with coronary artery disease [103]. Niacin may inhibit vascular inflammation and protect against endothelial dysfunction independent of changes in plasma lipid levels [104]. However, even in amounts as low as 50–100 mg, individuals may experience flushing or headache. Higher doses (over 500 mg daily) of niacin should be taken only under the direct supervision of a physician, because of the concern for hepatic dysfunction or hyperuricemia.

Vitamins B6, B12, and folate: Supplementation with these B vitamins reduces homocysteine levels. In addition, low levels of the NOS cofactor tetrahydrobiopterin can be restored with B vitamins, in particular folate [105, 106]. The results of the Nurses' Health Study indicated that women who consume more B vitamins (either from food or vitamins) have half the risk of heart disease. However, large randomized clinical trials of B vitamins, to reduce plasma levels of homocysteine and thereby reduce MACE, have been disappointing [107].

Antioxidants

Vitamins C and E: Oxidative stress plays a role in endothelial vasodilator dysfunction and in the initiation and progression of atherosclerosis. In experimental atherosclerosis, the endothelium produces excessive amounts of superoxide anion [108, 109]. Subsequently, superoxide anion rapidly combines with NO to form peroxynitrite anion, which itself can interfere with cell signaling [110].

In preclinical studies, antioxidants enhanced NO bioactivity [111]. Furthermore, antioxidants preserve the activity of tetrahydrobiopterin [112], a necessary cofactor for NOS. Finally, we have shown that antioxidants can protect the oxidant-sensitive enzyme DDAH [63, 64, 113], and thereby reduce the accumulation of ADMA and N^G-methyl-L-arginine, the endogenous NOS inhibitors. Small clinical studies indicate that administration of Vitamin E or C to patients with hypercholesterolemia or diabetes mellitus can restore endothelial vasodilator function [114, 115].

These observations led to large randomized clinical trials of antioxidant vitamin therapy to reduce cardiovascular events. The Cambridge Heart Antioxidant Study (CHAOS) studied approximately 2000 people with evidence of atherosclerosis. Participants who took 400–800 IU of Vitamin E daily enjoyed a reduced risk of heart attack and death from heart disease. However, larger randomized clinical trials have been compellingly negative [116]. In particular, the Heart Outcomes Prevention Evaluation (HOPE) study followed 10,000 patients for about 4.5 years. Subjects at high risk for cardiovascular events received 400 IU of Vitamin E or placebo daily, in a Latin square design with ramipril 5 mg or placebo. Whereas ramipril reduced MACE by about 25 %, vitamin E was not superior to placebo [117, 118].

Subsequently, it was suggested that a combination of antioxidants would be more effective than high doses of Vitamin E or C alone. However, Greg Brown and colleagues [119] found that a combination of antioxidant vitamins (β-carotene 12.5 mg BID, vitamin C 500 mg BID, vitamin E 400 IU BID, and selenium 50 µg BID) blunt the beneficial effects of statins to increase HDL cholesterol and induce regression of coronary artery disease. It seems better to obtain a variety of antioxidants through generous and frequent servings of fruit and vegetables, rather than supplemental vitamins.

Soy Protein and Phytoestrogens

Soy phytoestrogens, also called isoflavones, such as genistein and daidzein have antioxidant and anti-inflammatory effects. Furthermore, in postmenopausal women with high cholesterol, we found that soy isoflavones (genistein, daidzein, and glycitein) 50 mg once daily improve vasodilation, in the absence of an effect on the lipid profile [120, 121]. However, enthusiasm for supplemental phytoestrogens should be muted in the absence of large randomized clinical trials. Furthermore, caution is advisable in view of the disappointing results of estrogen supplementation in women with cardiovascular disease [122, 123]. Nevertheless, because it contains fiber and a complex assortment of phytonutrients, moderate consumption of soy (i.e., to substitute for red meat) seems reasonable.

Arginine

We observed that supplemental L-arginine could improve endothelium-dependent vasodilation in hypercholesterolemic animals and humans [124–126]. Because NO could also inhibit processes that are involved in atherosclerosis, such as platelet aggregation and vascular inflammation, we reasoned that dietary arginine might have an antiatherogenic effect. Indeed, we showed that L-arginine supplementation had an antiatherogenic effect probably through inhibiting platelet aggregation and monocyte adhesion in hypercholesterolemic animals and humans [127–130].

ADMA, the NOS antagonist: A likely explanation for the effect of supplemental L-arginine to enhance NO synthesis was provided by Vallance et al. [131], who demonstrated that endogenous methylarginines could antagonize the NOS pathway. The methylarginines include monomethylarginine (MMA), asymmetric dimethylarginine (ADMA), and symmetric dimethylarginine (SDMA), which are derived from the hydrolysis of proteins containing methylated arginine residues [132]. Both ADMA and MMA compete with L-arginine for binding to the NOS enzyme, whereas SDMA does not. Their antagonism of NOS is reversed by increasing the concentration of L-arginine [133].

Whereas 20 % of the clearance of these compounds is through renal excretion, about 80 % is due to the action of the cytoplasmic enzyme DDAH [134, 135], expressed ubiquitously as two isoforms [136]. This enzyme is inhibited by oxidative stress, possibly because it expresses a sulfhydryl group in its catalytic site [137]. We find that the enzyme activity of DDAH is impaired in hypercholesterolemia, diabetes mellitus, and hyperhomocysteinemia [113, 134, 138, 139]. In addition, ADMA is a predictor of the severity of vascular diseases and is an independent predictor of MACE and mortality [140–142].

Reversing the effects of ADMA with l-arginine supplementation: Short-term administration of L-arginine improves endothelial vasodilator function and relieves symptoms in patients with coronary artery disease and congestive heart failure [143–145]. We studied the potential benefit of L-arginine supplementation (3 g/day) versus placebo on endothelial vasodilator function and functional capacity in 220 patients with PAD [146]. After 6 months, those patients in the supplemented group had higher plasma L-arginine levels. However, measures of NO availability (including flow-mediated vasodilation, vascular compliance, plasma and urinary nitrogen oxides, and plasma citrulline formation) were reduced or not improved compared with placebo. Furthermore, the improvement in absolute claudication distance was significantly less in the arginine-treated group (28 % vs. 12 %; $p < 0.05$). Thus, in patients with PAD, long-term administration of L-arginine does not appear to be useful and may even induce a form of "arginine tolerance" or dysfunction of the NOS pathway.

Dietary Supplements and Interventions that Improve Cardiovascular Risk Factors

Hypolipidemic agents: By improving cardiovascular risk factors, some nutritional supplements may improve endothelial function. These include plant-based products containing sterols [147]. *Phytosterols* include sitosterol and campesterol which differ structurally from cholesterol by a methyl or ethyl group in their side chains. These structural differences cause them to be poorly absorbed. They associate with cholesterol and bile salts to increase cholesterol excretion. Clinical trials of medical foods that are enriched in plant sterols indicate they can reduce LDL cholesterol levels 10–20 % [148, 149]. However, these products could reduce the absorption of fat-soluble vitamins. Accordingly, their use should probably be reserved for hypercholesterolemic adults in the secondary prevention of vascular events.

Red yeast rice contains HMG coA reductase inhibitors. In a small clinical trial, red yeast rice (2.4 g daily) for 8 weeks reduced total cholesterol by 17 % [150]. However, like all dietary supplements, there is heterogeneity in product quality. A UCLA study showed that only one of nine red yeast rice supplements in health food stores contained all the monacolins that lower cholesterol [150]. Seven of nine brands contained a small amount of a toxic by-product of the fermentation process. Red yeast rice should not be combined with erythromycin, other statin drugs, fibrates, or high-dose niacin, except under medical supervision, as serious side effects (such as myositis or hepatic dysfunction) are theoretically possible. Currently, the FDA has imposed a ban on red yeast rice products in the United States.

Table 9.2 Dietary supplements that *might* mitigate disorders associated with cardiovascular disease and *might* enhance endothelial function

Agents which may improve the lipid profile	
Red yeast rice	Garlic (alliin)
Guggulsterone	Phytosterol (sitosterol, campesterol, stigmasterol)
Policosanol	Omega-3 fatty acid
Vitamin B_3	Soy protein (phytoestrogen)
Agents which may be useful in diabetes mellitus	
Chromium	α-Lipoic acid
Agents which may reduce blood pressure	
Dietary salt substitutes (potassium, magnesium, and lysine)	
Garlic	

In general, large randomized clinical trials are not available to support these dietary interventions

There are a number of other dietary supplements that appear to improve the lipid profile and that may therefore enhance endothelial vasodilator function (Table 9.2). These include alliin (a component of garlic), guggulsterone, and policosanol [151, 152]. It is most important to understand that the level of evidence for any of these dietary supplements is weak in comparison to FDA-approved hypolipidemic drugs. Large randomized clinical trials for primary and secondary prevention strongly support the use of FDA-approved cholesterol-lowering drugs such as statins.

For controlling blood sugar or blood pressure: *Chromium* has been theorized to enhance the action of insulin [153]. Severe chromium deficiency, as has been seen during total parenteral nutrition, is associated with elevated glucose, insulin, and lipid levels as well as CNS disturbance and peripheral neuropathy. Milder chromium deficiency may be a cause of insulin resistance. However, the results of randomized clinical trials have been mixed [154]. Another supplement proposed to be useful in diabetes mellitus is *α-lipoic acid*, which in one clinical trial reduced the symptoms secondary to diabetic neuropathy [155]. Lipoic acid may also improve endothelial function and reduce markers of inflammation in the metabolic syndrome [156].

Dietary salt substitutes (e.g., the combination of potassium, magnesium, and lysine) can reduce blood pressure by substituting for sodium chloride. The use of these agents in hypertensive individuals should be under medical supervision of physician, particularly so if the individual is taking medication that can reduce potassium excretion (i.e., aldactone, angiotensin-converting enzyme inhibitors, angiotensin receptor antagonists). In addition to its hypolipidemic effect, garlic preparations also have a mild antihypertensive effect, possibly due to enhancement of the NOS pathway [157].

NO generating lozenge: Studies using a patented formulation (US patents 8,303,995, 8,298,589, 8,435,5708 8,962,038 9,119,823 & 9,241,999) in the form of an orally disintegrating tablet found that it could modify cardiovascular risk factors in patients over the age of 40, significantly reduce triglycerides, and reduce blood pressure [158]. This same lozenge was used in a pediatric patient with argininosuccinic aciduria and significantly reduced his blood pressure when prescription medications were ineffective [159]. A more recent clinical trial using the nitrite lozenge reveals that a single lozenge can significantly reduce blood pressure, dilate blood vessels, improve endothelial function and arterial compliance in hypertensive patients [160]. Furthermore in a study of pre-hypertensive patients (BP >120/80 < 139/89), administration of one lozenge twice daily leads to a significant reduction in blood pressure (12 mmHg systolic and 6 mmHg diastolic) after 30 days [161]. Although it appears this dietary supplement is safe and effective in short term studies, long term studies are needed.

Conclusion

There are a wide variety of dietary interventions and supplements that can improve the function of the NOS pathway. However, dietary supplements that improve human endothelial function in the short term do not necessarily have long-term benefit on vascular function and cardiovascular events. Indeed, some of these supplements may even have adverse effects when used long term. Furthermore, dietary supplements may have adverse interactions with pharmacotherapies (e.g., reports of increased bleeding in patients taking aspirin and fish oil supplementation; salt substitutes and ACEIs causing hyperkalemia). In addition, dietary supplements are often of heterogeneous quality. Accordingly, before enthusiastic adoption of a particular dietary supplement, the cautious physician will demand evidence for efficacy and safety from large randomized clinical trials and will recommend supplements distributed by brand name vendors. With respect to various diets, the informed physician will recommend the Mediterranean diet for individuals with, or at risk of, cardiovascular disease, as this diet has the strongest support for a regimen that enhances endothelial vasodilator function and reduces cardiovascular events. The collective body of information provided in the coming chapters suggests that nitrite and nitrate content of certain foods in combination with antioxidants and systems promoting their reduction back to NO may need to be considered as other dietary factors of potential benefit.

References

1. Cooke JP, Rossitch E, Andon NA, Loscalzo J, Dzau VJ. Flow activates an endothelial potassium channel to release an endogenous nitrovasodilator. J Clin Invest. 1991;88(5):1663–71.
2. Nishida K, Harrison DG, Navas JP, et al. Molecular cloning and characterization of the constitutive bovine aortic endothelial cell nitric oxide synthase. J Clin Invest. 1992;90(5):2092–6.
3. Wong WT, Wong SL, Tian XY, Huang Y. Endothelial dysfunction: the common consequence in diabetes and hypertension. J Cardiovasc Pharmacol. 2010;55(4):300–7.
4. Taddei S, Virdis A, Ghiadoni L, et al. Age-related reduction of NO availability and oxidative stress in humans. Hypertension. 2001;38(2):274–9.
5. Gerhard M, Roddy MA, Creager SJ, Creager MA. Aging progressively impairs endothelium-dependent vasodilation in forearm resistance vessels of humans. Hypertension. 1996;27(4):849–53.
6. Egashira K, Inou T, Hirooka Y, et al. Effects of age on endothelium-dependent vasodilation of resistance coronary artery by acetylcholine in humans. Circulation. 1993;88(1):77–81.
7. Creager MA, Cooke JP, Mendelsohn ME, et al. Impaired vasodilation of forearm resistance vessels in hypercholesterolemic humans. J Clin Invest. 1990;86(1):228–34.
8. Cooke JP. Asymmetrical dimethylarginine: the Uber marker? Circulation. 2004;109(15):1813–8.
9. Stamler J, Mendelsohn ME, Amarante P, et al. N-acetylcysteine potentiates platelet inhibition by endothelium-derived relaxing factor. Circ Res. 1989;65(3):789–95.
10. Tsao PS, McEvoy LM, Drexler H, Butcher EC, Cooke JP. Enhanced endothelial adhesiveness in hypercholesterolemia is attenuated by L-arginine. Circulation. 1994;89(5):2176–82.
11. Tsao PS, Lewis NP, Alpert S, Cooke JP. Exposure to shear stress alters endothelial adhesiveness: role of nitric oxide. Circulation. 1995;92(12):3513–9.
12. Tsao PS, Buitrago R, Chan JR, Cooke JP. Fluid flow inhibits endothelial adhesiveness: nitric oxide and transcriptional regulation of VCAM-1. Circulation. 1996;94(7):1682–9.
13. Tsao PS, Wang B, Buitrago R, Shyy JYJ, Cooke JP. Nitric oxide regulates monocyte chemotactic protein-1. Circulation. 1997;96(3):934–40.
14. Cooke JP, Singer AH, Tsao P, Zera P, Rowan RA, Billingham ME. Antiatherogenic effects of L-arginine in the hypercholesterolemic rabbit. J Clin Invest. 1992;90(3):1168–72.
15. von der Leyen HE, Gibbons GH, Morishita R, et al. Gene therapy inhibiting neointimal vascular lesion: in vivo transfer of endothelial cell nitric oxide synthase gene. Proc Natl Acad Sci. 1995;92(4):1137–41.
16. Candipan RC, Wang B, Buitrago R, Tsao PS, Cooke JP. Regression or progression: dependency on vascular nitric oxide. Arterioscler Thromb Vasc Biol. 1996;16(1):44–50.
17. Schachinger V, Britten MB, Zeiher AM. Prognostic impact of coronary vasodilator dysfunction on adverse long-term outcome of coronary heart disease. Circulation. 2000;101(16):1899–906.

18. Suwaidi JA, Hamasaki S, Higano ST, Nishimura RA, Holmes DR, Lerman A. Long-term follow-Up of patients with mild coronary artery disease and endothelial dysfunction. Circulation. 2000;101(9):948–54.
19. Gokce N, Keaney JF, Hunter LM, et al. Predictive value of noninvasivelydetermined endothelial dysfunction for long-term cardiovascular events in patients with peripheral vascular disease. J Am Coll Cardiol. 2003;41(10):1769–75.
20. Brownlee M. The pathobiology of diabetic complications: a unifying mechanism. Diabetes. 2005;54(6):1615–25.
21. Hink U, Li H, Mollnau H, et al. Mechanisms underlying endothelial dysfunction in diabetes mellitus. Circ Res. 2001;88(2):e14–22.
22. Musicki B, Kramer MF, Becker RE, Burnett AL. Inactivation of phosphorylated endothelial nitric oxide synthase (Ser-1177) by O-GlcNAc in diabetes-associated erectile dysfunction. Proc Natl Acad Sci. 2005;102(33):11870–5.
23. Wells L. Glycosylation of nucleocytoplasmic proteins: signal transduction and O-GlcNAc. Science. 2001;291(5512):2376–8.
24. Musicki B, Kramer MF, Becker RE, Burnett AL. Age-related changes in phosphorylation of endothelial nitric oxide synthase in the rat penis. J Sex Med. 2005;2(3):347–57.
25. Musicki B, Liu T, Strong T, et al. Low-fat diet and exercise preserve eNOS regulation and endothelial function in the penis of early atherosclerotic pigs: a molecular analysis. J Sex Med. 2008;5(3):552–61.
26. Ceriello A. Evidence for an independent and cumulative effect of postprandial hypertriglyceridemia and hyperglycemia on endothelial dysfunction and oxidative stress generation: effects of short- and long-term simvastatin treatment. Circulation. 2002;106(10):1211–8.
27. Jensen-Urstad KJ, Reichard PG, Rosfors JS, Lindblad LE, Jensen-Urstad MT. Early atherosclerosis is retarded by improved long-term blood glucose control in patients with IDDM. Diabetes. 1996;45(9):1253–8.
28. Ghiadoni L, Magagna A, Versari D, et al. Different effect of antihypertensive drugs on conduit artery endothelial function. Hypertension. 2003;41(6):1281–6.
29. Plotnick GD, Corretti MC, Vogel RA, Hesslink R, Wise JA. Effect of supplemental phytonutrients on impairment of the flow-mediated brachialartery vasoactivity after a single high-fat meal. J Am Coll Cardiol. 2003; 41(10):1744–9.
30. Fard A, Tuck CH, Donis JA, et al. Acute elevations of plasma asymmetric dimethylarginine and impaired endothelial function in response to a high-fat meal in patients with type 2 diabetes. Arterioscler Thromb Vasc Biol. 2000;20(9):2039–44.
31. Keogh JB. Flow-mediated dilatation is impaired by a high-saturated fat diet but not by a high-carbohydrate diet. Arterioscler Thromb Vasc Biol. 2005;25(6):1274–9.
32. Jenkins D, Wolever T, Rao AV, et al. Effect on blood lipids of very high intakes of fiber in diets Low in saturated fat and cholesterol. N Engl J Med. 1993;329(1):21–6.
33. de Roos NM, Bots ML, Katan MB. Replacement of dietary saturated fatty acids by trans fatty acids lowers serum HDL cholesterol and impairs endothelial function in healthy Men and women. Arterioscler Thromb Vasc Biol. 2001;21(7):1233–7.
34. Mensink RP, Katan MB. Effect of dietary trans fatty acids on high-density and Low-density lipoprotein cholesterol levels in healthy subjects. N Engl J Med. 1990;323(7):439–45.
35. Sébédio JL, Vermunt SHF, Chardigny JM, et al. The effect of dietary trans α-linolenic acid on plasma lipids and platelet fatty acid composition: the TransLinE study. Eur J Clin Nutr. 2000;54(2):104–13.
36. Fraser GE. A possible protective effect of nut consumption on risk of coronary heart disease. Arch Intern Med. 1992;152(7):1416.
37. Dyerberg J. Eicosapentaenoic acid and prevention of thrombosis and atherosclerosis? Lancet. 1978; 312(8081):117–9.
38. Mori TA, Watts GF, Burke V, Hilme E, Puddey IB, Beilin LJ. Differential effects of eicosapentaenoic acid and docosahexaenoic acid on vascular reactivity of the forearm microcirculation in hyperlipidemic, overweight men. Circulation. 2000;102(11):1264–9.
39. Abeywardena M. Longchain n−3 polyunsaturated fatty acids and blood vessel function. Cardiovasc Res. 2001;52(3):361–71.
40. Funk CD, Powell WS. Release of prostaglandins and monohydroxy and trihydroxy metabolites of linoleic and arachidonic acids by adult and fetal aortae and ductus arteriosus. J Biol Chem. 1985;260(12):7481–8.
41. Fleischhauer FJ, Yan W-D, Fischell TA. Fish oil improves endothelium-dependent coronary vasodilation in heart transplant recipients. J Am Coll Cardiol. 1993;21(4):982–9.
42. Tushuizen ME, Nieuwland R, Scheffer PG, Sturk A, Heine RJ, Diamant M. Two consecutive high-fat meals affect endothelial-dependent vasodilation, oxidative stress and cellular microparticles in healthy men. J Thromb Haemost. 2006;4(5):1003–10.
43. Protein quality evaluation. Joint FAO/WHO. FAO Food and Nutrition paper. 1991;51:1-66.
44. Boger RH, Bode-Boger SM, Szuba A, et al. Asymmetric Di: its role in hypercholesterolemia. Circulation. 1998;98(18):1842–7.

45. Lockheart MSK, Steffen LM, Rebnord HM, et al. Dietary patterns, food groups and myocardial infarction: a case–control study. Br J Nutr. 2007;98(02):380.

46. Jenkins DJ, Kendall CW, Vuksan V, et al. Soluble fiber intake at a dose approved by the US Food and Drug Administration for a claim of health benefits: serum lipid risk factors for cardiovascular disease assessed in a randomized controlled crossover trial. Am J Clin Nutr. 2002;75(5):834–9.

47. Stein JH, Keevil JG, Wiebe DA, Aeschlimann S, Folts JD. Purple grape juice improves endothelial function and reduces the susceptibility of LDL cholesterol to oxidation in patients with coronary artery disease. Circulation. 1999;100(10):1050–5.

48. Widlansky ME, Hamburg NM, Anter E, et al. Acute EGCG supplementation reverses endothelial dysfunction in patients with coronary artery disease. J Am Coll Nutr. 2007;26(2):95–102.

49. Yung LM, Tian XY, Wong WT, et al. Chronic cranberry juice consumption restores cholesterol profiles and improves endothelial function in ovariectomized rats. Eur J Nutr. 2013;52(3):1145–55.

50. Flammer AJ, Hermann F, Sudano I, et al. Dark chocolate improves coronary vasomotion and reduces platelet reactivity. Circulation. 2007;116(21):2376–82.

51. Djoussé L, Ellison RC, McLennan CE, et al. Acute effects of a high-fat meal with and without red wine on endothelial function in healthy subjects. Am J Cardiol. 1999;84(6):660–4.

52. Fitzpatrick DF, Hirschfield SL, Coffey RG. Endothelium-dependent vasorelaxing activity of wine and other grape products. Am J Physiol. 1993;265(2 Pt 2):H774–8.

53. Cooper HA, Exner DV, Domanski MJ. Light-to-moderate alcohol consumption and prognosis in patients with left ventricular systolic dysfunction. J Am Coll Cardiol. 2000;35(7):1753–9.

54. Ajani UA, Gaziano JM, Lotufo PA, et al. Alcohol consumption and risk of coronary heart disease by diabetes status. Circulation. 2000;102(5):500–5.

55. Imhof A, Woodward M, Doering A, et al. Overall alcohol intake, beer, wine, and systemic markers of inflammation in western Europe: results from three MONICA samples (Augsburg, Glasgow, Lille). Eur Heart J. 2004;25(23):2092–100.

56. Sacco RL, Elkind M, Boden-Albala B. The protective effect of moderate alcohol consumption on ischemic stroke. Clin Cornerstone. 1999;1(6):56.

57. Vogel RA, Corretti MC, Plotnick GD. Effect of a single high-Fat meal on endothelial function in healthy subjects. Am J Cardiol. 1997;79(3):350–4.

58. Lundman P, Eriksson M, Schenck-Gustafsson K, Karpe F, Tornvall P. Transient triglyceridemia decreases vascular reactivity in young, healthy men without risk factors for coronary heart disease. Circulation. 1997;96(10):3266–8.

59. Yasuda S. Intensive treatment of risk factors in patients with type-2 diabetes mellitus is associated with improvement of endothelial function coupled with a reduction in the levels of plasma asymmetric dimethylarginine and endogenous inhibitor of nitric oxide synthase. Eur Heart J. 2005;27(10):1159–65.

60. Engler MM. Antioxidant vitamins C and E improve endothelial function in children with hyperlipidemia: Endothelial Assessment of Risk from Lipids in Youth (EARLY) Trial. Circulation. 2003;108(9):1059–63.

61. Plotnick GD. Effect of antioxidant vitamins on the transient impairment of endothelium-dependent brachial artery vasoactivity following a single high-Fat meal. JAMA. 1997;278(20):1682.

62. Raitakari OT, Adams MR, McCredie RJ, Griffiths KA, Stocker R, Celermajer DS. Oral vitamin C and endothelial function in smokers: short-term improvement, but no sustained beneficial effect. J Am Coll Cardiol. 2000;35(6):1616–21.

63. Stuhlinger MC. Endothelial dysfunction induced by hyperhomocyst(e)inemia: role of asymmetric dimethylarginine. Circulation. 2003;108(8):933–8.

64. Stuhlinger MC, Tsao PS, Her JH, Kimoto M, Balint RF, Cooke JP. Homocysteine impairs the nitric oxide synthase pathway: role of asymmetric dimethylarginine. Circulation. 2001;104(21):2569–75.

65. de Lorgeril M, Renaud S, Salen P, et al. Mediterranean alpha-linolenic acid-rich diet in secondary prevention of coronary heart disease. Lancet. 1994;343(8911):1454–9.

66. Burr ML, Gilbert JF, Holliday RM, et al. Effects of changes in fat, fish, and fibre intakes on death and myocardial reinfarction: Diet and Reinfarction Trial (DART). Lancet. 1989;334(8666):757–61.

67. Vogel RA, Corretti MC, Plotnick GD. The postprandial effect of components of the Mediterranean diet on endothelial function. J Am Coll Cardiol. 2000;36(5):1455–60.

68. Pérez-Jiménez F, Castro P, López-Miranda J, et al. Circulating levels of endothelial function are modulated by dietary monounsaturated fat. Atherosclerosis. 1999;145(2):351–8.

69. Fuentes F. Mediterranean and low-fat diets improve endothelial function in hypercholesterolemic men. Ann Intern Med. 2001;134(12):1115.

70. Lundberg JO, Feelisch M, Björne H, Jansson EÅ, Weitzberg E. Cardioprotective effects of vegetables: is nitrate the answer? Nitric Oxide. 2006;15(4):359–62.

71. Hord NG, Tang Y, Bryan NS. Food sources of nitrates and nitrites: the physiologic context for potential health benefits. Am J Clin Nutr. 2009;90(1):1–10.

72. Maruthur NM, Wang NY, Appel LJ. Lifestyle interventions reduce coronary heart disease risk: results from the PREMIER trial. Circulation. 2009;119(15):2026–31.
73. Appel LJ, Champagne CM, Harsha DW, et al. Effects of comprehensive lifestyle modification on blood pressure control: main results of the PREMIER clinical trial. JAMA. 2003;289(16):2083–93.
74. Larsen FJ, Ekblom B, Sahlin K, Lundberg JO, Weitzberg E. Effects of dietary nitrate on blood pressure in healthy volunteers. N Engl J Med. 2006;355(26):2792–3.
75. Weigle DS, Breen PA, Matthys CC, et al. A high-protein diet induces sustained reductions in appetite, ad libitum caloric intake, and body weight despite compensatory changes in diurnal plasma leptin and ghrelin concentrations. Am J Clin Nutr. 2005;82(1):41–8.
76. Fung TT, Stampfer MJ, Manson JE, Rexrode KM, Willett WC, Hu FB. Prospective study of major dietary patterns and stroke risk in women. Stroke. 2004;35(9):2014–9.
77. Ornish D. Intensive lifestyle changes for reversal of coronary heart disease. JAMA. 1998;280(23):2001.
78. Lefevre M, Champagne CM, Tulley RT, Rood JC, Most MM. Individual variability in cardiovascular disease risk factor responses to low-fat and low-saturated-fat diets in men: body mass index, adiposity, and insulin resistance predict changes in LDL cholesterol. Am J Clin Nutr. 2005;82(5):957–63. quiz 1145-6.
79. Archer SL, Stamler J, Moag-Stahlberg A, et al. Association of dietary supplement use with specific micronutrient intakes among middle-aged American Men and women: the INTERMAP study. J Am Diet Assoc. 2005;105(7):1106–14.
80. Connor WE, DeFrancesco CA, Connor SL. N-3 fatty acids from fish oil. Ann NY Acad Sci. 1993;683(1 Dietary Lipid):16–34.
81. Sacks FM, Stone PH, Gibson CM, Silverman DI, Rosner B, Pasternak RC. Controlled trial of fish oil for regression of human coronary atherosclerosis. J Am Coll Cardiol. 1995;25(7):1492–8.
82. Rivellese AA, Maffettone A, Iovine C, et al. Long-term effects of fish oil on insulin resistance and plasma lipoproteins in NIDDM patients with hypertriglyceridemia. Diabetes Care. 1996;19(11):1207–13.
83. Taccone-Gallucci M, Manca-di-Villahermosa S, Battistini L, Stuffler RG, Tedesco M, Maccarrone M. N-3 PUFAs reduce oxidative stress in ESRD patients on maintenance HD by inhibiting 5-lipoxygenase activity. Kidney Int. 2006;69(8):1450–4.
84. An WS, Kim HJ, Cho KH, Vaziri ND. Omega-3 fatty acid supplementation attenuates oxidative stress, inflammation, and tubulointerstitial fibrosis in the remnant kidney. Am J Physiol Renal Physiol. 2009;297(4):F895–903.
85. Svegliati-Baroni G, Candelaresi C, Saccomanno S, et al. A model of insulin resistance and nonalcoholic steatohepatitis in rats. Am J Pathol. 2006;169(3):846–60.
86. Casós K, Zaragozá MC, Zarkovic N, et al. A fish oil-rich diet reduces vascular oxidative stress in apoE –/– mice. Free Radic Res. 2010;44(7):821–9.
87. Virtanen JK, Mursu J, Voutilainen S, Tuomainen TP. Serum long-chain n-3 polyunsaturated fatty acids and risk of hospital diagnosis of atrial fibrillation in men. Circulation. 2009;120(23):2315–21.
88. Leaf A. Prevention of fatal arrhythmias in high-risk subjects by fish Oil n-3 fatty acid intake. Circulation. 2005;112(18):2762–8.
89. Djoussé L, Rautaharju PM, Hopkins PN, et al. Dietary linolenic acid and adjusted QT and JT intervals in the national heart, lung, and blood institute family heart study. J Am Coll Cardiol. 2005;45(10):1716–22.
90. Marchioli R. Early protection against sudden death by n-3 polyunsaturated fatty acids after myocardial infarction: time-course analysis of the results of the Gruppo Italiano per lo Studio della Sopravvivenza nell'Infarto Miocardico (GISSI)-Prevenzione. Circulation. 2002;105(16):1897–903.
91. Armah Christopher K, Jackson Kim G, Doman I, James L, Cheghani F, Minihane AM. Fish oil fatty acids improve postprandial vascular reactivity in healthy men. Clin Sci. 2008;114(11):679–86.
92. Goode GK, Garcia S, Heagerty AM. Dietary supplementation with marine fish Oil improves in vitro small artery endothelial function in hypercholesterolemic patients: a double-blind placebo-controlled study. Circulation. 1997;96(9):2802–7.
93. McVeigh GE, Brennan GM, Johnston GD, et al. Dietary fish oil augments nitric oxide production or release in patients with Type 2 (non-insulin-dependent) diabetes mellitus. Diabetologia. 1993;36(1):33–8.
94. Morgan DR, Dixon LJ, Hanratty CG, et al. Effects of dietary omega-3 fatty acid supplementation on endothelium-dependent vasodilation in patients with chronic heart failure. Am J Cardiol. 2006;97(4):547–51.
95. Holm T. Omega-3 fatty acids improve blood pressure control and preserve renal function in hypertensive heart transplant recipients. Eur Heart J. 2001;22(5):428–36.
96. Hu FB, Cho E, Rexrode KM, Albert CM, Manson JE. Fish and long-chain omega-3 fatty acid intake and risk of coronary heart disease and total mortality in diabetic women. Circulation. 2003;107(14):1852–7.
97. Hamilton MC, Hites RA, Schwager SJ, Foran JA, Knuth BA, Carpenter DO. Lipid composition and contaminants in farmed and wild salmon. Environ Sci Technol. 2005;39(22):8622–9.
98. van Beelen VA, Spenkelink B, Mooibroek H, et al. An n-3 PUFA-rich microalgal oil diet protects to a similar extent as a fish oil-rich diet against AOM-induced colonic aberrant crypt foci in F344 rats. Food Chem Toxicol. 2009;47(2):316–20.

99. El Abed MM, Marzouk B, Medhioub MN, Helal AN, Medhioub A. Microalgae: a potential source of polyunsaturated fatty acids. Nutr Health. 2008;19(3):221–6.
100. Lee JM, Robson MD, Yu LM, et al. Effects of high-dose modified-release nicotinic acid on atherosclerosis and vascular function: a randomized, placebo-controlled, magnetic resonance imaging study. J Am Coll Cardiol. 2009;54(19):1787–94.
101. Blankenhorn DH, Azen SP, Crawford DW, et al. Effects of colestipol-niacin therapy on human femoral atherosclerosis. Circulation. 1991;83(2):438–47.
102. Cashin-Hemphill L. Beneficial effects of colestipol-niacin on coronary atherosclerosis. JAMA. 1990;264(23):3013.
103. Warnholtz A, Wild P, Ostad MA, et al. Effects of oral niacin on endothelial dysfunction in patients with coronary artery disease: results of the randomized, double-blind, placebo-controlled INEF study. Atherosclerosis. 2009;204(1):216–21.
104. Wu BJ, Yan L, Charlton F, Witting P, Barter PJ, Rye KA. Evidence that niacin inhibits acute vascular inflammation and improves endothelial dysfunction independent of changes in plasma lipids. Arterioscler Thromb Vasc Biol. 2010;30(5):968–75.
105. Shirodaria C, Antoniades C, Lee J, et al. Global improvement of vascular function and redox state with Low-dose folic acid: implications for folate therapy in patients with coronary artery disease. Circulation. 2007;115(17):2262–70.
106. Doshi SN, McDowell IFW, Moat SJ, et al. Folate improves endothelial function in coronary artery disease: an effect mediated by reduction of intracellular superoxide? Arterioscler Thromb Vasc Biol. 2001;21(7):1196–202.
107. Stampfer MJ, Hu FB, Manson JE, Rimm EB, Willett WC. Primary prevention of coronary heart disease in women through diet and lifestyle. N Engl J Med. 2000;343(1):16–22.
108. Bonthu S, Heistad DD, Chappell DA, Lamping KG, Faraci FM. Atherosclerosis, vascular remodeling, and impairment of endothelium-dependent relaxation in genetically altered hyperlipidemic mice. Arterioscler Thromb Vasc Biol. 1997;17(11):2333–40.
109. Laursen JB, Somers M, Kurz S, et al. Endothelial regulation of vasomotion in ApoE-deficient mice: implications for interactions between peroxynitrite and tetrahydrobiopterin. Circulation. 2001;103(9):1282–8.
110. Beckman JS, Beckman TW, Chen J, Marshall PA, Freeman BA. Apparent hydroxyl radical production by peroxynitrite: implications for endothelial injury from nitric oxide and superoxide. Proc Natl Acad Sci. 1990;87(4):1620–4.
111. Förstermann U. Oxidative stress in vascular disease: causes, defense mechanisms and potential therapies. Nat Clin Pract Cardiovasc Med. 2008;5(6):338–49.
112. Baker TA, Milstien S, Katusic ZS. Effect of vitamin C on the availability of tetrahydrobiopterin in human endothelial cells. J Cardiovasc Pharmacol. 2001;37(3):333–8.
113. Lin KY, Ito A, Asagami T, et al. Impaired nitric oxide synthase pathway in diabetes mellitus: role of asymmetric dimethylarginine and dimethylarginine dimethylaminohydrolase. Circulation. 2002;106(8):987–92.
114. Hirashima O, Kawano H, Motoyama T, et al. Improvement of endothelial function and insulin sensitivity with vitamin C in patients with coronary spastic angina. J Am Coll Cardiol. 2000;35(7):1860–6.
115. Schindler TH, Nitzsche EU, Munzel T, et al. Coronary vasoregulation in patients with various risk factors in response to cold pressor testing. J Am Coll Cardiol. 2003;42(5):814–22.
116. Stephens NG, Parsons A, Brown MJ, et al. Randomised controlled trial of vitamin E in patients with coronary disease: Cambridge Heart Antioxidant Study (CHAOS). Lancet. 1996;347(9004):781–6.
117. Dagenais GR, Yusuf S, Bourassa MG, et al. Effects of ramipril on coronary events in high-risk persons: results of the heart outcomes prevention evaluation study. Circulation. 2001;104(5):522–6.
118. Sleight P, Yusuf S, Pogue J, Tsuyuki R, Diaz R, Probstfield J. Blood-pressure reduction and cardiovascular risk in HOPE study. Lancet. 2001;358(9299):2130–1.
119. Cheung MC, Zhao XQ, Chait A, Albers JJ, Brown BG. Antioxidant supplements block the response of HDL to simvastatin-niacin therapy in patients with coronary artery disease and low HDL. Arterioscler Thromb Vasc Biol. 2001;21(8):1320–6.
120. Lissin LW, Cooke JP. Phytoestrogens and cardiovascular health. J Am Coll Cardiol. 2000;35(6):1403–10.
121. Lissin LW, Oka R, Lakshmi S, Cooke JP. Isoflavones improve vascular reactivity in post-menopausal women with hypercholesterolemia. Vasc Med. 2004;9(1):26–30.
122. Tempfer CB, Froese G, Heinze G, Bentz E-K, Hefler LA, Huber JC. Side effects of phytoestrogens: a meta-analysis of randomized trials. Am J Med. 2009;122(10):939–46.e9.
123. Vincent A, Fitzpatrick LA. Soy isoflavones: are they useful in menopause? Mayo Clin Proc. 2000;75(11):1174–84.
124. Cooke JP, Andon NA, Girerd XJ, Hirsch AT, Creager MA. Arginine restores cholinergic relaxation of hypercholesterolemic rabbit thoracic aorta. Circulation. 1991;83(3):1057–62.
125. Rossitch E, Alexander E, Black PM, Cooke JP. L-arginine normalizes endothelial function in cerebral vessels from hypercholesterolemic rabbits. J Clin Invest. 1991;87(4):1295–9.
126. Creager MA, Gallagher SJ, Girerd XJ, Coleman SM, Dzau VJ, Cooke JP. L-arginine improves endothelium-dependent vasodilation in hypercholesterolemic humans. J Clin Invest. 1992;90(4):1248–53.

127. Theilmeier G, Chan JR, Zalpour C, et al. Adhesiveness of mononuclear cells in hypercholesterolemic humans is normalized by dietary L-arginine. Arterioscler Thromb Vasc Biol. 1997;17(12):3557–64.

128. Tsao PS, Theilmeier G, Singer AH, Leung LL, Cooke JP. L-arginine attenuates platelet reactivity in hypercholesterolemic rabbits. Arterioscler Thromb Vasc Biol. 1994;14(10):1529–33.

129. Wang B-Y, Candipan RC, Arjomandi M, Hsiun PTC, Tsao PS, Cooke JP. Arginine restores nitric oxide activity and inhibits monocyte accumulation after vascular injury in hypercholesterolemic rabbits. J Am Coll Cardiol. 1996;28(6):1573–9.

130. Wang B-Y, Singer AH, Tsao PS, Drexler H, Kosek J, Cooke JP. Dietary arginine prevents atherogenesis in the coronary artery of the hypercholesterolemic rabbit. J Am Coll Cardiol. 1994;23(2):452–8.

131. Vallance P, Leone A, Calver A, Collier J, Moncada S. Accumulation of an endogenous inhibitor of nitric oxide synthesis in chronic renal failure. Lancet. 1992;339(8793):572–5.

132. Leiper J, Murray-Rust J, McDonald N, Vallance P. S-nitrosylation of dimethylarginine dimethylaminohydrolase regulates enzyme activity: further interactions between nitric oxide synthase and dimethylarginine dimethylaminohydrolase. Proc Natl Acad Sci. 2002;99(21):13527–32.

133. Boger RH, Bode-Boger SM, Brandes RP, et al. Dietary L-arginine reduces the progression of atherosclerosis in cholesterol-fed rabbits: comparison with lovastatin. Circulation. 1997;96(4):1282–90.

134. Ito A, Tsao PS, Adimoolam S, Kimoto M, Ogawa T, Cooke JP. Novel mechanism for endothelial dysfunction: dysregulation of dimethylarginine dimethylaminohydrolase. Circulation. 1999;99(24):3092–5.

135. Leiper J, Nandi M, Torondel B, et al. Disruption of methylarginine metabolism impairs vascular homeostasis. Nat Med. 2007;13(2):198–203.

136. Leiper JM, Maria JS, Chubb A, et al. Identification of two human dimethylarginine dimethylaminohydrolases with distinct tissue distributions and homology with microbial arginine deiminases. Biochem J. 1999;343(1):209–14.

137. Hong L, Fast W. Inhibition of human dimethylarginine dimethylaminohydrolase-1 by S-nitroso-L-homocysteine and hydrogen peroxide. J Biol Chem. 2007;282(48):34684–92.

138. Rodionov RN, Dayoub H, Lynch CM, et al. Overexpression of dimethylarginine dimethylaminohydrolase protects against cerebral vascular effects of hyperhomocysteinemia. Circ Res. 2009;106(3):551–8.

139. Sydow K, Mondon CE, Schrader J, Konishi H, Cooke JP. Dimethylarginine dimethylaminohydrolase overexpression enhances insulin sensitivity. Arterioscler Thromb Vasc Biol. 2008;28(4):692–7.

140. Siroen MPC, van Leeuwen PAM, Nijveldt RJ, Teerlink T, Wouters PJ, Van den Berghe G. Modulation of asymmetric dimethylarginine in critically ill patients receiving intensive insulin treatment: a possible explanation of reduced morbidity and mortality? Crit Care Med. 2005;33(3):504–10.

141. Nicholls SJ, Wang Z, Koeth R, et al. Metabolic profiling of arginine and nitric oxide pathways predicts hemodynamic abnormalities and mortality in patients with cardiogenic shock after acute myocardial infarction. Circulation. 2007;116(20):2315–24.

142. Furuki K, Adachi H, Enomoto M, et al. Plasma level of asymmetric dimethylarginine (ADMA) as a predictor of carotid intima-media thickness progression: six-year prospective study using carotid ultrasonography. Hypertens Res. 2008;31(6):1185–9.

143. Quyyumi AA, Dakak N, Diodati JG, Gilligan DM, Panza JA, Cannon RO. Effect of l-arginine on human coronary endothelium-dependent and physiologic vasodilation. J Am Coll Cardiol. 1997;30(5):1220–7.

144. Chin-Dusting JPF, Kaye DM, Lefkovits J, Wong J, Bergin P, Jennings GL. Dietary supplementation with L-arginine fails to restore endothelial function in forearm resistance arteries of patients with severe heart failure. J Am Coll Cardiol. 1996;27(5):1207–13.

145. Parnell Melinda M, Holst Diane P, Kaye DM. Augmentation of endothelial function following exercise training is associated with increased L-arginine transport in human heart failure. Clin Sci. 2005;109(6):523–30.

146. Wilson AM, Harada R, Nair N, Balasubramanian N, Cooke JP. L-arginine supplementation in peripheral arterial disease: no benefit and possible harm. Circulation. 2007;116(2):188–95.

147. Weingärtner O, Lütjohann D, Ji S, et al. Vascular effects of diet supplementation with plant sterols. J Am Coll Cardiol. 2008;51(16):1553–61.

148. Miettinen TA, Railo M, Lepäntalo M, Gylling H. Plant sterols in serum and in atherosclerotic plaques of patients undergoing carotid endarterectomy. J Am Coll Cardiol. 2005;45(11):1794–801.

149. Blair SN, Capuzzi DM, Gottlieb SO, Nguyen T, Morgan JM, Cater NB. Incremental reduction of serum total cholesterol and low-density lipoprotein cholesterol with the addition of plant stanol ester-containing spread to statin therapy. Am J Cardiol. 2000;86(1):46–52.

150. Heber D, Lembertas A, Lu Q-Y, Bowerman S, Go VLW. An analysis of nine proprietary Chinese red yeast rice dietary supplements: implications of variability in chemical profile and contents. J Alt Complement Med. 2001;7(2):133–9.

151. Turner B, Mølgaard C, Marckmann P. Effect of garlic (Allium sativum) powder tablets on serum lipids, blood pressure and arterial stiffness in normo-lipidaemic volunteers: a randomised, double-blind, placebo-controlled trial. Br J Nutr. 2004;92(04):701.

152. Berthold HK, Unverdorben S, Degenhardt R, Bulitta M, Gouni-Berthold I. Effect of policosanol on lipid levels among patients with hypercholesterolemia or combined hyperlipidemia. JAMA. 2006;295(19):2262.
153. Urberg M, Zemel MB. Evidence for synergism between chromium and nicotinic acid in the control of glucose tolerance in elderly humans. Metabolism. 1987;36(9):896–9.
154. Ali A, Ma Y, Reynolds J, Wise J, Inzucchi S, Katz D. Chromium effects on glucose tolerance and insulin sensitivity in persons at risk for diabetes mellitus. Endo Pract. 2011;17(1):16–25.
155. Ziegler D, Gries FA. Lipoic acid in the treatment of diabetic peripheral and cardiac autonomic neuropathy. Diabetes. 1997;46 Suppl 2:S62–S6.
156. Sola S, Mir MQ, Cheema FA, et al. Irbesartan and lipoic acid improve endothelial function and reduce markers of inflammation in the metabolic syndrome: results of the Irbesartan and Lipoic Acid in Endothelial Dysfunction (ISLAND) study. Circulation. 2005;111(3):343–8.
157. Sobenin IA, Andrianova IV, Fomchenkov IV, Gorchakova TV, Orekhov AN. Time-released garlic powder tablets lower systolic and diastolic blood pressure in men with mild and moderate arterial hypertension. Hypertens Res. 2009;32(6):433–7.
158. Zand J, Lanza F, Garg HK, Bryan NS. All-natural nitrite and nitrate containing dietary supplement promotes nitric oxide production and reduces triglycerides in humans. Nutr Res. 2011;31(4): 262–9.
159. Nagamani SC, Campeau PM, Shchelochkov OA, Premkumar MH, Guse K, Brunetti-Pierri N, Chen Y, Sun Q, Tang Y, Palmer D, Reddy AK, Li L, Slesnick TC, Feig DI, Caudle S, Harrison D, Salviati L, Marini JC, Bryan NS, Erez A, Lee B. Nitric-oxide supplementation for treatment of long-term complications in argininosuccinic aciduria. Am J Hum Genet. 2012;90(5):836–46.
160. Houston M, Hays L. Acute effects of an oral nitric oxide supplement on blood pressure, endothelial function, and vascular compliance in hypertensive patients. J Clin Hypertens (Greenwich). 2014;16(7):524–9.
161. Biswas OS, Gonzalez VR, Schwarz ER. Effects of an oral nitric oxide supplement on functional capacity and blood pressure in adults with prehypertension. J Cardiovasc Pharmacol Ther. 2015;20(1):52–8.

Chapter 10
Dietary Flavonoids as Modulators of NO Bioavailability in Acute and Chronic Cardiovascular Diseases

Matthias Totzeck and Tienush Rassaf

Key Points

- Coronary heart disease and acute myocardial infarction remain the leading causes of death worldwide.
- Diet compositions have gained much attention in prevention from chronic cardiovascular diseases.
- Flavonoids—found in high concentration in green tea, red wine, and cocoa—are naturally occurring polyphenols.
- Improvement of diseases leading to arteriosclerosis (diabetes, hypertension) has been demonstrated in several randomized clinical trials under treatment with flavonoids.
- Cytoprotection by flavonoids has been largely related to modulation of nitric oxide bioavailability.
- Experimental evidence also suggests that flavonoids can alleviate myocardial damage in acute events, e.g., an acute myocardial infarction.
- Randomized clinical trials investigating the role of dietary interventions with flavonoids in acute myocardial infarctions have yet to be conducted.

Keywords Polyphenols • Wine • Cocoa • Chocolate • Coronary artery disease • Hypertension

Endothelial Dysfunction, Atherosclerosis, and Myocardial Infarction: Role of NO

Protection via dietary flavonoids has been largely related to an improvement of NO bioavailability. An imbalanced NO homeostasis, in turn, is a key element in the development of chronic cardiovascular diseases. Additionally, experimental data suggest an important role for NO in protection from myocardial damage during acute cardiovascular events (e.g., an acute myocardial infarction).

M. Totzeck, M.D. (✉) • T. Rassaf, M.D., Ph.D.
Department of Cardiology, West-German Heart and Vessel Center, University Hospital Essen,
University of Duisburg-Essen, Hufelandstrasse 55, Essen 45147, Germany
e-mail: matthiastotzeck@me.com; Tienush.rassaf@med.uni-duesseldorf.de

N.S. Bryan, J. Loscalzo (eds.), *Nitrite and Nitrate in Human Health and Disease*, Nutrition and Health,
DOI 10.1007/978-3-319-46189-2_10, © Springer International Publishing AG 2017

NO in Chronic Cardiovascular Diseases

Endothelial dysfunction precedes chronic cardiovascular disease and is caused by several risk factors, including aging, hypercholesterolemia, hypertension, diabetes mellitus, smoking, obesity, chronic systemic inflammation, hyperhomocysteinemia, and a family history of premature atherosclerosis. An intact endothelium, however, is required for the maintenance of vascular tone and architecture, blood fluidity, and antithrombotic protection. NO signaling plays a major role in maintaining these functions by regulation of vascular tone and smooth muscle cell proliferation, white blood cell adhesion, and platelet function.

Endothelial NO synthases (eNOS) produce NO from the amino acid L-arginine and several cofactors [1]. Disturbance in the NOS-dependent NO pathway is a key element in the development of endothelial dysfunction, which leads to the onset and progression of atherosclerosis. In endothelial dysfunction, reduced NO bioavailability is caused by an inhibition of eNOS function rather than by lower eNOS protein expression [2]. One of the events leading to altered eNOS activity has been related to decreased levels of the cofactor, (6R)-5,6,7,8-tetrahydro-L-biopterin, and has been termed "eNOS uncoupling" [3]. Under these pathological conditions, eNOS produce potentially harmful reactive oxygen species (ROS, e.g., superoxide) rather than NO. The extent of endothelial dysfunction relates to the risk of patients for developing chronic cardiovascular diseases and acute ischemic events, e.g., myocardial infarction and stroke.

Another key element in the maintenance of endothelial function is endothelial progenitor cells (EPCs, also referred to as circulating angiogenic cells). EPCs participate in the repair of vessel injury and neovascularization [4–6]. NO modulates the recruitment of EPCs, and balanced NO homeostasis is, therefore, required for appropriate interaction between resident endothelial cells and circulating EPCs. Imbalanced NO levels may thus contribute to impaired vascular regenerative processes, for example, in patients with chronic ischemic heart disease [7,8].

NO produced from eNOS can react with hemoglobin in the circulation to undergo oxidation to nitrate. Owing to the immense concentrations of hemoglobin in blood, signaling by NO has therefore been considered to occur only at the site of NO production. Indeed, evidence suggests that free hemoglobin is an effective NO scavenger that can attenuate NO bioavailability and NO-related cardiovascular functions; in patients with end-stage renal disease, impaired vascular function occurs, in part, due to hemodialysis-related release of free hemoglobin into the circulation [9]. Thus, the question of how NO bioavailability is preserved throughout the cardiovascular system is an important one for vascular homeostasis. In vivo, several biochemical mechanisms exist, allowing NO to be transported as NO itself, to react with plasma compounds to form nitrosospecies, and to be oxidized to bioactive nitrite [10–13]. Like endocrine hormones, these mechanisms ensure NO bioavailability and signaling throughout the body. Thus, the complexity of the circulating NO pool opens new, considerable pharmacological and therapeutical possibilities in the diagnosis and therapy of cardiovascular diseases.

In order to detect endothelial (dys)function and to evaluate the impact of dietary interventions on NO-mediated vascular functions, several invasive and noninvasive techniques have been utilized. The measurement of the so-called flow-mediated dilation (FMD) has been used in numerous interventional trials to assess endothelial function. In FMD, the relative diameter change (in percent) of the brachial artery from pre-ischemia (baseline) to reactive hyperemia is measured via high-resolution ultrasound. Ischemia is induced via inflation of a blood pressure cuff around the forearm, distal to the part of the artery assessed by ultrasound. During the ischemic period, the forearm vasculature vasodilates, and the final pressure release from the cuff consequently leads to increased flow in the conduit brachial artery. This is accompanied by an enhanced vessel wall shear stress, which stimulates eNOS to produce NO, which in turn causes vascular smooth muscle relaxation and accompanying arterial dilation. FMD is, therefore, generally regarded to be the gold standard in the characterization of endothelial NO function in vivo. Notably, the extent of endothelial function detected via FMD also correlates with the function in the coronary conduit arteries.

NO in Acute Cardiovascular Events

Acute myocardial infarction caused by rupture of an atherosclerotic plaque and subsequent occlusion of a large coronary artery by thrombus is the leading cause of death worldwide. Upon the onset of an acute myocardial infarction, the immediate and successful recanalization of the occluded coronary artery is the optimal therapy, which leads to an infarct size reduction and improves the prognosis of the patient. The process of restoring blood flow to the ischemic myocardium, however, can induce injury itself. This phenomenon, termed ischemia/reperfusion (I/R) injury, reduces the beneficial effects of myocardial reperfusion [14]. Apart from acute impairment of left ventricular (LV) function, large myocardial infarctions lead to a pathologic remodeling of the LV, limiting prognosis and quality of life.

Four phenomena, namely, myocardial stunning, the no-reflow phenomenon, arrhythmias, and lethal reperfusion injury, contribute to the extent and impact of the ultimate myocardial I/R injury [15–20]. The key mediators of these patho-biological mechanisms are ROS, deranged calcium levels (Ca^{2+}), pH, the mitochondrial permeability transition pore (mPTP), metabolic changes, and the immune responses. For a full summary on the mechanisms of myocardial I/R injury, the reader is referred to relevant, recent reviews [15,21,22].

As demonstrated in numerous experimental studies, myocardial I/R injury can be mitigated. Two of the most powerful protective mechanisms in this context are ischemic preconditioning (IPC) and ischemic postconditioning (PostC), which are capable of reducing final infarct size by ~30–60% of the myocardium at risk [21–23]. However, although numerous animal model-based studies showed much-reduced I/R injury, the respective translational clinical trials failed to demonstrate the same benefit in humans. Several of these experimental studies point to a protective role of the NO pool during myocardial I/R [24–27]. Administration of nitrite in an attempt to modulate I/R injury leads to an elevation of the circulating NO pool and exerts protective tissue effects in a mouse model of I/R injury in vivo [24,26,28]. The protective effects derive from a hemeprotein-dependent reduction of nitrite to NO, as demonstrated in mice without myoglobin [26]; NO derived from nitrite reduction via cardiac myoglobin reversibly modulates mitochondrial electron transport, thus decreasing reperfusion-derived oxidative stress and inhibiting cellular apoptosis leading to a smaller final infarct size. Notably, translational approaches to test nitrite for its clinical relevance in patients are still under investigation.

In summary, disturbed NO homeostasis is a hallmark of the development of chronic cardiovascular disease and plays a key role in acute ischemic events, such as an acute myocardial infarction. However, the involvement of NO in all of the mechanisms described above also points to the possibility of therapeutic interventions that modulate NO bioavailability, e.g., through diets enriched with flavonoids.

Biochemistry of Flavonoids

Flavonoids represent a subclass of polyphenols, which are widely distributed in plants [29]. They are synthesized as micronutrients in the secondary metabolism of plants to protect against microbes, fungi, and insects, or to serve as pigments [30]. The term, "flavonoid," has been used to describe a group of compounds that can be divided into six subclasses (Fig. 10.1). More precisely, flavonoids are ketone-containing compounds (flavanols, flavones, and isoflavones), whereas non-ketone polyphenols are referred to as flavanoids (e.g., flavan-3-ols) [31]. All compounds are formed in a metabolic pathway through a series of enzymatic modifications yielding numerous end products. Key dietary products with high flavonoid content are green tea, red wine, and products made from cocoa beans. The latter contains relatively high amounts of flavan-3-ols also known as flavanols [32]. Flavanols (e.g., catechin or (−)-epicatechin) or their oligomers, the so-called proanthocyanidins, are among the

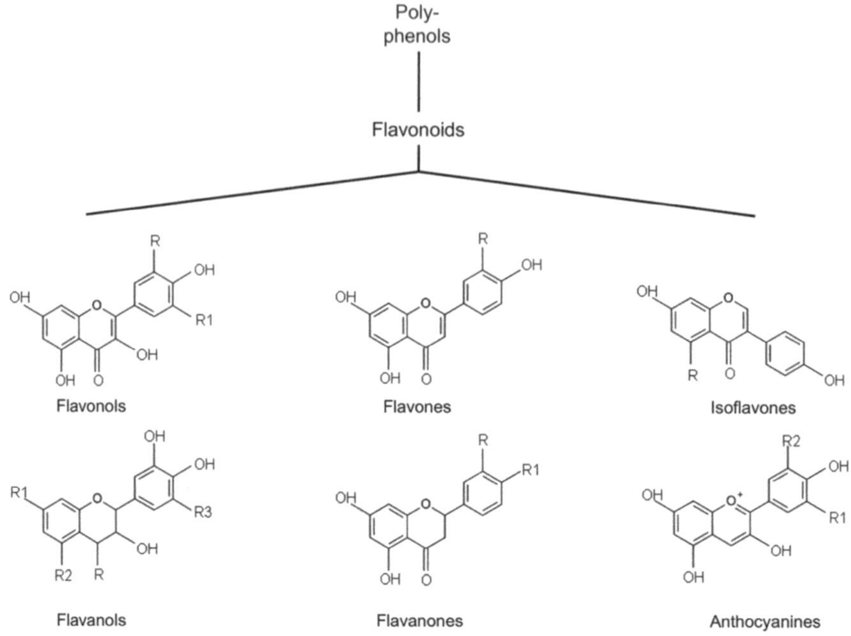

Fig. 10.1 Biochemistry of flavonoids. Metabolic processes in plants form flavonoid compounds that are classified into six groups according to their chemical structures

best-characterized substances in the field of micronutrients. As certain foods rich in flavonoids may contribute to cardiovascular health and protect from the onset of diseases, specific dietary recommendations are under intensive debate. Several steps will have to be taken before official guidelines can be issued. In a first step, certain foods will have to be evaluated in terms of their precise flavonoid content and the average population-based consumption. Environmental changes, as well as nutrition processing, may also influence total flavonoid levels. Moreover, methods for the determination of polyphenolic contents are avidly discussed currently, as recent studies have yielded divergent results. The US Department of Agriculture has, however, published a database with flavonoid contents of selected foods. On the basis of these data, the total actual daily intake of flavanols can be calculated. The next step for the development of recommendations is the assessment of flavonoid bio-metabolism. Despite their high concentrations in certain foods and beverages, several flavonoids show poor bioavailability due to chemical alterations in the digestive tract and liver metabolism. One example is the metabolism of catechin, which is already to a large degree metabolized in the small intestine [33,34].

Flavonoids and Cardiovascular Diseases

Epidemiological data suggest an inverse relation between the intake of certain foods and beverages and the incidence of cardiovascular disease. Current literature databases account for several thousand original publications and reviews examining this topic. In this context, epidemiological studies notoriously hold one major well-known flaw: the lack of demonstration of a cause and effect relationship. Moreover, research involving dietary interventions has to be conducted in a multiple step approach. International authorities in this field have, therefore, established the "International Conference on Polyphenols and Health." In its statutes, criteria for both clinical trials and basic research with flavonoid intervention were defined in detail (for further information, please go to www.polyphenolsandhealth.org.uk/). Interventional trials are generally preferred; they need to be well controlled and conducted over a considerable length of time. Control substances should be well balanced as, for example, cocoa-derived products have a distinct taste and contain numerous other ingredients (e.g., theobromine and caffeine) [35].

Several interventional studies have revealed acute and chronic improvements of vascular function as measured by FMD, circulating NO pool, and EPC mobilization after the ingestion of flavanol-rich foods or beverages, such as red wine [36], purple grape juice [37], tea [38], and cocoa-based products (Table 10.1) [7,39–43]. Even in optimally treated type 2 diabetic patients, beneficial health effects of flavanol-rich cocoa could be detected [39]. The latter study also identified that effects exerted by flavanols cannot only be found following prolonged ingestion (chronic FMD effects), as even an "acute-on-chronic" effect was described. The "positive" impact on FMD together with an enhanced circulating NO pool (nitroso species, nitrite) and a positive recruitment of EPCs provides strong evidence that the modulation of NO bioavailability is a key feature of diets rich in flavonoids.

Although the studies mentioned above have investigated the impact of a dietary intervention on chronic cardiovascular risk factors, the impact of flavonoids on acute ischemic events, such as I/R injury, can only be deduced from animal model-based studies to date. Can we, however, easily translate diet-based experimental studies and assume a similar benefit for the corresponding patient cohort? Given that mice and rats in comparison to humans show several physiological differences in terms of eating habits, nutrition, digestion, metabolism, and immune system, the existing studies have to be interpreted very carefully in this regard.

Table 10.1 Human interventional trials with flavanoid-rich food

Human interventional trials			
Disease	Treatment	Outcome	References
Healthy volunteers	Cocoa drink	PAT↑	Fisher et al. [76]
	Cocoa drink	FMD↑	Heiss et al. [42]
	Cocoa drink	FMD↑	Schroeter et al. [43]
	Chocolate	FMD↑	Grassi et al. [77]
	Chocolate	FMD↑	Engler et al. [78]
	Chocolate	FMD↑	Vlachopoulos et al. [79]
	Black tea	FMD↑	Grassi et al. [80]
	Red wine	FMD↑	Agewall et al. [36]
	Apple puree	NO↑ platelet reactivity↑	Gasper et al. [54]
Cardiovascular risk factors	Cocoa drink	FMD↑	Heiss et al. [40,41]
	Chocolate/cocoa drink	BP↓, FMD↑	Faridi et al. [81]
Coronary artery disease	Black tea	FMD↑	Duffy et al. [38]
	Grape juice	FMD↑	Stein et al. [37]
	Cocoa drink	FMD↑, EPC↑	Heiss et al. [7]
	Cocoa drink	MP↓	Horn et al. [82]
Hypertension	Chocolate	BP↓	Taubert et al. [83]
	Chocolate	BP↓	Taubert et al. [84]
	Chocolate	BP↓, FMD↑	Grassi et al. [85]
Hypercholesterolemia	Black tea	FMD↑	Hodgson et al. [86]
Diabetes	Cocoa drink	FMD↑	Balzer et al. [39]
Overweight	Dark Chocolate	FMD↑	Grassi et al. [87]
Heart failure	Dark chocolate	Vasodilation↑, Art. stiffness↓	West et al. [88]
	Chocolate	FMD↑, platelet adhesion↓	Flammer et al. [89]

PAT peripheral arterial tone measurement; *FMD* flow-mediated dilation measurement; *BP* blood pressure; *EPC* endothelial progenitor cells (mobilization), *MP* microparticles associated with coronary artery disease

In order to interpret data from rodent models of I/R, interspecies differences between mice, rats, and humans in absorption, metabolism, plasma concentration, and tissues distribution must be considered. The interspecies differences in the metabolism of *Vitis vinifera* (red wine) derived resveratrol may serve as an example: while humans metabolize resveratrol almost entirely into the glucuronidated form, the most abundant metabolite found in rat plasma is the sulfated form [44]. Resveratrol treatment has been shown to produce cardioprotection from lethal I/R injury in rats [45–49]; this benefit, however, may not be observed in humans owing to the differences in resveratrol metabolism pointed out above.

The studies on flavonoids and their role in protecting the myocardium in the event of I/R injury make use of a variety of different compounds that are found, for example, in red wine, green tea, cocoa beans, and vegetables. Even synthetic flavanol-based substances with questionable importance for dietary interventions were used in an attempt to modulate I/R injury induced by ROS [50,51]. Notably, the food formulations used in the animal models were inconsistent; although some studies used the purified compound, others relied on extracts (green tea) or the whole fruit (flesh from grapes), the latter being closer to the flavonoids' natural distribution and formulation. In summary, a total of about 60 experimental studies using flavonoids to modulate I/R suggest that these compounds also exert protection in acute ischemic events, a hypothesis, which has yet to be proven in clinical intervention trials.

Protection Mechanisms of Dietary Flavonoids

The in vivo effects of flavonoids strongly depend on the substances' pharmacodynamic and pharmacokinetic properties. Although numerous intervention trials using FMD strongly suggest involvement of NO, the exact mechanisms of protection by dietary flavonoids are still elusive. One of the key questions remaining in this context is whether flavonoids improve the bioavailability of NO via increased production or via inhibition of the metabolic inactivation of NO.

Flavonoids and the Nitrate–Nitrite–NO Pathway

In a recent in vitro study in human coronary artery endothelial cells, Ramirez-Sanchez et al. provided evidence that (−)-epicatechin, the most abundant flavonoid in cocoa, can activate eNOS together with the necessary involvement of phosphatidylinositol 3-kinase [52]. Additionally, Steffen et al. showed that the treatment of endothelial cells with (−)-epicatechin elevated intracellular levels of NO and cyclic guanosine monophosphate [53]. Recently, a study by Gasper and coworkers detected an increase in total NO metabolites following cocoa intake [54]. Further investigations revealed that (−)-epicatechin inhibited NADPH oxidase with a consequent reduction in superoxide anion production, which is known to scavenge NO. The inhibition of the NADPH oxidase, thus, increases bioavailability of NO rather than affecting NO synthesis. These results may explain the acute effects on endothelial function 1–2 h after the intake of a single flavanol-containing drink [40]. Prolonged intervention with flavanol-enriched drinks resulted in an increase in baseline levels of FMD [42]. In contrast to the acute impact, which can be related to the inhibition of NO metabolism, the chronic effects may be due to an increase in endothelial NOS expression in the vascular endothelium. Balanced NO homeostasis is required for a proper interaction between the endothelium and EPCs, which, in turn, participate in vascular regenerative processes. A very recent study in patients with coronary artery disease has shown that the treatment with flavanols over 30 days not only improves endothelial function as measured by FMD and the circulation NO pool [10,41], but also significantly increased circulating

EPC in these patients. The mobilization of EPC may serve as an additional explanation for the effects of a long-term dietary flavonoid intake on endothelial function and bio-repair.

The exact mechanisms of flavonoids on NO metabolism are currently unresolved, although, in light of the findings above, a strong interaction with eNOS activity can be assumed. Flavonoids might, however, interact with NO homeostasis in at least two other ways: (1) by increasing nitrite, which in turn exerts cytoprotection and (2) by direct bioconversion of nitrite to NO. As stated above, clinical intervention strategies demonstrated an increase in nitrite following the consumption of drinks enriched with flavanols [39,41,43]. Plasma nitrite either derives from nutritional sources or via endogenous decomposition of NO. It is no longer regarded to be an inert oxidative product. Experimental data suggest that nitrite exerts cytoprotection in a wide variety of pathologies, first and foremost during I/R of the heart [24,26]. These effects have been mainly related to nitrite bioconversion to NO. As they lead to increased nitrite levels, some of the protective effects from flavonoids may be related to this modulation in NO homeostasis. Nitrite reduction to bioactive NO can be mediated by deoxygenated hemoglobin [55] or myoglobin [27] in vivo. Notably, Gago et al. demonstrated that this might also occur directly via red wine polyphenols in the gut without the need of endogenous conversion mechanisms [56]. Following nitrite conversion, NO diffuses over the gut wall and induces muscle relaxation [57]. Taken together, the effects on the NO homeostasis maybe far reaching, ranging from increased NO production via eNOS and bioconversion of increased levels of nitrite and a modulated NO decomposition.

Cellular Targets of Flavonoids in I/R

Although the effects on endothelial and chronic cardiovascular function are mainly related to improved NO bioavailability, the mechanisms of protection from I/R injury by flavonoids remain unresolved. The processes during I/R injury are structurally complex and have been avidly investigated over the past two decades. As only a small number of studies on flavonoid protection from I/R injury have been published to date, a common mechanism or signaling cascade through flavonoid intervention remains unknown.

A considerable impact on oxidative stress is a common feature of most of the flavonoids in experimental trials on I/R injury. ROS formation during I/R may lead to cell damage, e.g., via interaction with lipids of the cell membrane. Possible direct ROS-reducing effects of flavonoids and especially flavanols have been discussed in detail, but the exact mechanisms remain unresolved. Likewise, the direct or indirect influence by dietary flavonoids on the formation of NO remains elusive. The tea polyphenol, epigallocatechin-3-gallate, applied via perfusate in a Langendorff heart model in guinea pigs decreased oxygen radicals, which was paralleled by an increase of NO [58]. Interestingly, eNOS inhibition with L-NAME completely abolished the positive effects of this green tea-derived flavonoid [59]. However, in vivo effects after oral administration of the substance may be considerably different due to the intestinal metabolism of epigallocatechin-3-gallate to form chemically altered and then absorbed substances—a process absent when using Langendorff heart preparations. Decreased ROS levels were also detected for procyanidin [60], osajin pomiferin [61], proanthocyanidin [62], anthocyanin [63], and (−)-epicatechin [64]. Soy-isoflavone is the only substance that does not exert an ROS-reducing effect on the myocardium [65]. Increased NO levels were also detected following an intervention with proanthocyanidin in cell cultures of cardiomyocytes [66]. Controversial effects were published for quercetin and resveratrol. Rabbits treated with quercetin undergoing ischemia and reperfusion had less eNOS and inducible NOS mRNA and protein expression compared with control [67], while another group reported that cardioprotective effects by resveratrol were completely abolished in inducible NOS-deficient mice [68]. The metabolism of quercetin is complex. After oral intake of quercetin or quercetin glucosides, an extensive biotransformation takes place in the gut and small intestine before absorption of the active species. In the blood, orally ingested quercetin appears almost entirely in conjugated forms after phase-II metabolism [69]. These conjugates, and not purified and commercially available quercetin, may be responsible for its beneficial effects.

Only a few studies focussed on the molecular analysis of signaling proteins involved in I/R responses. Myocardial protection strategies induce a specific signaling pathway that ultimately leads to an inhibition of mPTP opening. In this context, Akt plays a central role; it can phosphorylate a number of proteins, such as eNOS, GSK-3β, and Bcl-2, and modify their (enzymatic) activities [21]. Protection from I/R by NO is closely linked to the role of Akt. Inhibition of either Akt or NOS decreased NO levels and led to higher cell death in proanthocyanidin-treated cardiomyocytes [66]. GSK-3β is phosphorylated by a number of proteins, including Akt or Erk. Phosphorylation of GSK-3β leads to its inactivation. High levels of phosphorylated GSK-3β appear to exert an anti-apoptotic effect [70]. Application of resveratrol led to an increase in phosphorylated GSK-3β. Notably, the soluble guanylyl cyclase inhibitor, 1-H [7,35,40] oxadiazolo[4,3-a]quinoxalin-1-one (ODQ), completely abolished the effect indicating an important involvement of the NO pathway. In this study, increased levels of GSK-3β were also found in the mitochondria of isolated cardiomyocytes along with a decrease in mitochondrial swelling. The authors deduced that, via phosphorylation of GSK-3β and its translocation, opening of mPTP is inhibited and cell damage decreased [71]. mPTP opening also induces Ca^{2+} overload, which was significantly decreased in epicatechin-3-gallate-treated hearts [58]. Signaling in apoptosis partly depends on signaling proteins Bax and Bcl-2. Although increased levels of Bax induce cell death, high levels of Bcl-2 appear to be anti-apoptotic [21]. Several authors have described an increase in Bcl-2 and a decrease in Bax and therefore anti-apoptotic effect for flavonoid-treated hearts [49,51,72–74]. The role of the signaling protein JNK is controversial. It plays a key role in phosphorylating numerous other downstream mediators. Treatments with extracts containing proanthocyanidin do not only decrease ROS and apoptosis but also show decreased JNK and c-jun levels [75].

Conclusion

The cardiovascular effects of flavonoids have been extensively investigated over the past decades. Epidemiological data in concert with results from intervention trials suggest that flavonoids might contribute to the prevention of the onset and progression of chronic cardiovascular diseases largely via modulation of NO bioavailability. Despite numerous trials showing beneficial effects for vascular function and cardiovascular risk factors, studies that examine primary clinical endpoints, such as survival or pain-free walking distance in peripheral arterial disease, have not been published to date.

Although a number of studies have examined their impact on chronic diseases, little is known about the effect of flavonoids on acute ischemic events, e.g., a myocardial I/R injury. These effects must be deduced from studies in animal models. This leads quite readily to difficulties in the interpretation in terms of the relevance for clinical practice: (1) the respective studies were carried out in rodents with considerably different metabolisms compared with humans; (2) in most cases, they used purified compounds or even synthetic flavonoids; (3) they lacked the proper balanced control substances, and (4) they were conducted in some cases using isolated organ models, thus completely bypassing metabolism. Future animal studies with flavonoid intervention will, therefore, need to focus on the aspects of absorption/metabolism as necessary basis for translational studies in humans. In the past, many animal studies with substances tested positive for cardioprotection from I/R injury failed when translated into the clinical practice [15]. Nevertheless, the heterogeneous group of flavonoids has the capability to ameliorate myocardial I/R injury in animal model-based experimental studies. If proven true in human studies, functional foods containing the right amounts of flavonoids along with nitrite and/or nitrate might serve as a powerful tool in primary and secondary cardiovascular prevention.

References

1. Moncada S, Higgs A. The L-arginine-nitric oxide pathway. N Engl J Med. 1993;329(27):2002–12.
2. Li H, Wallerath T, Munzel T, Forstermann U. Regulation of endothelial-type NO synthase expression in pathophysiology and in response to drugs. Nitric Oxide. 2002;7(3):149–64.
3. Thum T, Fraccarollo D, Schultheiss M, Froese S, Galuppo P, Widder JD, et al. Endothelial nitric oxide synthase uncoupling impairs endothelial progenitor cell mobilization and function in diabetes. Diabetes. 2007;56(3):666–74.
4. Werner N, Kosiol S, Schiegl T, Ahlers P, Walenta K, Link A, et al. Circulating endothelial progenitor cells and cardiovascular outcomes. N Engl J Med. 2005;353(10):999–1007.
5. Dimmeler S, Zeiher AM. Vascular repair by circulating endothelial progenitor cells: the missing link in atherosclerosis? J Mol Med. 2004;82(10):671–7.
6. Shantsila E, Watson T, Lip GY. Endothelial progenitor cells in cardiovascular disorders. J Am Coll Cardiol. 2007;49(7):741–52.
7. Heiss C, Jahn S, Taylor M, Real WM, Angeli FS, Wong ML, et al. Improvement of endothelial function with dietary flavanols is associated with mobilization of circulating angiogenic cells in patients with coronary artery disease. J Am Coll Cardiol. 2010;56(3):218–24.
8. Dimmeler S. Regulation of bone marrow-derived vascular progenitor cell mobilization and maintenance. Arterioscler Thromb Vasc Biol. 2010;30(6):1088–93.
9. Meyer C, Heiss C, Drexhage C, Kehmeier ES, Balzer J, Muhlfeld A, et al. Hemodialysis-induced release of hemoglobin limits nitric oxide bioavailability and impairs vascular function. J Am Coll Cardiol. 2010;55(5):454–9.
10. Rassaf T, Bryan NS, Kelm M, Feelisch M. Concomitant presence of N-nitroso and S-nitroso proteins in human plasma. Free Radic Biol Med. 2002;33:1590–6.
11. Rassaf T, Bryan NS, Maloney RE, Specian V, Kelm M, Kalyanaraman B, et al. NO adducts in mammalian red blood cells: too much or too little? Nat Med. 2003;9:481–2.
12. Rassaf T, Kleinbongard P, Preik M, Dejam A, Gharini P, Lauer T, et al. Plasma nitrosothiols contribute to the systemic vasodilator effects of intravenously applied NO: experimental and clinical Study on the fate of NO in human blood. Circ Res. 2002;91(6):470–7.
13. Elrod JW, Calvert JW, Gundewar S, Bryan NS, Lefer DJ. Nitric oxide promotes distant organ protection: evidence for an endocrine role of nitric oxide. Proc Natl Acad Sci U S A. 2008;105(32):11430–5.
14. Piper HM, Garcia-Dorado D, Ovize M. A fresh look at reperfusion injury. Cardiovasc Res. 1998;38(2):291–300.
15. Yellon DM, Hausenloy DJ. Myocardial reperfusion injury. N Engl J Med. 2007;357(11):1121–35.
16. Braunwald E, Kloner RA. The stunned myocardium: prolonged, postischemic ventricular dysfunction. Circulation. 1982;66(6):1146–9.
17. Bolli R, Marban E. Molecular and cellular mechanisms of myocardial stunning. Physiol Rev. 1999;79(2):609–34.
18. Ito H. No-reflow phenomenon and prognosis in patients with acute myocardial infarction. Nat Clin Pract Cardiovasc Med. 2006;3(9):499–506.
19. Manning AS, Hearse DJ. Reperfusion-induced arrhythmias: mechanisms and prevention. J Mol Cell Cardiol. 1984;16(6):497–518.
20. Heusch G. Stunning–great paradigmatic, but little clinical importance. Basic Res Cardiol. 1998;93(3):164–6.
21. Murphy E, Steenbergen C. Mechanisms underlying acute protection from cardiac ischemia-reperfusion injury. Physiol Rev. 2008;88(2):581–609.
22. Skyschally A, Schulz R, Heusch G. Pathophysiology of myocardial infarction: protection by ischemic pre- and postconditioning. Herz. 2008;33(2):88–100.
23. Zhao ZQ, Corvera JS, Halkos ME, Kerendi F, Wang NP, Guyton RA, et al. Inhibition of myocardial injury by ischemic postconditioning during reperfusion: comparison with ischemic preconditioning. Am J Physiol Heart Circ Physiol. 2003;285(2):H579–88.
24. Duranski MR, Greer JJ, Dejam A, Jaganmohan S, Hogg N, Langston W, et al. Cytoprotective effects of nitrite during in vivo ischemia-reperfusion of the heart and liver. J Clin Invest. 2005;115(5):1232–40.
25. Hamid SA, Totzeck M, Drexhage C, Thompson I, Fowkes RC, Rassaf T, et al. Nitric oxide/cGMP signalling mediates the cardioprotective action of adrenomedullin in reperfused myocardium. Basic Res Cardiol. 2010;105:257–66.
26. Hendgen-Cotta UB, Merx MW, Shiva S, Schmitz J, Becher S, Klare JP, et al. Nitrite reductase activity of myoglobin regulates respiration and cellular viability in myocardial ischemia-reperfusion injury. Proc Natl Acad Sci U S A. 2008;105(29):10256–61.
27. Rassaf T, Flogel U, Drexhage C, Hendgen-Cotta U, Kelm M, Schrader J. Nitrite reductase function of deoxymyoglobin: oxygen sensor and regulator of cardiac energetics and function. Circ Res. 2007;100(12):1749–54.
28. Bryan NS, Calvert JW, Elrod JW, Gundewar S, Ji SY, Lefer DJ. Dietary nitrite supplementation protects against myocardial ischemia-reperfusion injury. Proc Natl Acad Sci U S A. 2007;104(48):19144–9.

29. Scalbert A, Williamson G. Dietary intake and bioavailability of polyphenols. J Nutr. 2000;130(8S Suppl):2073S–85.
30. Galeotti F, Barile E, Lanzotti V, Dolci M, Curir P. Quantification of major flavonoids in carnation tissues (*Dianthus caryophyllus*) as a tool for cultivar discrimination. Z Naturforsch C. 2008;63(3–4):161–8.
31. Ververidis F, Trantas E, Douglas C, Vollmer G, Kretzschmar G, Panopoulos N. Biotechnology of flavonoids and other phenylpropanoid-derived natural products. Part I: Chemical diversity, impacts on plant biology and human health. Biotechnol J. 2007;2(10):1214–34.
32. Williamson G, Manach C. Bioavailability and bioefficacy of polyphenols in humans. II. Review of 93 intervention studies. Am J Clin Nutr. 2005;81(1 Suppl):243S–55.
33. Donovan JL, Crespy V, Oliveira M, Cooper KA, Gibson BB, Williamson G. (+)-Catechin is more bioavailable than (−)-catechin: relevance to the bioavailability of catechin from cocoa. Free Radic Res. 2006;40(10):1029–34.
34. Donovan JL, Crespy V, Manach C, Morand C, Besson C, Scalbert A, et al. Catechin is metabolized by both the small intestine and liver of rats. J Nutr. 2001;131(6):1753–7.
35. Heiss C, Kelm M. Chocolate consumption, blood pressure, and cardiovascular risk. Eur Heart J. 2010;31(13):1554–6.
36. Agewall S, Wright S, Doughty RN, Whalley GA, Duxbury M, Sharpe N. Does a glass of red wine improve endothelial function? Eur Heart J. 2000;21(1):74–8.
37. Stein JH, Keevil JG, Wiebe DA, Aeschlimann S, Folts JD. Purple grape juice improves endothelial function and reduces the susceptibility of LDL cholesterol to oxidation in patients with coronary artery disease. Circulation. 1999;100(10):1050–5.
38. Duffy SJ, Keaney Jr JF, Holbrook M, Gokce N, Swerdloff PL, Frei B, et al. Short- and long-term black tea consumption reverses endothelial dysfunction in patients with coronary artery disease. Circulation. 2001;104(2):151–6.
39. Balzer J, Rassaf T, Heiss C, Kleinbongard P, Lauer T, Merx MW, et al. Sustained benefits in vascular function through flavanol-containing cocoa in mediated diabetic patients: a double-masked, randomized, controlled trial. J Am Coll Cardiol. 2008;51:2141–9.
40. Heiss C, Kleinbongard P, Dejam A, Perre S, Schroeter H, Sies H, et al. Acute consumption of flavanol-rich cocoa and the reversal of endothelial dysfunction in smokers. J Am Coll Cardiol. 2005;46(7):1276–83.
41. Heiss C, Dejam A, Kleinbongard P, Schewe T, Sies H, Kelm M. Vascular effects of cocoa rich in flavan-3-ols. JAMA. 2003;290(8):1030–1.
42. Heiss C, Finis D, Kleinbongard P, Hoffmann A, Rassaf T, Kelm M, et al. Sustained increase in flow-mediated dilation after daily intake of high-flavanol cocoa drink over 1 week. J Cardiovasc Pharmacol. 2007;49:74–80.
43. Schroeter H, Heiss C, Balzer J, Kleinbongard P, Keen CL, Hollenberg NK, et al. (−)-Epicatechin mediates beneficial effects of flavanol-rich cocoa on vascular function in humans. Proc Natl Acad Sci U S A. 2006;103(4):1024–9.
44. Yu C, Shin YG, Chow A, Li Y, Kosmeder JW, Lee YS, et al. Human, rat, and mouse metabolism of resveratrol. Pharm Res. 2002;19(12):1907–14.
45. Hattori R, Otani H, Maulik N, Das DK. Pharmacological preconditioning with resveratrol: role of nitric oxide. Am J Physiol Heart Circ Physiol. 2002;282(6):H1988–95.
46. Gurusamy N, Lekli I, Mukherjee S, Ray D, Ahsan MK, Gherghiceanu M, et al. Cardioprotection by resveratrol: a novel mechanism via autophagy involving the mTORC2 pathway. Cardiovasc Res. 2010;86:103–12.
47. Lekli I, Szabo G, Juhasz B, Das S, Das M, Varga E, et al. Protective mechanisms of resveratrol against ischemia-reperfusion-induced damage in hearts obtained from Zucker obese rats: the role of GLUT-4 and endothelin. Am J Physiol Heart Circ Physiol. 2008;294(2):H859–66.
48. Mokni M, Limam F, Elkahoui S, Amri M, Aouani E. Strong cardioprotective effect of resveratrol, a red wine polyphenol, on isolated rat hearts after ischemia/reperfusion injury. Arch Biochem Biophys. 2007;457(1):1–6.
49. Das S, Cordis GA, Maulik N, Das DK. Pharmacological preconditioning with resveratrol: role of CREB-dependent Bcl-2 signaling via adenosine A3 receptor activation. Am J Physiol Heart Circ Physiol. 2005;288(1):H328–35.
50. Wang S, Dusting GJ, May CN, Woodman OL. 3′, 4′-Dihydroxyflavonol reduces infarct size and injury associated with myocardial ischaemia and reperfusion in sheep. Br J Pharmacol. 2004;142(3):443–52.
51. Wang S, Fei K, Xu YW, Wang LX, Chen YQ. Dihydroxyflavonol reduces post-infarction left ventricular remodeling by preventing myocyte apoptosis in the non-infarcted zone in goats. Chin Med J (Engl). 2009;122(1):61–7.
52. Ramirez-Sanchez I, Maya L, Ceballos G, Villarreal F. (−)-Epicatechin activation of endothelial cell endothelial nitric oxide synthase, nitric oxide, and related signaling pathways. Hypertension. 2010;55:1398–405.
53. Steffen Y, Schewe T, Sies H. (−)-Epicatechin elevates nitric oxide in endothelial cells via inhibition of NADPH oxidase. Biochem Biophys Res Commun. 2007;359(3):828–33.
54. Gasper A, Hollands W, Casgrain A, Saha S, Teucher B, Dainty JR, Venema DP, et al. Consumption of both low and high (−)-epicatechin apple puree attenuates platelet reactivity and increases plasma concentrations of nitric oxide metabolites: a randomized controlled trial. Arch Biochem Biophys. 2014;559:29–37.
55. Cosby K, Partovi KS, Crawford JH, Patel RP, Reiter CD, Martyr S, et al. Nitrite reduction to nitric oxide by deoxy-hemoglobin vasodilates the human circulation. Nat Med. 2003;9(12):1498–505.
56. Gago B, Lundberg JO, Barbosa RM, Laranjinha J. Red wine-dependent reduction of nitrite to nitric oxide in the stomach. Free Radic Biol Med. 2007;43(9):1233–42.

57. Rocha BS, Gago B, Barbosa RM, Laranjinha J. Diffusion of nitric oxide through the gastric wall upon reduction of nitrite by red wine: physiological impact. Nitric Oxide. 2010;22(3):235–41.
58. Hirai M, Hotta Y, Ishikawa N, Wakida Y, Fukuzawa Y, Isobe F, et al. Protective effects of EGCg or GCg, a green tea catechin epimer, against postischemic myocardial dysfunction in guinea-pig hearts. Life Sci. 2007;80(11):1020–32.
59. Potenza MA, Marasciulo FL, Tarquinio M, Tiravanti E, Colantuono G, Federici A, et al. EGCG, a green tea poly-phenol, improves endothelial function and insulin sensitivity, reduces blood pressure, and protects against myocar-dial I/R injury in SHR. Am J Physiol Endocrinol Metab. 2007;292(5):E1378–87.
60. Chang WT, Shao ZH, Vanden Hoek TL, McEntee E, Mehendale SR, Li J, et al. Cardioprotective effects of grape seed proanthocyanidins, baicalin and wogonin: comparison between acute and chronic treatments. Am J Chin Med. 2006;34(2):363–5.
61. Florian T, Necas J, Bartosikova L, Klusakova J, Suchy V, Naggara EB, et al. Effects of prenylated isoflavones osajin and pomiferin in premedication on heart ischemia-reperfusion. Biomed Pap Med Fac Univ Palacky Olomouc Czech Repub. 2006;150(1):93–100.
62. Pataki T, Bak I, Kovacs P, Bagchi D, Das DK, Tosaki A. Grape seed proanthocyanidins improved cardiac recovery during reperfusion after ischemia in isolated rat hearts. Am J Clin Nutr. 2002;75(5):894–9.
63. Toufektsian MC, de Lorgeril M, Nagy N, Salen P, Donati MB, Giordano L, et al. Chronic dietary intake of plant-derived anthocyanins protects the rat heart against ischemia-reperfusion injury. J Nutr. 2008;138(4):747–52.
64. Yamazaki KG, Romero-Perez D, Barraza-Hidalgo M, Cruz M, Rivas M, Cortez-Gomez B, et al. Short- and long-term effects of (−)-epicatechin on myocardial ischemia-reperfusion injury. Am J Physiol Heart Circ Physiol. 2008;295(2):H761–7.
65. Suparto IH, Williams JK, Fox JL, Yusuf JT, Sajuthi D. Effects of hormone therapy and dietary soy on myocardial ischemia/reperfusion injury in ovariectomized atherosclerotic monkeys. Menopause. 2008;15(2):256–63.
66. Shao ZH, Wojcik KR, Dossumbekova A, Hsu C, Mehendale SR, Li CQ, et al. Grape seed proanthocyanidins protect cardiomyocytes from ischemia and reperfusion injury via Akt-NOS signaling. J Cell Biochem. 2009;107(4):697–705.
67. Wan LL, Xia J, Ye D, Liu J, Chen J, Wang G. Effects of quercetin on gene and protein expression of NOX and NOS after myocardial ischemia and reperfusion in rabbit. Cardiovasc Ther. 2009;27(1):28–33.
68. Imamura G, Bertelli AA, Bertelli A, Otani H, Maulik N, Das DK. Pharmacological preconditioning with resvera-trol: an insight with iNOS knockout mice. Am J Physiol Heart Circ Physiol. 2002;282(6):H1996–2003.
69. Graf BA, Ameho C, Dolnikowski GG, Milbury PE, Chen CY, Blumberg JB. Rat gastrointestinal tissues metabolize quercetin. J Nutr. 2006;136(1):39–44.
70. Tong H, Imahashi K, Steenbergen C, Murphy E. Phosphorylation of glycogen synthase kinase-3beta during preconditioning through a phosphatidylinositol-3-kinase-dependent pathway is cardioprotective. Circ Res. 2002;90(4):377–9.
71. Xi J, Wang H, Mueller RA, Norfleet EA, Xu Z. Mechanism for resveratrol-induced cardioprotection against reperfusion injury involves glycogen synthase kinase 3beta and mitochondrial permeability transition pore. Eur J Pharmacol. 2009;604(1–3):111–6.
72. Ling H, Lou Y. Total flavones from *Elsholtzia* blanda reduce infarct size during acute myocardial ischemia by inhibiting myocardial apoptosis in rats. J Ethnopharmacol. 2005;101(1–3):169–75.
73. Hao Y, Sun Y, Xu C, Jiang X, Sun H, Wu Q, et al. Improvement of contractile function in isolated cardiomyocytes from ischemia-reperfusion rats by ginkgolide B pretreatment. J Cardiovasc Pharmacol. 2009;54(1):3–9.
74. Ji ES, Yue H, Wu YM, He RR. Effects of phytoestrogen genistein on myocardial ischemia/reperfusion injury and apoptosis in rabbits. Acta Pharmacol Sin. 2004;25(3):306–12.
75. Sato M, Bagchi D, Tosaki A, Das DK. Grape seed proanthocyanidin reduces cardiomyocyte apoptosis by inhibiting ischemia/reperfusion-induced activation of JNK-1 and C-JUN. Free Radic Biol Med. 2001;31(6):729–37.
76. Fisher ND, Hughes M, Gerhard-Herman M, Hollenberg NK. Flavanol-rich cocoa induces nitric-oxide-dependent vasodilation in healthy humans. J Hypertens. 2003;21(12):2281–6.
77. Grassi D, Lippi C, Necozione S, Desideri G, Ferri C. Short-term administration of dark chocolate is followed by a significant increase in insulin sensitivity and a decrease in blood pressure in healthy persons. Am J Clin Nutr. 2005;81(3):611–4.
78. Engler MB, Engler MM, Chen CY, Malloy MJ, Browne A, Chiu EY, et al. Flavonoid-rich dark chocolate improves endothelial function and increases plasma epicatechin concentrations in healthy adults. J Am Coll Nutr. 2004;23(3):197–204.
79. Vlachopoulos C, Aznaouridis K, Alexopoulos N, Economou E, Andreadou I, Stefanadis C. Effect of dark chocolate on arterial function in healthy individuals. Am J Hypertens. 2005;18(6):785–91.
80. Grassi D, Mulder TP, Draijer R, Desideri G, Molhuizen HO, Ferri C. Black tea consumption dose-dependently improves flow-mediated dilation in healthy males. J Hypertens. 2009;27(4):774–81.
81. Faridi Z, Njike VY, Dutta S, Ali A, Katz DL. Acute dark chocolate and cocoa ingestion and endothelial function: a randomized controlled crossover trial. Am J Clin Nutr. 2008;88(1):58–63.

82. Horn P, Amabile N, Angeli FS, Sansone R, Stegemann B, Kelm M, Springer ML, et al. Dietary flavanol intervention lowers the levels of endothelial microparticles in coronary artery disease patients. Br J Nutr. 2014;111(7):1245–52.

83. Taubert D, Berkels R, Roesen R, Klaus W. Chocolate and blood pressure in elderly individuals with isolated systolic hypertension. JAMA. 2003;290(8):1029–30.

84. Taubert D, Roesen R, Lehmann C, Jung N, Schomig E. Effects of low habitual cocoa intake on blood pressure and bioactive nitric oxide: a randomized controlled trial. JAMA. 2007;298(1):49–60.

85. Grassi D, Necozione S, Lippi C, Croce G, Valeri L, Pasqualetti P, et al. Cocoa reduces blood pressure and insulin resistance and improves endothelium-dependent vasodilation in hypertensives. Hypertension. 2005;46(2):398–405.

86. Hodgson JM, Puddey IB, Burke V, Watts GF, Beilin LJ. Regular ingestion of black tea improves brachial artery vasodilator function. Clin Sci (Lond). 2002;102(2):195–201.

87. Grassi D, Desideri G, Necozione S, Ruggieri F, Blumberg JB, Stornello M, Ferri C. Protective effects of flavanol-rich dark chocolate on endothelial function and wave reflection during acute hyperglycemia. Hypertension. 2012;60(3):827–32.

88. West SG, McIntyre MD, Piotrowski MJ, Poupin N, Miller DL, Preston AG, Wagner P, Groves LF, Skulas-Ray AC. Effects of dark chocolate and cocoa consumption on endothelial function and arterial stiffness in overweight adults. Br J Nutr. 2014;111(4):653–61.

89. Flammer AJ, Sudano I, Wolfrum M, Thomas R, Enseleit F, Periat D, Kaiser P, et al. Cardiovascular effects of flavanol-rich chocolate in patients with heart failure. Eur Heart J. 2012;33(17):2172–80.

90. Heiss C, Lauer T, Dejam A, Kleinbongard P, Hamada S, Rassaf T, et al. Plasma nitroso compounds are decreased in patients with endothelial dysfunction. J Am Coll Cardiol. 2006;47:573–9.

Chapter 11
Nitrite and Nitrate in Human Breast Milk: Implications for Development

Pamela D. Berens and Nathan S. Bryan

Key Points

- Breast milk is nature's most perfect food with essential nutrients for the health and development of babies.
- Breast milk is enriched in both nitrite and nitrate, and the ratio of these anions changes as the composition of milk changes.
- Exposure rates of infants consuming colostrum from breast milk reach nearly 1 mg/kg, which exceeds the ADI for nitrite 14-fold.
- Breast milk contains higher concentrations of nitrite and nitrate than most commercially available formulas.
- The presence of nitrite and nitrate in breast milk suggests an essential physiological function for the growth and development of the neonate.

Keywords Nursing • Infants • Formula • Health disparity • Colostrum • Neonate

Introduction

Breast milk serves as the primary source of calories and nutrients in infants and children. Human breast milk is recommended as the exclusive food for the first 6 months of life, and continuing, along with safe, nutritious complementary foods, up through 2 years [1, 2]. Breast milk is nature's most perfect food. In fact, the U.S. Centers for Disease Control in 2010 noted that breast milk is widely acknowledged as the most complete form of nutrition for infants, with a range of benefits for infants' health, growth, immunity, and development. Breast milk is a unique nutritional source that cannot adequately be replaced by any other food, including infant formula. It remains superior to infant formula from the

P.D. Berens, M.D.
Department of Obstetrics, Gynecology, and Reproductive Sciences, McGovern Medical School at the University of Texas Health Science Center at Houston, 6431 Fannin St., MSB 3.116, Houston, TX 77030, USA
e-mail: Pamela.D.Berens@uth.tmc.edu

N.S. Bryan, Ph.D. (✉)
Department of Molecular and Human Genetics, Baylor College of Medicine, One Baylor Plaza, BCM225, Houston, TX 77030, USA
e-mail: Nathan.bryan@bcm.edu

N.S. Bryan, J. Loscalzo (eds.), *Nitrite and Nitrate in Human Health and Disease*, Nutrition and Health, DOI 10.1007/978-3-319-46189-2_11, © Springer International Publishing AG 2017

perspective of the overall health of both mother and child. Infants are fragile and susceptible to disease, partly because their immune system is not fully developed early on and they are dependent upon nutrients supplied by the nursing mother. They must be treated with special care and given adequate nourishment. Infant formulas have progressed over the years and are able to mimic a few of the nutritional components of breast milk, but formula cannot hope to duplicate the vast and constantly changing array of essential nutrients and immunological properties in human milk that vary over time. Furthermore, there may be many nutrients present in human breast milk that are still unrecognized and, therefore, lacking in infant formulas. Discovery of unrecognized nutrients may allow improvement in artificial formula developments and/or provide new insight into beneficial properties of certain molecules or nutrients. This chapter invokes a role for nitrite and nitrate as important nutrients in breast milk.

Human milk is known to confer significant nutritional and immunological benefits for the infant [3–5]. According to the CDC National Immunization Survey in 2012, 80 % of infants were breastfed in the early postpartum period, 51 % were at 6 months, and 29 % at 12 months of age. Among infants born in 2012, only 43 and 22 % were exclusively breastfed through 3 months and 6 months of age, respectively [6]. These data also suggest that approximately 19 % of breastfeeding infants will receive formula supplementation in the first 2 days of life. Nearly 29 % of breastfeeding infants have supplement by 3 months and 35 % have formula supplement before 6 months. These data suggest that, in addition to breast milk, other forms of milk or complementary foods may be significant nutrient sources for infants. Therefore, it is critical to understand the potential dietary exposures to nitrate and nitrite from these milk sources to assess potential health benefits and risks associated with their consumption, as well as to identify differences in these nutrients.

The relevance of dietary sources of nitrites and nitrates for human health is in its ascendancy. Historically, health risks due to nitrates in groundwater have been associated with a risk of methemoglobinemia (blue-baby syndrome) in children [7]. Infants less than six months of age may be exposed to excess nitrates in bacterially contaminated well water that reduces nitrate to nitrite [8]. Infants consuming excess nitrite experience blue baby syndrome due to nitrite-mediated oxidation of ferric (Fe^{2+}) iron in oxyhemoglobin that leads to hypoxia and cyanosis and the resulting blue color [9, 10]. Under physiological conditions, methemoglobin reduction is accomplished primarily by red cell-reduced nicotinamide adenine dinucleotide reductase so efficiently that there is <1 % of methemoglobin in the circulating blood of healthy adults. Infants younger than 6 months old are particularly susceptible to nitrate-induced methemoglobinemia because of their low stomach acid production, large numbers of nitrite-reducing bacteria, the relatively easy oxidation of fetal hemoglobin, and immaturity of the methemoglobin reductase system [11, 12]. As such, an American Academy of Pediatrics consensus panel concluded that all prenatal and well-infant visits should include questions about the home water supply; if the water source is a private well, the water should be tested for nitrate [7]. The panel concluded that infants fed commercially prepared infant foods are generally not at risk of nitrate poisoning, but that home-prepared infant foods from vegetables (e.g., spinach, beets, green beans, squash, carrots) should be avoided until infants are 3 months or older. Breastfed infants are not at risk of excessive nitrate exposure from mothers who ingest water with high nitrate content (up to 100 ppm nitrate nitrogen) as nitrate concentration is thought not to increase significantly in the breast milk under these circumstances [7].

It is noteworthy that the few human nitrate and nitrite exposure studies, including in children and adults, have not produced methemoglobinemia. Infants exposed to 175–700 mg nitrate per day did not experience methemoglobin levels above 7.5 %, suggesting that nitrate alone is not causative for methemoglobinemia [13]. A more recent randomized three-way crossover study exposed healthy volunteer adults to single doses of sodium nitrite that ranged from 150 to 190 mg per volunteer to 290–380 mg per volunteer [14]. Observed methemoglobin concentrations were 12.2 % for volunteers receiving the higher dose of nitrite and 4.5 % for those receiving the lower dose. Recent nitrite infusion studies of up to 110 µg/kg/min for 5 min induced methemoglobin concentrations of only 3.2 % in adults [15]. These data have led investigators to propose alternative explanations for the observed methemoglobinemia in infants, including gastroenteritis and associated iNOS-mediated production of NO induced by bacteria-contaminated water [16, 17]. Experts have questioned the veracity of the evidence supporting the hypothesis that nitrates and nitrites are toxic for healthy post-infant populations [9, 18, 19]. Although the previously held hypothesis of nitrite toxicity appears biologically plausible, conclusive scientific evidence supporting the hypothesis has not been documented [16, 17]. These findings highlight serious, but context-specific, risks associated with nitrite overexposure in infants. Potential health risks due to excessive nitrite and nitrate consumption in these specific population subgroups led to regulatory limits on permissible concentration of nitrate in drinking water (50 mg nitrate/L in the European Union, 44 mg/L in the United States) in accordance with World Health Organization recommendations first established in 1970 and reaffirmed in 2004 [20]. The US Environmental Protection Agency limits human exposure to inorganic nitrates to 0.10 mg/L (or 10 ppm nitrate nitrogen) and nitrites to 1 ppm nitrite nitrogen [21]. As a result, it is recommended that baby formulas and foods be made with nitrite and nitrate free water in order to reduce the risk of blue baby syndrome. However, there may be a low level of these anions that provide benefit to the infant without causing harm.

Breast Milk Confers Benefits to Infants Early in Life

What is indisputable is the established health benefits of breast milk. Human breast milk is the optimal source of infant nutrition and contains all the proteins, lipids, carbohydrates, and trace elements required for infant development [22]. The World Health Organization (WHO) recommends exclusive breastfeeding for the first 6 months of life and, thereafter, the introduction of appropriate solid foods with continuation of breastfeeding through 2 years and beyond. The American Academy of Pediatrics [23] recommends that mothers breastfeed for at least the first year of a child's life and continue until they both feel they are ready to stop. In the first 6 months, the baby should be nourished exclusively by breast milk. The AAP asserts that breast milk has the perfect balance of nutrients for the infant. It is enough sustenance along with vitamin D supplementation for approximately the first 6 months of life and should follow as the child's staple throughout the first year. Breast milk contains essential nutrients to foster growth and development and promote optimal health.

Studies have demonstrated a number of important health benefits to breastfeeding for both mother and infant. It provides a number of infant health advantages beginning at birth and continuing throughout a child's life. In fact, a large number of the health problems today's children face might be decreased, or even prevented, by breastfeeding the infant exclusively for at least the first 6 months of life [24]. The longer the mother breastfeeds, the more likely her child will acquire the health benefits of breastfeeding. Human breast milk acts to develop the immune system of infants such that those who are breastfed exclusively for six months have a more developed immune system than those who are not exclusively breastfed [25, 26]. Many studies show that breastfeeding strengthens the immune system. During nursing, the mother passes antibodies to the child, which helps the child resist diseases and helps improve the normal immune response to certain vaccines. Breastfeeding has also been shown to reduce the risks of the infant developing asthma and allergies [27] as well as childhood leukemia [28]. Cardiovascular disease risk is also reduced through the reduction of obesity, blood pressure, and cholesterol [5, 29].

Having been breastfed as an infant provides benefit later in life. Additional benefits may become apparent as the child grows older. Breastfed children are less likely to contract a number of diseases later in life, including type I diabetes mellitus, multiple sclerosis, heart disease, and cancer before the age of 15. In fact breastfed babies have been shown to have a small reduction in blood pressure later in life [30]. Breastfeeding during infancy is also associated with a reduction in risk of ischemic cardiovascular disease later in life [31]. What is it about breast milk that allows the health-promoting properties? The current theory is that immunoglobulins passed from mother to baby are responsible for these effects, but many of these benefits extend beyond the immune system. There is clear evidence now that nitrite and/or nitrate in breast milk may provide many of the aforementioned benefits of breastmilk.

Changes in Intestinal Microflora of the Developing Infant

Prior to uncomplicated birth, the fetus is sterile, and the first encounter with the microbial world begins during delivery. This exposure will vary depending on cesarean or vaginal delivery. During vaginal birth, bacterial colonization of a previously germ-free gut begins with organisms that are derived from the vagina, skin, and rectum of the mother. Vaginal microflora are some of the first colonizers in the gastrointestinal tract of newborns [32]. Naturally delivered babies experience a period of 2–3 days in which bacteria invading and reproducing within infant's gut are predominately aerobic bacteria such as *Enterobacteriaceae*, streptococci, and staphylococci [33]. These potentially pathogenic bacteria may not appear to be beneficial to the health of babies but research has shown that the metabolism of these early bacteria are believed to be positive factors in preparing for colonization by beneficial enteric bacteria such as bifidobacteria, lactobacilli, and other anaerobic bacteria that appear to reach the gut after 2–3 days [34]. After bacterial transmission from the mother to the baby during delivery, the mother provides a continual source of bacteria from oral and skin microbes through suckling, kissing, and caressing. These bacteria have been shown to be able to create a reduced environment favorable to anaerobic bacteria, which colonize the gut at the end of week 1 taking over from the aerobic bacteria.

An additional source of bacteria for breastfed neonates is mother's milk which contains up to 10^9 microbes/L in healthy mothers [35]. The most common bacteria colonized from this mode of transmission are staphylococci, streptococci, corynebacteria, lactobacilli, micrococci, propionibacteria, and bifidobacteria, as well as lactobacilli [36]. After the first 2 weeks of life, stable microflora is established and maintained. This microflora environment is dependent upon breast versus formula feeding. Supplementation of a breastfed baby with formula milk induces a rapid shift in the bacterial pattern of a breastfed baby [34]. This relationship between the microflora composition and the health of the baby is critical and thought to provide routes of metabolism and nutrients for the developing baby similar to vitamin K metabolism in adults. We will present data later in the chapter showing this bacterial symbiosis in the conversion of nitrate and nitrite to NO production.

Changes in Milk Characteristics over Time

The quality and characteristics of breast milk change over time. This change in composition is thought to be due partly to the changing gut microflora in the baby and the changing metabolic demands as the baby grows. Colostrum, the early milk made by mothers' breasts, is usually present after the fifth or sixth month of pregnancy. Colostrum has a yellow color, is thick in consistency, and is high in protein and low in fat and sugar. Colostrum is designed to meet a newborn's special needs. The protein content is three times higher than that of mature milk. It is rich in antibodies being passed from the mother. Breast milk will change and increase in quantity about 48–72 h after giving birth [37]. In addition to the macronutrients, vitamins, minerals, and water in human milk, specific components contribute to the immunocompetence of the infant, including secretory IgA, IgG, and IgM, lactoferrin, lysozyme, cytokines, and exosome-derived microRNAs [38–41]. These antibodies protect the baby and act as a natural laxative, helping the baby pass the first stool, called meconium. Once the baby is born, colostrum is present in small but gradually increasing amounts for the first 3 days to match the small size of the baby's stomach. Most babies do not need additional nutrition during this time. This precious colostrum should be given to the newborn as soon as possible after giving birth. It is recommended that mothers breastfeed at least 8–12 times per 24 h so the baby receives this valuable milk. Milk will typically increase in quantity by 72 h after giving birth, although it may take longer depending on when breastfeeding began and how frequently breastfeeding occurs.

Once a more mature milk supply is established, there is variation in the milk that occurs during feedings. When breastfeeding begins, the first milk the baby receives is called foremilk. It is thin and watery with a light blue tinge. Foremilk has a higher water content needed to satisfy the baby's thirst. Hind-milk is released after several minutes of nursing. It is similar in texture to cream and has the highest concentration of fat. The hind-milk has a relaxing effect on the baby. Hind-milk helps the baby feel satisfied and gain weight. Interestingly, several research studies have demonstrated relatively high concentrations of nitrite and nitrate in human breast milk. Ohta et al. found high concentrations of nitrite and nitrate (166–1246 μM) in the breast milk of Japanese mothers from days 1 to 8 [42]. We recently found that the ratio of nitrite and nitrate in breast milk changes as the composition of milk changes whereby colostrum contains the highest amount of nitrite (~10 μM) (Fig. 11.1) [43]. Iizuka et al. report a range of 19.3–82.4 μM with average concentration of nitrite in colostrum as 43.6 ± 5.3 μM and nitrate concentrations ranging from 86.6 to 278.7 μM with an average of 180.6 ± 63.1 μM [44]. In this particular study, when breast milk was ingested there was production of NO gas in the stomach that was absent in formula-fed babies, who have much lower concentrations of nitrite and nitrate [44]. The nitrite consumed by the infant enters the acidic environment of the stomach leading to the generation of NO in the gastric lumen. The effects of NO in adult stomachs are documented relative to gastric mucosal integrity and blood flow from the reduction of nitrate to nitrite to NO. Prior to the birthing process, the gastrointestinal tract of the infant is sterile, and it is rapidly colonized by bacteria originating from the mother and the environment, as discussed earlier [45, 46].

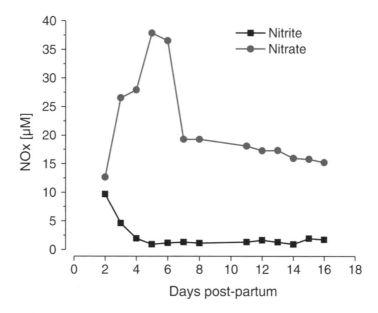

Fig. 11.1 A longitudinal study in the same nursing mothers from multiple collections over 16 days. Data are average of single donations from two nursing mothers for 16 days

Reduction of nitrate to nitrite requires the commensal bacteria that normally reside in the body since humans do not have a functional nitrate reductase enzyme. However in newborn infants, this pathway has not developed since they are born sterile without any nitrate-reducing bacteria. Therefore, breast milk high in nitrite relative to nitrate overcomes the natural deficiency early in life. This is the ratio found in colostrum [43]. At later stages of development, nitrate becomes the predominant anion when symbiosis exists with the colonized bacteria. This concept is supported by the data in Fig. 11.1 showing early post-partum colostrum enriched in nitrite. The nitrite concentration decreases concomitantly with nitrate increases during the time at which the gut becomes colonized by bacteria. There appears to be a complementary system whereby the nitrite in breast milk is supplied early on to overcome the inability to reduce nitrate. Colostrum is designed to meet a newborn's special needs. Our data suggest that the ratio of nitrite to nitrate also changes in order to meet the changing metabolic demands on the infant. Therefore, part of its role may be to supply a sufficient amount of nitrite in the early post-partum period until the gut microflora are established to then utilize nitrate as the primary substrate. Indeed, the pattern of nitrate and nitrite composition of human milk suggests that, as observed by others, these components also serve immunomodulatory and gastroprotective roles. In adults, the provision of nitrite, resulting from reduction by commensal lingual microbiota, is constant because of enterosalivary circulation of nitrate [47]. Interruption of the reduction of nitrate to nitrite has been shown to prevent a rise in plasma nitrite, hinder vasodilation with resulting increased blood pressure, and increased platelet aggregation [48]. Therefore, it is reasonable to surmise that nitrite must be supplied through human milk to the newborn in order to derive the resulting vascular, immunological, and gastroprotective benefits of NO to which it is metabolized.

Nitrite and Nitrate Anions in Breast Milk versus Formula and Cows' Milk

Currently, there are stringent regulations on the amount of nitrite and nitrate allowed in our food supply and drinking water. These regulations may not be applicable to human milk. There have been several studies showing high concentrations of nitrite and nitrate in breast milk. Cekmen et al. found extremely high concentrations of nitrite in breast milk of healthy mothers that was reduced in pre-eclampsia patients [49]. Total nitrite levels were 56.09 ± 11.18 μM vs. 82.20 ± 12.01 μM, $P < 0.05$, in colostrum of pre-eclamptics and controls, respectively. The level of total nitrite was 37.75 ± 12.10 μM vs. 53.28 ± 10.25 μM, $P < 0.05$, in 30th-day milk of these same patients [49], consistent with our data showing the highest concentrations of nitrite in colostrum that declines over time (Fig. 11.1). Breast milk from certain mothers contained the highest nitrite concentration of any food or beverage product tested. Human breast milk contains high concentrations of nitrate and nitrite in the early postpartum period, and the relative concentrations of the two anions change from colostrum to transition milk to mature milk. The concentrations are much higher than that found in commercial baby formulas. When comparing freshly expressed human breast milk at day 3 to Enfamil baby formula, human breast milk contained almost 20 times higher nitrite than that contained in Enfamil formula (Fig. 11.2). It is interesting to note that formula had much less nitrite but actually higher concentrations of nitrate than breast milk.

Table 11.1 contains nitrate and nitrite concentrations in a representative sample of colostrum, artificial formulas for infants and children, as well as samples of cow and soy milk. As shown below, colostrum contained the highest concentration of nitrite of any of the other milk products, and Silk Soy Vanilla milk contained the highest nitrate concentration. All of the pediatric formulas, with the exception of Bright Beginning Soy formula, had very low (barely detectable) levels of nitrite.

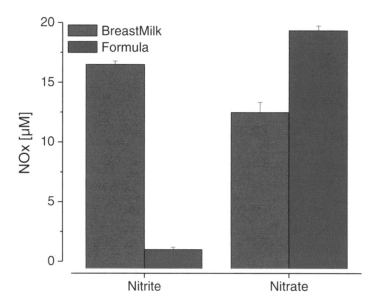

Fig. 11.2 Comparison of 3-day-old breast milk to Enfamil Baby Formula. Data are average ± SEM of $n = 3$ samples of each

Table 11.1 Nitrate and nitrite concentrations from infant and pediatric formula milk products

Milk product	Nitrite (mg/100 mL)	Nitrate (mg/100 mL)
Colostrum	0.08	0.19
Boost Kid Essentials Lactose Free®	0.007	0.26
Bright Beginnings Soy Pediatric®	0.03	0.16
Similac®	0.00005	0.05
Pregestimil®—Protein Hydrolysate	0.00005	0.12
Enfamil EnfaCare®	0.00005	0.10
Nestle Carnation Instant Breakfast Plus® (Lactose Free)	0.00005	0.08
TwoCal HN® with FOS-Milk Based Hydrolasate	0.00005	0.18
Nutramigen® AA Elemental	0.00005	0.02
Market Pantry 2% Cow's Milk	0.0002	0.23
Silk® Soy Vanilla Milk	0.0036	3.48
Silk® Soy Egg Nog	0.013	0.34

Source of Nitrite in Breast Milk

The origins of nitrate and nitrite in human milk are not known. Maternal nitrate and nitrite intakes are not reflected in nitrate and nitrite composition of human milk [7]. The data supporting this conclusion are sparse. One study published on this topic demonstrated that women who consumed water with a nitrate concentration of up to 100 mg/L did not produce milk with elevated nitrate levels [50]. Physiological production of nitrate and nitrite in tissues is dynamic; it is dependent on the mammalian nitrate reductase activity of enzymes, such as xanthine oxidoreductase and a host of other redundant systems [51–54], local conditions such as hypoxia [55, 56], and acidosis. Another hypothesis is that tissue production of nitrite and nitrate may be the result of bacterial nitrate reductase activity. Other work has demonstrated an important role for xanthine oxidoreductase (XO) in milk in the generation of NO_x in milk. It is worth noting that XO activity of human milk, while generally very much lower than that of cow's milk [57], is exceptionally high in the first few weeks postpartum [58]. This is precisely the period when antibacterial activity toward potential pathogens is required in the neonatal gut, it coincides with particularly high levels of nitrite [44], and it correlates with the highest levels of XO-dependent NO generation found in human milk [59]. In the early postpartum stage, nitrite may exert critical antimicrobial activity. The presence of high levels of XO in breast milk, along with the high nitrite levels and low oxygen tension, allows for the generation of peroxynitrite, a potent bactericidal agent, from NO [60]. These proposed antimicrobial actions of nitrite and peroxynitrite may help fight infections in the peripartum period. It is also biologically plausible that the high concentration of nitrite in colostrum may be the byproduct of NO oxidation resulting from its production during the initiation of lactation by nitric oxide synthase (NOS) [61, 62]. NOS-derived NO, using L-arginine as a substrate, may assist in the "let down" reflex at the initiation of lactation [61, 63]. After initiation of lactation, there may be less NO required, and, hence, less nitrite produced by NO oxidation, to begin the flow of milk from the breast tissue. This may be one potential explanation as to why nitrite concentrations drop significantly in transitional and mature milk.

Levels of Exposure of Nitrite and Nitrate from Breast Milk

The published data on nitrite and nitrate content of human breast milk reveal that substantial exposure to the anions occurs in newborn infants. This becomes interesting in terms of levels of exposure based on nitrite ingestion relative to body weight in the infant (over 1 mg/kg). The Joint Food and Agricultural Organization/World Health Organization has set the Acceptable Daily Intake (ADI) for the nitrate ion at 3.7 mg/kg body weight and for the nitrite ion at 0.06 mg/kg body weight [64]. Normal daily infant breastmilk intake is 2–3 oz per pound per day or 200 mL/kg. That translates into 14–21 oz or 400–630 mL of milk for a typical 7 lb (3 kg) infant or more simply ~200 mL/kg. Taking an average of 50 µmoles/L concentration of nitrite or 2300 µg/L, total daily nitrite exposure to nursing infants is roughly 1.2 mg/kg or 20 times higher than the acceptable daily intake for nitrite. Among the milk samples analyzed from Table 11.1, nitrate intakes were highest from pediatric formulas (Boost Kids Essential® Lactose Free and Bright Beginnings Soy Pediatric®) and lowest from infant formulas (Similac® and Pregestimil®). The highest nitrate exposure of the milk samples tested, relative to the WHO ADI standard, is from Silk® Soy Vanilla, which could result in intakes of 104 % of this standard. Boost Kids Essential® Lactose Free intake could result in 83 % of the WHO ADI for nitrate. Regarding potential nitrite exposures, Bright Beginnings Soy Pediatric® intake could result in 383 % and Boost Kids Essential® Lactose Free intake could result in 83 % of the WHO ADI standard. Human milk colostrum samples had the highest nitrite concentration of any milk product tested. The small amount produced by the breast and, therefore, consumed by the infant still translates to 42 % of WHO ADI at 100 mL intake.

The ability to exceed WHO ADI limits with usual intake levels of single foods, such as breast milk, spinach [65], or a dessicated vegetable supplement [66], indicates that these regulatory limits may not be applicable to all situations, such as food sources of nitrate and nitrite, for children and adults. This latter suggestion is warranted based on consideration of intake estimates of nitrite and nitrate from other food sources. We have found that a DASH diet pattern with high-nitrate food choices exceeds the WHO ADI for nitrate by 550 % for a hypothetical 60 kg adult. If this hypothetical adult were to consume, in addition to the high-nitrate DASH diet, three cups of the soy milk analyzed herein, the concentration of dietary nitrate from these foods (1248 mg total or 20.8 mg/kg) would exceed WHO ADI levels by 562 % for a 60 kg person. We have recently presented compelling data for consideration of nitrite and nitrate as indispensable nutrients [67].

Implications for Health

The nitrate and nitrite content of milk products demonstrate that these foods are significant contributors to total dietary nitrate and nitrite intakes. The data presented in this chapter suggest that human milk provides a dietary source for nitrite prior to the establishment of lingual and gastrointestinal microbiota. Once the microbiota are established, these commensal organisms are capable of reducing dietary nitrate, via enterosalivary circulation, to nitrite, and support gastrointestinal, immune, and cardiovascular health. The significant concentration of nitrate and nitrite in bovine milk demonstrated here indicates that these conclusions may be applicable across mammalian species. When we compare freshly express human breast milk at day 3 to Enfamil baby formula, human breast milk contains almost 20 times more nitrite. Studies have demonstrated a number of important health benefits to breastfeeding [5]. It provides a number of health advantages beginning at birth and continuing throughout a child's life.

In addition to the established cardiovascular and immunological benefits of NO derived from breast milk nitrite, there is growing evidence related to its anticancer activity, as well. A review of the evidence for an association between breastfed infants and certain cancers suggests that children who are never breastfed or are breastfed short-term have a higher risk than those breastfed for ≥6 months of developing Hodgkin's disease [68], but not non-Hodgkin's lymphoma or acute lymphoblastic leukemia [69]. Human milk contains an extensive array of antimicrobial activity and appears to stimulate early development of the infant immune system. Artificially fed infants negotiate exposure to infectious agents without the benefits of this immunological armamentarium and do not do as well as breastfed infants in resisting infection. Thus, human milk may make the breastfed infant better able to negotiate future carcinogenic insults by modulating the interaction between infectious agents and the developing infant immune system or by directly affecting the long-term development of the infant immune system. Part of this response may be mediated by the nitrite and nitrate content of breast milk.

Necrotizing enterocolitis (NEC) is the leading cause of death from gastrointestinal disease in the premature infant, and is gradually increasing in frequency [70]. The defining pathological feature of NEC is the presence of patchy areas of ischemia and necrosis of the small and large intestine [71]. Although prematurity is the leading risk factor for NEC development, breast milk administration has been identified as the most important protective strategy [72]. Yazji et al. recently demonstrated that administration of sodium nitrate and nitrite within the formula significantly reduced NEC severity in mouse models of NEC. The reduction in NEC severity after the administration of nitrite correlated with a marked improvement in the degree of perfusion of the intestinal microcirculation, in fact to an extent comparable to that of breastfed mice [73]. These findings demonstrate that nitrite and nitrate in breast milk or supplemented in formula can restore intestinal perfusion and attenuate the severity of NEC, and also provide mechanistic insights to explain the protective effects of breast milk in this devastating disease.

Future research is required to establish definitively the role of nitrite and nitrate in formula-fed versus breastfed infants. The dietary and physiological determinants of milk nitrate and nitrite concentrations remain to be established. However, the presence of nitrate and nitrite in human milk provides indirect evidence for a physiological benefit of dietary nitrite for the protection of the gastrointestinal tract in the neonate prior to the establishment of commensal bacteria in the mouth and gut. The temporal relationship between the provision of nitrite in human milk and the development of commensal microbiota capable of reducing dietary nitrate to nitrite supports an hypothesis that milk nitrite may supply this component in the immediate term after birth. One conclusion that can be made from these exposure estimates is that humans are adapted to receive dietary nitrite and nitrate from birth and, therefore, they may not pose a significant risk at levels naturally found in certain foods. In fact, the absence of an essential nutrient, namely nitrite in baby formulas, may be involved in many of the health disparities in formula-fed babies, including necrotizing enterocolitis, infections, poor nutrient absorption, and even increased health risks later in life. If claims of nitrite and nitrate in promoting carcinogenesis were warranted, we would expect a higher incidence of cancer in breastfed infants compared to formula-fed infants, a result which has not been reported. In fact, there is a recognized and large disparity in the health of breastfed versus formula-fed babies favoring breastfeeding. The current theory holds that this is due to immunoglobulins transferred from mother to baby which are responsible for the benefits of breast milk. A case can be made for nitrite as an essential molecule in the development and immune function of infants, as well. Nature has devised a perfect system to nourish and foster the growth and development of nursing babies. Nitrite appears to be one of those indispensable nutrients [67].

References

1. Heinig MJ. The American Academy of Pediatrics recommendations on breastfeeding and the use of human milk. J Hum Lact. 1998;14(1):2–3.
2. Gartner LM, Morton J, Lawrence RA, Naylor AJ, O'Hare D, Schanler RJ, et al. Breastfeeding and the use of human milk. Pediatrics. 2005;115(2):496–506.
3. Hoddinott P, Tappin D, Wright C. Breast feeding. BMJ. 2008;336(7649):881–7.
4. James DC, Lessen R. Position of the American Dietetic Association: promoting and supporting breastfeeding. J Am Diet Assoc. 2009;109(11):1926–42.
5. Ip S, Chung M, Raman G, Chew P, Magula N, DeVine D, et al. Breastfeeding and maternal and infant health outcomes in developed countries. Evidence Report/Technology Assessment. Rockville, MD: Tufts-New England Medical Center Evidence-Based Practice Center, under Contract No. 290-02-00222007 Contract No.: 07-E007.
6. Centers for Disease Control and Prevention (CDC). Breastfeeding trends and updated national health objectives for exclusive breastfeeding--United States, birth years 2000-2004. MMWR Morb Mortal Wkly Rep. 2007;56(30):760–3.
7. Greer FR, Shannon M. Infant methemoglobinemia: the role of dietary nitrate in food and water. Pediatrics. 2005;116(3):784–6.
8. Johnson CJ, Kross BC. Continuing importance of nitrate contamination of groundwater and wells in rural areas. Am J Ind Med. 1990;18(4):449–56.
9. McKnight GM, Duncan CW, Leifert C, Golden MH. Dietary nitrate in man: friend or foe? Br J Nutr. 1999;81(5):349–58.
10. Fan AM, Steinberg VE. Health implications of nitrate and nitrite in drinking water: an update on methemoglobinemia occurrence and reproductive and developmental toxicity. Regul Toxicol Pharmacol. 1996;23(1 Pt 1):35–43.
11. Dusdieker LB, Getchell JP, Liarakos TM, Hausler WJ, Dungy CI. Nitrate in baby foods. Adding to the nitrate mosaic. Arch Pediatr Adolesc Med. 1994;148(5):490–4.
12. Kross BC, Ayebo AD, Fuortes LJ. Methemoglobinemia: nitrate toxicity in rural America. Am Fam Physician. 1992;46(1):183–8.
13. Cornblath MaH AF. Methhemoglobinaemia in young infants. J Pediatr. 1948;33:421–5.
14. Kortboyer J, Olling, M, Zeilmaker, MJ The oral bioavailability of sodium nitrite investigated in healthy adult volunteers. Bilthoven: National Institute of Public Health and the Environment; 1997.
15. Dejam A, Hunter CJ, Tremonti C, Pluta RM, Hon YY, Grimes G, et al. Nitrite infusion in humans and nonhuman primates: endocrine effects, pharmacokinetics, and tolerance formation. Circulation. 2007;116(16):1821–31.
16. L'Hirondel JL, Avery AA, Addiscott T. Dietary nitrate: where is the risk? Environ Health Perspect. 2006;114(8):A458–9; author reply A9-61.
17. Powlson DS, Addiscott TM, Benjamin N, Cassman KG, de Kok TM, van Grinsven H, et al. When does nitrate become a risk for humans? J Environ Qual. 2008;37(2):291–5.
18. L'Hirondel JL. Nitrate and man: Toxic, Harmless or Beneficial? Wallingford, UK: CABI Publishing; 2001.
19. Authority EFS. Nitrate in vegetables: scientific opinion of the panel on contaminants in the food chain. EFSA J. 2008;289:1–79.
20. World Health Organization. Recommendations; nitrate and nitrite. In: Guidelines for drinking water quality. 3rd edn. Geneva: WHO; 2004.
21. National Primary Drinking Water Regulations: Final Rule, 40., CFR parts 141–143 (1991).
22. Newton ER. Breastmilk: the gold standard. Clin Obstet Gynecol. 2004;47(3):632–42.
23. Albanes D, Heinonen OP, Taylor PR, Virtamo J, Edwards BK, Rautalahti M, et al. Alpha-Tocopherol and beta-carotene supplements and lung cancer incidence in the alpha-tocopherol, beta-carotene cancer prevention study: effects of base-line characteristics and study compliance. J Natl Cancer Inst. 1996;88(21):1560–70.

24. Bartick M, Reinhold A. The burden of suboptimal breastfeeding in the United States: a pediatric cost analysis. Pediatrics. 2010;125(5):e1048–56.
25. Hawkes JS, Neumann MA, Gibson RA. The effect of breast feeding on lymphocyte subpopulations in healthy term infants at 6 months of age. Pediatr Res. 1999;45(5 Pt 1):648–51.
26. Hawkes JS, Gibson RA. Lymphocyte subpopulations in breast-fed and formula-fed infants at six months of age. Adv Exp Med Biol. 2001;501:497–504.
27. Fulhan J, Collier S, Duggan C. Update on pediatric nutrition: breastfeeding, infant nutrition, and growth. Curr Opin Pediatr. 2003;15(3):323–32.
28. Shu XO, Linet MS, Steinbuch M, Wen WQ, Buckley JD, Neglia JP, et al. Breast-feeding and risk of childhood acute leukemia. J Natl Cancer Inst. 1999;91(20):1765–72.
29. Singhal A. Early nutrition and long-term cardiovascular health. Nutr Rev. 2006;64(5 Pt 2):S44–9; discussion S72–91.
30. Martin RM, Gunnell D, Smith GD. Breastfeeding in infancy and blood pressure in later life: systematic review and meta-analysis. Am J Epidemiol. 2005;161(1):15–26.
31. Rich-Edwards JW, Stampfer MJ, Manson JE, Rosner B, Hu FB, Michels KB, et al. Breastfeeding during infancy and the risk of cardiovascular disease in adulthood. Epidemiology. 2004;15(5):550–6.
32. Brook I, Barrett CT, Brinkman 3rd CR, Martin WJ, Finegold SM. Aerobic and anaerobic bacterial flora of the maternal cervix and newborn gastric fluid and conjunctiva: a prospective study. Pediatrics. 1979;63(3):451–5.
33. Morelli L. Postnatal development of intestinal microflora as influenced by infant nutrition. J Nutr. 2008;138(9):1791S–5.
34. Favier CF, Vaughan EE, De Vos WM, Akkermans AD. Molecular monitoring of succession of bacterial communities in human neonates. Appl Environ Microbiol. 2002;68(1):219–26.
35. West PA, Hewitt JH, Murphy OM. Influence of methods of collection and storage on the bacteriology of human milk. J Appl Bacteriol. 1979;46(2):269–77.
36. Martin R, Langa S, Reviriego C, Jiminez E, Marin ML, Xaus J, et al. Human milk is a source of lactic acid bacteria for the infant gut. J Pediatr. 2003;143(6):754–8.
37. Neville MC. Physiology of lactation. Clin Perinatol. 1999;26(2):251–79. v.
38. Hamosh M. Bioactive factors in human milk. Pediatr Clin North Am. 2001;48(1):69–86.
39. Garofalo R. Cytokines in human milk. J Pediatr. 2010;156(2 Suppl):S36–40.
40. Kosaka N, Izumi H, Sekine K, Ochiya T. microRNA as a new immune-regulatory agent in breast milk. Silence. 2010;1(1):7.
41. Admyre C, Johansson SM, Qazi KR, Filen JJ, Lahesmaa R, Norman M, et al. Exosomes with immune modulatory features are present in human breast milk. J Immunol. 2007;179(3):1969–78.
42. Ohta N, Tsukahara H, Ohshima Y, Nishii M, Ogawa Y, Sekine K, et al. Nitric oxide metabolites and adrenomedullin in human breast milk. Early Hum Dev. 2004;78(1):61–5.
43. Hord NG, Ghannam JS, Garg HK, Berens PD, Bryan NS. Nitrate and nitrite content of human, formula, bovine, and soy milks: implications for dietary nitrite and nitrate recommendations. Breastfeed Med. 2011;6(6):393–9.
44. Iizuka T, Sasaki M, Oishi K, Uemura S, Koike M, Shinozaki M. Non-enzymatic nitric oxide generation in the stomachs of breastfed neonates. Acta Paediatr. 1999;88(10):1053–5.
45. Fujita K, Murono K. Nosocomial acquisition of Escherichia coli by infants delivered in hospitals. J Hosp Infect. 1996;32(4):277–81.
46. Mandar R, Mikelsaar M. Transmission of mother's microflora to the newborn at birth. Biol Neonate. 1996;69(1):30–5.
47. Petersson J, Carlstrom M, Schreiber O, Phillipson M, Christoffersson G, Jagare A, et al. Gastroprotective and blood pressure lowering effects of dietary nitrate are abolished by an antiseptic mouthwash. Free Radic Biol Med. 2009;46(8):1068–75.
48. Webb AJ, Patel N, Loukogeorgakis S, Okorie M, Aboud Z, Misra S, et al. Acute blood pressure lowering, vasoprotective, and antiplatelet properties of dietary nitrate via bioconversion to nitrite. Hypertension. 2008;51(3):784–90.
49. Cekmen MB, Balat A, Balat O, Aksoy F, Yurekli M, Erbagci AB, et al. Decreased adrenomedullin and total nitrite levels in breast milk of preeclamptic women. Clin Biochem. 2004;37(2):146–8.
50. Dusdieker LB, Stumbo PJ, Kross BC, Dungy CI. Does increased nitrate ingestion elevate nitrate levels in human milk? Arch Pediatr Adolesc Med. 1996;150(3):311–4.
51. Jansson EA, Huang L, Malkey R, Govoni M, Nihlen C, Olsson A, et al. A mammalian functional nitrate reductase that regulates nitrite and nitric oxide homeostasis. Nat Chem Biol. 2008;4(7):411–7.
52. Dalsgaard T, Simonsen U, Fago A. Nitrite-dependent vasodilation is facilitated by hypoxia and is independent of known NO-generating nitrite reductase activities. Am J Physiol Heart Circ Physiol. 2007;292(6):H3072–8.
53. Li H, Cui H, Kundu TK, Alzawahra W, Zweier JL. Nitric oxide production from nitrite occurs primarily in tissues not in the blood: critical role of xanthine oxidase and aldehyde oxidase. J Biol Chem. 2008;283(26):17855–63.
54. Feelisch M, Fernandez BO, Bryan NS, Garcia-Saura MF, Bauer S, Whitlock DR, et al. Tissue processing of nitrite in hypoxia: an intricate interplay of nitric oxide-generating and -scavenging systems. J Biol Chem. 2008; 283(49):33927–34.

55. Zweier JL, Wang P, Samouilov A, Kuppusamy P. Enzyme-independent formation of nitric oxide in biological tissues. Nat Med. 1995;1(8):804–9.
56. Bryan NS, Calvert JW, Elrod JW, Gundewar S, Ji SY, Lefer DJ. Dietary nitrite supplementation protects against myocardial ischemia-reperfusion injury. Proc Natl Acad Sci U S A. 2007;104(48):19144–9.
57. Abadeh S, Killacky J, Benboubetra M, Harrison R. Purification and partial characterization of xanthine oxidase from human milk. Biochim Biophys Acta. 1992;1117(1):25–32.
58. Brown AM, Benboubetra M, Ellison M, Powell D, Reckless JD, Harrison R. Molecular activation-deactivation of xanthine oxidase in human milk. Biochim Biophys Acta. 1995;1245(2):248–54.
59. Stevens CR, Millar TM, Clinch JG, Kanczler JM, Bodamyali T, Blake DR. Antibacterial properties of xanthine oxidase in human milk. Lancet. 2000;356(9232):829–30.
60. Hancock JT, Salisbury V, Ovejero-Boglione MC, Cherry R, Hoare C, Eisenthal R, et al. Antimicrobial properties of milk: dependence on presence of xanthine oxidase and nitrite. Antimicrob Agents Chemother. 2002;46(10):3308–10.
61. Iizuka T, Sasaki M, Oishi K, Uemura S, Koike M, Minatogawa Y. Nitric oxide may trigger lactation in humans. J Pediatr. 1997;131(6):839–43.
62. Iizuka T, Sasaki M, Oishi K, Uemura S, Koike M. The presence of nitric oxide synthase in the mammary glands of lactating rats. Pediatr Res. 1998;44(2):197–200.
63. Lacasse P, Farr VC, Davis SR, Prosser CG. Local secretion of nitric oxide and the control of mammary blood flow. J Dairy Sci. 1996;79(8):1369–74.
64. Authority EFS. Nitrate in vegetables: scientific opinion of the panel on contaminants in the food chain. EFSA J. 2008;689:1–79.
65. Lundberg JO, Feelisch M, Bjorne H, Jansson EA, Weitzberg E. Cardioprotective effects of vegetables: is nitrate the answer? Nitric Oxide. 2006;15(4):359–62.
66. Hord NG, Tang Y, Bryan NS. Food sources of nitrates and nitrites: the physiologic context for potential health benefits. Am J Clin Nutr. 2009;90(1):1–10.
67. Bryan NS, Ivy JL. Inorganic nitrite and nitrate: evidence to support consideration as dietary nutrients. Nutr Res. 2015;35(8):643–54.
68. Havens JR, Strathdee SA, Fuller CM, Ikeda R, Friedman SR, Des Jarlais DC, et al. Correlates of attempted suicide among young injection drug users in a multi-site cohort. Drug Alcohol Depend. 2004;75(3):261–9.
69. Davis MK. Review of the evidence for an association between infant feeding and childhood cancer. Int J Cancer Suppl. 1998;11:29–33.
70. Neu J, Walker WA. Necrotizing enterocolitis. N Engl J Med. 2011;364(3):255–64.
71. Dominguez KM, Moss RL. Necrotizing enterocolitis. Clin Perinatol. 2012;39(2):387–401.
72. Gephart SM, McGrath JM, Effken JA, Halpern MD. Necrotizing enterocolitis risk: state of the science. Adv Neonatal Care. 2012;12(2):77–87; quiz 8–9.
73. Yazji I, Sodhi CP, Lee EK, Good M, Egan CE, Afrazi A, et al. Endothelial TLR4 activation impairs intestinal microcirculatory perfusion in necrotizing enterocolitis via eNOS-NO-nitrite signaling. Proc Natl Acad Sci U S A. 2013;110(23):9451–6.

Chapter 12
Regulation of Dietary Nitrate and Nitrite: Balancing Essential Physiological Roles with Potential Health Risks

Norman G. Hord and Melissa N. Conley

Key Points

- US and European Union regulatory limits on nitrates in drinking water are necessary to limit environmental pollution known as eutrophication.
- Health concerns of excessive nitrate and nitrite consumption have driven regulatory actions due to perceived risk of methemoglobinemia in infants and gastrointestinal cancer risk in adults.
- The World Health Organization's Acceptable Daily Intake recommendations for nitrate can be exceeded by normal daily intakes of single foods and recommended dietary patterns, such as the DASH diet.
- Inconsistent positions on the health risks and benefits of foods containing nitrates and nitrites by the International Agency for Research on Cancer (IARC) and the European Food Safety Authority (EFSA) may contribute to confusion for consumers; regulators must take the opportunity to clarify and expand upon these positions in order to provide coherent dietary guidance.
- The established vasoprotective, blood pressure lowering, and antiplatelet aggregation properties of nitrite alone, or of nitrite originating from dietary nitrate, requires a new regulatory paradigm that incorporates the concepts of physiological deficiency, sufficiency, and excess.
- There is a need to engage an independent panel of experts from academia, industry, and governmental and non-governmental sectors to undertake the first comprehensive, systematic review of the potential health risks and benefits of food sources of nitrates and nitrites.
- U.S. Institute of Medicine's Dietary Reference Intake paradigm may be a useful guide to the development of coherent dietary nitrate and nitrite intake recommendations.

Keywords Accepted daily intake • Dietary recommendations • World Health Organization • Food safety • Risk benefit evaluation

Introduction

Current regulatory limits on nitrate concentrations in drinking water are necessary to reduce the risk of eutrophication; dietary intake limits for nitrate and nitrite based on potential risk of carcinogenicity and methemoglobinemia have more equivocal scientific support. The World Health Organization

N.G. Hord, Ph.D., M.P.H., R.D. (✉) • M.N. Conley, M.S.
School of Biological and Population Health Sciences, Oregon State University, Corvallis, OR, USA
e-mail: norman.hord@oregonstate.edu; melisanicole@gmail.com

N.S. Bryan, J. Loscalzo (eds.), *Nitrite and Nitrate in Human Health and Disease*, Nutrition and Health, DOI 10.1007/978-3-319-46189-2_12, © Springer International Publishing AG 2017

intake limits for nitrate are exceeded by normal daily intakes of single foods, such as soya milk and spinach, as well as recommended dietary patterns such as the Dietary Approaches to Stop Hypertension diet. Inconsistent regulatory positions on dietary exposures to nitrate and nitrite from processed meats and vegetables may be a source of public concern about the quality of dietary recommendations from health regulators. There is a great necessity to undertake a multidisciplinary, independent, and systematic review of the potential health benefits and risks of dietary nitrates and nitrites. Regulatory bodies must consider all available data on the physiological roles of nitrate and nitrite in order to derive rational bases for dietary recommendations. The development of a more sophisticated regulatory paradigm that acknowledges physiological states of nitrate and nitrite deficiency, adequacy and excess, as exemplified by the Institute of Medicine's Dietary Reference Intake paradigm, may be needed. This paradigm should also consider the relevant metabolic interactions between food sources of nitrates and nitrites in specific physiological and pathophysiological contexts. These efforts could improve dietary guidance regarding nitrate and nitrite intakes that result in public health benefits.

> Everything should be made as simple as possible, but not simpler.
> —Albert Einstein's comment on the philosophy of parsimony of William of Occam, English philosopher and Franciscan friar (c. 1285–1349).

Nitrate and Nitrite in Biology

Nitrate enters the food chain through plant foods via the action of lightning and soil bacteria. In addition to being a required nutrient for plants, it is an approved food additive [1]. Nitrate from vegetables is the major dietary source of this chemical while surface and ground water is a minor contributor. Nitrate intakes from vegetables are determined by the type of vegetable consumed, the levels of nitrate in the vegetables (including the nitrate content of fertilizer), the amount of vegetables consumed, and the level of nitrate in the water supply [2]. Nitrate concentration is the highest in the leaves whereas lower concentrations occur in seeds or tubers [3]. Nitrite exposure is mainly from endogenous nitrate conversion via a process known as enterosalivary circulation [4, 5]. Endogenous reduction of nitrate to nitrite is a source of nitric oxide (NO) and related metabolites (referred to as NO_x) in tissues and, in hypoxia, the vasculature [6, 7].

Regulatory Limits on Nitrate in Drinking Water

Environmental contamination with excess nitrate from fertilizer use is a persistent and growing problem. Nitrate and nitrite are naturally occurring ions that serve as nutrients for plants via fixation by soil bacteria. The enhancement of nitrogen fixation by the provision of nitrate-containing fertilizers has surpassed the amount that occurs naturally [8]. The resulting contamination of ground and surface water with excess nitrate by these agricultural practices is a global concern. Regulatory positions that address this concern are exemplified by the European Union's Nitrate Directive. The Nitrates Directive aims to protect water quality across Europe by preventing nitrates from agricultural sources polluting ground and surface waters and by promoting the use of good farming practices [9]. This type of pollution is referred to as eutrophication and is characterized by the excessive development of certain types of algae disturbs the aquatic ecosystems and becomes a health risk for animals and humans. The primary cause of eutrophication is an excessive concentration of plant nutrients, including nitrates, originating from agricultural practices or sewage treatment. While it is beyond the scope of this chapter to discuss these environmental issues, the contribution of contaminated drinking water sources to potential health effects of dietary nitrate and nitrite is highly relevant.

The permissible concentration of nitrate in drinking water is 50 mg nitrate/L in the European Union and 44 mg/L in the United States in accordance with World Health Organization recommendations first established in 1970 and reaffirmed in 2004 [10]. The harmony of US and European Union nitrate and nitrite regulations is betrayed by a growing body of evidence that ascribes essential functions in vascular and tissue homeostasis and immune function to these dietary constituents. Whereas accidental toxic exposures of nitrates and nitrites have occurred [11, 12], the health risks due to excessive nitrate and nitrite consumption, such as methemoglobinemia, appear only in specific subgroups of the population.

Experts have questioned the veracity of the evidence supporting the hypothesis that nitrates and nitrites are toxic for healthy adolescent and adult populations due to their ability to cause methemoglobinemia [13–16]. Many scientists now interpret the available data as evidence that the condition is caused by bacterial infection-induced enteritis rather than nitrate [15, 16]. Thus, it appears that the biologically plausible hypothesis of nitrite toxicity with regard to methemoglobinemia has essentially transformed a plausible hypothesis into sacrosanct dogma [13], despite the lack of proof [14, 16].

The excessive concentration of nitrate in drinking water, typically a minor contributor to dietary nitrate and nitrite concentrations, must be considered a serious health concern, particularly for infants [17]. Even so, the Society of Agricultural and Biological Engineers has, in parallel to those made this volume for nitrate and nitrite exposures from foods, called for a more rational approach to setting exposure limits on nitrogen-containing effluents in wastewater treatment [18]. To wit,

> Considering that definitive evidence of nitrate health risks is conspicuously lacking, a more rational approach to setting effluent limits for waste treatment systems is needed, one that considers costs/benefits and recognizes factors that act to limit nitrogen buildup in groundwater. Such factors include nitrogen removal by soil microorganisms, and aquifer hydrogeology [18].

Regulators must integrate data from potential exposures from drinking water with those of food sources of nitrate and nitrite. The balance of this chapter will address the potential health concerns regarding dietary nitrate and nitrite from the context of food sources and the physiological contexts of the actions of their metabolic products.

Regulatory Limits on Dietary Nitrate and Nitrite Intakes

The US Environmental Protection Agency limits human exposure to inorganic nitrates to 0.10 mg/L (or 10 ppm nitrate nitrogen) and nitrites to 1 ppm nitrite nitrogen [19]. The Joint Food and Agricultural Organization/World Health Organization has set the acceptable daily intake (ADI) for the nitrate ion at 3.7 mg/kg body wt and for the nitrite ion at 0.06 mg/kg body wt [1]. Likewise, Environmental Protection Agency has set a Reference Dose for nitrate of 1.6 mg nitrate nitrogen/kg body wt/day (equivalent to 7.0 mg nitrate ion/kg body wt/day).

Dietary Intake Estimates of Nitrate and Nitrite

The mean intake estimates for nitrate and nitrite from food in the United States and Europe vary from ~40–100 to 31–185 mg/day, respectively [20, 21]. Nitrite intakes vary from 0 to 20 mg/day. Nitrate intakes from sources other than vegetables, including drinking water and cured meats, have been estimated to average 35–44 mg/person/day for a 60-kg human [1]. On the basis of a conservative recommendation to consume 400 g of different fruits and vegetables per day at median nitrate concentrations, the dietary concentration of nitrate would be ~157 mg/day [1]. In the European Union, fruit consumption (average nitrate concentration: 10 mg/kg FW) constitutes more than half of the recommended intake of 400 g, nitrate intakes are estimated to be ~81–106 mg/day before additional nitrate losses from washing, peeling, and/or cooking are taken into consideration. In a convenience sample of foods analyzed for nitrate and nitrite content, two dietary patterns based upon the Dietary Approaches to Stop Hypertension that featured either low- or high-nitrate foods were found to contain between 174 mg nitrate/0.41 mg nitrite and 1,222 mg nitrate/0.35 mg nitrite [22]. It is noteworthy that the bioavailability of dietary nitrate is 100 % [23].

Dietary Intakes in the Context of WHO ADI Levels

The appreciation of four facts regarding human exposures to nitrate and nitrite casts concern over current regulatory limits on nitrate and nitrite consumption. First, it is possible to approach or exceed WHO ADI limits with usual intake levels of single foods, such as colostrum (at 100 mL intake in a newborn infant delivering 42 % of the WHO ADI intake limit), soya milk (750 mL intake for a hypothetical 6.8 kg infant yields 104 % of the WHO ADI intake limit) [24], spinach [25], or a dessicated vegetable supplement [22]. Second, recommended dietary intakes of vegetables and fruits, such as a Dietary Approaches to Stop Hypertension pattern with high nitrate food choices, exceed the World Health Organization's Acceptable Daily Intake for nitrate by 550 % for a 60-kg adult [22]. Third, for adults consuming the recommended intakes of vegetables and fruits, the origin of over 80 % of dietary nitrate and nitrite, the concentration of nitrate in saliva, via enterosalivary circulation, can reach up to three times the concentration in most global regulatory limits for drinking water [26]. Fourth, provision of dietary nitrate, as beetroot juice [27], dietary nitrate [28], or in a traditional Japanese dietary pattern [29], is effective in lowering blood pressure in humans. These facts indicate that WHO intake limits may not reflect optimal nitrate and nitrite concentrations from foods that confer health benefits. If nitrates and nitrites act as nutrients, it is likely that they do so to bolster the reserve of nitrite-derived NO_x metabolites required for optimal functioning through periods of physiological stress (e.g., hypoxia and acidosis) and diseases characterized by endothelial dysfunction [30–33].

Potential Negative Health Effects of Dietary Nitrate and Nitrite Exposures

Epidemiological and clinical studies have associated nitrate and nitrite consumption with increased risk of gastrointestinal cancers, thyroid dysfunction and thyroid cancer [34], chronic obstructive pulmonary disease in women [35] as well as other conditions [36]. Nitrate and nitrite exposures have been associated with gastrointestinal cancer risk through the consumption of cured and processed meats [37, 38]. A recent meta-analysis has challenged this conclusion; these results conclude that available epidemiological evidence is not sufficient to support a clear and unequivocal independent positive association between processed meat consumption and colorectal cancer risk [39]. Nitrates added to meats serve as antioxidants, develop flavor, and stabilize the red color in meats but must be converted to nitrite to exert these actions. Sodium nitrite is used as a colorant, flavor enhancer, and antimicrobial agent in cured and processed meats. Nitrate and nitrite use in meat products, including bacon, bologna, corned beef, hot dogs, luncheon meats, sausages, and canned and cured meat and hams is subject to limits put forth in Food and Drug Administration (FDA) and US Department of Agriculture (USDA) regulations. These regulations can be found in the Code of Federal Regulations (CFR) (21CFR 170.60, 172.170, and 172.175 for FDA and 9CFR 318.7 for USDA regulations, respectively). The use of nitrites in bacon must be accompanied by the use of either sodium erythorbate or sodium ascorbate (vitamin C), antioxidants that inhibit the nitrosation effect of nitrites on secondary amines [40]. The use of these antioxidants, along with lower nitrate and nitrite levels in processed meats, has lowered residual nitrite levels in cured meat products in the US by ~80 % since the mid-1970s [41].

While there exists little evidence that consumption of nitrates or nitrites is associated with risk of gastrointestinal cancers [16], the chemistry of gastric digestion of meals containing nitrite, dietary fat, and vitamin C may promote the formation of carcinogenic N-nitrosamines (NOCs) [42, 43]. The generation of nitrite-derived NO is maximal at the gastroesophageal junction where salivary nitrite first meets gastric acid. This chemistry is modeled using a physiologically relevant two-phase (aqueous and lipid) in vitro model system. Acid-catalyzed nitrosation was found to be enhanced by vitamin C in the presence of lipid. Thus, the presence of lipid, in the microenvironment of the stomach, converts vitamin C from inhibiting to promoting acid nitrosation. In the presence of 10 % lipid (a food matrix component for processed meats), the presence of vitamin C increased the formation of nitrosodimethylamine, nitrosodiethylamine, and N-nitrosopiperidine 8-, 60-, and 140-fold, respectively [42]. This effect is attributable to the ability of vitamin C to assist in the generation of NO in the aqueous phase, which enables the regeneration of nitrosating species by reacting with oxygen in the lipid phase. This concept has been supported by observations in a carcinogen-induced rat model of colorectal cancer in which processed meat consumption enhanced the number of preneoplastic lesions in the colon [44]. Taken together, these observations provide a biologically plausible mechanism for the observed association between processed meat consumption and gastrointestinal cancer risk [45]. It has been postulated that gastric formation of NOCs may be inhibited by nutrients and other components of vegetables and fruit [46]. In this two-phase model system, it has been shown that certain polyphenols block the formation of NOCs [47]. Clearly, more research is needed to address the potential interactions among food constituents to affect cancer risk.

Potential Health Benefits of Dietary Nitrate and Nitrite

Data from the laboratories of Drs. Jon Lundberg, Mark Gladwin, Jay Zweier, Nathan Bryan, and others have formed the modern basis for proposed health benefits of dietary nitrates and nitrites [**48–50**]. The nitrate–nitrite–NO pathway has been demonstrated to serve as a backup system to ensure NO supply in situations when the endogenous L-arginine/NO synthase pathway is dysfunctional [**51**]. This redundant system of NO production in tissues has important implications for cardiovascular, gastrointestinal, and immune function related to the provision of dietary nitrate and nitrite. As nitrite-dependent NO generation has been shown to play critical physiological and pathological roles, and is controlled by oxygen tension, pH, reducing substrates and nitrite levels [**52**], it is necessary to balance these contexts in a modern regulatory framework that acknowledges a physiological requirement for nitrate and nitrite supplied by dietary means.

Addressing Context-Specific Health Effects of Food Sources of Nitrate and Nitrite

Two recent advisories have been issued regarding the potential health effects of food sources of dietary nitrates and nitrites. The International Agency for Research on Cancer (IARC) considered the carcinogenic risk associated with nitrate and nitrite consumption [53]. This thorough review concluded that the carcinogenicity from drinking water and food sources of nitrates and nitrites were not strongly supported by the epidemiological evidence. This group considered a combination of positive and negative results from epidemiological and animal studies that supported a positive, coherent association with gastrointestinal cancer risk in conditions that favor endogenous formation of N-nitroso compounds. They reported the strongest associations were recorded in individuals with high nitrite and low vitamin C intake, a combination that promotes these nitrosation reactions. The presence of factors in foods, such as flavonoids in vegetables, may account for the lack of association with cancer risk for nitrate in food. The lack of association for nitrate in drinking water was attributed to low exposure in the available studies. They did not rule out the potential for nitrate in drinking water to result in endogenous nitrosation because water can be consumed if it was consumed in the absence of nitrosation inhibitors. Overall, the IARC group concluded that "ingested nitrate or nitrite under conditions that result in endogenous nitrosation is probably carcinogenic to humans (group 2A)" [53].

The European Food Safety Authority (EFSA) considered the potential health effects of dietary nitrate and nitrite consumption from vegetables because these foods are recommended for health benefits and are the primary dietary source of nitrate [1]. This group concluded that the diets of a small proportion of certain EU Member States that consume primarily high amounts of leafy vegetables, the ADI consumption limits for nitrate could be exceeded. Like other expert groups, EFSA concluded that epidemiological studies do not suggest that nitrate intake from diet or drinking water is associated with increased cancer risk. Furthermore, this group considers the evidence that high intake of nitrite might be associated with increased cancer risk to be equivocal.

The EFSA panel concluded their analysis of the risk and benefits of exposure to nitrate from vegetables by stating:

> Overall, the estimated exposures to nitrate from vegetables are unlikely to result in appreciable health risks, therefore the recognized beneficial effects of consumption of vegetables prevail. The Panel recognized that there are occasional circumstances, e.g., unfavorable local/home production conditions for vegetables which constitute a large part of the diet, or individuals with a diet high in vegetables such as rucola which need to be assessed on a case-by-case basis [1].

The fact that the IARC and EFSA recommendations were the result of a review process that considered similar bodies of evidence is noteworthy. More striking is the fact that, in spite of generally similar conclusions, the scope or purview of each group was limited. The IARC group did not consider other potential health effects of nitrate and nitrite and the EFSA group considered only those associated with nitrates in vegetables. While these excellent reports serve their intended functions, they highlight the need for a broader examination of the potential health benefits and risks of dietary nitrates and nitrites.

Determinants of Regulatory Paradigm Change

Beyond the ascription of risks and benefits of dietary nitrates and nitrites, a recent review concluded that these dietary constituents, based on their demonstrated physiological functions, be considered nutrients [22, 54]. Analogous to *all* essential or indispensable nutrients, intake of excess nitrate and nitrite exposure is, in specific contexts, associated with an increased risk of negative health outcomes. A set of dietary reference intake (DRI) categories are set by the Food and Nutrition Board of the National Academy of Sciences for essential nutrients to clearly define, where possible, the contexts in which intakes are deficient, safe, or potentially excessive. These DRI categories include the recommended dietary allowance (RDA), adequate intake (AI), tolerable upper level intake (TUL), and estimated average requirement (EAR) [55]. The process of setting DRIs for nutrients considers a broad range of physiological factors, including nutritional status and potential toxicities. Rational methodologies such as these, including the consideration of normal dietary consumption patterns of nitrate and nitrite-containing foods, have not been applied in setting exposure limits or in considering the potential health benefits of dietary nitrates and nitrites. Consideration of dietary nitrates and nitrites as nutrients, and the determination of physiological concentrations considered low, sufficient or excessive will require a consensus among researchers, health professionals, and regulators.

Time to Relax Regulatory Limits on Dietary Nitrate and Nitrite?

While there are compelling indications that dietary nitrate or nitrite may reduce cardiovascular disease risk, it has been suggested that the data are insufficient to relax standards for nitrate in drinking water and foods [56]. Of course, the lack of awareness of the association between nitrate and nitrite and blood pressure means that the necessary tools employed by clinical researchers, epidemiologists, and trialists to assess the association between nitrate/nitrite concentrations in foods and specific health biomarkers are not widely available. This means that in clinical trials, such as those completed to assess the efficacy of the DASH diet, nitrate and nitrite concentrations in foods were, because they were not considered causally related to blood pressure, not measured. Because these unmeasured factors likely contributed to the hypotensive effects of the DASH dietary pattern [22], dietary nitrate and nitrite would be considered confounding factors. This characterization is apropos for studies attributing cardiovascular benefits to vegetable intake [57], plant-based diets [58], and Mediterranean diet interventions for the secondary prevention of cardiovascular disease [59]. Dietary nitrates and nitrites may, in part, contribute to the salutary effects of vegetable and fruit consumption. It is hoped that dietary concentrations of these effect modifiers are measured or estimated and reported in epidemiological and clinical studies of cardiovascular risk.

Lack of inclusion of food nitrate and nitrite concentrations in standard food databases (e.g., USDA National Nutrient Database for Standard Reference) is another obstacle to the development of a solid epidemiological basis for quantifying cardiovascular and other health benefits of dietary nitrates and nitrites in human populations. As such, the development and availability of a database of food nitrate/nitrite concentrations would encourage more thorough investigations of hypotheses associating dietary nitrate/nitrite and specific health outcomes. Work is under way to establish this database at the U.S. Department of Agriculture's Nutrient Data Laboratory (Hord 2016; unpublished comment).

The compelling results of clinical studies demonstrate great potential for the treatment and prevention of cardiovascular diseases including ischemia–reperfusion (IR) injury and hypertension by dietary means [60] (Kapil, Khambata et al. 2015, unpublished). It is incumbent upon regulators to carry out a comprehensive, systematic, and independent review of all available evidence of health effects of dietary nitrate and nitrite. An Institute of Medicine (IOM)-type process would support the scholarly nature of this effort in an independent institution. This effort would be synergistic with the recent IOM report "A Population-Based Policy and Systems Change Approach to Prevent and Control Hypertension" [61]. Among other public health strategies, this report highlights population-based strategies to reduce the prevalence of hypertension. These strategies include behavioral and lifestyle interventions such as reducing sodium intake, increasing consumption of fruits and vegetables, and increasing physical activity. Many scientists understand that consideration of diet-based therapies emphasizing nitrate and nitrite intakes could benefit individuals with hypertension.

Conclusion

After decades of being subject to regulatory limits on dietary nitrate and nitrite based upon the poor practice of causal inference, the public deserves cohesive regulations that reflect the physiological necessity of nitrate and nitrite while accounting for contexts in which these dietary substances may produce health risks. The necessary work of bringing together experts from disparate scientific disciplines to craft meaningful dietary recommendations for nitrate and nitrite intakes will be a boon for public health.

References

1. European Food Safety Authority. Nitrate in vegetables: scientific opinion of the panel on contaminants in the food chain. EFSA J. 2008;689:1–79.
2. Pennington J. Dietary exposure models for nitrates and nitrites. Food Control. 1998;9:385–95.
3. Santamaria P. Nitrate in vegetables; content, toxicity and EC regulation. J Sci Food Agric. 2005;86(1):10–7.
4. Spiegelhalder B, Eisenbrand G, Preussmann R. Influence of dietary nitrate on nitrite content of human saliva: possible relevance to in vivo formation of N-nitroso compounds. Food Cosmet Toxicol. 1976;14(6):545–8.
5. Petersson J, Carlstrom M, Schreiber O, Phillipson M, Christoffersson G, Jagare A, et al. Gastroprotective and blood pressure lowering effects of dietary nitrate are abolished by an antiseptic mouthwash. Free Radic Biol Med. 2009;46(8):1068–75.
6. Jansson EA, Huang L, Malkey R, Govoni M, Nihlen C, Olsson A, et al. A mammalian functional nitrate reductase that regulates nitrite and nitric oxide homeostasis. Nat Chem Biol. 2008;4(7):411–7.
7. van Faassen EE, Bahrami S, Feelisch M, Hogg N, Kelm M, Kim-Shapiro DB, et al. Nitrite as regulator of hypoxic signaling in mammalian physiology. Med Res Rev. 2009;29(5):683–741.
8. Camargo JA, Alonso A. Ecological and toxicological effects of inorganic nitrogen pollution in aquatic ecosystems: a global assessment. Environ Int. 2006;32(6):831–49.
9. Monteny GJ, The EU. Nitrates Directive: a European approach to combat water pollution from agriculture. ScientificWorldJournal. 2001;1 Suppl 2:927–35.
10. World Health Organization. Recommendations; nitrate and nitrite. In: Guidelines for drinking water quality. 3rd ed. Geneva: WHO; 2004. p. 417–20.
11. Center for Disease Control and Prevention (CDC). Methemoglobinemia following unintentional ingestion of sodium nitrite–New York. MMWR Morb Mortal Wkly Rep. 2002;51(29):639–42.
12. Butler AR, Feelisch M. Therapeutic uses of inorganic nitrite and nitrate: from the past to the future. Circulation. 2008;117(16):2151–9.
13. McKnight GM, Duncan CW, Leifert C, Golden MH. Dietary nitrate in man: friend or foe? Br J Nutr. 1999;81(5):349–58.
14. L'Hirondel JL, Avery AA, Addiscott T. Dietary nitrate: where is the risk? Environ Health Perspect. 2006;114(8):A458–9; author reply A459–61.
15. L'hirondel J, L'hirondel JL. Nitrate and man: toxic, harmless or beneficial? Wallingford, UK: CABI; 2001.
16. Powlson DS, Addiscott TM, Benjamin N, Cassman KG, de Kok TM. vanGrinsven H, et al. When does nitrate become a risk for humans? J Environ Qual. 2008;37(2):291–5.
17. Greer FR, Shannon M. Infant methemoglobinemia: the role of dietary nitrate in food and water. Pediatrics. 2005;116(3):784–6.
18. Churchill JJ. Re-examining "nitrate toxicity": a call for a more rational approach to effluent limits for nitrogen in decentralized wastewater treatment. In: Paper presented at Eleventh Individual and Small Community Sewage Systems Conference Proceedings, Warwick; 2007
19. National Primary Drinking Water Regulations: Final Rule, 40. Federal Register. 1991;CFR Parts 141–143(56 (20)):3526–97.
20. Mensinga TT, Speijers GJ, Meulenbelt J. Health implications of exposure to environmental nitrogenous compounds. Toxicol Rev. 2003;22(1):41–51.
21. Gangolli SD, van den Brandt PA, Feron VJ, Janzowsky C, Koeman JH, et al. Nitrate, nitrite and N-nitroso compounds. Eur J Pharmacol. 1994;292(1):1–38.
22. Hord NG, Tang Y, Bryan NS. Food sources of nitrates and nitrites: the physiologic context for potential health benefits. Am J Clin Nutr. 2009;90(1):1–10.
23. van Velzen AG, Sips AJ, Schothorst RC, Lambers AC, Meulenbelt J. The oral bioavailability of nitrate from nitrate-rich vegetables in humans. Toxicol Lett. 2008;181(3):177–81.
24. Hord NG, Ghannam J, Garg H, Tang YP, Bryan NS. Nitrate and nitrite content of human, formula, bovine and soy milks: implications for dietary nitrite and nitrate recommendations Breastfeed Med. 2010 [Epub ahead of print].
25. Lundberg JO, Feelisch M, Bjorne H, Jansson EA, Weitzberg E. Cardioprotective effects of vegetables: is nitrate the answer? Nitric Oxide. 2006;15(4):359–62.
26. Li H, Thompson I, Carter P, Whitely A, Bailey M, Leifert C, et al. Salivary nitrate–an ecological factor in reducing oral acidity. Oral Microbiol Immunol. 2007;22(1):67–71.
27. Webb AJ, Patel N, Loukogeorgakis S, Okorie M, Aboud Z, Misra S, et al. Acute blood pressure lowering, vasoprotective, and antiplatelet properties of dietary nitrate via bioconversion to nitrite. Hypertension. 2008;51(3):784–90.
28. Kapil V, Milsom AB, Okorie M, Miliki-Toyserkani S, Akram F, Rehman F, et al. Inorganic nitrate supplementation lowers blood pressure in humans: role for nitrite-derived NO. Hypertension. 2010;56(2):274–81.

29. Sobko T, Marcus C, Govoni M, Kamiya S. Dietary nitrate in Japanese traditional foods lowers diastolic blood pressure in healthy volunteers. Nitric Oxide. 2010;22(2):136–40.
30. Stokes KY, Dugas TR, Tang Y, Garg H, Guidry E, Bryan NS. Dietary nitrite prevents hypercholesterolemic microvascular inflammation and reverses endothelial dysfunction. Am J Physiol Heart Circ Physiol. 2009;296(5):H1281–8.
31. Garg HK, Bryan NS. Dietary sources of nitrite as a modulator of ischemia/reperfusion injury. Kidney Int. 2009;75(11):1140–4.
32. Elrod JW, Calvert JW, Gundewar S, Bryan NS, Lefer DJ. Nitric oxide promotes distant organ protection: evidence for an endocrine role of nitric oxide. Proc Natl Acad Sci U S A. 2008;105(32):11430–5.
33. Lundberg JO, Gladwin MT, Ahluwalia A, Benjamin N, Bryan NS, Butler A, et al. Nitrate and nitrite in biology, nutrition and therapeutics. Nat Chem Biol. 2009;5(12):865–9.
34. Ward MH, Kilfoy BA, Weyer PJ, Anderson KE, Folsom AR, Cerhan JR. Nitrate intake and the risk of thyroid cancer and thyroid disease. Epidemiology. 2010;21(3):389–95.
35. Jiang R, Camargo Jr CA, Varraso R, Paik DC, Willett WC, Barr RG. Consumption of cured meats and prospective risk of chronic obstructive pulmonary disease in women. Am J Clin Nutr. 2008;87(4):1002–8.
36. Panesar NS. Downsides to the nitrate-nitrite-nitric oxide pathway in physiology and therapeutics? Nat Rev Drug Discov. 2008;7(8):710. author reply 710.
37. Santarelli RL, Pierre F, Corpet DE. Processed meat and colorectal cancer: a review of epidemiologic and experimental evidence. Nutr Cancer. 2008;60(2):131–44.
38. Norat T, Bingham S, Ferrari P, Slimani N, Jenab M, Mazuir M, et al. Meat, fish, and colorectal cancer risk: the European Prospective Investigation into cancer and nutrition. J Natl Cancer Inst. 2005;97(12):906–16.
39. Alexander DD, Miller AJ, Cushing CA, Lowe KA. Processed meat and colorectal cancer: a quantitative review of prospective epidemiologic studies. Eur J Cancer Prev. 2010;19(5):328–41.
40. Rao GS, Osborn JC, Adatia MR. Drug-nitrite interactions in human saliva: effects of food constituents on carcinogenic N-nitrosamine formation. J Dent Res. 1982;61(6):768–71.
41. Cassens RG. Residual nitrite in cured meat. Food Technol. 1997;51:53–5.
42. Combet E, Paterson S, Iijima K, Winter J, Mullen W, Crozier A, et al. Fat transforms ascorbic acid from inhibiting to promoting acid-catalysed N-nitrosation. Gut. 2007;56(12):1678–84.
43. Combet E, Preston T, McColl KE. Development of an in vitro system combining aqueous and lipid phases as a tool to understand gastric nitrosation. Rapid Commun Mass Spectrom. 2010;24(5):529–34.
44. Santarelli RL, Vendeuvre JL, Naud N, Tache S, Gueraud F, Viau M, et al. Meat processing and colon carcinogenesis: cooked, nitrite-treated, and oxidized high-heme cured meat promotes mucin-depleted foci in rats. Cancer Prev Res (Phila). 2010;3(7):852–64.
45. World Cancer Research Fund, American Institute of Cancer Research. Food, nutrition, physical activity, and the prevention of cancer: a global perspective. Second Expert Report; 2007.
46. de Kok TM, Engels LG, Moonen EJ, Kleinjans JC. Inflammatory bowel disease stimulates formation of carcinogenic N-nitroso compounds. Gut. 2005;54(5):731.
47. Combet E, El Mesmari A, Preston T, Crozier A, McColl KE. Dietary phenolic acids and ascorbic acid: influence on acid-catalyzed nitrosative chemistry in the presence and absence of lipids. Free Radic Biol Med. 2010;48(6):763–71.
48. Lundberg JO, Govoni M. Inorganic nitrate is a possible source for systemic generation of nitric oxide. Free Radic Biol Med. 2004;37(3):395–400.
49. Dejam A, Hunter CJ, Schechter AN, Gladwin MT. Emerging role of nitrite in human biology. Blood Cells Mol Dis. 2004;32(3):423–9.
50. Bryan NS, Fernandez BO, Bauer SM, Garcia-Saura MF, Milsom AB, Rassaf T, et al. Nitrite is a signaling molecule and regulator of gene expression in mammalian tissues. Nat Chem Biol. 2005;1(5):290–7.
51. Lundberg JO, Weitzberg E. NO generation from inorganic nitrate and nitrite: role in physiology, nutrition and therapeutics. Arch Pharm Res. 2009;32(8):1119–26.
52. Zweier JL, Li H, Samouilov A, Liu X. Mechanisms of nitrite reduction to nitric oxide in the heart and vessel wall. Nitric Oxide. 2010;22(2):83–90.
53. Grosse Y, Baan R, Straif K, Secretan B, El Ghissassi F, Cogliano V. Carcinogenicity of nitrate, nitrite, and cyanobacterial peptide toxins. Lancet Oncol. 2006;7(8):628–9.
54. Bryan, Ivy Inorganic nitrite and nitrate: Evidence to support consideration as dietary nutrients Nutr Res. 2015;35(8):643–54.
55. Otten JJ, Hellwig J, Meyers LD. editors. Dietary reference intakes: the essential guide to nutrient requirements. Washington, DC: National Academy Press, Food and Nutrition Board, Institute of Medicine, National Academy of Sciences; 2006.
56. Katan MB. Nitrate in foods: harmful or healthy? Am J Clin Nutr. 2009;90(1):11–2.
57. Joshipura KJ, Hung HC, Li TY, Hu FB, Rimm EB, Stampfer MJ, et al. Intakes of fruits, vegetables and carbohydrate and the risk of CVD. Public Health Nutr. 2009;12(1):115–21.

58. Jenkins DJ, Wong JM, Kendall CW, Esfahani A, Ng VW, Leong TC, et al. The effect of a plant-based low-carbohydrate ("Eco-Atkins") diet on body weight and blood lipid concentrations in hyperlipidemic subjects. Arch Intern Med. 2009;169(11):1046–54.
59. de Lorgeril M, Salen P, Martin JL, Monjaud I, Delaye J, Mamelle N. Mediterranean diet, traditional risk factors, and the rate of cardiovascular complications after myocardial infarction: final report of the Lyon Diet Heart Study. Circulation. 1999;99(6):779–85.
60. Kapil V, et al. Dietary nitrate provides sustained blood pressure lowering in hypertensive patients: a randomized, phase 2, double-blind, placebo-controlled study. Hypertension. 2015;65(2):320–7.
61. Institute of Medicine. A population-based policy and systems change approach to prevent and control hypertension. Washington, DC: The National Academy of Sciences; 2010.

Part III
Nitrite and Nitrate in Therapeutics and Disease

Chapter 13
Nitric Oxide Signaling in Health and Disease

Nathan S. Bryan and Jack R. Lancaster Jr.

Key Points

- Nitric oxide is a key signaling molecule in the cardiovascular, immune, and nervous systems.
- There are three enzymes that produce NO from L-arginine, two are constitutive and produce low amounts of NO for short periods of time, the other inducible and can produce high levels of NO for prolonged periods.
- Nitric oxide reacts primarily with transition metals and other free radicals.
- A number of diseases are associated with dysregulation of NO production.
- The nitric oxide pathway is a key target for development of new therapies.

Key words Soluble guanylyl cyclase • Cyclic GMP • Nitrite • Nitrate • Heme iron nitric oxide synthase • EDRF • Nitric oxide synthase • L-arginine • Nitrosothiols • Inflammation • Free radical

Introduction

The discovery in the 1980s of the mammalian biosynthesis of nitric oxide ($^{\bullet}$NO, nitrogen monoxide) and its roles in the immune [1–5], cardiovascular [6–9], and nervous [10] systems established a startling new paradigm in the history of cellular signaling mechanisms. Prior to this discovery, it was essentially inconceivable that cells would intentionally produce a toxic molecule as a messenger; $^{\bullet}$NO is a common air pollutant, a constituent of cigarette smoke, and a toxic gas, which appears in the exhaust of motor cars and jet airplanes, causes acid rain, and destroys the ozone layer. Amazingly, despite this nasty reputation, it is now known that $^{\bullet}$NO is one of a family of reactive signaling molecules, which includes both reactive nitrogen and reactive oxygen species [11, 12]. This is, in fact, a hallmark

The original version of this chapter was revised. An erratum to this chapter can be found at DOI 10.1007/978-3-319-46189-2_23

N.S. Bryan, Ph.D. (✉)
Department of Molecular and Human Genetics, Baylor College of Medicine, One Baylor Plaza, BCM225, Houston, TX 77030, USA
e-mail: Nathan.bryan@bcm.edu

J.R. LancasterJr. , Ph.D.
Departments of Pharmacology & Chemical Biology, Medicine, and Surgery, University of Pittsburgh School of Medicine, 200 Lothrop St., Pittsburgh, PA 15261, USA
e-mail: doctorno@pitt.edu

N.S. Bryan, J. Loscalzo (eds.), *Nitrite and Nitrate in Human Health and Disease*, Nutrition and Health, DOI 10.1007/978-3-319-46189-2_13, © Springer International Publishing AG 2017

example of the propensity of nature to seek out and exquisitely utilize the unique properties of unusual molecules. Its best recognized role, for which the 1998 Nobel Prize in Physiology or Medicine was awarded to Robert Furchgott, Louis Ignarro, and Ferid Murad, is as a signal molecule in the vasculature and specifically in the control of blood pressure. In addition to this role, $\cdot NO$ is one of the most important signaling molecules in the body and is involved in virtually every organ system where it is responsible for modulating an astonishing variety of effects. Indeed, there have been over 150,000 publications on $\cdot NO$, more than half of which have appeared in the last 12 years. $\cdot NO$ has been shown to be involved in and affect (just to list a few major examples) neurotransmission, memory, stroke, glaucoma and neural degeneration, pulmonary hypertension, penile erection, angiogenesis, wound healing, atherogenesis, inflammation such as arthritis, nephritis, colitis, autoimmune diseases (diabetes, inflammatory bowel disease), invading pathogens, tumors, asthma, tissue transplantation, septic shock, platelet aggregation and blood coagulation, sickle cell disease, gastrointestinal motility, hormone secretion, gene regulation, hemoglobin delivery of oxygen, stem cell proliferation and differentiation, and bronchodilation.

Here, we will focus on the physiology/pathophysiology of the three major organ systems originally documented for the importance of $\cdot NO$ as a signal, the cardiovascular, immune, and nervous systems, and understandings relevant to nitrite/nitrate as particularly related to the cardiovascular system. Other important pathologies where $\cdot NO$ also plays critical roles, including cancer [13], sepsis [14], diabetes [15], and lung [16] and kidney injury [17], are covered elsewhere.

Enzymatic NO Synthesis

The first $\cdot NO$ synthase (NOS) to be isolated and the gene sequenced was the enzyme from rat cerebellum, by Bredt and Snyder in 1991 [18, 19]. Since then, the genes for the two other major NOS isoforms have been identified and cloned, revealing that all three enzymes perform the same basic enzymatic reaction. These heme- and flavin-containing enzymes utilize electrons from NADPH and produce $\cdot NO$ by the mixed-function oxidation of one of the two equivalent guanidine nitrogen atoms of the amino acid L-arginine (Fig. 13.1) [20, 21]. The three major isoforms are most commonly referred to as the neuronal ("nNOS" or NOS1), inducible ("iNOS" or NOS2), and endothelial

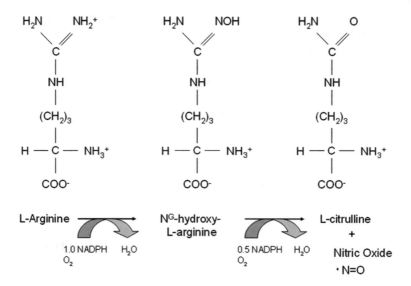

Fig. 13.1 Mechanism of $\cdot NO$ synthesis from L-arginine by NOS

("eNOS" or NOS3), classified according to their initially identified location and function. nNOS and eNOS enzymes are constitutive, cytosolic, and Ca^{2+}/calmodulin dependent, and release ·NO for short time periods in response to receptor or physical stimulation. The ·NO released by these enzymes acts as a signaling mechanism underlying several physiological responses depending on when and where it is produced and producing relatively small amounts of ·NO. The other enzyme type (iNOS) is induced after activation of macrophages, endothelial cells, and a number of other cells by endotoxin and pro-inflammatory cytokines, and once expressed, synthesizes NO for long periods of time and at higher levels than the constitutive enzymes. Furthermore, the inducible form is Ca^{2+} independent since calmodulin is already bound to the enzyme. It is now appreciated that eNOS is found in other cells and tissues beside the endothelium; nNOS is found in other cells than neurons; iNOS is found constitutively in some tissues; and there are inducible forms of both eNOS and nNOS, adding confusion to the nomenclature as it was first described. In addition, the activity and subcellular location of all three enzymes is regulated by a variety of different posttranslational modifications, especially eNOS [22]. Finally, there is evidence of NOS associated with erythrocytes [23] and also mitochondria [24, 25], although these associations are less well characterized.

Interconversions Between ·NO and Nitrite/Nitrate

Formation of Nitrite/Nitrate from ·NO

The major pathway for ·NO metabolism is the stepwise oxidation to nitrite and nitrate [26]. Quantitatively, the major mechanism for ·NO metabolism in man in vivo is via reaction with oxyhemoglobin [27]. In plasma or other physiological fluids or buffers, ·NO is oxidized almost completely to nitrite, where it remains stable for several hours. ·NO and nitrite are rapidly oxidized to nitrate in whole blood. The half-life of nitrite (NO_2^-) in human blood is about 110 s when infused systemically [26] or about 90–120 min when taken orally [28]. Nitrate (NO_3^-), on the other hand, has a circulating half life of 5–8 h. For years, both nitrite and nitrate have been used as surrogate markers of ·NO production in biological tissues. During fasting conditions with low intake of nitrite/nitrate, enzymatic ·NO formation from NOS accounts for the majority of nitrite [29], and plasma nitrite reflects eNOS activity [30]. On the basis of these and other studies, it was believed that ·NO is acutely terminated by oxidation to nitrite and nitrate; however, as described later, it is now appreciated that nitrite and/or nitrate can recapitulate ·NO physiology.

Formation of ·NO from Nitrite/Nitrate

There are several reports of the biological production of ·NO from nitrite which predate the discovery in mammals. The first identification of enzymatic synthesis of ·NO was in the 1970s where it was shown that it is produced in plants by the enzymatic reduction of NO_2^- as an intermediate in nitrogen fixation [31, 32], although the resulting heme-nitrosyl most probably does not release free ·NO. In 1983, formation of ·NO from nitrite was proposed as the mechanism whereby nitrite prevents botulism [33]. ·NO is a central intermediate in microbial denitrification [34].

It is now recognized that there are non-NOS sources of ·NO [35]. The first description of such ·NO formation was in 1994 when two independent groups demonstrated ·NO detection from acidified nitrite in the stomach [36, 37], which is predictable based on well-established chemistry of acidified nitrite but which may not apply to other, less acidic bodily compartments [38]. Nitrate per se is not metabolized by mammals; however, commensal facultative bacteria in the oral cavity can convert nitrate to nitrite and is a significant source of circulating nitrite. As described in more detail later, there are multiple mechanisms for the reductive conversion of nitrite to ˙NO that may be therapeutically effective.

Biochemical Actions and Targets of ˙NO

Reactivity and Diffusivity of ˙NO

Although commonly characterized as a "highly reactive radical," the biochemistry of ˙NO is surprisingly simple since it is determined by one major principle: ˙NO undergoes chemistry that serves to stabilize its unpaired electron. There are major consequences of this chemistry: (1) ˙NO interacts almost exclusively with only two species, molecules that also contain an unpaired electron (thus allowing the electrons to pair); and (2) transition metals to which ˙NO binds, thereby stabilizing the unpaired electron owing to special interactions with the electrons on metals in their d orbitals. As a result of these two phenomena, in general ˙NO essentially does not react with molecules with covalent bonds (all electrons paired) that do not contain metal ions.

Being an extremely small, uncharged molecule with limited reactivity, ˙NO is highly diffusible in biological milieu; at first glance, these properties would seem to violate the concept of specificity required for a biological signal and was initially a startling (and controversial) concept. Indeed, modeling studies predicted that with sustained ˙NO production (such as from iNOS), a gradient of ˙NO will extend great distances from the source, depending on the half-life of ˙NO [39, 40]. However, the actions of ˙NO in some systems are quite spatially confined, and, in fact, are highly localized to the immediate environment surrounding NOS [41]. The explanation for this confinement of a highly diffusible molecule is unclear, but could be explained by rapid reactivity at the source, e.g., with superoxide [42].

Intracellular ˙NO Targets: Transition Metals

The interaction between ˙NO and its primary protein target, the enzyme, soluble guanylyl cyclase (cGC), which mediates target cell responses such as vascular smooth muscle relaxation and platelet inhibition, has been well characterized although controversies remain [43, 44]. After entering the target cell, ˙NO binds to the heme moiety of sGC and activates the enzyme by inducing a conformational change that displaces iron out of the plane of the porphyrin ring [45]. sGC then catalyzes the production of cyclic GMP from GTP to elevate cyclic GMP. Cyclic GMP then triggers a cascade of intracellular events that culminate in a reduction in calcium-dependent vascular smooth muscle tone by inactivating myosin light chain kinase or MLCK [46]. MLCK normally phosphorylates the regulatory set of myosin light chains. This phosphorylation event activates cross-bridge cycling and initiates contraction. cGMP modulates MLCK activity by activating a cGMP-dependent protein kinase that phosphorylates MLCK. Phosphorylation of MLCK diminishes its affinity for calmodulin and, as a consequence, decreases the phosphorylation of myosin light chain, which, in turn, stabilizes the inactive form of myosin. In this manner, cGMP may induce vasorelaxation by indirectly decreasing myosin light chain-dependent myosin activation.

In addition to the heme iron of sGC, there are three other major intracellular iron targets of ˙NO. As mentioned previously, the reaction of ˙NO with intraerythrocytic hemoglobin (either oxygenated or deoxygenated) is quantitatively the most important in vivo reaction of ˙NO [27], the major product of which is nitrate. A very minor reaction comparatively is the formation of S-NO-hemoglobin, which is S-nitrosated at a specific cysteine residue (cysβ93), and which has been proposed as a major mechanism of blood flow and tissue oxygenation modulation in response to hypoxia [47, 48]. This controversial proposal is discussed in more detail later.

Another target of ˙NO is the mitochondrion, which dates back to 1979 [49]. At relatively low levels, relevant to ˙NO signaling, ˙NO binds to the same mitochondrial site(s) as O_2, thus exhibiting competitive kinetics with O_2 [50]. This competition may have critically important consequences in intracellular signaling events, including the generation of reactive oxygen species (see later), as well as major influences on cellular and tissue distribution of ˙NO and of O_2, and possible contributions to tissue hypoxic vasodilation [40, 51].

Intracellular nonheme iron (NHI) is also a target for ˙NO. One effect of such ˙NO binding is to produce iron–˙NO complexes, which are visible by electron paramagnetic resonance (EPR) spectroscopy. Observation of these signals was important in developing concepts of cellular actions of ˙NO in the early stages of the mammalian ˙NO field [52, 53]. It was initially believed that the origin of the NHI under these conditions is mitochondrial iron–sulfur centers, which are targets of ˙NO under conditions of high inflammatory levels of ˙NO, but quantitative and kinetic evidence now appears to support the so-called chelatable iron pool as the iron source [54].

Intracellular ˙NO Targets: Radicals

˙NO reacts at near-diffusion controlled rates with other radicals, and these reactions can have either damaging or protective effects due to three consequences: (1) removal of ˙NO, (2) removal of the radical with which ˙NO reacts, and (3) formation of products that may itself also be very reactive [55]. Arguably the most important ˙NO/radical reaction (certainly the most discussed in the literature, especially with reference to the effects of ˙NO in disease conditions) is the extremely rapid reaction with the superoxide anion radical ($O_2^{˙-}$). $O_2^{˙-}$ is perhaps the "canonical" reactive oxygen species; the discovery in 1969 of the enzyme that detoxifies it (superoxide dismutase) by McCord and Fridovich is responsible for the birth of the entire field [56]. So it might seem that the ˙NO + $O_2^{˙-}$ reaction would only be protective, but, in fact, the two other effects of the reaction are in principle damaging because it decreases the beneficial activities of ˙NO and the product (peroxynitrite, ONOO−) is potentially an even more damaging ROS than $O_2^{˙-}$ [57, 58]. There are several well-established sources of $O_2^{˙-}$ that are relevant to the modulation of ˙NO signaling, as described later.

Current Aspects of ˙NO in Health and Disease: A Janus-Faced Molecule

The Cardiovascular System

The endothelium is the thin layer of cells that line the interior surface of blood vessels, forming an interface between circulating blood in the lumen and the rest of the vessel wall. Endothelial cells line the entire circulatory system, from the heart to the smallest capillary. Endothelial dysfunction is defined by the loss of normal biochemical processes carried out by the endothelium, and is a hallmark for a plethora of vascular diseases, and often leads to atherosclerosis. Endothelial dysfunction is very common, for example, in patients with diabetes mellitus, hypertension, or other chronic pathophysiological conditions with a substantial cardiovascular component. Abnormal vasodilation as an important component of endothelial dysfunction arises from variations in blood flow observed in patients with atherosclerosis compared with healthy subjects [59]. In healthy subjects, activation of eNOS causes vasodilation in both muscular conduit vessels and resistance arterioles. In contrast, in subjects with atherosclerosis, similar stimulation yields attenuated vasodilation in peripheral vessels and causes paradoxical vasoconstriction in coronary arteries, thus indicating a decrease in the availability of bioactive 'NO [60, 61]. Interestingly, endothelial dysfunction can be demonstrated in patients with risk factors for atherosclerosis in the absence of atherosclerosis itself [62, 63]. In addition, feeding healthy volunteers a high-fat meal leads to endothelial vasodilator dysfunction in a time span of just a few hours [64]. Experimental and clinical studies provide evidence that detective endothelial 'NO function is not only associated with all major cardiovascular risk factors, such as hyperlipidemia, diabetes, hypertension, smoking, and the severity of established atherosclerosis, but also has a profound predictive value for future atherosclerotic disease progression [59]. This concept is illustrated in Fig. 13.2. Thus, the dysfunctional eNOS/'NO pathway (including both 'NO formation and also disappearance) provides an ideal target for therapeutic or preventive intervention once 'NO homeostasis is better understood in atherosclerosis.

Potential consequences or manifestations of endothelial dysfunction include increased vascular contraction to vasoconstrictors, such as endothelin-1, thromboxanes, and serotonin [65]; enhanced

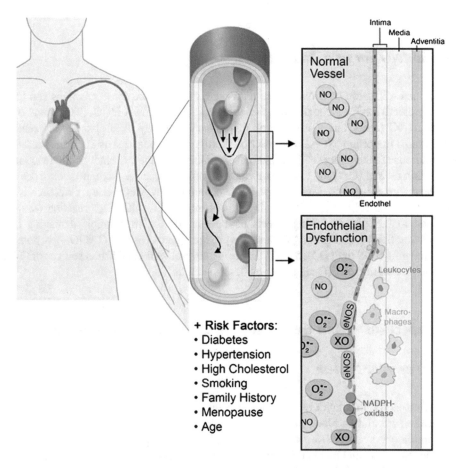

Fig. 13.2 The innermost lining of the blood vessel is a single layer of endothelial cells. Healthy endothelial cells produce nitric oxide to combat disease and regulate blood pressure. Endothelial dysfunction is defined by the inability of the vessels to transduce 'NO activity, either by inability to produce it from L-arginine or inactivation by superoxide ($O_2^{\bullet-}$) [104]

thrombus formation; and exacerbated smooth muscle cell (SMC) proliferation and migration [66]. Decreased ˙NO production may be the result of oxidized LDL-mediated displacement of eNOS from plasmalemmal caveolae, thereby inhibiting acetylcholine-induced activation of the enzyme [67]. Moreover, decreased ˙NO increases the tendency for lesion progression by enhancing vascular smooth muscle proliferation and migration, augmenting platelet activation and thrombosis, possibly participating in intravascular neovascularization, and favoring adverse lipid modification [68]. Once lesions have developed, endothelial dysfunction may exacerbate the risk of developing clinical events. Impaired endothelium may abnormally reduce vascular perfusion, produce factors that decrease plaque stability, and augment the thrombotic response to plaque rupture [69]. Augmentation of ˙NO or restoration of NOS function seems a logical means by which to inhibit atherosclerosis. However, overexpression of endothelial NOS accelerates lesion formation in apoE-deficient mice [70] demonstrating that enhanced NOS-derived ˙NO may not always be beneficial, or, perhaps, a reflection of local uncoupling of eNOS in the plaque leading to superoxide generation by eNOS. Supplementation with tetrahydrobiopterin (BH_4) reduced the lesion size to those seen in Apo E knockout mice revealing the requirement of enzyme cofactors and consistent with BH_4's role in preventing enzymatic uncoupling (see later). Even with BH_4 supplementation, there was still no effect on lesion development.

The physiological effects of ˙NO extend well beyond the vascular endothelium. Radomski et al. [71] have shown that human platelets contain a NOS that is activated when platelets are stimulated to aggregate. Thus, platelets themselves also have the enzymatic capacity to synthesize ˙NO with both a constitutive and inducible form of NOS identified in human megakaryoblasts [72]. NOS activity increases with platelet activation, and this response appears to modulate platelet aggregation, thereby potentially limiting the self-amplification of platelet thrombus formation in vivo [73, 74]. It was also reported early on that human neutrophils inhibit platelet aggregation by releasing an ˙NO-like factor [75]. These antithrombotic properties of the endothelium may be a consequence of the synergistic action of ˙NO and prostacyclin. Radomski et al. [76] have shown the synergistic antiaggregatory effects of ˙NO and prostacyclin on platelets. ˙NO and prostacyclin may act in concert to oppose local vasospasm or thrombus formation at sites where platelets aggregate and the coagulation cascade is activated. It has also been proposed that the antiplatelet effects of endothelium-derived ˙NO may prevent thromboembolic events during administration of potent prostacyclin inhibitors such as aspirin [77]. In this regard, ˙NO acts as an anti-inflammatory molecule.

Endothelial dysfunction in the setting of cardiovascular risk factors has been shown to be, at least in part, dependent on the production of ROS, such as superoxide, and the subsequent decrease in vascular availability of nitric oxide. ROS production has been demonstrated to occur in the endothelial cell and also within the media and adventitia, all of which may impair ˙NO signaling within vascular tissue to endothelium-dependent but also endothelium-independent vasodilators [59]. More recent experimental but also clinical studies point to the pathophysiological importance of xanthine oxidase, vascular NADPH oxidases, mitochondria, and uncoupled endothelial nitric oxide synthase as significant enzymatic superoxide sources. These phenomena are described later.

The mitochondrial respiratory chain can be a major source of superoxide. During aerobic metabolism, the oxidation/reduction energy of mitochondrial electron transport is converted to the high-energy phosphate bond of ATP via a multicomponent electron transfer complex. Molecular oxygen serves as the final electron acceptor for cytochrome c oxidase (complex IV), the terminal component of the respiratory chain, and is ultimately reduced to water (H_2O). Up to 1–4 % of O_2 even under normal physiological conditions may be incompletely reduced, resulting in $O_2{}^{\cdot}-$ formation, mainly at complex I (NADH coenzyme Q reductase) and complex III (ubiquinol Cyt c reductase) [78]. Increased mitochondrial $O_2{}^{\cdot}-$ generation can be enhanced in certain conditions such as conditions of metabolic perturbation, hypoxia–reoxygenation, and ischemia–reperfusion, where the enhanced $O_2{}^{\cdot}-$ is at least partially responsible for an increase in endothelial permeability [79].

Xanthine oxidoreductase (XOR) is a ubiquitous metalloflavoprotein found in one of two interconvertible yet functionally distinct forms, namely, xanthine dehydrogenase (XD), which is constitutively expressed in vivo, and XO, which is generated by the posttranslational modification of XD [80, 81]. Functionally, both XD and XO catalyze oxidation of hypoxanthine to xanthine and xanthine to urate. However, whereas XD requires NAD^+ as an electron acceptor, XO instead requires the reduction of molecular O_2, thereby generating $O_2{}^{\cdot}-$. The conversion of XD to XO occurs either through reversible thiol oxidation of sulfhydryl residues on XD or via irreversible proteolytic cleavage of a segment of XD during hypoxia, ischemia, or in the presence of various proinflammatory mediators [80], e.g., tumor necrosis factor-α (TNF-α). Dysfunctional or uncoupled NOSs are also a source of $O_2{}^{\cdot}-$. The essential NOS cofactor, tetrahydrobiopterin (BH_4), appears to have a key role in regulating NOS function by "coupling" the reduction of molecular O_2 to L-arginine oxidation as well as maintaining the stability of NOS dimers [81]. Thus, BH_4 availability may be a crucial factor in the balance between ˙NO and $O_2{}^{\cdot}-$ generation by eNOS. Furthermore, BH_4 itself is highly susceptible to oxidative degradation, and the initial oxidative loss of BH_4 in response to increased ROS production by NADPH oxidases (see later) has been shown to amplify oxidative stress through the resulting loss of ˙NO production and increased NOS-dependent $O_2{}^{\cdot}-$ generation [82]. In addition to increased catabolism or degradation, another reason for BH_4 depletion may be its reduced synthesis or incomplete reduction of BH_2 back to BH_4.

In recent years, it has become apparent that endothelial cells and other nonphagocytic cells constitutively express an $O_2{}^{\cdot}-$-generating enzyme analogous to the phagocyte NADPH oxidase of neutrophils [83]. All the classical neutrophil oxidase components are expressed in endothelial cells, but the enzyme nevertheless exhibits several major differences from the neutrophil oxidase; for example, it continuously generates a low level of $O_2{}^{\cdot}-$ even in unstimulated cells, although its activity can be further increased by several agonists; and a substantial proportion of the $O_2{}^{-\cdot}$ generated by the enzyme is produced intracellularly, whereas neutrophil oxidase $O_2{}^{-\cdot}$ generation occurs mainly in the extracellular compartment.

Normal physiological processes are continuously generating superoxide and other oxygen radicals that can quickly and effectively inactivate ˙NO. Fortunately there are a number of antioxidant systems, both enzymatic and nonenzymatic, that help to limit the amount of ROS produced to preserve ˙NO activity. It appears the production of ˙NO is a war of attrition. There are many circumstances in the production pathway that can diminish its output, but there are also many physiological factors that can quickly inactivate ˙NO once it has been successfully produced.

ROS such as superoxide are produced in abundance by the dysfunctional endothelium, and limitation of ROS generation increases the availability of ˙NO. For this reason antioxidant therapy with vitamin C and cholesterol-lowering therapy with statins (HMG-CoA reductase inhibitors) improve endothelial function, at least acutely [59]. An alternative approach to increasing levels of bioactive ˙NO and to improving endothelial function is to increase the synthesis of ˙NO. Enhanced synthesis of ˙NO can be achieved by increased availability of agonists that stimulate release of ˙NO from the endothelial cells, like bradykinin, assuming a healthy endothelium and sufficient cofactor to supply the NOS enzymes. Another straightforward approach to increasing ˙NO synthesis is to provide additional substrate (L-arginine) to the endothelial cell [84]. However, in some patients this may not be a beneficial strategy. L-arginine supplementation in postinfarct and PAD patients causes these patients to worsen and actually increases mortality for as yet unclear reasons [85, 86]. Furthermore, L-arginine takes part in protein synthesis, endocrine functions, wound healing, and erectile function. It is not regarded as an essential amino acid as the adult human is able to synthesize L-arginine de novo via the urea cycle. In adults, the synthesis of L-arginine results in L-citrulline, a by-product of glutamine metabolism in the gut and in the liver. Citrulline is excreted into the circulation and is reabsorbed in the kidney and converted to L-arginine. L-citrulline is reformed if L-arginine is shunted through the NOS pathway and is itself shunted through the partial urea cycle to regenerate L-arginine illustrating the requirement of argininosuccinate synthase (ASA) and argininosuccinate lyase (ASL) in the NOS pathway [87]. This protein complex, including ASA, ASL, and NOS, provides a spatial and temporal molecular shuttling of critical NO substrate and cofactors to allow for more efficient NO production. This protein complex also reveals that cells use intracellular pools of L-citrulline, which is then converted to L-arginine, rather than utilizing extracellular L-arginine to make NO. The dietary application of L-arginine is the basic determinant of the L-arginine level in plasma, as the biosynthesis of L-arginine is not able to balance inadequate intake or deficiency. Providing supplementation of substrate to individuals with inadequate ˙NO, therefore, has been suggested as a rational approach to increase ˙NO production by the ˙NO synthase [84], provided the enzyme is functional and the protein complex described earlier is intact [87].

Erythrocytes as Delivery Agents for ˙NO

The phenomenon of hypoxic vasodilation, whereby peripheral vessels dilate under decreased O_2 conditions, has been known for more than a century but the mechanism underlying it is not clear; in particular it is not clear as to what the "sensor" is that detects hypoxia and initiates the response. In 1995 Ellsworth et al. [88] proposed that the sensor is intraerythrocytic hemoglobin, which responds by increasing release of ATP to induce vasodilation. Shortly thereafter, Jia et al. [47] advanced a surprising alternative mechanism for the hemoglobin signal: rather than simply being an irreversible sink for ˙NO, hemoglobin within the red blood cell actually reacts with the ˙NO in the lung (where O_2 levels are high) forming S-nitrosohemoglobin (SNOHb) and releases it in vascular beds where O_2 is low (hypoxia). Important evidence for this hypothesis was the report that the levels of SNOHb are higher in the arterial than the venous circulation, implying liberation of ˙NO upon transit through the hypoxic region (called the A/V transit). Gladwin et al. [89] subsequently found no change in SNOHb upon A/V transit; however, there was significant consumption of plasma nitrite upon induction of hypoxia by exercise, and subsequent work suggested that deoxygenation of hemoglobin induces its reaction with nitrite to produce ˙NO. These three mechanisms of hypoxic vasodilation (ATP, SNOHb, and nitrite) are highly controversial and may in fact be interrelated [90, 91]. What appears to be less controversial is the critical and essential role of NO in the delivery of oxygen to the periphery [92].

Independent of its validity as the mechanism for hypoxic vasodilation, a series of recent studies have shown that nitrite administration is remarkably salutary for a variety of clinically significant applications, including myocardial infarction, stroke, hypertension, angiogenesis, and organ transplantation [93]. In addition, a report by Kleinbongard et al. [94] demonstrates that plasma nitrite levels progressively decrease with increasing cardiovascular risk burden indicating a reduction in nitric oxide produced. Risk factors considered include age, hypertension, smoking, and hypercholesterolemia, conditions all known for reduced availability of ˙NO. Although a correlation exists in plasma, it is not known whether the situation is mirrored in the heart or other tissue at risk for ischemic injury or disease. If so, tissue nitrite may serve as an index of risk, and restoring tissue nitrite may act as a first line of defense for protecting organs from ischemic and/or I/R injury. Since a substantial portion of steady-state nitrite concentrations in blood and tissue are derived from dietary sources, modulation of nitrite intake may provide a first line of defense for cardiovascular disease [93].

The Immune System

˙NO is also generated by macrophages and neutrophils as part of the human immune response. ˙NO is toxic to bacteria and other human pathogens. It is the inducible isoform of NOS that is responsible for macrophage ˙NO production. Inducible NOS has been found in many cell types, including macrophages [5], and is immunologically activated by exposure to bacterial endotoxin or pro-inflammatory cytokines, such as interleukin-1, or interferon-gamma [95, 96] and tumor necrosis factor. The presence of iNOS message or protein can serve as a biomarker for inflammation in tissues. The iNOS protein may remain present for several days [97]. At these high concentrations and flux rates, ˙NO is cytotoxic and plays a key role in the immune response of macrophages to bacteria and other pathogens. Antimicrobial activity and ˙NO production parallel tumor necrosis factor activity, and a strong correlation exists between antimicrobial activity and production of L-Arg-derived ˙NO by cytokine-activated cells observed during in vitro studies [98]. The precise mechanism of ˙NO-mediated bactericidal and tumoricidal activity is unknown, but these observations suggest that macrophage ˙NO production contributes to nonspecific immunity. ˙NO from activated macrophages may be responsible for the profound loss of vascular tone seen in septic patients [99]. It is this relative overproduction of ˙NO and the subsequent vasodilation that are thought to mediate ˙NO's pathophysiological role during sepsis and multiorgan failure during hypovolemia and hypoxia. Despite the rapid progress in our understanding of the complex physiological and pathophysiological processes involving ˙NO, uncertainties remain with regard to the critical cellular targets of ˙NO cytotoxicity, the relative importance of different ˙NO redox states and carrier molecules, and the importance of the ˙NO antimicrobial system in human phagocytes. Ultimately, the immunoregulatory and vasoregulatory activities of ˙NO may prove to be just as important as its antimicrobial properties during infection.

The Nervous System

In the central nervous system, ˙NO is a neurotransmitter that underpins several functions, including the formation of memory. As in other organ systems, this ˙NO pathway may also play a role in the pathology of the central nervous system. The NOS isoform in the nervous system is activated by glutamate acting on N-methyl-D-aspartate receptors. In a matter of seconds, the glutamate-induced increase in intracellular calcium concentration activates NOS via the calcium/calmodulin interaction as previously described. Under most circumstances, eNOS and nNOS are constitutive in the sense that their activation does not require new enzyme synthesis. However, both forms of NOS are inducible in that new enzyme synthesis occurs primarily under conditions of traumatic or pathological insult. The calcium influx that accompanies prolonged NMDA receptor activation is associated with degeneration of the neurons through a mechanism(s) that involves ˙NO, but is still not precisely clear [100]. Thus, the dichotomy of both the protective and deleterious actions of ˙NO is again revealed in the nervous system.

In the periphery, there is a widespread network of nerves, previously recognized as nonadrenergic and noncholinergic (NANC), that operate through a ˙NO-dependent mechanism to mediate some forms of neurogenic vasodilation and regulate various gastrointestinal, respiratory, and genitourinary tract functions, as well as autonomic innervation of smooth muscle in the gastrointestinal tract, the pelvic viscera, the airways, and other systems [101]. NOS has been detected in the gastric mucosa, and ˙NO appears to play a role in protecting the gastric mucosa during physiologic stress by acting as an endogenous vasodilator and thus supporting mucosal blood flow [102]. The exact mechanism of ˙NO's protective effect is unclear, but may relate to vasodilation, inhibition of platelet aggregation in the gastric microvasculature, or a protective effect on the epithelial cells themselves [103]. However, alternative means to enhance ˙NO in the periphery and stomach will certainly provide benefit to a number of conditions.

Conclusion

Nitric oxide research has expanded rapidly in the past 30 years, and the roles of •NO in physiology and pathology have been extensively studied. The pathways of •NO synthesis, signaling, and metabolism in vascular biological systems have been and continue to be a major area of research. As a gas (in the pure state and under standard temperature and pressure) and free radical with an unshared electron, •NO participates in various biological processes. Understanding its production, regulation, and molecular targets will be essential for the development of new therapies for various pathological conditions characterized by an imbalanced production and metabolism of •NO.

References

1. Wagner D, Tannenbaum S. Enhancement of nitrate biosynthesis by *Escherichia coli* lipopolysaccharide. In: Magee PN, editor. Nitrosamines and human cancer. New York: Cold Spring Harbor Press; 1982. p. 437–43.
2. Hegesh E, Shiloah J. Blood nitrates and infantile methemoglobinemia. Clin Chim Acta. 1982;125(2):107–15.
3. Stuehr DJ, Marletta MA. Mammalian nitrate biosynthesis: mouse macrophages produce nitrite and nitrate in response to Escherichia coli lipopolysaccharide. Proc Natl Acad Sci U S A. 1985;82(22):7738–42.
4. Hibbs Jr JB, Taintor RR, Vavrin Z. Macrophage cytotoxicity: role for L-arginine deiminase and imino nitrogen oxidation to nitrite. Science. 1987;235(4787):473–6.
5. Marletta MA, Yoon PS, Iyengar R, Leaf CD, Wishnok JS. Macrophage oxidation of L-arginine to nitrite and nitrate: nitric oxide is an intermediate. Biochemistry. 1988;27(24):8706–11.
6. Arnold WP, Mittal CK, Katsuki S, Murad F. Nitric oxide activates guanylate cyclase and increases guanosine 3′:5′-cyclic monophosphate levels in various tissue preparations. Proc Natl Acad Sci U S A. 1977; 74(8):3203–7.
7. Ignarro LJ, Buga GM, Wood KS, Byrns RE, Chaudhuri G. Endothelium-derived relaxing factor produced and released from artery and vein is nitric oxide. Proc Natl Acad Sci U S A. 1987;84:9265–9.
8. Furchgott RF, Zawadzki JV. The obligatory role of endothelial cells in the relaxation of arterial smooth muscle by acetycholine. Nature. 1980 27 nov 1980;288(5789):373-6.
9. Palmer RMJ, Ferrige AG, Moncada S. Nitric oxide release accounts for the biological activity of endothelium-derived relaxing factor. Nature. 1987 11 June 1987;327(6122):524-6.
10. Garthwaite J, Charles SL, Chess-Williams R. Endothelium-derived relaxing factor release on activation of NMDA receptors suggests role as intercellular messenger in the brain. Nature. 1988;336(6197):385–8.
11. Forman HJ, Maiorino M, Ursini F. Signaling functions of reactive oxygen species. Biochemistry. 2010; 49(5):835–42.
12. Hill BG, Dranka BP, Bailey SM, Lancaster Jr JR, Darley-Usmar VM. What part of NO don't you understand? Some answers to the cardinal questions in nitric oxide biology. J Biol Chem. 2010;285(26):19699–704.
13. Fukumura D, Kashiwagi S, Jain RK. The role of nitric oxide in tumour progression. Nat Rev Cancer. 2006;6(7):521–34.
14. Fernandes D, Assreuy J. Nitric oxide and vascular reactivity in sepsis. Shock. 2008;30 Suppl 1:10–3.
15. Capellini VK, Celotto AC, Baldo CF, Olivon VC, Viaro F, Rodrigues AJ, et al. Diabetes and vascular disease: basic concepts of nitric oxide physiology, endothelial dysfunction, oxidative stress and therapeutic possibilities. Curr Vasc Pharmacol. 2010;8(4):526–44.
16. Mehta S. The effects of nitric oxide in acute lung injury. Vascul Pharmacol. 2005;43(6):390–403.
17. Baylis C. Nitric oxide deficiency in chronic kidney disease. Am J Physiol Renal Physiol. 2008;294(1):F1–9.
18. Bredt DS, Snyder SH. Isolation of nitric oxide synthetase, a calmodulin-requiring enzyme. Proc Natl Acad Sci U S A. 1990;87(2):682–5.
19. Bredt DS, Hwang PM, Glatt CE, Lowenstein C, Reed RR, Snyder SH. Cloned and expressed nitric oxide synthase structurally resembles cytochrome P-450 reductase. Nature. 1991;351(6329):714–8.
20. Li H, Poulos TL. Structure-function studies on nitric oxide synthases. J Inorg Biochem. 2005;99(1):293–305.
21. Daff S. NO synthase: structures and mechanisms. Nitric Oxide. 2010;23(1):1–11.
22. Michel T, Vanhoutte PM. Cellular signaling and NO production. Pflugers Arch. 2010;459(6):807–16.
23. Ozuyaman B, Grau M, Kelm M, Merx MW, Kleinbongard P. RBC NOS: regulatory mechanisms and therapeutic aspects. Trends Mol Med. 2008;14(7):314–22.

24. Finocchietto PV, Franco MC, Holod S, Gonzalez AS, Converso DP, Antico Arciuch VG, et al. Mitochondrial nitric oxide synthase: a masterpiece of metabolic adaptation, cell growth, transformation, and death. Exp Biol Med (Maywood). 2009;234(9):1020–8.
25. Lacza Z, Pankotai E, Busija DW. Mitochondrial nitric oxide synthase: current concepts and controversies. Front Biosci (Landmark Ed). 2009;14:4436-43.
26. Kelm M. Nitric oxide metabolism and breakdown. Biochim Biophys Acta. 1999;1411:273–89.
27. Wennmalm A, Benthin G, Edlund A, Kieler-Jensen N, Lundin S, Petersson AS, et al. Nitric oxide synthesis and metabolism in man. Ann N Y Acad Sci. 1994;714:158–64.
28. Greenway FL, Predmore BL, Flanagan DR, Giordano T, Qiu Y, Brandon A, et al. Single-dose pharmacokinetics of different oral sodium nitrite formulations in diabetes patients. Diabetes Technol Ther. 2012; 14(7):552–60.
29. Rhodes P, Leone AM, Francis PL, Struthers AD, Moncada S, Rhodes PM. The L-arginine:nitric oxide pathway is the major source of plasma nitrite in fasted humans. Biochem Biophys Res Commun. 1995;209(2):590–6.
30. Kleinbongard P, Dejam A, Lauer T, Rassaf T, Schindler A, Picker O, et al. Plasma nitrite reflects constitutive nitric oxide synthase activity in mammals. Free Radic Biol Med. 2003;35(7):790–6.
31. Aparicio PJ, Knaff DB, Malkin R. The role of an iron-sulfur center and siroheme in spinach nitrite reductase. Arch Biochem Biophys. 1975;169(1):102–7.
32. Lancaster JR, Vega JM, Kamin H, Orme-Johnson NR, Orme-Johnson WH, Krueger RJ, et al. Identification of the iron-sulfur center of spinach ferredoxin-nitrite reductase as a tetranuclear center, and preliminary EPR studies of mechanism. J Biol Chem. 1979;254(4):1268–72.
33. Reddy D, Lancaster Jr JR, Cornforth DP. Nitrite inhibition of Clostridium botulinum: electron spin resonance detection of iron-nitric oxide complexes. Science. 1983;221(4612):769–70.
34. Goretski J, Hollocher TC. Trapping of nitric oxide produced during denitrification by extracellular hemoglobin. J Biol Chem. 1988;263(5):2316–23.
35. Lundberg JO, Weitzberg E. NO-synthase independent NO generation in mammals. Biochem Biophys Res Commun. 2010;396(1):39–45.
36. Lundberg JO, Weitzberg E, Lundberg JM, Alving K. Intragastric nitric oxide production in humans: measurements in expelled air. Gut. 1994;35(11):1543–6.
37. Benjamin N, O'Driscoll F, Dougall H, Duncan C, Smith L, Golden M, et al. Stomach NO synthesis. Nature. 1994;368(6471):502.
38. Butler AR, Ridd JH. Formation of nitric oxide from nitrous acid in ischemic tissue and skin. Nitric Oxide. 2004;10(1):20–4.
39. Lancaster Jr JR. A tutorial on the diffusibility and reactivity of free nitric oxide. Nitric Oxide. 1997;1(1):18–30.
40. Thomas DD, Liu X, Kantrow SP, Lancaster JRJ. The biological lifetime of nitric oxide: implications for the perivascular dynamics of NO and O_2. Proc Natl Acad Sci U S A. 2001;98:355–60.
41. Hare JM. Nitric oxide and excitation-contraction coupling. J Mol Cell Cardiol. 2003;35(7):719–29.
42. Tziomalos K, Hare JM. Role of xanthine oxidoreductase in cardiac nitroso-redox imbalance. Front Biosci (Landmark Ed). 2009;14:237-62.
43. Derbyshire ER, Marletta MA. Biochemistry of soluble guanylate cyclase. Handb Exp Pharmacol. 2009;191:17–31.
44. Garthwaite J. New insight into the functioning of nitric oxide-receptive guanylyl cyclase: physiological and pharmacological implications. Mol Cell Biochem. 2010;334(1-2):221–32.
45. Ignarro LJ. Signal transduction mechanisms involving nitric oxide. Biochem Pharmacol. 1991;41(4):485–90.
46. Francis SH, Busch JL, Corbin JD, Sibley D. cGMP-dependent protein kinases and cGMP phosphodiesterases in nitric oxide and cGMP action. Pharmacol Rev. 2010;62(3):525–63.
47. Jia L, Bonaventura C, Bonaventura J, Stamler JS. S-nitrosohaemoglobin: a dynamic activity of blood involved in vascular control. Nature. 1996;380:221–6.
48. Allen BW, Stamler JS, Piantadosi CA. Hemoglobin, nitric oxide and molecular mechanisms of hypoxic vasodilation. Trends Mol Med. 2009;15(10):452–60.
49. Stevens TH, Brudvig GW, Bocian DF, Chan SI. Structure of cytochrome a3-Cua3 couple in cytochrome c oxidase as revealed by nitric oxide binding studies. Proc Natl Acad Sci U S A. 1979;76(7):3320–4.
50. Brunori M, Giuffre A, Sarti P. Cytochrome c oxidase, ligands and electrons. J Inorg Biochem. 2005;99(1):324–36.
51. Erusalimsky JD, Moncada S. Nitric oxide and mitochondrial signaling: from physiology to pathophysiology. Arterioscler Thromb Vasc Biol. 2007;27(12):2524–31.
52. Lancaster Jr JR, Hibbs Jr JB. EPR demonstration of iron-nitrosyl complex formation by cytotoxic activated macrophages. Proc Natl Acad Sci U S A. 1990;87(3):1223–7.
53. Pellat C, Henry Y, Drapier JC. IFN-gamma-activated macrophages: detection by electron paramagnetic resonance of complexes between L-arginine-derived nitric oxide and non-heme iron proteins. Biochem Biophys Res Commun. 1990;166(1):119–25.

54. Toledo Jr JC, Bosworth CA, Hennon SW, Mahtani HA, Bergonia HA, Lancaster Jr JR. Nitric oxide-induced conversion of cellular chelatable iron into macromolecule-bound paramagnetic dinitrosyliron complexes. J Biol Chem. 2008;283(43):28926–33.

55. Bryan NS, Rassaf T, Maloney RE, Rodriguez CM, Saijo F, Rodriguez JR, et al. Cellular targets and mechanisms of nitros(yl)ation: an insight into their nature and kinetics in vivo. Proc Natl Acad Sci U S A. 2004;101(12):4308–13.

56. McCord JM, Fridovich I. Superoxide dismutase. An enzymic function for erythrocuprein (hemocuprein). J Biol Chem. 1969;244(22):6049–55.

57. Beckman JS, Beckman TW, Chen J, Marshall PA, Freeman BA. Apparent hydroxyl radical production by peroxynitrite: implications for endothelial injury from nitric oxide and superoxide. Proc Natl Acad Sci U S A. 1990;87(4):1620–4.

58. Pacher P, Beckman JS, Liaudet L. Nitric oxide and peroxynitrite in health and disease. Physiol Rev. 2007;87(1):315–424.

59. Forstermann U. Nitric oxide and oxidative stress in vascular disease. Pflugers Arch. 2010;459(6):923–39.

60. Lieberman EH, Gerhard MD, Uehata A, Selwyn AP, Ganz P, Yeung AC, et al. Flow-induced vasodilation of the human brachial artery is impaired in patients <40 years of age with coronary artery disease. Am J Cardiol. 1996;78(11):1210–4.

61. Ludmer PL, Selwyn AP, Shook TL, Wayne RR, Mudge GH, Alexander RW, et al. Paradoxical vasoconstriction induced by acetylcholine in atherosclerotic coronary arteries. N Engl J Med. 1986;315(17):1046–51.

62. Creager MA, Cooke JP, Mendelsohn ME, Gallagher SJ, Coleman SM, Loscalzo J, et al. Impaired vasodilation of forearm resistance vessels in hypercholesterolemic humans. J Clin Invest. 1990;86(1):228–34.

63. Celermajer DS, Sorensen KE, Georgakopoulos D, Bull C, Thomas O, Robinson J, et al. Cigarette smoking is associated with dose-related and potentially reversible impairment of endothelium-dependent dilation in healthy young adults. Circulation. 1993;88(5 Pt 1):2149–55.

64. Esposito K, Nappo F, Giugliano F, Giugliano G, Marfella R, Giugliano D. Effect of dietary antioxidants on post-prandial endothelial dysfunction induced by a high-fat meal in healthy subjects. Am J Clin Nutr. 2003;77(1):139–43.

65. Lamping K, Faraci F. Enhanced vasoconstrictor responses in eNOS deficient mice. Nitric Oxide. 2003;8(4):207–13.

66. Huang PL. Endothelial nitric oxide synthase and endothelial dysfunction. Curr Hypertens Rep. 2003;5(6):473–80.

67. Blair A, Shaul PW, Yuhanna IS, Conrad PA, Smart EJ. Oxidized low density lipoprotein displaces endothelial nitric-oxide synthase (eNOS) from plasmalemmal caveolae and impairs eNOS activation. J Biol Chem. 1999;274(45):32512–9.

68. Loscalzo J. Nitric oxide insufficiency, platelet activation, and arterial thrombosis. Circ Res. 2001;88(8):756–62.

69. Faxon DP, Fuster V, Libby P, Beckman JA, Hiatt WR, Thompson RW, et al. Atherosclerotic vascular disease conference: Writing Group III: pathophysiology. Circulation. 2004;109(21):2617–25.

70. Ozaki M, Kawashima S, Yamashita T, Hirase T, Namiki M, Inoue N, et al. Overexpression of endothelial nitric oxide synthase accelerates atherosclerotic lesion formation in apoE-deficient mice. J Clin Invest. 2002;110(3):331–40.

71. Radomski MW, Palmer RM, Moncada S. An L-arginine/nitric oxide pathway present in human platelets regulates aggregation. Proc Natl Acad Sci U S A. 1990;87(13):5193–7.

72. Sase K, Michel T. Expression of constitutive endothelial nitric oxide synthase in human blood platelets. Life Sci. 1995;57(22):2049–55.

73. Freedman JE, Loscalzo J, Barnard MR, Alpert C, Keaney JF, Michelson AD. Nitric oxide released from activated platelets inhibits platelet recruitment. J Clin Invest. 1997;100(2):350–6.

74. Freedman JE, Sauter R, Battinelli EM, Ault K, Knowles C, Huang PL, et al. Deficient platelet-derived nitric oxide and enhanced hemostasis in mice lacking the NOSIII gene. Circ Res. 1999;84(12):1416–21.

75. Salvemini D, de Nucci G, Gryglewski RJ, Vane JR. Human neutrophils and mononuclear cells inhibit platelet aggregation by releasing a nitric oxide-like factor. Proc Natl Acad Sci U S A. 1989;86(16):6328–32.

76. Radomski MW, Palmer RM, Moncada S. The anti-aggregating properties of vascular endothelium: interactions between prostacyclin and nitric oxide. Br J Pharmacol. 1987;92(3):639–46.

77. Bednar MM, Gross CE, Howard DB, Russell SR, Thomas GR. Nitric oxide reverses aspirin antagonism of t-PA thrombolysis in a rabbit model of thromboembolic stroke. Exp Neurol. 1997;146(2):513–7.

78. Lambert AJ, Brand MD. Reactive oxygen species production by mitochondria. Methods Mol Biol. 2009;554:165–81.

79. Pearlstein DP, Ali MH, Mungai PT, Hynes KL, Gewertz BL, Schumacker PT. Role of mitochondrial oxidant generation in endothelial cell responses to hypoxia. Arterioscler Thromb Vasc Biol. 2002;22(4):566–73.

80. Meneshian A, Bulkley GB. The physiology of endothelial xanthine oxidase: from urate catabolism to reperfusion injury to inflammatory signal transduction. Microcirculation. 2002;9(3):161–75.

81. Vasquez-Vivar J, Kalyanaraman B, Martasek P. The role of tetrahydrobiopterin in superoxide generation from eNOS: enzymology and physiological implications. Free Radic Res. 2003;37(2):121–7.

82. Landmesser U, Dikalov S, Price SR, McCann L, Fukai T, Holland SM, et al. Oxidation of tetrahydrobiopterin leads to uncoupling of endothelial cell nitric oxide synthase in hypertension. J Clin Invest. 2003;111(8):1201–9.

83. Lambeth JD. NOX enzymes and the biology of reactive oxygen. Nat Rev Immunol. 2004;4(3):181–9.

84. Boger RH. L-Arginine therapy in cardiovascular pathologies: beneficial or dangerous? Curr Opin Clin Nutr Metab Care. 2008;11(1):55–61.

85. Schulman SP, Becker LC, Kass DA, Champion HC, Terrin ML, Forman S, et al. L-arginine therapy in acute myocardial infarction: the vascular interaction with age in myocardial infarction (VINTAGE MI) randomized clinical trial. JAMA. 2006;295(1):58–64.

86. Wilson AM, Harada R, Nair N, Balasubramanian N, Cooke JP. L-arginine supplementation in peripheral arterial disease: no benefit and possible harm. Circulation. 2007;116(2):188–95.

87. Erez A, Nagamani SC, Shchelochkov OA, Premkumar MH, Campeau PM, Chen Y, et al. Requirement of argininosuccinate lyase for systemic nitric oxide production. Nat Med. 2011;17(12):1619–26.

88. Ellsworth ML. The red blood cell as an oxygen sensor: what is the evidence. Acta Physiol Scand. 2000;168:551–9.

89. Gladwin MT, Shelhamer JH, Schechter AN, Pease-Fye ME, Waclawiw MA, Panza JA, et al. Role of circulating nitrite and S-nitrosohemoglobin in the regulation of regional blood flow in humans. Proc Natl Acad Sci U S A. 2000;97(21):11482–7.

90. Robinson JM, Lancaster Jr JR. Hemoglobin-mediated, hypoxia-induced vasodilation via nitric oxide: mechanism(s) and physiologic versus pathophysiologic relevance. Am J Respir Cell Mol Biol. 2005;32(4):257–61.

91. Jensen FB. The dual roles of red blood cells in tissue oxygen delivery: oxygen carriers and regulators of local blood flow. J Exp Biol. 2009;212(Pt 21):3387–93.

92. Zhang R, Hess DT, Qian Z, Hausladen A, Fonseca F, Chaube R, et al. Hemoglobin betaCys93 is essential for cardiovascular function and integrated response to hypoxia. Proc Natl Acad Sci U S A. 2015;112(20):6425–30.

93. Bryan NS, Ivy JL. Inorganic nitrite and nitrate: evidence to support consideration as dietary nutrients. Nutr Res. 2015;35(8):643–54.

94. Kleinbongard P, Dejam A, Lauer T, Jax T, Kerber S, Gharini P, et al. Plasma nitrite concentrations reflect the degree of endothelial dysfunction in humans. Free Radic Biol Med. 2006;40(2):295–302.

95. Stuehr DJ, Marletta MA. Induction of nitrite/nitrate synthesis in murine macrophages by BCG infection, lymphokines, or interferon-gamma. J Immunol. 1987;139(2):518–25.

96. Ding AH, Nathan CF, Stuehr DJ. Release of reactive nitrogen intermediates and reactive oxygen intermediates from mouse peritoneal macrophages. Comparison of activating cytokines and evidence for independent production. J Immunol. 1988;141(7):2407–12.

97. Xie QW, Cho HJ, Calaycay J, Mumford RA, Swiderek KM, Lee TD, et al. Cloning and characterization of inducible nitric oxide synthase from mouse macrophages. Science. 1992;256(5054):225–8.

98. Nathan CF, Hibbs Jr JB. Role of nitric oxide synthesis in macrophage antimicrobial activity. Curr Opin Immunol. 1991;3(1):65–70.

99. Cauwels A. Nitric oxide in shock. Kidney Int. 2007;72(5):557–65.

100. Garthwaite G, Garthwaite J. Differential dependence on Ca^{2+} of N-methyl-D-aspartate and quisqualate neurotoxicity in young rat hippocampal slices. Neurosci Lett. 1989;97(3):316–22.

101. Zhang J, Snyder SH. Nitric oxide in the nervous system. Annu Rev Pharmacol Toxicol. 1995;35:213–33.

102. Whittle BJ, Boughton-Smith NK, Moncada S. Biosynthesis and role of the endothelium-derived vasodilator, nitric oxide, in the gastric mucosa. Ann N Y Acad Sci. 1992;664:126–39.

103. Calatayud S, Barrachina D, Esplugues JV. Nitric oxide: relation to integrity, injury, and healing of the gastric mucosa. Microsc Res Tech. 2001;53(5):325–35.

104. Munzel T, Sinning C, Post F, et al. Pathophysiology, diagnosis and prognostic implications of endothelial dysfunction. Ann Med. 2008;40(3):180–96.

Chapter 14
Inhaled Nitric Oxide

Kenneth D. Bloch, Andrea U. Steinbicker, Lisa Lohmeyer, and Rajeev Malhotra

Key Points

- Development of inhaled NO as a selective pulmonary vasodilator.
- Overview of how inhaled NO elicits its effects is provided, as well as alternate approaches to increasing pulmonary NO and cGMP concentrations.
- Consideration of the clinical applications of inhaled NO.
- Discussion of the methods required for the safe administration of NO gas.
- Potential applications of inhaled NO to the treatment of systemic vascular diseases.

Keywords Inhalative therapy • Pulmonary hypertension • Methemoglobinemia • Second messengers • cGMP

Introduction

In 1998, the Nobel Prize in Physiology or Medicine was awarded to jointly to Robert Furchgott, Louis Ignarro, and Ferid Murad for their discoveries concerning "nitric oxide as a signaling molecule in the cardiovascular system." In 1977, Ferid Murad and colleagues proposed that nitroglycerin and related

K.D. Bloch, M.D.
Department of Anesthesia, Critical Care, and Pain Medicine, Anesthesia Center for Critical Care
Research, Cardiovascular Research Center, Massachusetts General Hospital, Harvard Medical School,
Boston, MA 02114, USA
e-mail: kdbloch@partners.org

A.U. Steinbicker, M.D., M.P.H.
Department of Anesthesiology, Intensive Care and Pain Medicine, University Hospital Muenster,
Albert-Schweitzer-Campus 1, Building A1, Muenster 48149, Germany
e-mail: andrea.steinbicker@ukmuenster.de

L. Lohmeyer, M.D.
Department of Medicine I: Hematology, Oncology and Stem-Cell Transplantation, University Medical
Center Freiburg, Hugstetter Strasse 49, Freiburg im Breisgau, Baden-Württemberg 79106, Germany
e-mail: lisa.isbell@uniklinik-freiburg.de

R. Malhotra, M.D., M.S. (✉)
Department of Medicine, Cardiology Division, Massachusetts General Hospital,
55 Fruit Street Yawkey 5700, Boston, MA 02114, USA
e-mail: rmalhotra@mgh.harvard.edu

N.S. Bryan, J. Loscalzo (eds.), *Nitrite and Nitrate in Human Health and Disease*, Nutrition and Health,
DOI 10.1007/978-3-319-46189-2_14, © Springer International Publishing AG 2017

vasodilator compounds act by generating nitric oxide (NO). In 1980, Furchgott and Zawadski described endothelium-derived relaxing factor, which, 10 years later, Ignarro and colleagues reported was equivalent to NO.

Nitric oxide is a free radical with a short half-life in biological fluids. It is rapidly oxidized to nitrite and nitrate, but can also react with several types of intracellular targets including superoxide leading to the production of the strong oxidant, peroxynitrite. Nitric oxide or NO metabolites can nitrosylate thiol-containing amino acids, thereby modulating the activities of a variety of enzymes, including caspases. Nitric oxide also reacts with heme-containing proteins including hemoglobin, mitochondrial cytochromes, and soluble guanylyl cyclase (sGC). The rapid reaction of NO with oxyhemoglobin (containing Fe^{2+}-heme) leads to the formation of methemoglobin (containing Fe^{3+}-heme) and nitrate. Nitric oxide inhibits electron transport by competing with oxygen for heme moieties in mitochondrial cytochrome enzymes.

Binding of NO to the heme moiety in sGC activates the enzyme leading to the synthesis of cGMP from GTP. Cyclic GMP reacts with a variety of intracellular targets including phosphodiesterases, cyclic nucleotide-gated ion channels, and cGMP-dependent protein kinases. Cyclic GMP activation of cGMP-dependent protein kinases modulates a broad range of biological processes including vascular relaxation, inhibition of platelet and leukocyte activation, and modulation of gastrointestinal motility and neurotransmission. The actions of cGMP are limited by phosphodiesterases (PDEs), some of which are cGMP-specific (e.g., PDE5).

Nitric oxide-generating drugs, including nitroglycerin and sodium nitroprusside, have long been used by clinicians to treat patients with a broad spectrum of cardiovascular diseases including angina pectoris and systemic hypertension. The application of NO-donor compounds to the treatment of patients with pulmonary hypertension is severely limited by the systemic vasodilator effects of these agents [1]. Moreover, in patients with lung injury and ventilation-perfusion mismatch, systemic administration of NO-donor compounds reduces hypoxic pulmonary vasoconstriction leading to a deterioration of oxygenation (due to enhanced perfusion of poor-ventilated lung regions).

A variety of novel drugs have been developed that target cGMP metabolism including PDE5 inhibitors for male erectile dysfunction and pulmonary hypertension. More recently, agents which sensitize sGC to NO and/or directly activate the enzyme are being evaluated as treatments for pulmonary and systemic hypertension [2, 3].

In this chapter, we describe the development of inhaled NO as a therapeutic strategy to increase pulmonary vascular NO and cGMP concentrations. Inhaled NO induces pulmonary vasodilation without the side effects of systemic hypotension or hypoxemia. An overview of how inhaled NO elicits its effects is provided, and alternate approaches to increasing pulmonary NO and cGMP concentrations are considered. Several of the clinical applications of inhaled NO are presented, as is a discussion of the methods required for the safe administration of NO gas. Finally, we present experimental and clinical observations highlighting the potential applications of inhaled NO to the treatment of systemic vascular diseases.

Inhaled NO Is a Selective Pulmonary Vasodilator

Pulmonary-selective therapies are useful in a variety of disease states. For example, inhaled β-adrenergic agents and glucocorticoids are highly efficacious for treatment of asthma while avoiding the important side-effects associated with these drugs when they are administered systemically. Could inhalation of NO gas or a NO-donor compound be used to treat pulmonary hypertension without causing systemic hypotension or impairing ventilation-perfusion matching?

Prior to 1990, NO gas was viewed, almost exclusively, as a poison. Nitric oxide is generated by combustion engines and in chemical synthesis reactions used in industry. Nitric oxide levels can reach

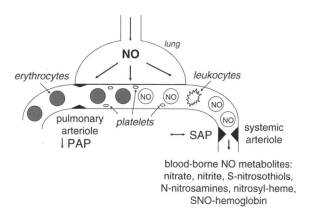

Fig. 14.1 Inhaled NO is a selective pulmonary vasodilator with actions on the systemic vasculature. A schematic of an alveolar-capillary unit is presented highlighting the ability of inhaled NO to dilate pulmonary arterioles and reduce pulmonary artery pressure (PAP). Inhaled NO does not dilate systemic arterioles or alter systemic arterial pressure (SAP) under normal conditions. Inhaled NO does have systemic effects, which are described in the text and which may be mediated by circulating cells exposed to NO in the lungs and blood-borne NO metabolites (Adapted with permission from reference [110])

1000 parts per million (ppm) in tobacco smoke [4]. Nitric oxide reacts with oxygen in the atmosphere to form nitrogen dioxide (NO_2), one of the greenhouse gases. NO_2, in turn, can react with water to form nitric acid, which causes lung injury. These considerations and others led the US Occupational Safety and Health Administration to establish a permissible exposure limit for NO of 25 ppm averaged over an 8-h work shift.

Zapol and Frostell hypothesized that inhalation of low concentrations of NO gas would be sufficient to relax pulmonary vascular smooth muscle [5]. Moreover, upon reaching the bloodstream, NO would be rapidly inactivated by oxyhemoglobin avoiding the systemic vasodilation seen with NO-donor compounds (Fig. 14.1). The methemoglobin formed in the reaction of NO with oxyhemoglobin would be reduced to hemoglobin by methemoglobin reductase present in erythrocytes. These investigators studied conscious instrumented adult lambs in which pulmonary hypertension was induced by systemic administration of the thromboxane analog, U46619, or by reducing the oxygen concentration in the inhaled gas mixture. By minimizing the time during which NO was mixed with oxygen prior to delivery to the animal, levels of inhaled NO_2 were maintained below 5 ppm. Breathing NO rapidly reduced pulmonary artery pressure (PAP) in a dose-dependent manner; inhalation of 80 ppm nearly completely reversed the pulmonary hypertension. The pulmonary vasodilator effects of breathing NO were rapidly reversible. Interestingly in the absence of pulmonary vasoconstriction, breathing NO did not alter pulmonary vascular resistance (PVR). Most importantly, breathing up to 80 ppm NO did not alter systemic blood pressure and did not cause lung injury. Moreover, breathing 80 ppm NO for up to 3 h did not significantly increase methemoglobin levels.

Mechanisms Responsible for the Pulmonary Vasodilator Effects of Breathing NO

As noted above, NO signals via both cGMP-dependent and cGMP-independent mechanisms. Several lines of evidence suggest that inhaled NO induces pulmonary vasodilation via a cGMP-dependent mechanism. In early studies, it was reported that breathing NO led to an increase in plasma cGMP levels [6]. Differences in arterial and venous cGMP levels suggested that the lungs were net producers of cGMP during NO inhalation [7].

Strategies designed to decrease cGMP metabolism by inhibiting PDE5 augmented the pulmonary vasodilator effects of inhaled NO in animal models [8] and in patients with pulmonary hypertension [9, 10]. In addition, administration of an agent that sensitizes sGC to NO, BAY41-2272, augmented the pulmonary vasodilator effects of inhaled NO [11]. As noted above, pulmonary vasodilator effects of breathing NO do not persist after gas administration is discontinued. On the other hand, during BAY41-2272 infusion, the duration of pulmonary vasodilation after discontinuing NO was greater than the duration of the pulmonary vasodilation in the absence of BAY 41-2272. These findings suggest that it may be possible to develop strategies for pulmonary vasodilation that require only intermittent administration of NO thereby reducing the amount of gas needed and potentially improving patient convenience.

sGC is a heterodimer composed of an α subunit and a β subunit. Although the mammalian genome has two α subunits and two β subunits, only α1β1 and α2β1 appear to be active, and α1β1 is thought to be the predominant isoform in the heart and vasculature. Janssens and his team reported that inhaled NO cannot induce pulmonary vasodilation in mice deficient in the sGC α1β1 isoform [12].

Taken together, these results strongly support the concept that inhaled NO vasodilates the pulmonary vasculature via its ability to stimulate cGMP synthesis by sGC. The clinical application of the combination of inhaled NO with PDE5 inhibitors or sGC activators remained relatively unexplored until recently. Concomitant administration of inhaled NO along with a PD5 inhibitor was more effective than either alone in protecting the heart from myocardial ischemia–reperfusion injury in mice [13]. Animal models also reveal that breathing NO during CPR markedly improved resuscitation success, 7-day neurological outcomes, and survival in a rat model of VF-induced cardiac arrest and CPR [14]. These results have been corroborated in large animal models of prolonged cardiac arrest and resuscitation [15]. These results support the beneficial effects of NO inhalation after cardiac arrest and CPR. It will be interesting to see how these data translate to humans.

Alternative Strategies to Increase Pulmonary NO and/or cGMP Concentrations Using Inhaled Drugs

Inhalation of NO-donor compounds is effective for treatment of systemic vascular diseases. For example, inhaled nitroglycerin is used to treat chest pain in patients with angina pectoris. In animals with pulmonary hypertension, aerosol inhalation of nitroglycerin [16, 17] or sodium nitroprusside [18] reduced PAP, but higher doses also lowered systemic blood pressure. In small clinical series, inhalation of nitroprusside improved oxygenation in hypoxic neonates [19] and reduced PAP and PVR in children with congenital heart disease [20].

Stamler and colleagues have explored the pulmonary vascular effects of inhaling an S-nitroso thiol, O-nitrosoethanol, which is resistant to reaction with oxygen or superoxide [21]. Inhalation of O-nitrosoethanol induced selective pulmonary vasodilation in hypoxic piglets. In a small non-randomized clinical study, these investigators reported that O-nitrosoethanol inhalation improved oxygenation and systemic hemodynamics in hypoxemic newborns [22].

Gladwin and colleagues have studied the therapeutic applications of nitrite as a NO reservoir (reviewed in reference [23]). Nitrite can be converted to NO by a variety of enzymes including hemoglobin and xanthine oxidoreductase. Hunter and colleagues reported that inhalation of high concentrations of nitrite reduced PAP in newborn lambs with hypoxia- and U46619-induced pulmonary vasoconstriction [24]. More recently, these investigators reported that intermittent administration of a low-dose nitrite aerosol could attenuate pulmonary vascular remodeling in rodent models of pulmonary arterial hypertension [25]. More recently inhaled nebulized nitrite reverses hemolysis-induced pulmonary vasoconstriction in newborn lambs without blood participation [26]. Inhaled nitrite limits organ injury and inflammation in animal models of hemorrhagic shock [27]. More recently, the safety of inhaled nitrite was tested in human subjects. Nebulized sodium nitrite was well tolerated following 6 days of every 8 h administration up to 90 mg, producing significant increases in circulating Hb(Fe)-NO, R-SNO, and FENO. Pulmonary absorption of nitrite was rapid and complete, and plasma exposure dose was proportional through the maximum tolerated dose of 90 mg, without accumulation following repeated inhalation. At higher dosage levels, dose-limiting toxicities were orthostasis (observed at 120 mg) and hypotension with tachycardia (at 176 mg), but venous methemoglobin did not exceed 3.0 % at any time in any subject. Neither the tolerability nor pharmacokinetics of nitrite was impacted by conditions of mild hypoxia, or co-administration with sildenafil, supporting the safe use of inhaled nitrite in the clinical setting of PAH.

A variety of other strategies using inhaled molecules have been developed to increase pulmonary cGMP concentrations. Studying awake lambs with U46619-induced pulmonary hypertension, Ichinose and colleagues demonstrated that inhalation of the PDE5 inhibitors, zaprinast or sildenafil, can selectively dilate the pulmonary vasculature and potentiate the effects of inhaled NO [28, 29]. More recently, Evgenov and colleagues studied the pulmonary vasodilator effects of inhaling sGC stimulators encapsulated in a biodegradable microparticle [30]. They observed that inhalation of microparticles containing BAY 41-8543 induced sustained pulmonary vasodilation in conscious sheep and could markedly potentiate the effects of inhaled NO. Of note, the ability of PDE5 inhibitors and sGC activators to potentiate the pulmonary vasodilator effects of NO appeared to be greater when the agents were delivered as an inhaled aerosol than when they were administered as an intravenous infusion.

Clinical Applications

Over the past 20 years, the application of inhaled NO to a wide variety of pulmonary and systemic disorders has been explored in experimental models and clinical trials. In this chapter, we highlight the application of inhaled NO to treat hypoxia in the newborn, to prevent the development of bronchopulmonary dysplasia (BPD) in premature infants, to improve matching of ventilation and perfusion, and to unload the right ventricle.

Persistent pulmonary hypertension of the newborn—Infants with PPHN typically present shortly after birth with respiratory distress and cyanosis, but with a structurally normal heart (reviewed in reference [31]). In these infants, the pulmonary vasculature fails to dilate normally with the first few breaths after birth, and high pulmonary vascular resistance leads to right-to-left shunting of deoxygenated blood via the ductus arterious and/or foramen ovale and results in severe hypoxemia. The systemic hypoxemia in PPHN often fails to improve with conventional therapies (e.g., supplemental oxygen), and acidosis and further pulmonary vasoconstriction may develop [32]. Extracorporeal membrane oxygenation (ECMO) is commonly required to rescue these infants.

The incidence of PPHN is estimated at 0.2 % of live-born term or near-term infants (≥34 weeks gestation). Persistent pulmonary hypertension of the newborn may be associated with normal lung parenchyma and excessively muscularized pulmonary vessels (idiopathic PPHN). In other cases, hypoplastic pulmonary vasculature, such as that associated with congenital diaphragmatic hernia, may cause pulmonary hypertension to persist after birth. In addition, PPHN is often associated with other conditions in which perinatal pulmonary vasodilation is inhibited including pulmonary diseases and sepsis.

Two pilot studies reported in the Lancet in 1992 demonstrated that breathing NO could increase oxygenation in severely hypoxemic newborns with PPHN without causing systemic hypotension [33, 34]. Randomized multicenter controlled studies confirmed that breathing NO can increase systemic oxygen levels in term and near-term babies with hypoxemia and pulmonary hypertension and reduce the need for ECMO (reviewed in references [35, 36]). It is important to note that breathing NO did not improve survival in PPHN patients in these trials likely because ECMO was used as a rescue therapy for infants with persistent hypoxemia. Moreover, inhaled NO did not improve survival or reduce ECMO use in infants with hypoxemia associated with congenital diaphragmatic hernia [37]. Hypoxemia in term and near-term infants with pulmonary hypertension is the only indication for which treatment with inhaled NO is currently approved by the US Food and Drug Administration.

Bronchopulmonary dysplasia—Bronchopulmonary dysplasia is a chronic respiratory disease seen in infants who were born prematurely and experience lung injury due to oxygen therapy and mechanical ventilation. Pathological findings are characterized by alveolar hypoplasia with reduced surface area available for gas exchange and abnormal pulmonary vascular development, which sometimes leads to pulmonary hypertension. In experimental animal models of BPD, breathing NO appeared to enhance lung growth [38–40] and decrease pulmonary vascular disease. An early report by Kinsella and colleagues suggested that breathing NO did not increase the survival of severely hypoxic premature newborns at risk for BPD [41]. However, in a single center randomized controlled trial, Schreiber and colleagues observed that breathing NO decreased the incidence of BPD and mortality in pre-term infants [42]. Multiple multicenter randomized controlled trials have followed up on these findings with mixed results. Comparison of these trials is challenging because of differences in the patient populations studied, as well as the dose, duration, and timing of NO administration. In two trials, early treatment of severely hypoxemic premature infants with inhaled NO did not reduce the incidence of BPD and tended to increase the frequency of severe intraventricular cerebral hemorrhage (IVH) [43, 44]. However, in subgroup analyses, it appeared that NO inhalation by larger infants (greater than 1 kg birth weight) reduced the incidence of BPD and IVH. Moreover, in another multicenter clinical trial in which older infants (7–21 days of age) at increased risk of developing BPD were studied, breathing NO also reduced the incidence of BPD [45]: On the other hand, in a recently completed trial of premature infants with mild to moderate respiratory distress, inhaled NO did not prevent BPD and did not reduce brain injury [46]. Whether or not inhaled NO is beneficial in subgroups of premature infants remains to be determined.

Improving matching of ventilation and perfusion with inhaled NO—In acute respiratory distress syndrome (ARDS) and acute lung injury (ALI), perfusion of poorly ventilated lung regions leads to hypoxemia that often is resistant to supplemental oxygen. By virtue of its method of administration, inhaled NO is selectively delivered to well-ventilated portions of the lung where it vasodilates the vasculature and improves matching of ventilation with perfusion and, thereby, enhances oxygenation. In 1993, Roissant and colleagues studied 9 patients with severe ARDS using the multiple inert-gas-elimination technique [47]. They reported that inhaled NO improved oxygenation and reduced PAP in these patients associated with a reduction in intrapulmonary shunting. These investigators later reported that the ability of inhaled NO to improve systemic oxygenation in patients with ALI did not require a reduction in mean PAP [48]. Only very low doses of NO were required to improve oxygenation, and concentrations higher than 10 ppm could worsen oxygenation, possible by spilling over to poorly ventilated lung regions. However, despite improving systemic oxygenation, inhaled NO did not improve outcomes in patient with ARDS in multiple randomized clinical trials (reviewed in reference [49]). Moreover, there was an increased risk of renal dysfunction in patients who were randomized to receive inhaled NO. Nonetheless, some clinicians use inhaled NO as a rescue therapy in patients with severe hypoxemia.

Unloading the right ventricle with inhaled NO—In adults and children undergoing cardiac surgery and cardiac transplantation, inhaled NO is frequently used to treat pulmonary hypertension and improve cardiac output in the perioperative period [50, 51]. Preoperative evaluation of pulmonary vasoreactivity is useful for identifying patients with pulmonary hypertension who are likely to respond to inhaled NO in the perioperative period (see "Evaluation of pulmonary vasoreactivity with inhaled NO" below).

Right ventricular failure in the perioperative period after placement of a left ventricular assist device is associated with increased mortality [52, 53]. Frequently, placement of a right ventricular assist device (RVAD) is required to maintain LVAD flow. Nitric oxide inhalation reduces PVR and can augment LVAD flow without causing systemic hypotension [54, 55]. In a small clinical trial of LVAD patients with elevated PVR, who were randomly assigned to treatment with and without inhaled NO, Argenziano and colleagues observed that inhaled NO reduced mean PAP and augmented LVAD flow [56]. In patients with RV failure after LVAD placement clinicians routinely administer inhaled NO prior to implantation of a RVAD. However, a randomized clinical trial demonstrating that inhaled NO decreases the necessity for RVAD placement has yet to be reported.

In a subgroup of patients presenting with ST elevation myocardial infarction (MI) and occlusion of the right coronary artery, RV myocardial infarction (RVMI) can cause RV failure, resulting in cardiogenic shock without pulmonary edema. Prompt revascularization is the most effective means to treat cardiogenic shock due to RVMI [57]. However, in patients, who present late in the course of their MI or in whom RV marginal coronary arteries cannot be fully revascularized, cardiogenic shock may require hemodynamic support with inotropic agents, intra-aortic balloon counterpulsation, or even RVAD placement. Several case reports [58–60] and one report of 13 patients [61] demonstrated that breathing NO can improve cardiac index in patients with cardiogenic shock due to RVMI. Moreover, in the subset of RVMI patients with hypoxemia due to right-to-left shunting via a patent foramen ovale, breathing NO reduces shunting and improves oxygenation [59, 61].

Evaluation of pulmonary vasoreactivity with inhaled NO—Acute pulmonary vasodilator testing (APVT) is performed in clinical practice for three primary indications. Most commonly, APVT is used in patients with World Health Organization (WHO) Group I pulmonary arterial hypertension [62] to identify those patients with greatest potential for sustained benefit to oral calcium-channel blocker (CCB) therapy and to evaluate clinical prognosis [63]. World Health Organization Group I PAH patients have a resting mean pulmonary arterial pressure (mPAP) $\geq$25 mmHg at rest with a pulmonary capillary wedge pressure of $\leq$15 mmHg (to exclude patients with pulmonary venous hypertension) and a PVR $\geq$3 Wood units. This group includes patients with idiopathic (IPAH) and familial (FPAH) PAH and patients with associated forms of PAH (APAH; e.g., pulmonary hypertension in conjunction with connective tissue disease, portal hypertension, HIV, drugs and toxins, etc.). Inhaled NO is the preferred agent of choice during APVT for PAH, since it is relatively well-tolerated, is selective for the pulmonary circulation, and has rapid onset and offset.

A positive response to APVT in PAH is currently defined as a decrease in mean PAP by at least 10 mmHg to an absolute level less than 40 mmHg without a decrease in cardiac output. The use of CCBs as a vasodilator in APVT should be avoided since a fraction of non-responders will develop severe adverse effects including shock and severe systemic hypotension [64]. On the other hand, the degree of vasoreactive responsiveness to inhaled NO during APVT is predictive of long-term benefit to CCB therapy in PAH [64, 65]. Although alternative agents such as intravenous epoprostenol or intravenous adenosine can be used, the acute administration of these agents has been associated with systemic hypotension and other adverse effects [66, 67].

APVT with inhaled NO has also been used in patients with congenital heart disease (CHD) and pulmonary hypertension undergoing preoperative evaluation for either corrective surgery or cardiac transplantation [68]. The presence of fixed PH is a predictor of perioperative morbidity and mortality in CHD. Although exact criteria have not been prospectively validated, many centers use a PVR less than 10–14 Wood units, a pulmonary:systemic vascular resistance (PVR:SVR) ratio less than or equal to 0.66, and a PVR:SVR ratio less than or equal to 0.33 during vasodilator administration as thresholds for determining operability for surgical correction and identifying patients with improved surgical prognosis. Preoperative APVT with a combination of inhaled NO and oxygen is more sensitive in identifying appropriate operative candidates than the use of oxygen alone in vasodilator testing [69].

APVT is also used to assess peri-operative risk in patients with acquired heart disease who undergoing evaluation for heart transplantation. Elevated resting PVR and transpulmonary gradient are independent predictors of increased post-transplant mortality [70–72]. Patients with fixed PH have a poorer postoperative prognosis than do patients who exhibit reversibility with APVT [73]. Although no thresholds have been prospectively validated, typical values used to determine candidacy for heart transplantation include a PVR less than or equal to 6 Wood units at rest or less than 3 Wood units with maximal vasodilation [74]. Patients exceeding these threshold values can be considered for combined heart-lung transplantation. Inhaled NO has been used for APVT in the preoperative evaluation for cardiac transplantation [75–77], but operators must proceed with caution because NO inhalation can increase LV filling pressures in the presence of LV dysfunction and result in pulmonary edema [75, 78, 79].

Methods Required for Safe Administration of Inhaled NO

NO reacts with oxygen to form the pollutant nitrogen dioxide (NO_2). The amount of NO_2 formed depends on the concentrations of NO and oxygen and for how long the two gases are mixed. Exposure of the lung and respiratory epithelium to NO_2 can cause pulmonary injury. Moreover, because NO is unstable when mixed with oxygen, it must be shipped and stored in nitrogen.

During the clinical development of inhaled NO, new technologies and equipment were devised to enable the reliable and continuous delivery of low concentrations of NO gas and the mixture of NO with oxygen (at concentrations dictated by the needs of the patient) via endotracheal tube, tight-fitting face mask, or nasal canula. Additional considerations included the requirement that the NO delivery system be portable to enable transport of patients breathing NO. Commercially available equipment minimizes the time during which oxygen and NO are mixed prior to administration to the patient and monitors the levels of NO, oxygen, and NO_2 administered to the patient in real-time. Clinicians who administer NO gas to patients need to be aware of several "side-effects" associated with breathing NO including methemoglobinemia, inhibition of platelet function, and increases in pulmonary capillary wedge pressure (in the setting of LV dysfunction; see "Evaluation of pulmonary vasoreactivity with inhaled NO" above). Clinicians also need to be cognizant of the rebound pulmonary hypertension that occurs if inhaled NO is abruptly discontinued.

Methemoglobinemia—When NO reacts with oxyhemoglobin to form nitrate, methemoglobin is formed. Methemoglobin does not bind oxygen, impairing the ability of the blood to deliver oxygen from the lungs to the periphery. Methemoglobin is rapidly converted to ferrous hemoglobin by methemoglobin reductase in erythrocytes. Methemoglobin levels greater than 5 % are typically seen only in patients who breathe 80 ppm NO [80] but are also rarely seen in infants breathing lesser concentrations [81].

Bleeding risk: Nitric oxide is well known to reduce the ability of platelets to adhere and become activated. Inhaled NO was reported to increase bleeding time in rabbits and people [82] and to reduce thrombosis after thrombolysis in dogs [83]. The risk of bleeding has been a particular concern in studies of the use of inhaled NO in premature infants who are already at increased risk of intracranial bleeding. However, recent studies of premature infants treated with inhaled NO have not shown an increased risk of IVH [46].

Rebound pulmonary hypertension: In the mid-1990s, two groups reported that pulmonary hypertension could develop as patients were weaned from breathing NO [84, 85]. Using an experimental lamb model, Black and Fineman and their colleagues reported that prolonged NO inhalation led to increased endothelin 1 levels and reduced NO synthase activity, both of which contributed to the pulmonary vascular rebound observed when inhaled NO was discontinued [86]. Administration of PDE5 inhibitors was found to attenuate the rebound seen when withdrawing inhaled NO [87, 88]. In general, however, pulmonary vascular rebound can be prevented by slowly weaning the concentration of inhaled NO.

Extrapulmonary Effects of Breathing NO

Initial studies by Zapol and colleagues suggested that the effects of breathing NO were limited to the lung, as reflected by the absence of systemic vasodilation [5]. However, in 1993, Hogman et al. reported that breathing NO could inhibit platelet function as reflected by a prolongation of the bleeding time [82]. These observations suggest that at least some of the inhaled NO reaching the bloodstream escaped inactivation by hemoglobin.

Although the impact of inhaled NO on platelet function remains controversial [89–91], Hogman's report led other research teams to identify additional systemic effects of breathing NO. For example, Lee and colleagues reported that breathing NO (80 ppm) for 2 weeks could reduce neointima formation in balloon-injured rat carotid arteries [92]. Semigran and colleagues demonstrated that breathing 80 ppm NO, but not 20 ppm, enhanced coronary artery patency in a canine model of coronary thrombosis after thrombolysis [83]. These investigators went on to show that PDE5 inhibitors markedly augment the ability of inhaled NO to prevent thrombosis [93].

Fox-Robichaud and colleagues explored the impact of inhaled NO on leukocyte function [94]. Studying a feline model of intestinal ischemia–reperfusion injury, these investigators reported that breathing 80 ppm NO, but not 20 ppm, could decrease leukocyte adherence and activation and maintain mesenteric blood flow.

Hataishi and colleagues extended these observations by studying the impact of breathing NO on cardiac ischemia–reperfusion injury [95]. Studying mice subjected to transient coronary artery ligation and reperfusion, they reported that breathing NO for 24 h (beginning during ischemia) could decrease MI size. Similar to the findings of Fox-Robichaud et al., Hataishi and his team reported that breathing 40 or 80 ppm NO was cardioprotective, whereas breathing 20 ppm NO was not. Liu and colleagues explored the impact of breathing NO on cardiac ischemia–reperfusion in a large animal model [96]. They observed that breathing 80 ppm NO reduced MI size in the hearts of pigs subjected to transient coronary artery occlusion, whereas intravenous administration of nitroglycerin did not. Of note, in both the mouse and pig models, breathing NO during cardiac ischemia and reperfusion reduced accumulation of leukocytes in the heart and importantly did not cause systemic hypotension. Taken together, these observations suggest that concentrations of NO required to elicit effects in the systemic circulation are greater than those required to modulate pulmonary vascular tone.

Promising results using inhaled NO in experimental models of ischemia–reperfusion injury have led to several small-scale clinical trials. Lang and colleagues reported that breathing 80 ppm NO reduced muscle inflammation in the lower extremities of patients undergoing limb ischemia (for regional anesthesia) and reperfusion [97]. These investigators went on to report that breathing NO reduced ischemia–reperfusion injury in liver transplantation associated with a reduction in hepatocyte apoptosis and improved synthetic function [98]. Although both of these studies were encouraging, they were small (~10 patients in each experimental group). Additional multicenter randomized controlled trials will be necessary to determine whether inhaled NO can improve outcomes in patients experiencing systemic organ ischemia–reperfusion injury.

Fox-Robichaud and colleagues reported that breathing NO elicited its greatest effects under conditions characterized by reduced NO levels, such as those caused by inhibiting NO synthase or induced by ischemia and reperfusion [94]. Studied performed by Gladwin and colleagues supported this hypothesis by demonstrating the breathing NO could increase forearm blood flow during regional blockade of NO synthesis [99, 100]. In contrast, Hataishi et al. failed to detect systemic vasodilation in response to inhaled NO in mice deficient in the endothelial NO synthase isoform (NOS3) [101].

In another model characterized by a systemic deficiency of NO, Gladwin and colleagues found that hemolysis induced systemic vasoconstriction via the ability of cell-free hemoglobin to scavenge NO generated by the endothelium. They reported that breathing 80 ppm NO dramatically reduced systemic blood pressure in dogs with systemic vasoconstriction induced by hemolysis associated with nearly complete oxidation of cell-free hemoglobin to methemoglobin [102]. These investigators hypothesized that oxidation of extracellular hemoglobin is required for the vasodilator effects of inhaled NO in hemolysis. This hypothesis was challenged by the findings of Yu et al. [103], who reported that pretreatment of mice with inhaled NO could prevent the vasoconstriction induced by the subsequent administration of cell-free hemoglobin without oxidizing the extracellular hemoglobin to methemoglobin. The observations of Yu and colleagues suggest that pretreatment with inhaled NO may enable administration of hemoglobin-based oxygen carriers as blood substitutes (reviewed in reference [104]).

How does inhaled NO elicit its extrapulmonary effects?—Because of the exceedingly short half-life of NO in biological fluids, it seems unlikely that NO reaching the bloodstream during NO inhalation will be transported to the periphery in an unmodified form. Several hypotheses have been proposed to account for the ability of inhaled NO to elicit systemic effects (Fig. 14.1). One hypothesis was that exposure of platelets and leukocytes to NO, as they transit the pulmonary circulation, would inhibit their function in the periphery. Fox-Robichaud et al. reported that breathing NO did not alter the activation of leukocytes exposed to an artificial surface suggesting that inhaled NO does not directly inactivate leukocytes [94].

An alternate hypothesis is that inhaled NO that reaches the circulation can react with plasma and/or cellular constituents to form stable NO metabolites such as nitrite and *S*-nitrosothiols. These stable NO metabolites, in turn, regenerate NO in the periphery. A variety of NO metabolites have been detected in animals and human beings breathing NO. NO can react with heme in hemoglobin forming nitrosyl-hemoglobin and can nitrosylate the cysteine-93 in the hemoglobin β unit to form SNO-hemoglobin. SNO-hemoglobin formation depends on oxygen concentrations: NO reacts with cysteine-93 in the presence of high oxygen concentrations in the lung, and SNO-hemoglobin releases NO under the low oxygen concentrations present in the periphery (reviewed in reference [105]). In fact, this mechanism has now been shown to be necessary and sufficient for oxygen delivery from hemoglobin in the periphery [106].

Breathing NO leads to a modest increase in the blood levels of nitrite and a more marked increase in nitrate levels [99]. Recent studies have demonstrated that administration of nitrite can reduce cardiac ischemia–reperfusion injury [107, 108] leading to the speculation that the cardioprotective effects of inhaled NO are attributable to nitrite.

Nitric oxide reacts with a broad spectrum of plasma components to produce *S*-nitrosothiols (RSNO) and *N*-nitrosamines (RNNO). Recently, Nagasaka and Fernandes and their colleagues performed a quantitative analysis of the NO metabolites formed in blood and tissues during NO breathing in mice [109]. They reported that NO inhalation led to the rapid accumulation of nitrate and nitrite in plasma and erythrocytes, as well as a dramatic increase in RSNO, RNNO, and nitrosyl-heme concentrations in erythrocytes (Fig. 14.2). These findings suggest that both blood compartments could contribute to the transport of NO metabolites to the periphery. During NO inhalation, NO metabolites accumulated in all of the tissues sampled, but tissue accumulation of NO metabolites during NO inhalation was organ-specific. Taken together, these observations support the hypothesis that inhaled NO elicits its systemic effects, at least in part, via transport of NO metabolites to the periphery.

Conclusions

Over 20 years ago, Frostell and Zapol learned that breathing low concentrations of a "poisonous" gas, NO, could reduce pulmonary vascular tone in a uniquely selective manner. Extensive basic and clinical research led to the rapid translation of the remarkably safe pulmonary vasodilator effects of breathing NO to benefit neonates with hypoxemia and pulmonary hypertension. On the other hand, randomized clinical trials have demonstrated that inhaled NO does not improve outcomes in most patients with ARDS, and the impact of inhaled NO on BPD is at best uncertain. The observations in

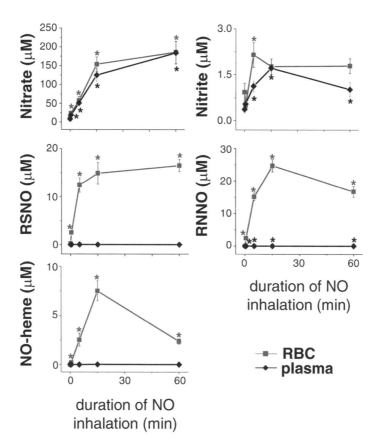

Fig. 14.2 Distribution and kinetics of accumulation of NO-metabolites in mice breathing nitric oxide (plasma and erythrocytes). Concentrations of NO-metabolites were measured in blood of mice breathing air supplemented with NO (80 ppm) for 0, 0.5, 5, 15, and 60 min ($n = 5–7$). Abbreviations: erythrocytes (RBC), nitrosyl-heme species (NO-heme), N-nitrosamines (RNNO), S-nitrosothiols (RSNO). $*P < 0.05$ vs. mice not breathing nitric oxide (Adapted with permission from reference [109])

animal models and small clinical trials that breathing NO can load the body with NO metabolites, such as nitrite and nitrate, sufficient to decrease ischemia–reperfusion injury without causing systemic hypotension are exciting and merit additional investigation.

Acknowledgements and Disclosures The authors acknowledge research funding support from the Deutsche Forschungsgemeinschaft DFG (A.U.S.), NHLBI 5 K08 HL111210 (R.M.), and a sponsored research agreement with Ikaria Inc. (K.D.B.).

The authors acknowledge the mentorship and support of Warren M. Zapol M.D. The authors thank Dr. Fumito Ichinose and Jesse D. Roberts Jr. for their critical review of this chapter.

The MGH has obtained patents relating to the use of inhaled nitric oxide and has licensed them to Ikaria Inc. and Linde Gas Therapeutics AB.

References

1. Packer M, Halperin JL, Brooks KM, Rothlauf EB, Lee WH. Nitroglycerin therapy in the management of pulmonary hypertensive disorders. Am J Med. 1984;76:67–75.
2. Stasch JP, Hobbs AJ (2009) NO-independent, haem-dependent soluble guanylate cyclase stimulators. Handb Exp Pharmacol. 2009;(191):277–308.

3. Schmidt HH, Schmidt PM, Stasch JP (2009) NO- and haem-independent soluble guanylate cyclase activators. Handb Exp Pharmacol. 2009;(191):309–39.
4. Norman V, Keith CH. Nitrogen oxides in tobacco smoke. Nature. 1965;205:915–6.
5. Frostell C, Fratacci MD, Wain JC, Jones R, Zapol WM. Inhaled nitric oxide. A selective pulmonary vasodilator reversing hypoxic pulmonary vasoconstriction. Circulation. 1991;83:2038–47.
6. Roberts Jr JD, Chen TY, Kawai N, Wain J, Dupuy P, et al. Inhaled nitric oxide reverses pulmonary vasoconstriction in the hypoxic and acidotic newborn lamb. Circ Res. 1993;72:246–54.
7. Rovira I, Chen TY, Winkler M, Kawai N, Bloch KD, et al. Effects of inhaled nitric oxide on pulmonary hemodynamics and gas exchange in an ovine model of ARDS. J Appl Physiol. 1994;76:345–55.
8. Ichinose F, Adrie C, Hurford WE, Zapol WM. Prolonged pulmonary vasodilator action of inhaled nitric oxide by Zaprinast in awake lambs. J Appl Physiol. 1995;78:1288–95.
9. Ivy DD, Ziegler JW, Kinsella JP, Wiggins JW, Abman SH. Hemodynamic effects of dipyridamole and inhaled nitric oxide in pediatric patients with pulmonary hypertension. Chest. 1998;114:17S.
10. Lepore JJ, Maroo A, Pereira NL, Ginns LC, Dec GW, et al. Effect of sildenafil on the acute pulmonary vasodilator response to inhaled nitric oxide in adults with primary pulmonary hypertension. Am J Cardiol. 2002;90:677–80.
11. Evgenov OV, Ichinose F, Evgenov NV, Gnoth MJ, Falkowski GE, et al. Soluble guanylate cyclase activator reverses acute pulmonary hypertension and augments the pulmonary vasodilator response to inhaled nitric oxide in awake lambs. Circulation. 2004;110:2253–9.
12. Vermeersch P, Buys E, Pokreisz P, Marsboom G, Ichinose F, et al. Soluble guanylate cyclase-alpha1 deficiency selectively inhibits the pulmonary vasodilator response to nitric oxide and increases the pulmonary vascular remodeling response to chronic hypoxia. Circulation. 2007;116:936–43.
13. Lux A, Pokreisz P, Swinnen M, Caluwe E, Gillijns H, et al. Concomitant phosphodiesterase 5-inhibition enhances myocardial protection of inhaled nitric oxide in ischemia-reperfusion injury. 2015. J Pharmacol Exp Ther.
14. Brucken A, Derwall M, Bleilevens C, Stoppe C, Gotzenich A, et al. Brief inhalation of nitric oxide increases resuscitation success and improves 7-day-survival after cardiac arrest in rats: a randomized controlled animal study. Crit Care. 2015;19:408.
15. Derwall M, Ebeling A, Nolte KW, Weis J, Rossaint R, et al. Inhaled nitric oxide improves transpulmonary blood flow and clinical outcomes after prolonged cardiac arrest: a large animal study. Crit Care. 2015;19:328.
16. Bando M, Ishii Y, Kitamura S, Ohno S. Effects of inhalation of nitroglycerin on hypoxic pulmonary vasoconstriction. Respiration. 1998;65:63–70.
17. Gong F, Shiraishi H, Kikuchi Y, Hoshina M, Ichihashi K, et al. Inhalation of nebulized nitroglycerin in dogs with experimental pulmonary hypertension induced by U46619. Pediatr Int. 2000;42:255–8.
18. Adrie C, Ichinose F, Holzmann A, Keefer L, Hurford WE, et al. Pulmonary vasodilation by nitric oxide gas and prodrug aerosols in acute pulmonary hypertension. J Appl Physiol. 1998;84:435–41.
19. Palhares DB, Figueiredo CS, Moura AJ. Endotracheal inhalatory sodium nitroprusside in severely hypoxic newborns. J Perinat Med. 1998;26:219–24.
20. Goyal P, Kiran U, Chauhan S, Juneja R, Choudhary M. Efficacy of nitroglycerin inhalation in reducing pulmonary arterial hypertension in children with congenital heart disease. Br J Anaesth. 2006;97:208–14.
21. Moya MP, Gow AJ, McMahon TJ, Toone EJ, Cheifetz IM, et al. S-nitrosothiol repletion by an inhaled gas regulates pulmonary function. Proc Natl Acad Sci U S A. 2001;98:5792–7.
22. Moya MP, Gow AJ, Califf RM, Goldberg RN, Stamler JS. Inhaled ethyl nitrite gas for persistent pulmonary hypertension of the newborn. Lancet. 2002;360:141–3.
23. Lundberg JO, Weitzberg E, Gladwin MT. The nitrate-nitrite-nitric oxide pathway in physiology and therapeutics. Nat Rev Drug Discov. 2008;7:156–67.
24. Hunter CJ, Dejam A, Blood AB, Shields H, Kim-Shapiro DB, et al. Inhaled nebulized nitrite is a hypoxia-sensitive NO-dependent selective pulmonary vasodilator. Nat Med. 2004;10:1122–7.
25. Zuckerbraun BS, Shiva S, Ifedigbo E, Mathier MA, Mollen KP, et al. Nitrite potently inhibits hypoxic and inflammatory pulmonary arterial hypertension and smooth muscle proliferation via xanthine oxidoreductase-dependent nitric oxide generation. Circulation. 2010;121:98–109.
26. Blood AB, Schroeder HJ, Terry MH, Merrill-Henry J, Bragg SL, et al. Inhaled nitrite reverses hemolysis-induced pulmonary vasoconstriction in newborn lambs without blood participation. Circulation. 2011;123:605–12.
27. Kautza B, Gomez H, Escobar D, Corey C, Ataya B, et al. Inhaled, nebulized sodium nitrite protects in murine and porcine experimental models of hemorrhagic shock and resuscitation by limiting mitochondrial injury. Nitric Oxide. 2015;51:7–18.
28. Ichinose F, Adrie C, Hurford WE, Bloch KD, Zapol WM. Selective pulmonary vasodilation induced by aerosolized zaprinast. Anesthesiology. 1998;88:410–6.
29. Ichinose F, Erana-Garcia J, Hromi J, Raveh Y, Jones R, et al. Nebulized sildenafil is a selective pulmonary vasodilator in lambs with acute pulmonary hypertension. Crit Care Med. 2001;29:1000–5.
30. Evgenov OV, Kohane DS, Bloch KD, Stasch JP, Volpato GP, et al. Inhaled agonists of soluble guanylate cyclase induce selective pulmonary vasodilation. Am J Respir Crit Care Med. 2007;176:1138–45.

31. Steinhorn RH. Neonatal pulmonary hypertension. Pediatr Crit Care Med. 2010;11:S79–84.

32. Rao S, Bartle D, Patole S. Current and future therapeutic options for persistent pulmonary hypertension in the newborn. Expert Rev Cardiovasc Ther. 2010;8:845–62.

33. Roberts JD, Polaner DM, Lang P, Zapol WM. Inhaled nitric oxide in persistent pulmonary hypertension of the newborn. Lancet. 1992;340:818–9.

34. Kinsella JP, Neish SR, Shaffer E, Abman SH. Low-dose inhalation nitric oxide in persistent pulmonary hypertension of the newborn. Lancet. 1992;340:819–20.

35. Konduri GG, Kim UO. Advances in the diagnosis and management of persistent pulmonary hypertension of the newborn. Pediatr Clin North Am. 2009;56:579–600, Table of Contents.

36. Finer NN, Barrington KJ. Nitric oxide for respiratory failure in infants born at or near term. Cochrane Database Syst Rev. 2001;18(4):CD000399.

37. The Neonatal Inhaled Nitric Oxide Study Group (NINOS). Inhaled nitric oxide and hypoxic respiratory failure in infants with congenital diaphragmatic hernia. Pediatrics. 1997;99:838–45.

38. Roberts Jr JD, Chiche JD, Weimann J, Steudel W, Zapol WM, et al. Nitric oxide inhalation decreases pulmonary artery remodeling in the injured lungs of rat pups. Circ Res. 2000;87:140–5.

39. Tang JR, Markham NE, Lin YJ, McMurtry IF, Maxey A, et al. Inhaled nitric oxide attenuates pulmonary hypertension and improves lung growth in infant rats after neonatal treatment with a VEGF receptor inhibitor. Am J Physiol Lung Cell Mol Physiol. 2004;287:L344–51.

40. McCurnin DC, Pierce RA, Chang LY, Gibson LL, Osborne-Lawrence S, et al. Inhaled NO improves early pulmonary function and modifies lung growth and elastin deposition in a baboon model of neonatal chronic lung disease. Am J Physiol Lung Cell Mol Physiol. 2005;288:L450–9.

41. Kinsella JP, Walsh WF, Bose CL, Gerstmann DR, Labella JJ, et al. Inhaled nitric oxide in premature neonates with severe hypoxaemic respiratory failure: a randomised controlled trial. Lancet. 1999;354:1061–5.

42. Schreiber MD, Gin-Mestan K, Marks JD, Huo D, Lee G, et al. Inhaled nitric oxide in premature infants with the respiratory distress syndrome. N Engl J Med. 2003;349:2099–107.

43. Van Meurs KP, Wright LL, Ehrenkranz RA, Lemons JA, Ball MB, et al. Inhaled nitric oxide for premature infants with severe respiratory failure. N Engl J Med. 2005;353:13–22.

44. Kinsella JP, Cutter GR, Walsh WF, Gerstmann DR, Bose CL, et al. Early inhaled nitric oxide therapy in premature newborns with respiratory failure. N Engl J Med. 2006;355:354–64.

45. Ballard RA, Truog WE, Cnaan A, Martin RJ, Ballard PL, et al. Inhaled nitric oxide in preterm infants undergoing mechanical ventilation. N Engl J Med. 2006;355:343–53.

46. Mercier JC, Hummler H, Durrmeyer X, Sanchez-Luna M, Carnielli V, et al. Inhaled nitric oxide for prevention of bronchopulmonary dysplasia in premature babies (EUNO): a randomised controlled trial. Lancet. 2010;376:346–54.

47. Rossaint R, Falke KJ, Lopez F, Slama K, Pison U, et al. Inhaled nitric oxide for the adult respiratory distress syndrome. N Engl J Med. 1993;328:399–405.

48. Gerlach H, Rossaint R, Pappert D, Falke KJ. Time-course and dose-response of nitric oxide inhalation for systemic oxygenation and pulmonary hypertension in patients with adult respiratory distress syndrome. Eur J Clin Invest. 1993;23:499–502.

49. Adhikari NK, Burns KE, Friedrich JO, Granton JT, Cook DJ, et al. Effect of nitric oxide on oxygenation and mortality in acute lung injury: systematic review and meta-analysis. BMJ. 2007;334:779.

50. Fullerton DA, Jones SD, Jaggers J, Piedalue F, Grover FL, et al. Effective control of pulmonary vascular resistance with inhaled nitric oxide after cardiac operation. J Thorac Cardiovasc Surg. 1996;111:753–62. discussion 762-753.

51. Beck JR, Mongero LB, Kroslowitz RM, Choudhri AF, Chen JM, et al. Inhaled nitric oxide improves hemodynamics in patients with acute pulmonary hypertension after high-risk cardiac surgery. Perfusion. 1999;14:37–42.

52. Matthews JC, Koelling TM, Pagani FD, Aaronson KD. The right ventricular failure risk score a pre-operative tool for assessing the risk of right ventricular failure in left ventricular assist device candidates. J Am Coll Cardiol. 2008;51:2163–72.

53. Drakos SG, Janicki L, Horne BD, Kfoury AG, Reid BB, et al. Risk factors predictive of right ventricular failure after left ventricular assist device implantation. Am J Cardiol. 2010;105:1030–5.

54. Hare JM, Shernan SK, Body SC, Graydon E, Colucci WS, et al. Influence of inhaled nitric oxide on systemic flow and ventricular filling pressure in patients receiving mechanical circulatory assistance. Circulation. 1997;95:2250–3.

55. Wagner F, Dandel M, Gunther G, Loebe M, Schulze-Neick I, et al. Nitric oxide inhalation in the treatment of right ventricular dysfunction following left ventricular assist device implantation. Circulation. 1997;96:II-291–6.

56. Argenziano M, Choudhri AF, Moazami N, Rose EA, Smith CR, et al. Randomized, double-blind trial of inhaled nitric oxide in LVAD recipients with pulmonary hypertension. Ann Thorac Surg. 1998;65:340–5.

57. Bowers TR, O'Neill WW, Grines C, Pica MC, Safian RD, et al. Effect of reperfusion on biventricular function and survival after right ventricular infarction. N Engl J Med. 1998;338:933–40.

58. Fujita Y, Nishida O, Sobue K, Ito H, Kusama N, et al. Nitric oxide inhalation is useful in the management of right ventricular failure caused by myocardial infarction. Crit Care Med. 2002;30:1379–81.

59. Fessler MB, Lepore JJ, Thompson BT, Semigran MJ. Right-to-left shunting through a patent foramen ovale in right ventricular infarction: improvement of hypoxemia and hemodynamics with inhaled nitric oxide. J Clin Anesth. 2003;15:371–4.

60. Valenti V, Patel AJ, Sciarretta S, Kandil H, Fabrizio B, et al. Use of inhaled nitric oxide in the treatment of right ventricular myocardial infarction. Am J Emerg Med. 2011;29(4):473.e3–5.

61. Inglessis I, Shin JT, Lepore JJ, Palacios IF, Zapol WM, et al. Hemodynamic effects of inhaled nitric oxide in right ventricular myocardial infarction and cardiogenic shock. J Am Coll Cardiol. 2004;44:793–8.

62. Kelm M, Feelisch M, Spahr R, Piper HM, Noack E, et al. Quantitative and kinetic characterization of nitric oxide and EDRF released from cultured endothelial cells. Biochem Biophys Res Commun. 1988;154:236–44.

63. McLaughlin VV, Archer SL, Badesch DB, Barst RJ, Farber HW, et al. ACCF/AHA 2009 expert consensus document on pulmonary hypertension a report of the American College of Cardiology Foundation Task Force on Expert Consensus Documents and the American Heart Association developed in collaboration with the American College of Chest Physicians; American Thoracic Society, Inc.; and the Pulmonary Hypertension Association. J Am Coll Cardiol. 2009;53:1573–619.

64. Sitbon O, Humbert M, Jagot JL, Taravella O, Fartoukh M, et al. Inhaled nitric oxide as a screening agent for safely identifying responders to oral calcium-channel blockers in primary pulmonary hypertension. Eur Respir J. 1998;12:265–70.

65. Sitbon O, Humbert M, Jais X, Ioos V, Hamid AM, et al. Long-term response to calcium channel blockers in idiopathic pulmonary arterial hypertension. Circulation. 2005;111:3105–11.

66. Badesch DB, Abman SH, Ahearn GS, Barst RJ, McCrory DC, et al. Medical therapy for pulmonary arterial hypertension: ACCP evidence-based clinical practice guidelines. Chest. 2004;126:35S–62.

67. Oliveira EC, Ribeiro AL, Amaral CF. Adenosine for vasoreactivity testing in pulmonary hypertension: a head-to-head comparison with inhaled nitric oxide. Respir Med. 2010;104:606–11.

68. Warnes CA, Williams RG, Bashore TM, Child JS, Connolly HM, et al. ACC/AHA 2008 Guidelines for the Management of Adults with Congenital Heart Disease: a report of the American College of Cardiology/American Heart Association Task Force on Practice Guidelines (writing committee to develop guidelines on the management of adults with congenital heart disease). Circulation. 2008;118:e714–833.

69. Balzer DT, Kort HW, Day RW, Corneli HM, Kovalchin JP, et al. Inhaled Nitric Oxide as a Preoperative Test (INOP Test I): the INOP Test Study Group. Circulation. 2002;106:I76–81.

70. Kirklin JK, Naftel DC, Kirklin JW, Blackstone EH, White-Williams C, et al. Pulmonary vascular resistance and the risk of heart transplantation. J Heart Transplant. 1988;7:331–6.

71. Erickson KW, Costanzo-Nordin MR, O'Sullivan EJ, Johnson MR, Zucker MJ, et al. Influence of preoperative transpulmonary gradient on late mortality after orthotopic heart transplantation. J Heart Transplant. 1990;9:526–37.

72. Murali S, Kormos RL, Uretsky BF, Schechter D, Reddy PS, et al. Preoperative pulmonary hemodynamics and early mortality after orthotopic cardiac transplantation: the Pittsburgh experience. Am Heart J. 1993;126:896–904.

73. Chen JM, Levin HR, Michler RE, Prusmack CJ, Rose EA, et al. Reevaluating the significance of pulmonary hypertension before cardiac transplantation: determination of optimal thresholds and quantification of the effect of reversibility on perioperative mortality. J Thorac Cardiovasc Surg. 1997;114:627–34.

74. Natale ME, Pina IL. Evaluation of pulmonary hypertension in heart transplant candidates. Curr Opin Cardiol. 2003;18:136–40.

75. Semigran MJ, Cockrill BA, Kacmarek R, Thompson BT, Zapol WM, et al. Hemodynamic effects of inhaled nitric oxide in heart failure. J Am Coll Cardiol. 1994;24:982–8.

76. Adatia I, Perry S, Landzberg M, Moore P, Thompson JE, et al. Inhaled nitric oxide and hemodynamic evaluation of patients with pulmonary hypertension before transplantation. J Am Coll Cardiol. 1995;25:1656–64.

77. Mahajan A, Shabanie A, Varshney SM, Marijic J, Sopher MJ. Inhaled nitric oxide in the preoperative evaluation of pulmonary hypertension in heart transplant candidates. J Cardiothorac Vasc Anesth. 2007;21:51–6.

78. Bocchi EA, Bacal F, Auler Junior JO, Carmone MJ, Bellotti G, et al. Inhaled nitric oxide leading to pulmonary edema in stable severe heart failure. Am J Cardiol. 1994;74:70–2.

79. Loh E, Stamler JS, Hare JM, Loscalzo J, Colucci WS. Cardiovascular effects of inhaled nitric oxide in patients with left ventricular dysfunction. Circulation. 1994;90:2780–5.

80. Weinberger B, Laskin DL, Heck DE, Laskin JD. The toxicology of inhaled nitric oxide. Toxicol Sci. 2001;59:5–16.

81. Hamon I, Gauthier-Moulinier H, Grelet-Dessioux E, Storme L, Fresson J, et al. Methaemoglobinaemia risk factors with inhaled nitric oxide therapy in newborn infants. Pediatr Res. 2015;77(3):472–6.

82. Hogman M, Frostell C, Arnberg H, Hedenstierna G. Bleeding time prolongation and NO inhalation. Lancet. 1993;341:1664–5.

83. Adrie C, Bloch KD, Moreno PR, Hurford WE, Guerrero JL, et al. Inhaled nitric oxide increases coronary artery patency after thrombolysis. Circulation. 1996;94:1919–26.

84. Miller OI, Tang SF, Keech A, Celermajer DS. Rebound pulmonary hypertension on withdrawal from inhaled nitric oxide. Lancet. 1995;346:51–2.
85. Atz AM, Adatia I, Wessel DL. Rebound pulmonary hypertension after inhalation of nitric oxide. Ann Thorac Surg. 1996;62:1759–64.
86. Oishi P, Grobe A, Benavidez E, Ovadia B, Harmon C, et al. Inhaled nitric oxide induced NOS inhibition and rebound pulmonary hypertension: a role for superoxide and peroxynitrite in the intact lamb. Am J Physiol Lung Cell Mol Physiol. 2006;290:L359–66.
87. Ivy DD, Kinsella JP, Ziegler JW, Abman SH. Dipyridamole attenuates rebound pulmonary hypertension after inhaled nitric oxide withdrawal in postoperative congenital heart disease. J Thorac Cardiovasc Surg. 1998;115:875–82.
88. Namachivayam P, Theilen U, Butt WW, Cooper SM, Penny DJ, et al. Sildenafil prevents rebound pulmonary hypertension after withdrawal of nitric oxide in children. Am J Respir Crit Care Med. 2006;174:1042–7.
89. Albert J, Norman M, Wallen NH, Frostell C, Hjemdahl P. Inhaled nitric oxide does not influence bleeding time or platelet function in healthy volunteers. Eur J Clin Invest. 1999;29:953–9.
90. Gries A, Herr A, Motsch J, Holzmann A, Weimann J, et al. Randomized, placebo-controlled, blinded and cross-matched study on the antiplatelet effect of inhaled nitric oxide in healthy volunteers. Thromb Haemost. 2000;83:309–15.
91. Beghetti M, Sparling C, Cox PN, Stephens D, Adatia I. Inhaled NO inhibits platelet aggregation and elevates plasma but not intraplatelet cGMP in healthy human volunteers. Am J Physiol Heart Circ Physiol. 2003;285:H637–42.
92. Lee JS, Adrie C, Jacob HJ, Roberts Jr JD, Zapol WM, et al. Chronic inhalation of nitric oxide inhibits neointimal formation after balloon-induced arterial injury. Circ Res. 1996;78:337–42.
93. Schmidt U, Han RO, DiSalvo TG, Guerrero JL, Gold HK, et al. Cessation of platelet-mediated cyclic canine coronary occlusion after thrombolysis by combining nitric oxide inhalation with phosphodiesterase-5 inhibition. J Am Coll Cardiol. 2001;37:1981–8.
94. Fox-Robichaud A, Payne D, Hasan SU, Ostrovsky L, Fairhead T, et al. Inhaled NO as a viable antiadhesive therapy for ischemia/reperfusion injury of distal microvascular beds. J Clin Invest. 1998;101:2497–505.
95. Hataishi R, Rodrigues AC, Neilan TG, Morgan JG, Buys E, et al. Inhaled nitric oxide decreases infarction size and improves left ventricular function in a murine model of myocardial ischemia-reperfusion injury. Am J Physiol Heart Circ Physiol. 2006;291:H379–84.
96. Liu X, Huang Y, Pokreisz P, Vermeersch P, Marsboom G, et al. Nitric oxide inhalation improves microvascular flow and decreases infarction size after myocardial ischemia and reperfusion. J Am Coll Cardiol. 2007;50:808–17.
97. Mathru M, Huda R, Solanki DR, Hays S, Lang JD. Inhaled nitric oxide attenuates reperfusion inflammatory responses in humans. Anesthesiology. 2007;106:275–82.
98. Lang Jr JD, Teng X, Chumley P, Crawford JH, Isbell TS, et al. Inhaled NO accelerates restoration of liver function in adults following orthotopic liver transplantation. J Clin Invest. 2007;117:2583–91.
99. Cannon 3rd RO, Schechter AN, Panza JA, Ognibene FP, Pease-Fye ME, et al. Effects of inhaled nitric oxide on regional blood flow are consistent with intravascular nitric oxide delivery. J Clin Invest. 2001;108:279–87.
100. Zuzak KJ, Schaeberle MD, Gladwin MT, Cannon 3rd RO, Levin IW. Noninvasive determination of spatially resolved and time-resolved tissue perfusion in humans during nitric oxide inhibition and inhalation by use of a visible-reflectance hyperspectral imaging technique. Circulation. 2001;104:2905–10.
101. Hataishi R, Zapol WM, Bloch KD, Ichinose F. Inhaled nitric oxide does not reduce systemic vascular resistance in mice. Am J Physiol Heart Circ Physiol. 2006;290:H1826–9.
102. Minneci PC, Deans KJ, Zhi H, Yuen PS, Star RA, et al. Hemolysis-associated endothelial dysfunction mediated by accelerated NO inactivation by decompartmentalized oxyhemoglobin. J Clin Invest. 2005;115:3409–17.
103. Yu B, Raher MJ, Volpato GP, Bloch KD, Ichinose F, et al. Inhaled nitric oxide enables artificial blood transfusion without hypertension. Circulation. 2008;117:1982–90.
104. Yu B, Bloch KD, Zapol WM. Hemoglobin-based red blood cell substitutes and nitric oxide. Trends Cardiovasc Med. 2009;19:103–7.
105. Lima B, Forrester MT, Hess DT, Stamler JS. S-nitrosylation in cardiovascular signaling. Circ Res. 2010;106:633–46.
106. Zhang R, Hess DT, Qian Z, Hausladen A, Fonseca F, et al. Hemoglobin betaCys93 is essential for cardiovascular function and integrated response to hypoxia. Proc Natl Acad Sci U S A. 2015;112:6425–30.
107. Duranski MR, Greer JJ, Dejam A, Jaganmohan S, Hogg N, et al. Cytoprotective effects of nitrite during in vivo ischemia-reperfusion of the heart and liver. J Clin Invest. 2005;115:1232–40.
108. Gonzalez FM, Shiva S, Vincent PS, Ringwood LA, Hsu LY, et al. Nitrite anion provides potent cytoprotective and antiapoptotic effects as adjunctive therapy to reperfusion for acute myocardial infarction. Circulation. 2008;117:2986–94.
109. Nagasaka Y, Fernandez BO, Garcia-Saura MF, Petersen B, Ichinose F, et al. Brief periods of nitric oxide inhalation protect against myocardial ischemia-reperfusion injury. Anesthesiology. 2008;109:675–82.
110. Bloch KD, Ichinose F, Roberts Jr JD, Zapol WM. Inhaled NO as a therapeutic agent. Cardiovasc Res. 2007;75:339–48.

Chapter 15
Pharmacology of Nitrovasodilators

Thomas Münzel and Andreas Daiber

Key Points

- Organic nitrates such as nitroglycerin, isosorbide mono- and dinitrate, and pentaerythrityl tetra-nitrate are potent vasodilators thereby improving clinical symptoms in patients with cardiovascular disease.
- The mechanism of vasodilation includes the release of ˙NO or a related compound, an activation of the soluble guanylyl cyclase, and the cGMP-dependent kinase I leading to decreases in intracellular calcium concentrations.
- GTN is bioactivated by the mitochondrial aldehyde dehydrogenase (ALDH-2).
- Nitrovasodilator therapy leads to side effects such as nitrate tolerance and endothelial dysfunction.
- These side effects are associated and/or secondary to stimulation of production of reactive oxygen and nitrogen species produced by the NADPH oxidase or an uncoupled nitric oxide synthase.
- Nitrovasodilator-induced vascular ROS production (mainly by GTN and ISMN) stimulates autocrine vascular endothelin-1 production associated with a supersensitivity to vasoconstrictors contributing to nitrate tolerance.
- Strategies to avoid tolerance or endothelial dysfunction include a nitrate-free interval, or cotherapy with antioxidants, AT-1 receptor blockers or ACE inhibitors, and the direct vasodilating compound hydralazine with potent ROS scavenging properties.

Keywords Nitroglycerin • Organic nitrates • Nitrate tolerance • Endothelial dysfunction • Mitochondrial aldehyde dehydrogenase • Mitochondrial reactive oxygen species • Peroxynitrite

Hemodynamics

Organic nitrates have beneficial hemodynamic effects in patients with coronary artery disease, acute and chronic congestive heart failure, and arterial hypertension. Organic nitrates dilate venous capacitance vessels, large and medium-sized coronary arteries, collaterals [1, 2], and also the aorta,

T. Münzel, M.D. (✉) • A. Daiber, Ph.D.
University Medical Center Mainz, Center for Cardiology, Building 605, Langenbeckstr. 1, Mainz 55131, Germany
e-mail: tmuenzel@uni-mainz.de

N.S. Bryan, J. Loscalzo (eds.), *Nitrite and Nitrate in Human Health and Disease*, Nutrition and Health, 195
DOI 10.1007/978-3-319-46189-2_15, © Springer International Publishing AG 2017

while coronary and peripheral arterioles with a diameter <100 μm have been demonstrated to be nitrate resistant [3, 4]. Thus, in the setting of stable angina, the preferential venodilation induced by anti-ischemic doses of nitrates results in venous pooling, and, therefore, preload reduction, leading to a reduction of left ventricular end-diastolic pressure and wall tension. The consequences are a reduction in myocardial workload and oxygen demand [1, 2]. Beyond these effects, in patients with coronary artery disease, nitrates have potent anti-ischemic effects that depend on the dilation of large epicardial coronary arteries and coronary collaterals. This leads to improved blood perfusion and, therefore, oxygen delivery to subendocardial regions mainly by increasing total coronary conductance. In addition, since arteriolar tone is largely unaffected by nitrates [3], coronary steal phenomena as well as reflex tachycardia are in general avoided. Nitrates also markedly improve left ventricular function in patients with acute and chronic congestive heart failure. Nitrates decrease the right atrial pressure with a redistribution of blood from the central circulation into larger capacitance veins. Nitrates cause also an unloading of the failing afterload-dependent ventricle by reducing the impedance to left ventricular ejection mainly via dilation of large capacitance arteries such as the aorta. The increase in compliance of the arterial vasculature, in turn, leads to a reduction in the magnitude, frequency, and velocity of reflected waves in the arterial circulation [5]. Thus, these hemodynamic afterload effects of nitrates cause an increase in cardiac output, a reduction in left ventricular filling pressure and wall tension, and also a reduction in mitral regurgitation, thereby shifting the stroke volume/left ventricular end-diastolic pressure relationship from a negative to a positive slope.

Cellular Mechanisms of Vasodilation by Organic Nitrates

The activation of the enzyme soluble guanylyl cyclase (sGC) by nitrate-derived nitric oxide ($^{\bullet}$NO) was identified as the principal mechanism of action of these drugs (for review see [6]). sGC activation leads to increased bioavailability of cyclic guanosine-3′,-5′-monophosphate (cGMP) and activation of cGMP-dependent protein kinases, such as the cGMP-dependent protein kinase I (cGK-I). The relaxation downstream to these processes requires Ca^{2+}-dependent and/or -independent mechanisms. cGK-I inhibits the inositol-1,4,5-trisphosphate [IP_3]-dependent calcium release mediated by phosphorylation of the IP_3 receptor-associated cGMP kinase substrate (IRAG), and activates the big calcium-activated potassium channel (BK_{Ca}) through phosphorylation, leading to hyperpolarization and reduced calcium influx. Furthermore, cGK-I activates the Ca^{2+}-ATPase-pump and, thereby, enhances the efflux of calcium to the extracellular space. Ca^{2+}-independent relaxation by cGK-I involves phosphorylation of the myosin-binding subunit [e.g., myosin phosphatase targeting subunit 1 (MYPT1)] (Fig. 15.1). Furthermore, cGK-I might phosphorylate, and thereby inhibit, the small GTP-binding protein RhoA, leading to decreased Rho-kinase (ROK) activity and conserved activity of myosin light chain phosphatase (MLCP), all of which is vasodilatory. ROK can also directly phosphorylate and increase the contractility of myosin light chain (MLC). cGK-I also induces a feedback mechanism (which lowers the intracellular cGMP concentration) by the phosphorylation and activation of phosphodiesterases. More recent studies also put emphasis on epigenetic regulation of nitrate-induced smooth muscle relaxation [7].

Organic Nitrate Metabolism and $^{\bullet}$NO Generation

The concept that the group of organic nitrates causes vasodilation in general by $^{\bullet}$NO release was challenged by independent studies demonstrating an almost 100-fold discrepancy between GTN-evoked $^{\bullet}$NO formation and vasodilation, whereas for ISMN and ISDN, a direct correlation between these parameters was shown [8]. In addition, a similar discrepancy between hemodynamic effects in

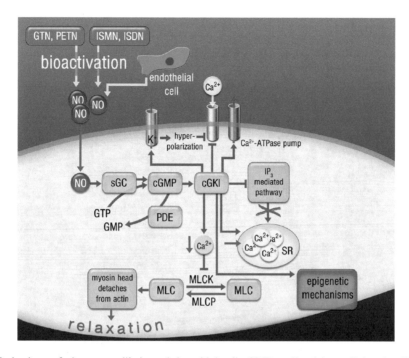

Fig. 15.1 Mechanisms of nitrate vasodilation: nitric oxide/cyclic GMP-mediated intracellular signaling leading to smooth muscle cell relaxation involve the activation of cGK-I and decrease in intracellular calcium levels (via inhibition of the IP3-receptor-regulated calcium channel, activation of potassium channels with subsequent inhibition of calcium channels, and activation of the calcium pump) as well as epigenetic mechanisms. Recent data appear to suggest that the biotransformation of GTN occurs in the mitochondria only when higher local concentrations are reached. *cGK-I* cGMP-dependent kinase, *ALDH-2* aldehyde dehydrogenase, *PDE* phosphodiesterase, *MLCP* myosin light-chain phosphatase, *MLCK* myosin light-chain kinase, *GTN* nitroglycerin, *PETN* pentaerythrityl tetranitrate, *ISMN, ISDN* isosorbide mono- and dinitrate. Adopted from [82] with permission © Oxford University Press. Reproduced from [86] with permission. Copyright © 2014 Elsevier Inc. All rights reserved

response to GTN administration and ˙NO release was reported [9], suggesting that the action of GTN is unrelated to its bioconversion to ˙NO. Thus, it appears that at least GTN-induced vasodilation might be mediated by a ˙NO-related species but not by ˙NO itself. A detailed discussion on the identity of the vasodilating species, which might be formed by organic nitrates, was provided in previous review articles but may comprise iron-nitrosyl or *S*-nitroso species [10, 11]. Support for this notion is provided by the significant increase in iron-nitrosyl or *S*-nitroso species can be seen some minutes after oral intake of GTN in human volunteers or animals [12–14].

Further evidence is provided by recent studies showing that so-called ˙NO donors, such as PETN and GTN, have substantially different effects on gene expression [15, 16]. Treatment with GTN resulted in a larger expression of cardiotoxic genes and inhibition of the expression of cardioprotective proteins, whereas PETN treatment enhances the expression of genes in the opposite direction. A possible explanation for this differential gene expression profile in response to both "˙NO donors" may be the release of different vasoactive molecules upon bioactivation [15].

Clinical Uses/Usefulness of Organic Nitrates and Nitrovasodilators

In general, treatment of patients with coronary artery disease, chronic congestive heart failure, or arterial hypertension with nitrates comprises organic nitrates and nitrovasodilators, such as GTN (mainly spray, capsule, patch, and infusion), PETN (tablet), ISMN (tablet), ISDN (tablet), nicorandil (tablet),

Commonly used organic nitrates					
Compound	Structure	Application	Dose [mg]	Time until action [min]	Duration of action [h]
Glyceryl trinitrate (GTN)		Spray	0.5	1	0.5
		Tablet	52 – 5	1 – 2	
		Patch	5 – 105 – 10	4 – 8	
Pentaerythrityl tetranitrate (PETN)		Tablet	50 – 80	10 – 20	8 – 12
Isosorbide dinitrate (ISDN)		Tablet	5	1 – 2	1
		Slow-release tablet	20 – 120	10 – 30	8 – 12
		Spray	5 – 20	10 – 20	4 – 6
Isosorbide-5-mononitrate (ISMN)	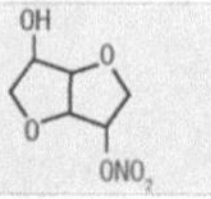	Tablet	20 – 40	10 – 30	1 – 2
Nicorandil	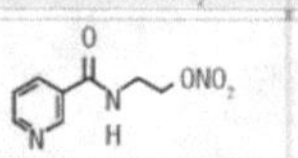	Tablet	10 – 20	30 – 60	2 – 4
Sodium nitroprusside (SNP)	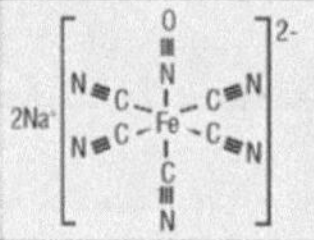	Infusion	0.001 – 0.003 mg/kg/min	1	0.1 – 0.2

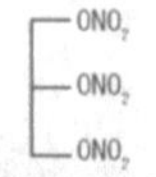

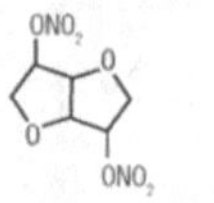

Fig. 15.2 Commonly used organic nitrates. Reproduced from [58] with permission of the publisher. Copyright © 2015, Mary Ann Liebert, Inc (permission not required)

and sodium nitroprusside (SNP, infusion) (Fig. 15.2). Although nitrates in general failed to improve prognosis in patients with coronary artery disease (CAD) and congestive heart failure (CHF) (except for the combination ISDN and hydralazine in patients with CHF), they can be considered for acute and long-term treatment in these patient groups. Actually the current guidelines for stable angina [17], acute coronary syndromes [18], acute and chronic congestive heart failure (nitrates alone for patients with CHF and angina and ISDN + hydralazine for patients with CHF alone) [19] and arterial hypertension [20] still recommend, the use of oral or intravenous organic nitrates.

Recently, Ambrosio et al. [21] demonstrated that treatment of patients with CAD with organic nitrates was associated with a shift away from ST-segment elevation myocardial infarction to non ST-segment elevation infarction, and that this shift was associated with a less pronounced release of cardiac markers. The authors speculated that this beneficial effect may be mediated at least in part by the positive effects of nitrates on ischemic preconditioning [21]. In the setting of heart failure and coronary artery disease, nitrates demonstrated an improvement in symptoms but mostly failed to demonstrate an improvement in prognosis in studies such as ISIS-4 [22] or GISSI-3 [23]. To the contrary, several retrospective studies (meta-analysis) performed in postinfarct patients [24–26], patients following percutaneous coronary intervention (PCI) and diabetes [27], or patients with vasospastic angina revealed that the nitrate use is associated with an unfavorable prognosis. In addition, more recent studies in patients with diabetes mellitus revealed that long-term therapy with ISMN leads to more cardiovascular events consisting out of cardiovascular revascularization, nonfatal MI, and cardiovascular death [24, 25]. Moreover, in patients with vasospastic angina a multicenter registry study from Japan just recently revealed that in patients treated with organic nitrates, the GTN use (patches) was associated with a significant increase in the incidence of major adverse cardiovascular events, including cardiac death, nonfatal MI, hospitalization due to unstable angina, and heart failure and appropriate implantable cardioverter defibrillator shocks [28].

PETN is an organic nitrate with completely different effects on the vasculature. PETN does not cause tolerance, it does not increase oxidative stress within the vasculature, and it does not cause endothelial dysfunction when treating patients with coronary artery disease [29]. The reason for this is that this nitrate stimulates the expression of heme oxygenase 1 (HO-1) in endothelial and smooth muscle cells, which in turn leads to enhanced production of one of the most potent antioxidants in our body, bilirubin [30–32]. It induces also the expression of ferritin, which, in turn, diminishes the levels of free iron and, therefore, the production of hydroxyl radicals ($\cdot$OH) via Fenton chemistry [31, 32]. Moreover, it also causes an increase in the production of carbon monoxide (CO), which, in turn, stimulates the sGC and thereby lowers vascular tone [30]. Indeed, we recently performed a large randomized, double-blind, placebo-controlled multicenter trial (*CLEOPATRA study*) to investigate the anti-ischemic efficacy of 80 mg of PETN (b.i.d., given in the morning and at midday) over placebo in patients with stable angina pectoris. Despite the fact that PETN therapy alone did not provide any additional benefit in unselected patients with known CAD, its administration in combination with modern anti-ischemic drugs could increase exercise tolerance in symptomatic patients with reduced exercise capacity [33].

Chemical Structure of Nitrates

In order to become a vasodilator, all organic nitrates need to undergo a 3-electron reduction leading to nitric oxide and the respective alcohol. It is still elusive whether under biological conditions this reduction involves inorganic or organic nitrite as an intermediate or not.

The chemical structure of the organic nitrates currently in clinical use differs markedly (Fig. 15.2). ISMN and ISDN have one or two $-ONO_2$ groups linked to the isosorbide sugar ring, respectively, whereas GTN has three $-ONO_2$ groups linked to the glycerol backbone; PETN has a more spherical shape due to linkage of 4 methylene–ONO_2 groups to the central carbon atom. These differences in their chemical structure result in quite distinct pharmacokinetic profiles, including different transporters for cellular uptake (for review see [11]). Of note, after oral application of PETN to humans, only the dinitrate and mononitrate metabolites (PEDN and PEMN) can be found in the blood [34], while PETN itself has never been quantified and its trinitrate metabolite (PETriN) has only been detected occasionally and in very low concentrations. This suggests that despite its high stability in aqueous media, PETN is rapidly bioconverted by intestinal microorganisms yielding the less effective mono- and dinitrate metabolites which are absorbed and produce their therapeutic effects [35–37]. This also reflects in different pharmacokinetic properties: in contrast to GTN (whose absorption and metabolism upon administration shows a peak at 30 min [38, 39]), PETN metabolites reach their maximum plasma concentration 2–3 h after oral administration [37].

SNP, although not classified as an organic nitrate, is highly water soluble and shows an extremely fast onset of vasodilation of less than 1 min upon infusion. SNP overdosing can result in cyanosis upon escape of cyanide ions from the complex. SNP is not available as an oral drug. Nicorandil is a hybrid molecule harboring a potassium channel opening function and an organic nitrate group. Importantly, no tolerance side effects were described for nicorandil in humans and additional pharmacological actions as well as cardioprotection on top of the nitrate function probably rely on the opening of mitochondrial ATP-sensitive potassium channels and induction of preconditioning and/or the combination of hyperpolarization and NO release.

Organic Nitrate Bioactivation

Organic nitrates need biotransformation in order to release a vasoactive molecule. In particular, GTN has been intensively studied. GTN concentration–relaxation curves show a biphasic response suggesting a high potency pathway (operative at clinically relevant GTN concentrations <1 μM and mediated by ALDH-2) and a low potency pathway (operative at suprapharmacological GTN concentrations >1 μM and mediated by other enzymes or low-molecular-weight reductants) (Fig. 15.3).

High Potency (Affinity) Pathway Involving the ALDH-2 in GTN Bioactivation

In 2002, Chen et al. identified the mitochondrial isoform of aldehyde dehydrogenases (ALDH-2) as a key enzyme in a clinically relevant (high affinity) bioactivation process of GTN [40]. This was later proven at the molecular level using ALDH-2 deficient (ALDH-2$^{-/-}$) mice with impaired relaxation in response to GTN but not SNP or ISDN [41]. In addition, the bioactivation of PETN and its trinitrate metabolite PETriN was impaired in ALDH-2$^{-/-}$ mice [42]. The ALDH-2 is well known from the alcoholism research field and accounts for the removal of toxic acetaldehyde upon ethanol consumption. Individuals with the East-Asian variant of the enzyme (e.g., ALDH2*2) with reduced activity show

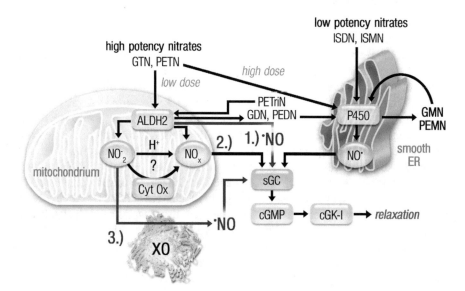

Fig. 15.3 Proposed mechanisms underlying bioactivation of organic nitrates. *Left*, Characterization of the bioactivation of high-potency nitrates such as nitroglycerin (glyceryl trinitrate [GTN]), pentaerythrityl tetranitrate (PETN), and pentaerythrityl trinitrate (PETriN) by mitochondrial aldehyde dehydrogenase (ALDH-2) when used at low, clinically relevant concentrations (<1 μmol/L). The reductase activity converts the organic nitrates to nitrite and the denitrated metabolite (1,2-glyceryl dinitrate, PETriN, or its dinitrate, PEDN). In general, there are three proposed mechanisms including nitrogen oxide formed via a reduction of nitrite (NO$_2^-$), nitric oxide, formed directly in response to interaction with the ALDH-2, and nitrite, released from the mitochondria, may be reduced by the xanthine oxidase in the cytoplasm to form ˙NO. *Right*, The bioactivation of low-potency nitrates such as isosorbide dinitrate (ISDN) and isosorbide-5-mononitrate (ISMN) but also GDN, PEDN, and their respective mononitrates GMN and PEMN by P450 enzyme(s) in the endoplasmic reticulum (ER), directly yielding nitric oxide. The latter mechanism also accounts for the high-potency nitrates when used at high concentrations (>1 μmol/L). Cyt Ox indicates cytochrome c oxidase. Adopted from [10]. Reproduced from [79] with permission of the publisher. Copyright © 2011, Wolters Kluwer Health

the so-called flushing syndrome in response to alcohol intake [43] but also displayed impaired GTN bioactivation [44, 45]. Of note, in 2008, the group of Mochly-Rosen identified another important therapeutic potential of ALDH-2 consisting of the anti-ischemic protection in a model of myocardial infarction, e.g., by using the synthetic activator Alda-1 [46].

The isolated ALDH-2 enzyme generates inorganic nitrite (NO_2^-) and 1,2-glyceryl dinitrate (1,2-GDN) from GTN, and the reaction is accelerated by NAD^+. During this reaction, a thionitrate intermediate will be formed from GTN in the active center of ALDH-2 [47, 48], with concomitant release of 1,2-GDN. The intermediate then reacts by nucleophilic attack of an adjacent second cysteine thiol group under formation of a disulfide bridge releasing nitrite as the leaving group. Nonspecific inhibitors of ALDH-2 (disulfiram, cyanamide, chloral hydrate, benomyl) and high substrate concentrations (acetaldehyde) attenuated the vasodilating activity of GTN and inhibited the organic nitrate reductase activity of ALDH-2 [40, 49–51]. The inactive thiol-oxidized enzyme can be reactivated by thiol donors like dithiothreitol and 2-mercapthoethanol. Dihydrolipoic acid serves as the natural reducing agent in animals [52, 53] and in humans [14]. Likewise, the thioredoxin/thioredoxin reductase system accounts for the reactivation of ALDH-2 [54]. The enzymatic activity can be further regulated by reaction of the disulfide bridge with glutathione (S-glutathionylation) [52]. There is also irreversible inhibition of the enzyme, probably via formation of sulfonic acid groups by oxidants, such as superoxide or peroxynitrite [52], which requires *de novo* synthesis of the ALDH-2. In principle, this concept is a revival of the Needleman "thiol theory," which already suggested an interaction of organic nitrates with the mitochondria (swelling and increased oxygen uptake) as well as a depletion of mitochondrial thiol pools in response to chronic GTN treatment [55].

Low Potency (Affinity) Pathway for Bioactivation of GTN Besides the ALDH-2

The low potency pathway leads to formation of measurable amounts of ˙NO in vascular tissues in vivo [56] and in vitro [8]. Therefore, ˙NO is a vasoactive principle of GTN applied at higher concentrations upon bioactivation by low-molecular-weight factors such as cysteine, *N*-acetyl-cysteine, thiosalicylic acid, and ascorbate, as well as enzymatic systems such as deoxyhemoglobin, deoxymyoglobin, cytochrome P450 (CYP), xanthine oxidase, glutathione-*S*-transferase, glyceraldehyde-3-phosphate dehydrogenase (GAPDH), or other ALDH isoforms (reviewed in [10, 11, 57, 58]). While ˙NO formation by deoxyhemoproteins and xanthine oxidase will be confined to tissues of low oxygen tension (deoxyhemoproteins), the CYP pathway may be operative at high GTN concentrations in liver, lung, and kidney [56]. Bioactivation by thiols or ascorbate requires high concentrations of these reductants (mM range) and GTN (μM range) questioning their physiological significance.

Side Effects of Chronic Nitroglycerin Therapy: Nitrate Tolerance and Nitrate-Induced Endothelial Dysfunction

The clinical introduction of organic nitrates at the end of the nineteenth century was soon followed by the observation that the hemodynamic and clinical effects of GTN, ISMN, and ISDN invariably wane upon continuous therapy. In the setting of coronary artery disease, nitrate tolerance has been demonstrated as the loss of effects on treadmill walking time and time of onset of angina [59]. In congestive heart failure, it has been described as the loss of hemodynamic effect of the administered nitrate [60] and in hypertension, it is evident as the rapid loss of the hypotensive effects of these drugs.

Table 15.1 Hypotheses proposed to explain the development of nitrate tolerance

Pseudotolerance
Activation of renin–angiotensin–aldosterone system
Increase in circulating catecholamine levels and catecholamine release rates
Increase in vasopressin levels
Volume expansion
Vascular tolerance
Impaired GTN biotransformation
Increased vascular superoxide production
Desensitization of the soluble guanylate cyclase
Increase in phosphodiesterase activity
Increased sensitivity to vasoconstrictors
Increased endothelin expression
GTN indicates glyceryl trinitrate

Reproduced from [58]. Copyright © 2015, Mary Ann Liebert, Inc. (permission not required)

Rather controversial data have been reported for tolerance to the antiplatelet effects of GTN. Early work showed that antiplatelet effects of GTN could be potentiated by *N*-acetylcysteine, and that *S*-nitroso-*N*-acetylcysteine was itself a direct platelet inhibitor acting via the cGMP pathway [61]. Preclinical studies demonstrated that tolerance is associated with a paradoxical activation of platelets [62] and another report showed that prior exposure to GTN, even in very low doses, induces tolerance to the antiaggregatory effects of the drug [62]. In contrast, other studies in both rats and humans have shown that platelet responsiveness is preserved despite hemodynamic tolerance [63, 64].

Another issue is so-called *nitrate resistance*, defined by the reduced effectiveness of organic nitrates in the setting of cardiovascular disease (and therefore increased oxidative stress). Nitrate resistance shares the same features as nitrate tolerance, but impaired nitrate effectiveness is not induced by prior nitrate therapy. For instance, McVeigh et al. reported reduced hemodynamic effects in diabetic patients and (as mentioned earlier) that GTN-induced inhibition of platelet aggregability is blunted in patients with coronary artery disease or diabetes [65]. To date, it remains unclear whether these different forms of reduced responsiveness to nitrates share common mechanisms (e.g., dysfunction of downstream 'NO signaling pathways) or should rather be considered as two distinct entities. The mechanism of nitrate bioactivation is a highly complex process. However, the mechanism of the development of nitrate tolerance is even more complex since it involves neurohormonal counter-regulation; expansion of plasma volume (collectively classified as pseudotolerance); and intrinsic vascular processes, defined as true tolerance (Table 15.1 and Fig. 15.4).

Pseudotolerance

The vasodilation evoked by intravenous, oral, and transdermal nitrate therapy causes an increase in catecholamine release, and plasma vasopressin levels, and increases in plasma renin activity and aldosterone levels (for review see [58]). Such activation of neurohormonal vasoconstrictor forces, which is not specific for GTN therapy and has also been observed in response to other vasodilators, has been demonstrated in patients with coronary artery disease, patients with heart failure, and healthy subjects [58]. In line with these data, long-term continuous transdermal GTN therapy has been associated with altered autonomic neural function, including impaired baroreflex activity and preponderance of sympathetic over parasympathetic tone in the regulation of heart rate [66]. A marked increase in intravascular volume, secondary to the transvascular shift of fluid and/or to aldosterone-mediated salt and water retention [67, 68], has also been observed in patients treated with GTN. Although these changes could attenuate the preload effect of GTN, evidence suggests that these mechanisms are not sufficient to explain fully the loss

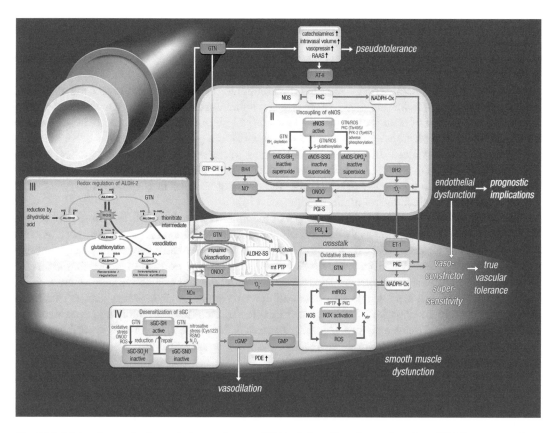

Fig. 15.4 Molecular mechanisms of nitrate tolerance. Within 1 day of continuous low-dose GTN therapy, neurohormonal counter-regulation consisting of increased catecholamine and vasopressin plasma levels, increased intravascular volume, and activation of the renin–angiotensin–aldosterone system (RAAS) reduces therapeutic efficacy (pseudotolerance). After 3 days, endothelial and smooth muscle dysfunction develops (vascular tolerance and cross-tolerance) by different mechanisms: (1) Increased endothelial and smooth muscle superoxide formation from NADPH oxidase activation by PKC and from the mitochondria; (2) direct inhibition of NOS activation by PKC; (3) uncoupling of endothelial NOS caused by limited BH_4 availability caused by peroxynitrite ($ONOO^-$)-induced oxidation of BH_4 and reduced expression of GTP-cyclohydrolase I (GTPCH-I); (4) Vasoconstrictor supersensitivity caused by increased smooth muscle PKC activity; (5) impaired bioactivation of GTN caused by inhibition of ALDH-2; (6) inhibition of smooth muscle soluble guanylate cyclase by superoxide and peroxynitrite; (7) increased inactivation of cGMP by PDE; (8) inhibition of prostacyclin synthase (PGI_2-S) by peroxynitrite, leading to reduced PGI_2 formation; and (9) nitrosylation of the soluble guanylyl cyclase, leading to a desensitization of the enzyme to $^{\cdot}NO$. Reproduced from [82] with permission of the publisher. Copyright © European Society of Cardiology/Oxford University Press

of nitrate effectiveness. For instance, there is a difference in the time frame of neurohormonal activation, plasma expansion, and development of tolerance [68]; furthermore, studies testing the effects of diuretics, beta-blockers, or ACE inhibitors did not invariably reverse or prevent tolerance. Thus, although the possible prognostic implications of these changes need to be acknowledged, other mechanisms of tolerance and a hypothesis that explained all these changes had to be found.

True Vascular Tolerance Mechanisms

True vascular tolerance is thought to be due to an inability of the vascular tissue to respond to organic nitrates in the absence of the neurohormonal environment. Thus, vessels from animals pretreated with nitrates demonstrated a blunted reaction in response, for example, to GTN. In the mid-1980s true vascular tolerance mechanisms comprised impaired GTN biotransformation, intracellular SH-group depletion, a desensitization of the ˙NO target enzyme sGC, as well as an increase in PDE activity.

The concept of nitroglycerin-induced depletion of thiol stores is mainly based on the observations by Needleman on the inhibition of thiol-dependent mitochondrial proteins in response to different organic nitrates [55]. This group also reported the importance of sulfhydryl groups for the enzymatic function of essential proteins and linked GTN-induced inhibition of these enzymes and loss of vasodilation to the depletion of thiol groups [69, 70]. Later, the loss of cellular thiol groups in response to nitroglycerin and the excessive formation of S-nitrosothiols as a potential mechanism of nitrate tolerance (impaired bioactivation) was favored [71, 72] and indirectly supported by clinical studies [73, 74]. The involvement of PDE enzymes in GTN tolerance was based on the observation that cGMP turnover was changed in nitrate-tolerant vessels [75] and the demonstration of increased expression of PDE (isoform 1A1) in an animal model of nitrate tolerance [76].

Within the last 30 years several new concepts of tolerance emerged such as increases in oxidative stress in the tolerant vasculature as well as an increase in sensitivity to vasoconstrictors, such as serotonin, phenylephrine, angiotensin II, and thromboxane A [77, 78] mainly due to increased autocrine endothelin-1 production (Fig. 15.4) [77, 79–81]. There is a growing body of evidence that increased RNOS (reactive nitrogen oxide species; superoxide and peroxynitrite) production in response to GTN therapy may be actually responsible for almost all that true tolerance phenomena such as decreased sGC responsiveness and impaired GTN biotransformation by the ALDH-2 simply due to oxidation of critical SH groups of these enzymes [11, 82]. Increased RNOS production will contribute to the induction of vascular endothelin-1 expression (within the endothelium, smooth muscle, and adventitia) [83, 84].

It is now clear that the group of organic nitrates is not a uniform NO-releasing group of compounds but rather quite heterogeneous. Nitrates differ markedly with regard to their mechanism of vasodilation, speed and degree of tolerance development, capacity to increase oxidative stress, and to cause the development of endothelial dysfunction (cross-tolerance) [85, 86].

GTN Increases Oxidative Stress and Causes Tolerance

In 1995, we proposed a new molecular mechanism for GTN tolerance and cross-tolerance [87]. With these studies we demonstrated that GTN markedly increase ROS (reactive oxygen species and also peroxynitrite) production in the vasculature and caused tolerance and endothelial dysfunction, all of which was corrected by the addition of liposomal superoxide dismutase, which dismutates $O_2^{˙-}$ to H_2O_2 and oxygen [87], a finding which was later confirmed in patient studies [65, 88–90]. GTN tolerance was also associated with increased markers of free radical-induced lipid peroxidation such as cytotoxic aldehydes and isoprostanes [91] and esterified 8-epi-prostaglandin $F_{2\alpha}$ ($PGF_{2\alpha}$) [92]. Several sources of oxidative stress have been shown to contribute significantly to nitrate tolerance, such as the NADPH oxidases [77, 81, 93] and the mitochondria respiratory chain (Fig. 15.4) [49, 94, 95].

Oxidative Stress Impairs GTN Biotransformation

The recognition of the role of a mitochondrial enzyme in the biotransformation of organic nitrates and of a role of mitochondrial oxidative stress in the development of tolerance provided an elusive link between two apparently separate hypotheses (reduced bioactivation versus ROS-mediated ˙NO scavenging or ROS-mediated inactivation of ˙NO signaling). This concept is essentially based on the concept that the oxidation of thiol groups may cause inhibition of several key enzymes (including ALDH-2 [52, 96] and sGC [97]) and, therefore, both reduced GTN biotransformation and inhibited ˙NO signal transduction [82]. In line with this view, treatment of tolerant animals with mitochondria-targeted antioxidants completely prevented or reversed GTN tolerance [98, 99], and heterozygous knockout of manganese-superoxide dismutase (MnSOD⁺/⁻ mice), which is located with the mitochondria, markedly aggravated tolerance development in response to GTN [95]. These data reconcile the bioactivation and oxidative stress hypotheses and provide an interesting clinical corollary of the original observations published by Needleman and Hunter, showing that incubation with high concentrations of nitrates induced swelling of isolated cardiac mitochondria, stimulated oxygen consumption, and uncoupled oxidative phosphorylation [55, 69, 70], being consistent with a mitochondrial source of nitrate-induced ROS. Subsequently, we could demonstrate that in vivo and in vitro GTN treatment increases RNOS (most likely peroxynitrite) formation in isolated mitochondria and simultaneously decreases bioactivation of GTN to the 1,2-GDN metabolite. Furthermore, the GTN-dependent activation of soluble guanylyl cyclase (sGC) was lost upon chemical depletion of mitochondrial proteins [49].

Importantly, however, these considerations apply to GTN tolerance, but most likely not to ISMN or ISDN tolerance because these drugs do not undergo mitochondrial metabolism [96]. Regardless of the exact mechanism by which GTN stimulates mitochondrial ROS production (e.g., premature release of partially reduced oxygen from mitochondrial complex I or III, initiation of lipid peroxidation, depolarization of mitochondrial membrane potential, mitochondrial swelling [11]), these observations suggest that oxidative stress may directly impair GTN biotransformation, either by oxidative inhibition of ALDH-2 or by depletion of essential repair cofactors such as lipoic acid [52]. Recent data obtained with purified ALDH-2 provide evidence that ALDH-2 could be even a source of GTN-triggered ROS formation [54]. The pathways leading to pseudo and true vascular tolerance in response to GTN are summarized in Fig. 15.3. A critical role for ALDH-2 in causing nitrate (GTN) tolerance was also provided in human studies where vascular tissue of GTN-treated patients showed impaired activity of the ALDH-2 in veins and mammary artery but also decreased expression of ALDH-2 [100].

Desensitization of the Soluble Guanylyl Cyclase

In the late 1980s, the desensitization of the sGC was suggested as a mechanism of tolerance [101, 102]. This desensitization is compatible with the evidence that patients treated with one nitrate also show reduced sensitivity to other ˙NO-dependent vasodilators (so-called cross-tolerance). Importantly, it has been shown that S-nitrosylation of sGC results in decreased responsiveness to ˙NO characterized by loss of ˙NO-stimulated sGC and cGK-I activity [103].

The authors extended this in vitro evidence to in vivo observations by demonstrating that development of nitrate tolerance and cross-tolerance by 3-day chronic GTN treatment correlates with S-nitrosylation and desensitization of sGC in tolerant tissues, and that tolerance was reversed by concomitant treatment with the sulfhydryl donor, N-acetylcysteine [97]. In line with this observation, our group just established that GTN-induced tolerance is partially prevented in rats by therapy with a sGC activator but not with a sGC stimulator [104], suggesting that oxidation of sGC is a critical event in causing desensitization of the enzyme (see also paragraph "sGC stimulators and activators"). Due to the fact that at least four cysteine residues were reported for S-nitros(yl)ation-dependent sGC desensitization (Cys78, (122, 243, 516)) that are distributed between the heterodimers of the enzyme, the enzyme is highly susceptible to oxidative and nitrosative stress. Oxidative inactivation and nitrosative desensitization of sGC were previously shown for superoxide [105], peroxynitrite [106], and iNOS induction under inflammatory conditions [107]. Inflammation is a new confirmed risk factor for cardiovascular disease [108, 109], and inactivation of sGC under these conditions is very likely.

Increased Sensitivity to Vasoconstrictors in Response to GTN and ISMN Therapy

GTN has also been shown to trigger a supersensitivity of the vasculature to vasoconstrictors such as angiotensin II, phenylephrine, and serotonin, a phenomenon that may markedly compromise the vasodilatory effects of nitrates [81]. Interestingly, it appears that increased autocrine levels of endothelin within the vasculature, with subsequent activation of phospholipase C (PLC) and protein kinase C (PKC), are responsible for this true tolerance mechanism [77].

These pathways (PLC and PKC) depend on increased intracellular calcium levels, activate the myosin light chain kinase (MLCK) leading to increased contractility of the myosin–actin filaments (Fig. 15.1), but also provide the link to cytosolic oxidative stress. Similarly, agonist-driven calcium-independent activation of the RhoA/ROK pathway contributes to vasoconstriction via inhibition of myosin light chain phosphatase (MLCP).

Importantly, this phenomenon of increased vascular sensitivity to vasoconstrictors is likely due to increased oxidative stress within endothelial and smooth muscle cells, since reactive oxygen species have been shown to increase the expression of endothelin, which in turn activates PKC [83, 84]. The increased sensitivity to vasoconstriction was shown for norepinephrine, KCl, serotonin, angiotensin II, and PKC activators, and was normalized following inhibition of PKC [77]. Studies in patients with coronary artery disease confirmed these observations [110].

In addition, ISMN treatment, the most frequently used oral organic nitrate worldwide, is associated with a strong increase in the expression of endothelin-1, mainly within the endothelial cell layer and the adventitia, and by increased sensitivity of the vasculature to vasoconstricting agents, such as phenylephrine and angiotensin II, as demonstrated previously in GTN-tolerant vessels [80]. Interestingly, incubation of inflammatory cells with ISMN activated the phagocytic NADPH oxidase and caused an oxidative burst, all of which was blocked in vitro by the endothelin receptor blocker bosentan and was normalized in vivo by gp91phox deficiency [80]. In aortic tissue, ISMN induced a similar increase in ROS (superoxide) formation, most likely by Nox2, which was prevented by the unspecific flavin-dependent oxidoreductase inhibitor apocynin, by gp91phox deficiency, as well as by the ET-receptor antagonist bosentan. Although these adverse effects of ISMN appear similar to the previous observations in response to GTN treatment, there are several fundamental differences. First, in contrast to GTN, ISMN is not bioactivated by mitochondrial ALDH-2, a process that leads to a marked increase in mitochondrial ROS production; and second, NADPH oxidase activation in response to ISMN is not dependent on the cross-talk between ROS-producing mitochondria and NADPH oxidase, and is an independent phenomenon triggered by unclear mechanisms. Thus, these findings may explain, at least in part, why therapy of postinfarct patients with ISMN leads to an increased rate of coronary events [26].

Epigenetic Mechanisms of GTN-Induced Tolerance

More recent studies demonstrated the involvement of epigenetic mechanisms in the development of nitrate tolerance. Colussi et al. observed that GTN administration increased cGMP production but also protein N^{ε} lysine acetylation in cultured rat smooth muscle cells, including myosin light chain phosphorylation and actomyosin formation [7]. These effects were abolished by pretreatment with GTN and restored by treatment with trichostatin. Ex vivo experiments performed on aortic rings where tolerance was induced in vivo with subcutaneous injections of GTN revealed that all proacetylation drugs studied caused a reversal of nitrate tolerance. In addition, the vasodilator response of

GTN was abolished by the histone acetylase inhibitor anacardic acid. These findings suggest that GTN therapy increases histone acetylase activity all of which is lost upon chronic treatment with GTN. These findings may also indicate that a combination therapy of epigenetic drugs with GTN may be a novel approach to prevent GTN tolerance development. In addition, epigenetic drugs may also affect the progression of atherosclerosis in general since more recent evidence linked DNA methylation to the progression of atherosclerosis [111].

Organic Nitrates Cause Endothelial Dysfunction in the Clinical Setting

Therapy with most organic nitrates in clinically used doses impairs responsiveness to stimuli for the release of endothelium-derived ·NO. This phenomenon, a consequence of endothelial dysfunction, has been observed in animal studies and in humans during prolonged GTN [89, 112, 113], ISMN [114], and ISDN therapy [115].

Endothelial dysfunction has also been observed in response to intermittent dosing of ISMN [114]. Importantly, no endothelial dysfunction was observed in response to long-term PETN treatment [29], which is likely due to the previously mentioned upregulation of the antioxidant enzyme HO-1 [30]. Compared with equipotent dosages of nicorandil, ISDN (40 mg/day) treatment for 3 months caused endothelial dysfunction [115].

Endothelial dysfunction in response to organic nitrates may be caused by different mechanisms including decreased expression and/or activity of the endothelial NOS or even uncoupling of the enzyme [116, 117]. In the setting of tolerance, the expression of the ·NO synthase is upregulated rather than downregulated suggesting pointing to a dysfunctional uncoupled eNOS enzyme [118]. Molecular mechanisms causing eNOS uncoupling in the setting of GTN tolerance include a deficiency of the eNOS cofactor tetrahydrobiopterin (BH4) [119, 120], a decrease in the expression of the BH4 synthesizing enzyme GTP-cyclohydrolase-1, and an increase in the expression of the BH2 recycling enzyme dihydrofolate reductase [121]. Likewise, in ISMN-treated animals, endothelial dysfunction was associated with decreased GTP-cyclohydrolase-1 expression [80]. Further pieces of evidence for a dysfunctional, uncoupled eNOS in response to nitrate treatment were demonstrated by changes in the eNOS phosphorylation pattern. GTN therapy markedly inhibited eNOS phosphorylation at Ser1177, which in turn leads to decreased basal and stimulated ·NO release of the enzyme [121]. GTN treatment also increases phosphorylation of eNOS at Thr495 [121] known to cause dysfunction and even uncoupling of the enzyme [122]. GTN therapy and ISMN therapy lead to an enhanced S-glutathionylation at one or more cysteine residues of the reductase domain contributing to uncoupling and, therefore, endothelial dysfunction [80, 121, 123]. There is also evidence for decreased activity and/or expression of prostacyclin synthase in response to therapy with GTN, which will act in concert with the NOS synthase to continuously lower vascular tone [124].

Further enzymatic superoxide sources activated by therapy with nitrovasodilators include NADPH oxidase(s) [87], the mitochondria [49, 96], and an uncoupled eNOS [80, 118, 121]. Previously we have reported on a cross-talk between mitochondrial and cytosolic RONS in a model of increased mitochondrial GTN-induced oxidative stress [125]. In this animal model, endothelial dysfunction and nitrate tolerance were dependent on the activation of both superoxide sources [125]. With these studies we could also show NADPH oxidase knockout mice developed tolerance but no endothelial dysfunction in response to nitroglycerin treatment, suggesting that this enzyme is a primary mechanism for nitrate-induced endothelial dysfunction [125, 129, 130]. The findings of this study are summarized and discussed in view of similar observations in different animal models in recent review articles [126, 127].

Strategies to Prevent Development of Nitrate Tolerance and Nitrate-Induced Endothelial Dysfunction

Tolerance toward the hemodynamic effects of GTN can be avoided by treatment regimens with nitrate-free intervals, sulfhydryl group donors, or antioxidants (for review see [58]). Since the neuro-hormonal counter-regulatory mechanisms triggered by the hypotensive effects of nitrate therapy may offset the direct vasodilatory effects of GTN, therapy with inhibitors of the RAAS may be beneficial. Indeed, concomitant treatment with the angiotensin II type 1 (AT-1) receptor blocker, e.g., losartan or telmisartan, normalized superoxide levels, and prevented tolerance and cross-tolerance to the endothelium-dependent vasodilator acetylcholine [81, 121]. Finally, a hybrid nitrate containing valsartan as a backbone showed no signs of tolerance or other side effects in vitro [128].

A combination therapy of ISDN with hydralazine was able to significantly improve the prognosis in patients with CHF in the V-HeFT I Trial as compared to vasodilator therapy with the alpha blocker prazosin [129] and to significantly improve exercise capacity compared to enalapril in the V-HeFT II trial [130]. In addition, the double blind, randomized African-American Heart Failure Trial (A-HeFT) demonstrated that the combination of ISDN and hydralazine was markedly effective in improving the composite endpoint of the trial which included death from any cause, a first hospitalization for heart failure, and quality of life measures [131]. Hydralazine has also been shown to be a highly potent scavenger of peroxynitrite [132]. Thus, the peroxynitrite scavenging effects of hydralazine may explain, at least in part, why this combination therapy is devoid of tolerance and why a combination therapy improves prognosis in patients with chronic CHF [133].

Further compounds that have been demonstrated to ameliorate nitrate tolerance include the beta-blocker carvedilol with antioxidant properties [134], statins [135], or endothelin receptor blockers, since ISMN and GTN therapy has been shown to stimulate autocrine endothelin-1 production within the vascular wall [80, 81]. Recently, sGC stimulators and activators were tested to what extent tolerance may be modified by these compounds. With these studies it was demonstrated that sGC activation but not sGC stimulation was able to restore vascular 'NO levels, to reduce oxidative stress, and, therefore, restore the activity of the cGMP-dependent kinase in tolerant tissue [104]. It is important to note that by using EPR technique we also demonstrated for the first time that in vivo treatment with GTN decreases significantly ambient vascular 'NO levels.

Pleiotropic Effects of Organic Nitrates

Organic nitrates are not just vasodilators but also have other pleiotropic effects. In particular, the effects of GTN on ischemic conditioning or the antioxidant properties of PETN by stimulating the activity and expression of the HO-1 are of clinical interest.

Short-term administration of GTN has been shown to reduce electrocardiographic evidence of myocardial ischemia during exercise stress and during angioplasty in patients with cardiovascular disease [136, 137], likely mediated by a stimulation of ROS production. The transient GTN-triggered ROS production [94] has two types of effects: (1) upon short-term GTN administration, it would cause preconditioning-based protection against ischemia reperfusion [138], which, however, can be blocked by antioxidant supplementation (contributing the phenomenon of the antioxidant paradox) [139]; (2) upon prolonged administration, via oxidative damage, accumulation of ROS would lead to the toxic effects of GTN, such as endothelial and autonomic dysfunction as described earlier. PETN conferred preconditioning effects also under chronic conditions [138] whereas GTN even induced a "tolerance" to the preconditioning effects, probably based on oxidative damage by GTN-induced

RNOS formation [85]. The loss of anti-ischemic protection by chronic GTN therapy was associated with inhibition of ALDH-2 [138]. Of note, ALDH-2 inhibition per se suppressed remote preconditioning in experimental models and humans [140].

The organic nitrate PETN and its trinitrate metabolite PETriN increase (in contrast to GTN) the activity and expression of the very potent antioxidant enzyme heme oxygenase 1 in vitro [31, 32, 57, 141] and in vivo [42, 142, 143]. The induction of HO-1 by PETN is likely being based on a direct activation of the HO-1 promoter region via Nrf2 (either directly via Keap1 or indirectly via ˙NO) as well as a stabilization of the HO-1 mRNA by ˙NO-triggered HuR binding, effects that are not shared at all by other nitrates [16, 80, 144]. The induction of extracellular SOD and glutathione peroxidase-1 [145] represents another important intrinsic antioxidant pathway activated by PETN in vivo treatment [146] and may explain its beneficial effects in hyperlipidemia [147, 148] and arterial hypertension [142]. PETN also induces heritable epigenetic changes that cause blood pressure reduction in female offspring of PETN-treated hypertensive rats via enhanced histone 3 lysine 27 acetylation and histone 3 lysine 4 trimethylation with subsequent induction of aortic endothelial nitric oxide synthase, mitochondrial superoxide dismutase, glutathione peroxidase 1, and heme oxygenase 1 [149].

Administration of organic nitrates also has profound effects on the number and function of circulating EPCs. Both PETN and ISDN increase the level of circulating EPCs in control animals but also in animals with decreased EPCs levels with ischemic cardiomyopathy [150]. In humans, GTN treatment increases apoptosis and decreases phenotypic differentiation, migration, and mitochondrial dehydrogenase activity of EPCs [151].

Conclusion and Future Perspectives for the Class of Nitrovasodilators

At present, organic nitrates should not be considered as a homogeneous, interchangeable class of ˙NO-releasing compounds since they display a considerable diversity [85]. Especially GTN induces clinical tolerance, oxidative stress, and endothelial dysfunction, side effects which are shared to more or less extent with the other nitrates, ISMN and ISDN. ISMN is more likely to induce endothelial dysfunction and increases oxidative stress via activation of the vascular/phagocytic NADPH oxidase. In contrast, PETN is obviously devoid of these adverse side effects via acting on the powerful antioxidant enzyme HO-1. Much less is known concerning ISDN and tolerance and endothelial dysfunction. GTN and PETN, both are bioactivated by the ALDH-2 and GTN but not PETN therapy causes simultaneously an inactivation of its own bioactivating pathway via mitochondrial ROS formation. Oxidative stress causes nitrate tolerance and endothelial dysfunction in response to organic nitrates although the extent and the cellular source of oxidative stress differ markedly between the ISMN, ISDN, and GTN. Oxidative stress induction may represent a major limitation for the clinical use of organic nitrates since increased production of reactive oxygen and nitrogen species (RNOS) formation per se is a hallmark and/or biomarker of cardiovascular disease [152, 153]. Thus, organic nitrate therapy could, therefore, even elicit harmful effects in concert with RNOS-associated cardiovascular disease [154]. A novel approach to avoid tolerance and endothelial dysfunction may be the development of hybrid molecules, e.g., the combination of a ˙NO-releasing moiety coupled with a nonsteroidal anti-inflammatory moiety (NSAIDs) in one molecule, such a nitro-aspirin. However, similar approaches with NO-naproxen have failed to gain FDA approval.

Acknowledgments We thank Margot Neuser for graphical support.

Conflict of Interest A.D. and T.M. received honoraries and research support from Actavis Deutschland GmbH (Langenfeld, Germany), the manufacturer of PETN.

References

1. Abrams J. A reappraisal of nitrate therapy. J Am Med Assoc. 1988;259:396–401.
2. Abrams J. The role of nitrates in coronary heart disease. Arch Intern Med. 1995;155(4):357–64.
3. Sellke FW, Myers PR, Bates JN, Harrison DG. Influence of vessel size on the sensitivity of porcine coronary microvessels to nitroglycerin. Am J Physiol. 1990;258(2 Pt 2):H515–20.
4. Sellke FW, Tomanek RJ, Harrison DG. L-cysteine selectively potentiates nitroglycerin-induced dilation of small coronary microvessels. J Pharmacol Exp Ther. 1991;258(1):365–9.
5. Zobel LR, Finkelstein SM, Carlyle PF, Cohn JN. Pressure pulse contour analysis in determining the effect of vasodilator drugs on vascular hemodynamic impedance characteristics in dogs. Am Heart J. 1980;100(1):81–8.
6. Munzel T, Feil R, Mulsch A, Lohmann SM, Hofmann F, Walter U. Physiology and pathophysiology of vascular signaling controlled by guanosine $3',5'$-cyclic monophosphate-dependent protein kinase [corrected]. Circulation. 2003;108(18):2172–83.
7. Colussi C, Scopece A, Vitale S, Spallotta F, Mattiussi S, Rosati J, et al. P300/CBP associated factor regulates nitroglycerin-dependent arterial relaxation by N{varepsilon}-lysine acetylation of contractile proteins. Arterioscler Thromb Vasc Biol. 2012;32(10):2435–43. Epub 2012/08/04.eng.
8. Kleschyov AL, Oelze M, Daiber A, Huang Y, Mollnau H, Schulz E, et al. Does nitric oxide mediate the vasodilator activity of nitroglycerin? Circ Res. 2003;93(9):e104–12.
9. Nunez C, Victor VM, Tur R, Alvarez-Barrientos A, Moncada S, Esplugues JV, et al. Discrepancies between nitroglycerin and NO-releasing drugs on mitochondrial oxygen consumption, vasoactivity, and the release of NO. Circ Res. 2005;97(10):1063–9.
10. Munzel T, Daiber A, Mulsch A. Explaining the phenomenon of nitrate tolerance. Circ Res. 2005;97(7):618–28.
11. Daiber A, Wenzel P, Oelze M, Munzel T. New insights into bioactivation of organic nitrates, nitrate tolerance and cross-tolerance. Clin Res Cardiol. 2008;97(1):12–20. Epub 2007/10/17.eng.
12. Janero DR, Bryan NS, Saijo F, Dhawan V, Schwalb DJ, Warren MC, et al. Differential nitros(yl)ation of blood and tissue constituents during glyceryl trinitrate biotransformation in vivo. Proc Natl Acad Sci U S A. 2004;101(48):16958–63.
13. Noack E, Feelisch M. Molecular mechanisms of nitrovasodilator bioactivation. Basic Res Cardiol. 1991;86 Suppl 2:37–50.
14. Wenzel P, Schulz E, Gori T, Ostad MA, Mathner F, Schildknecht S, et al. Monitoring white blood cell mitochondrial aldehyde dehydrogenase activity: implications for nitrate therapy in humans. J Pharmacol Exp Ther. 2009;330(1):63–71. Epub 2009/04/07.eng.
15. Pautz A, Rauschkolb P, Schmidt N, Art J, Oelze M, Wenzel P, et al. Effects of nitroglycerin or pentaerithrityl tetranitrate treatment on the gene expression in rat hearts: evidence for cardiotoxic and cardioprotective effects. Physiol Genomics. 2009;38(2):176–85. Epub 2009/05/07.eng.
16. Daiber A, Oelze M, Wenzel P, Bollmann F, Pautz A, Kleinert H. Heme oxygenase-1 induction and organic nitrate therapy: beneficial effects on endothelial dysfunction, nitrate tolerance, and vascular oxidative stress. Int J Hypertens. 2012;2012:842632. Pubmed Central PMCID: 3312327, Epub 2012/04/17.eng.
17. Task Force M, Montalescot G, Sechtem U, Achenbach S, Andreotti F, Arden C, et al. 2013 ESC guidelines on the management of stable coronary artery disease: the Task Force on the management of stable coronary artery disease of the European Society of Cardiology. Eur Heart J. 2013;34(38):2949–3003.
18. Hamm CW, Bassand JP, Agewall S, Bax J, Boersma E, Bueno H, et al. ESC guidelines for the management of acute coronary syndromes in patients presenting without persistent ST-segment elevation: The Task Force for the management of acute coronary syndromes (ACS) in patients presenting without persistent ST-segment elevation of the European Society of Cardiology (ESC). Eur Heart J. 2011;32(23):2999–3054.
19. McMurray JJ, Adamopoulos S, Anker SD, Auricchio A, Bohm M, Dickstein K, et al. ESC guidelines for the diagnosis and treatment of acute and chronic heart failure 2012: The Task Force for the Diagnosis and Treatment of Acute and Chronic Heart Failure 2012 of the European Society of Cardiology. Developed in collaboration with the Heart Failure Association (HFA) of the ESC. Eur J Heart Fail. 2012;14(8):803–69.
20. Mancia G, Fagard R, Narkiewicz K, Redon J, Zanchetti A, Bohm M, et al. 2013 ESH/ESC practice guidelines for the management of arterial hypertension. Blood Press. 2014;23(1):3–16.
21. Ambrosio G, Del Pinto M, Tritto I, Agnelli G, Bentivoglio M, Zuchi C, et al. Chronic nitrate therapy is associated with different presentation and evolution of acute coronary syndromes: insights from 52,693 patients in the Global Registry of Acute Coronary Events. Eur Heart J. 2010;31(4):430–8. Epub 2009/11/12.eng.
22. ISIS-4: a randomised factorial trial assessing early oral captopril, oral mononitrate, and intravenous magnesium sulphate in 58,050 patients with suspected acute myocardial infarction. ISIS-4 (Fourth International Study of Infarct Survival) Collaborative Group. Lancet. 1995;345(8951):669–85. Epub 1995/03/18.eng.
23. GISSI-3: effects of lisinopril and transdermal glyceryl trinitrate singly and together on 6-week mortality and ventricular function after acute myocardial infarction. Gruppo Italiano per lo Studio della Sopravvivenza nell'infarto Miocardico. Lancet. 1994;343(8906):1115–22. Epub 1994/05/07.eng.

24. Ishikawa K, Kanamasa K, Ogawa I, Takenaka T, Naito T, Kamata N, et al. Long-term nitrate treatment increases cardiac events in patients with healed myocardial infarction. Secondary Prevention Group. Jpn Circ J. 1996;60(10):779–88.
25. Kanamasa K, Hayashi T, Kimura A, Ikeda A, Ishikawa K. Long-term, continuous treatment with both oral and transdermal nitrates increases cardiac events in healed myocardial infarction patients. Angiology. 2002;53(4):399–408.
26. Nakamura Y, Moss AJ, Brown MW, Kinoshita M, Kawai C. Long-term nitrate use may be deleterious in ischemic heart disease: a study using the databases from two large-scale postinfarction studies. Multicenter Myocardial Ischemia Research Group. Am Heart J. 1999;138(3 Pt 1):577–85.
27. Yiu KH, Pong V, Siu CW, Lau CP, Tse HF. Long-term oral nitrate therapy is associated with adverse outcome in diabetic patients following elective percutaneous coronary intervention. Cardiovasc Diabetol. 2011;10:52. Pubmed Central PMCID: 3129297.
28. Takahashi J, Nihei T, Takagi Y, Miyata S, Odaka Y, Tsunoda R, et al. Prognostic impact of chronic nitrate therapy in patients with vasospastic angina: multicentre registry study of the Japanese coronary spasm association. Eur Heart J. 2015;36(4):228–37.
29. Schnorbus B, Schiewe R, Ostad MA, Medler C, Wachtlin D, Wenzel P, et al. Effects of pentaerythritol tetranitrate on endothelial function in coronary artery disease: results of the PENTA study. Clin Res Cardiol. 2010;99(2):115–24. Epub 2009/12/04.eng.
30. Wenzel P, Oelze M, Coldewey M, Hortmann M, Seeling A, Hink U, et al. Heme oxygenase-1: a novel key player in the development of tolerance in response to organic nitrates. Arterioscler Thromb Vasc Biol. 2007;27(8):1729–35.
31. Oberle S, Schwartz P, Abate A, Schroder H. The antioxidant defense protein ferritin is a novel and specific target for pentaerithrityl tetranitrate in endothelial cells. Biochem Biophys Res Commun. 1999;261(1):28–34.
32. Oberle S, Abate A, Grosser N, Vreman HJ, Dennery PA, Schneider HT, et al. Heme oxygenase-1 induction may explain the antioxidant profile of pentaerythrityl trinitrate. Biochem Biophys Res Commun. 2002;290(5):1539–44. Epub 2002/02/01.eng.
33. Munzel T, Meinertz T, Tebbe U, Schneider HT, Stalleicken D, Wargenau M, et al. Efficacy of the long-acting nitro vasodilator pentaerithrityl tetranitrate in patients with chronic stable angina pectoris receiving anti-anginal background therapy with beta-blockers: a 12-week, randomized, double-blind, placebo-controlled trial. Eur Heart J. 2014;35(14):895–903. Pubmed Central PMCID: 3977134.
34. Neurath GB, Dunger M. Blood levels of the metabolites of glyceryl trinitrate and pentaerythritol tetranitrate after administration of a two-step preparation. Arzneimittelforschung. 1977;27(2):416–9.
35. Seeling A, Lehmann J. NO-donors, part X: investigations on the stability of pentaerythrityl tetranitrate (PETN) by HPLC-chemoluminescence-N-detection (CLND) versus UV-detection in HPLC. J Pharm Biomed Anal. 2006;40(5):1131–6.
36. Haustein KO, Winkler U, Loffler A, Huller G. Absorption and bioavailability of pentaerithrityl-tetranitrate (PETN, Dilcoran 80). Int J Clin Pharmacol Ther. 1995;33(2):95–102.
37. Weber W, Michaelis K, Luckow V, Kuntze U, Stalleicken D. Pharmacokinetics and bioavailability of pentaerithrityl tetranitrate and two of its metabolites. Arzneimittelforschung. 1995;45(7):781–4.
38. Woodward AJ, Lewis PA, Rudman AR, Maddock J. Determination of nitroglycerin and its dinitrate metabolites in human plasma by high-performance liquid chromatography with thermal energy analyzer detection. J Pharm Sci. 1984;73(12):1838–40. Epub 1984/12/01.eng.
39. Sun JX, Piraino AJ, Morgan JM, Joshi JC, Cipriano A, Chan K, et al. Comparative pharmacokinetics and bioavailability of nitroglycerin and its metabolites from Transderm-Nitro, Nitrodisc, and Nitro-Dur II systems using a stable-isotope technique. J Clin Pharmacol. 1995;35(4):390–7.
40. Chen Z, Zhang J, Stamler JS. Identification of the enzymatic mechanism of nitroglycerin bioactivation. Proc Natl Acad Sci U S A. 2002;99(12):8306–11.
41. Chen Z, Foster MW, Zhang J, Mao L, Rockman HA, Kawamoto T, et al. An essential role for mitochondrial aldehyde dehydrogenase in nitroglycerin bioactivation. Proc Natl Acad Sci U S A. 2005;102(34):12159–64.
42. Wenzel P, Hink U, Oelze M, Seeling A, Isse T, Bruns K, et al. Number of nitrate groups determines reactivity and potency of organic nitrates: a proof of concept study in ALDH-2-/- mice. Br J Pharmacol. 2007;150:526–33.
43. Xiao Q, Weiner H, Crabb DW. The mutation in the mitochondrial aldehyde dehydrogenase (ALDH2) gene responsible for alcohol-induced flushing increases turnover of the enzyme tetramers in a dominant fashion. J Clin Invest. 1996;98(9):2027–32.
44. Mackenzie IS, Maki-Petaja KM, McEniery CM, Bao YP, Wallace SM, Cheriyan J, et al. Aldehyde dehydrogenase 2 plays a role in the bioactivation of nitroglycerin in humans. Arterioscler Thromb Vasc Biol. 2005;25(9):1891–5.
45. Li Y, Zhang D, Jin W, Shao C, Yan P, Xu C, et al. Mitochondrial aldehyde dehydrogenase-2 (ALDH2) Glu504Lys polymorphism contributes to the variation in efficacy of sublingual nitroglycerin. J Clin Invest. 2006;116(2):506–11.

46. Chen CH, Budas GR, Churchill EN, Disatnik MH, Hurley TD, Mochly-Rosen D. Activation of aldehyde dehydrogenase-2 reduces ischemic damage to the heart. Science. 2008;321(5895):1493–5. Epub 2008/09/13.eng.
47. Fung HL. Biochemical mechanism of nitroglycerin action and tolerance: is this old mystery solved? Annu Rev Pharmacol Toxicol. 2004;44:67–85.
48. Lang BS, Gorren AC, Oberdorfer G, Wenzl MV, Furdui CM, Poole LB, et al. Vascular bioactivation of nitroglycerin by aldehyde dehydrogenase-2: reaction intermediates revealed by crystallography and mass spectrometry. J Biol Chem. 2012;287(45):38124–34.
49. Sydow K, Daiber A, Oelze M, Chen Z, August M, Wendt M, et al. Central role of mitochondrial aldehyde dehydrogenase and reactive oxygen species in nitroglycerin tolerance and cross-tolerance. J Clin Invest. 2004;113(3):482–9. Pubmed Central PMCID: 324536.
50. Kollau A, Hofer A, Russwurm M, Koesling D, Keung WM, Schmidt K, et al. Contribution of aldehyde dehydrogenase to mitochondrial bioactivation of nitroglycerin. Evidence for activation of purified soluble guanylyl cyclase via direct formation of nitric oxide. Biochem J. 2005;385(Pt 3):769–77.
51. Zhang J, Chen Z, Cobb FR, Stamler JS. Role of mitochondrial aldehyde dehydrogenase in nitroglycerin-induced vasodilation of coronary and systemic vessels: an intact canine model. Circulation. 2004;110(6):750–5.
52. Wenzel P, Hink U, Oelze M, Schuppan S, Schaeuble K, Schildknecht S, et al. Role of reduced lipoic acid in the redox regulation of mitochondrial aldehyde dehydrogenase (ALDH-2) activity. Implications for mitochondrial oxidative stress and nitrate tolerance. J Biol Chem. 2007;282(1):792–9. Epub 2006/11/15.eng.
53. Dudek M, Bednarski M, Bilska A, Iciek M, Sokolowska-Jezewicz M, Filipek B, et al. The role of lipoic acid in prevention of nitroglycerin tolerance. Eur J Pharmacol. 2008;591(1–3):203–10. Epub 2008/07/12.eng.
54. Oelze M, Knorr M, Schell R, Kamuf J, Pautz A, Art J, et al. Regulation of human mitochondrial aldehyde dehydrogenase (ALDH-2) activity by electrophiles in vitro. J Biol Chem. 2011;286(11):8893–900. Epub 2011/01/22.eng.
55. Needleman P, Hunter Jr FE. Effects of organic nitrates on mitochondrial respiration and swelling: possible correlations with the mechanism of pharmacologic action. Mol Pharmacol. 1966;2(2):134–43.
56. Mulsch A, Bara A, Mordvintcev P, Vanin A, Busse R. Specificity of different organic nitrates to elicit NO formation in rabbit vascular tissues and organs in vivo. Br J Pharmacol. 1995;116(6):2743–9.
57. Daiber A, Munzel T, Gori T. Organic nitrates and nitrate tolerance—state of the art and future developments. Adv Pharmacol. 2010;60:177–227. Epub 2010/11/18.eng.
58. Daiber A, Munzel T. Organic nitrate therapy, nitrate tolerance, and nitrate-induced endothelial dysfunction: emphasis on redox biology and oxidative stress. Antioxid Redox Signal. 2015;23(11):899–942.
59. Parker JD, Parker JO. Nitrate therapy for stable angina pectoris. N Engl J Med. 1998;338(8):520–31.
60. Elkayam U, Kulick D, McIntosh N, Roth A, Hsueh W, Rahimtoola SH. Incidence of early tolerance to hemodynamic effects of continuous infusion of nitroglycerin in patients with coronary artery disease and heart failure. Circulation. 1987;76(3):577–84.
61. Loscalzo J. N-Acetylcysteine potentiates inhibition of platelet aggregation by nitroglycerin. J Clin Invest. 1985;76(2):703–8. Pubmed Central PMCID: 423881.
62. Fink B, Bassenge E. Association between vascular tolerance and platelet upregulation: comparison of nonintermittent administration of pentaerithrityltetranitrate and glyceryltrinitrate. J Cardiovasc Pharmacol. 2002;40(6):890–7. Epub 2002/11/27.eng.
63. Booth BP, Jacob S, Bauer JA, Fung HL. Sustained antiplatelet properties of nitroglycerin during hemodynamic tolerance in rats. J Cardiovasc Pharmacol. 1996;28(3):432–8.
64. Holmes AS, Chirkov YY, Willoughby SR, Poropat S, Pereira J, Horowitz JD. Preservation of platelet responsiveness to nitroglycerine despite development of vascular nitrate tolerance. Br J Clin Pharmacol. 2005;60(4):355–63. Pubmed Central PMCID: 1884829, Epub 2005/09/29.eng.
65. McVeigh GE, Hamilton P, Wilson M, Hanratty CG, Leahey WJ, Devine AB, et al. Platelet nitric oxide and superoxide release during the development of nitrate tolerance: effect of supplemental ascorbate. Circulation. 2002;106(2):208–13. Epub 2002/07/10.eng.
66. Gori T, Floras JS, Parker JD. Effects of nitroglycerin treatment on baroreflex sensitivity and short-term heart rate variability in humans. J Am Coll Cardiol. 2002;40(11):2000–5.
67. Parker JD, Farrell B, Fenton T, Cohanim M, Parker JO. Counter-regulatory responses to continuous and intermittent therapy with nitroglycerin. Circulation. 1991;84(6):2336–45.
68. Munzel T, Heitzer T, Kurz S, Harrison DG, Luhman C, Pape L, et al. Dissociation of coronary vascular tolerance and neurohormonal adjustments during long-term nitroglycerin therapy in patients with stable coronary artery disease. J Am Coll Cardiol. 1996;27(2):297–303.
69. Jakschik B, Needleman P. Sulfhydryl reactivity of organic nitrates: biochemical basis for inhibition of glyceraldehyde-P dehydrogenase and monoamine oxidase. Biochem Biophys Res Commun. 1973;53(2):539–44.
70. Needleman P, Jakschik B, Johnson Jr EM. Sulfhydryl requirement for relaxation of vascular smooth muscle. J Pharmacol Exp Ther. 1973;187(2):324–31.
71. Feelisch M, Noack E, Schroder H. Explanation of the discrepancy between the degree of organic nitrate decomposition, nitrite formation and guanylate cyclase stimulation. Eur Heart J. 1988;9(Suppl A):57–62.

72. Fung HL, Chung SJ, Chong S, Hough K, Kakami M, Kowaluk E. Cellular mechanisms of nitrate action. Z Kardiol. 1989;78 Suppl 2:14–7, discussion 64–7.
73. Boesgaard S, Aldershvile J, Poulsen HE. Preventive administration of intravenous N-acetylcysteine and development of tolerance to isosorbide dinitrate in patients with angina pectoris. Circulation. 1992;85(1):143–9. Epub 1992/01/01.eng.
74. Ghio S, de Servi S, Perotti R, Eleuteri E, Montemartini C, Specchia G. Different susceptibility to the development of nitroglycerin tolerance in the arterial and venous circulation in humans. Effects of N-acetylcysteine administration. Circulation. 1992;86(3):798–802.
75. Axelsson KL, Karlsson JO. Nitroglycerin tolerance in vitro: effect on cGMP turnover in vascular smooth muscle. Acta Pharmacol Toxicol (Copenh). 1984;55(3):203–10.
76. Kim D, Rybalkin SD, Pi X, Wang Y, Zhang C, Munzel T, et al. Upregulation of phosphodiesterase 1A1 expression is associated with the development of nitrate tolerance. Circulation. 2001;104(19):2338–43.
77. Munzel T, Giaid A, Kurz S, Stewart DJ, Harrison DG. Evidence for a role of endothelin 1 and protein kinase C in nitroglycerin tolerance. Proc Natl Acad Sci. 1995;92(11):5244–8.
78. Heitzer T, Munzel T. Lack of effect of tetrahydrobiopterin and vitamin C on endothelial dysfunction in patients with congestive heart failure. Circulation. 1998;98:I-318 (abstract).
79. Munzel T, Daiber A, Gori T. Nitrate therapy: new aspects concerning molecular action and tolerance. Circulation. 2011;123(19):2132–44. Epub 2011/05/18.eng.
80. Oelze M, Knorr M, Kroller-Schon S, Kossmann S, Gottschlich A, Rummler R, et al. Chronic therapy with isosorbide-5-mononitrate causes endothelial dysfunction, oxidative stress, and a marked increase in vascular endothelin-1 expression. Eur Heart J. 2013;34(41):3206–16.
81. Kurz S, Hink U, Nickenig G, Borthayre AB, Harrison DG, Munzel T. Evidence for a causal role of the renin-angiotensin system in nitrate tolerance. Circulation. 1999;99(24):3181–7.
82. Munzel T, Daiber A, Gori T. More answers to the still unresolved question of nitrate tolerance. Eur Heart J. 2013;34(34):2666–73.
83. Kahler J, Ewert A, Weckmuller J, Stobbe S, Mittmann C, Koster R, et al. Oxidative stress increases endothelin-1 synthesis in human coronary artery smooth muscle cells. J Cardiovasc Pharmacol. 2001;38(1):49–57.
84. Kahler J, Mendel S, Weckmuller J, Orzechowski HD, Mittmann C, Koster R, et al. Oxidative stress increases synthesis of big endothelin-1 by activation of the endothelin-1 promoter. J Mol Cell Cardiol. 2000;32(8):1429–37.
85. Gori T, Daiber A. Non-hemodynamic effects of organic nitrates and the distinctive characteristics of pentaerithrityl tetranitrate. Am J Cardiovasc Drugs. 2009;9(1):7–15. Epub 2009/01/31.eng.
86. Munzel T, Steven S, Daiber A. Organic nitrates: update on mechanisms underlying vasodilation, tolerance and endothelial dysfunction. Vascul Pharmacol. 2014;63(3):105–13.
87. Munzel T, Sayegh H, Freeman BA, Tarpey MM, Harrison DG. Evidence for enhanced vascular superoxide anion production in nitrate tolerance. A novel mechanism underlying tolerance and cross-tolerance. J Clin Invest. 1995;95(1):187–94.
88. Chirkov YY, Holmes AS, Chirkova LP, Horowitz JD. Nitrate resistance in platelets from patients with stable angina pectoris. Circulation. 1999;100(2):129–34.
89. Schulz E, Tsilimingas N, Rinze R, Reiter B, Wendt M, Oelze M, et al. Functional and biochemical analysis of endothelial (dys)function and NO/cGMP signaling in human blood vessels with and without nitroglycerin pretreatment. Circulation. 2002;105(10):1170–5.
90. Sage PR, de la Lande IS, Stafford I, Bennett CL, Phillipov G, Stubberfield J, et al. Nitroglycerin tolerance in human vessels: evidence for impaired nitroglycerin bioconversion. Circulation. 2000;102(23):2810–5.
91. Jurt U, Gori T, Ravandi A, Babaei S, Zeman P, Parker JD. Differential effects of pentaerythritol tetranitrate and nitroglycerin on the development of tolerance and evidence of lipid peroxidation: a human in vivo study. J Am Coll Cardiol. 2001;38(3):854–9.
92. McGrath LT, Dixon L, Morgan DR, McVeigh GE. Production of 8-epi prostaglandin F(2alpha) in human platelets during administration of organic nitrates. J Am Coll Cardiol. 2002;40(4):820–5.
93. Fukatsu A, Hayashi T, Miyazaki-Akita A, Matsui-Hirai H, Furutate Y, Ishitsuka A, et al. Possible usefulness of apocynin, an NADPH oxidase inhibitor, for nitrate tolerance: prevention of NO donor-induced endothelial cell abnormalities. Am J Physiol Heart Circ Physiol. 2007;293(1):H790–7.
94. Daiber A, August M, Baldus S, Wendt M, Oelze M, Sydow K, et al. Measurement of NAD(P)H oxidase-derived superoxide with the luminol analogue L-012. Free Radic Biol Med. 2004;36(1):101–11.
95. Daiber A, Oelze M, Sulyok S, Coldewey M, Schulz E, Treiber N, et al. Heterozygous deficiency of manganese superoxide dismutase in mice (Mn-SOD+/-): a novel approach to assess the role of oxidative stress for the development of nitrate tolerance. Mol Pharmacol. 2005;68(3):579–88.
96. Daiber A, Oelze M, Coldewey M, Bachschmid M, Wenzel P, Sydow K, et al. Oxidative stress and mitochondrial aldehyde dehydrogenase activity: a comparison of pentaerythritol tetranitrate with other organic nitrates. Mol Pharmacol. 2004;66(6):1372–82.

97. Sayed N, Kim DD, Fioramonti X, Iwahashi T, Duran WN, Beuve A. Nitroglycerin-induced S-nitrosylation and desensitization of soluble guanylyl cyclase contribute to nitrate tolerance. Circ Res. 2008;103(6):606–14. Pubmed Central PMCID: 2737267, Epub 2008/08/02.eng.
98. Esplugues JV, Rocha M, Nunez C, Bosca I, Ibiza S, Herance JR, et al. Complex I dysfunction and tolerance to nitroglycerin: an approach based on mitochondrial-targeted antioxidants. Circ Res. 2006;99(10):1067–75.
99. Garcia-Bou R, Rocha M, Apostolova N, Herance R, Hernandez-Mijares A, Victor VM. Evidence for a relationship between mitochondrial Complex I activity and mitochondrial aldehyde dehydrogenase during nitroglycerin tolerance: effects of mitochondrial antioxidants. Biochim Biophys Acta. 2012;1817(5):828–37.
100. Hink U, Daiber A, Kayhan N, Trischler J, Kraatz C, Oelze M, et al. Oxidative inhibition of the mitochondrial aldehyde dehydrogenase promotes nitroglycerin tolerance in human blood vessels. J Am Coll Cardiol. 2007;50(23):2226–32. Epub 2007/12/07.eng.
101. Mulsch A, Busse R, Bassenge E. Desensitization of guanylate cyclase in nitrate tolerance does not impair endothelium-dependent responses. Eur J Pharmacol. 1988;158(3):191–8.
102. Molina CR, Andresen JW, Rapoport RM, Waldman S, Murad F. Effect of in vivo nitroglycerin therapy on endothelium-dependent and independent vascular relaxation and cyclic GMP accumulation in rat aorta. J Cardiovasc Pharmacol. 1987;10(4):371–8.
103. Sayed N, Baskaran P, Ma X, van den Akker F, Beuve A. Desensitization of soluble guanylyl cyclase, the NO receptor, by S-nitrosylation. Proc Natl Acad Sci U S A. 2007;104(30):12312–7. Pubmed Central PMCID: 1940331, Epub 2007/07/20.eng.
104. Jabs A, Oelze M, Mikhed Y, Stamm P, Kroller-Schon S, Welschof P, et al. Effect of soluble guanylyl cyclase activator and stimulator therapy on nitroglycerin-induced nitrate tolerance in rats. Vascul Pharmacol. 2015;71:181–91.
105. Brune B, Schmidt KU, Ullrich V. Activation of soluble guanylate cyclase by carbon monoxide and inhibition by superoxide anion. Eur J Biochem. 1990;192(3):683–8.
106. Weber M, Lauer N, Mulsch A, Kojda G. The effect of peroxynitrite on the catalytic activity of soluble guanylyl cyclase. Free Radic Biol Med. 2001;31(11):1360–7.
107. Kessler P, Bauersachs J, Busse R, Schini-Kerth VB. Inhibition of inducible nitric oxide synthase restores endothelium-dependent relaxations in proinflammatory mediator-induced blood vessels. Arterioscler Thromb Vasc Biol. 1997;17(9):1746–55. Epub 1997/11/05.eng.
108. Keaney Jr JF. Immune modulation of atherosclerosis. Circulation. 2011;124(22):e559–60. Pubmed Central PMCID: 3233261.
109. McManus DD, Beaulieu LM, Mick E, Tanriverdi K, Larson MG, Keaney Jr JF, et al. Relationship among circulating inflammatory proteins, platelet gene expression, and cardiovascular risk. Arterioscler Thromb Vasc Biol. 2013;33(11):2666 73. Pubmed Central PMCID: 4289138.
110. Heitzer T, Just H, Meinertz T, Brockhoff C, Munzel T. Chronic angiotensin converting enzyme inhibition with captopril prevents nitroglycerin induced hypersensitivity to vasoconstrictors in patients with stable coronary artery disease. J Am Coll Cardiol. 1998;31:83–8.
111. Cooper MP, Keaney Jr JF. Epigenetic control of angiogenesis via DNA methylation. Circulation. 2011; 123(25):2916–8.
112. Gori T, Mak SS, Kelly S, Parker JD. Evidence supporting abnormalities in nitric oxide synthase function induced by nitroglycerin in humans. J Am Coll Cardiol. 2001;38(4):1096–101.
113. Caramori PR, Adelman AG, Azevedo ER, Newton GE, Parker AB, Parker JD. Therapy with nitroglycerin increases coronary vasoconstriction in response to acetylcholine. J Am Coll Cardiol. 1998;32(7):1969–74.
114. Thomas GR, DiFabio JM, Gori T, Parker JD. Once daily therapy with isosorbide-5-mononitrate causes endothelial dysfunction in humans: evidence of a free-radical-mediated mechanism. J Am Coll Cardiol. 2007;49(12):1289–95. Epub 2007/03/31.eng.
115. Sekiya M, Sato M, Funada J, Ohtani T, Akutsu H, Watanabe K. Effects of the long-term administration of nicorandil on vascular endothelial function and the progression of arteriosclerosis. J Cardiovasc Pharmacol. 2005;46(1):63–7.
116. Schulz E, Jansen T, Wenzel P, Daiber A, Munzel T. Nitric oxide, tetrahydrobiopterin, oxidative stress, and endothelial dysfunction in hypertension. Antioxid Redox Signal. 2008;10(6):1115–26. Epub 2008/03/07.eng.
117. Munzel T, Daiber A, Ullrich V, Mulsch A. Vascular consequences of endothelial nitric oxide synthase uncoupling for the activity and expression of the soluble guanylyl cyclase and the cGMP-dependent protein kinase. Arterioscler Thromb Vasc Biol. 2005;25(8):1551–7.
118. Munzel T, Li H, Mollnau H, Hink U, Matheis E, Hartmann M, et al. Effects of long-term nitroglycerin treatment on endothelial nitric oxide synthase (NOS III) gene expression, NOS III-mediated superoxide production, and vascular NO bioavailability. Circ Res. 2000;86(1):E7–12.
119. Gruhn N, Aldershvile J, Boesgaard S. Tetrahydrobiopterin improves endothelium-dependent vasodilation in nitroglycerin-tolerant rats. Eur J Pharmacol. 2001;416(3):245–9.
120. Gori T, Saunders L, Ahmed S, Parker JD. Effect of folic acid on nitrate tolerance in healthy volunteers: differences between arterial and venous circulation. J Cardiovasc Pharmacol. 2003;41(2):185–90.

121. Knorr M, Hausding M, Kroller-Schuhmacher S, Steven S, Oelze M, Heeren T, et al. Nitroglycerin-induced endothelial dysfunction and tolerance involve adverse phosphorylation and S-glutathionylation of endothelial nitric oxide synthase: beneficial effects of therapy with the AT1 receptor blocker telmisartan. Arterioscler Thromb Vasc Biol. 2011;31(10):2223–31. Epub 2011/07/16.eng.

122. Fleming I, Fisslthaler B, Dimmeler S, Kemp BE, Busse R. Phosphorylation of Thr(495) regulates Ca(2+)/calmodulin-dependent endothelial nitric oxide synthase activity. Circ Res. 2001;88(11):E68–75.

123. Chen CA, Wang TY, Varadharaj S, Reyes LA, Hemann C, Talukder MA, et al. S-glutathionylation uncouples eNOS and regulates its cellular and vascular function. Nature. 2010;468(7327):1115–8. Pubmed Central PMCID: 3370391.

124. Daiber A, Oelze M, Daub S, Steven S, Schuff A, Kroller-Schon S, et al. Vascular redox signaling, redox switches in endothelial nitric oxide synthase and endothelial dysfunction. In: Laher I, editor. Systems biology of free radicals and antioxidants. Berlin: Springer; 2014. p. 1177–211.

125. Wenzel P, Mollnau H, Oelze M, Schulz E, Wickramanayake JM, Muller J, et al. First evidence for a crosstalk between mitochondrial and NADPH oxidase-derived reactive oxygen species in nitroglycerin-triggered vascular dysfunction. Antioxid Redox Signal. 2008;10(8):1435–47. Epub 2008/06/05.eng.

126. Daiber A. Redox signaling (cross-talk) from and to mitochondria involves mitochondrial pores and reactive oxygen species. Biochim Biophys Acta. 2010;1797(6–7):897–906. Epub 2010/02/04.eng.

127. Schulz E, Wenzel P, Munzel T, Daiber A. Mitochondrial redox signaling: interaction of mitochondrial reactive oxygen species with other sources of oxidative stress. Antioxid Redox Signal. 2014;20(2):308–24. Pubmed Central PMCID: 3887453.

128. Knorr M, Hausding M, Schulz E, Oelze M, Rummler R, Schuff A, et al. Characterization of new organic nitrate hybrid drugs covalently bound to valsartan and cilostazol. Pharmacology. 2012;90(3–4):193–204.

129. Cohn JN, Archibald DG, Ziesche S, Franciosa JA, Harston WE, Tristani FE, et al. Effect of vasodilator therapy on mortality in chronic congestive heart failure. Results of a Veterans Administration Cooperative Study. N Engl J Med. 1986;314(24):1547–52.

130. Cohn JN, Johnson G, Ziesche S, Cobb F, Francis G, Tristani F, et al. A comparison of enalapril with hydralazine-isosorbide dinitrate in the treatment of chronic congestive heart failure. N Engl J Med. 1991;325(5):303–10.

131. Taylor AL, Ziesche S, Yancy C, Carson P, D'Agostino Jr R, Ferdinand K, et al. Combination of isosorbide dinitrate and hydralazine in blacks with heart failure. N Engl J Med. 2004;351(20):2049–57.

132. Daiber A, Oelze M, Coldewey M, Kaiser K, Huth C, Schildknecht S, et al. Hydralazine is a powerful inhibitor of peroxynitrite formation as a possible explanation for its beneficial effects on prognosis in patients with congestive heart failure. Biochem Biophys Res Commun. 2005;338(4):1865–74.

133. Hare JM. Nitroso-redox balance in the cardiovascular system. N Engl J Med. 2004;351(20):2112–4.

134. Watanabe H, Kakihana M, Ohtsuka S, Sugishita Y. Randomized, double-blind, placebo-controlled study of carvedilol on the prevention of nitrate tolerance in patients with chronic heart failure. J Am Coll Cardiol. 1998;32(5):1194–200.

135. Fontaine D, Otto A, Fontaine J, Berkenboom G. Prevention of nitrate tolerance by long-term treatment with statins. Cardiovasc Drugs Ther. 2003;17(2):123–8.

136. Leesar MA, Stoddard MF, Dawn B, Jasti VG, Masden R, Bolli R. Delayed preconditioning-mimetic action of nitroglycerin in patients undergoing coronary angioplasty. Circulation. 2001;103(24):2935–41.

137. Crisafulli A, Melis F, Tocco F, Santoboni UM, Lai C, Angioy G, et al. Exercise-induced and nitroglycerin-induced myocardial preconditioning improves hemodynamics in patients with angina. Am J Physiol Heart Circ Physiol. 2004;287(1):H235–42.

138. Lisi M, Oelze M, Dragoni S, Liuni A, Steven S, Luca MC, et al. Chronic protection against ischemia and reperfusion-induced endothelial dysfunction during therapy with different organic nitrates. Clin Res Cardiol. 2012;101(6):453–9.

139. Dragoni S, Gori T, Lisi M, Di Stolfo G, Pautz A, Kleinert H, et al. Pentaerythrityl tetranitrate and nitroglycerin, but not isosorbide mononitrate, prevent endothelial dysfunction induced by ischemia and reperfusion. Arterioscler Thromb Vasc Biol. 2007;27(9):1955–9. Epub 2007/07/21.eng.

140. Contractor H, Stottrup NB, Cunnington C, Manlhiot C, Diesch J, Ormerod JO, et al. Aldehyde dehydrogenase-2 inhibition blocks remote preconditioning in experimental and human models. Basic Res Cardiol. 2013;108(3):343.

141. Oberle S, Abate A, Grosser N, Hemmerle A, Vreman HJ, Dennery PA, et al. Endothelial protection by pentaerithrityl trinitrate: bilirubin and carbon monoxide as possible mediators. Exp Biol Med (Maywood). 2003;228(5):529–34.

142. Schuhmacher S, Wenzel P, Schulz E, Oelze M, Mang C, Kamuf J, et al. Pentaerythritol tetranitrate improves angiotensin II-induced vascular dysfunction via induction of heme oxygenase-1. Hypertension. 2010;55(4):897–904. Pubmed Central PMCID: 3080599, Epub 2010/02/17.eng.

143. Daiber A, Munzel T. Characterization of the antioxidant properties of pentaerithrityl tetranitrate (PETN)-induction of the intrinsic antioxidative system heme oxygenase-1 (HO-1). Methods Mol Biol. 2010;594:311–26. Epub 2010/01/15.eng.

144. Schuhmacher S, Oelze M, Bollmann F, Kleinert H, Otto C, Heeren T, et al. Vascular dysfunction in experimental diabetes is improved by pentaerithrityl tetranitrate but not isosorbide-5-mononitrate therapy. Diabetes. 2011;60(10):2608–16. Epub 2011/08/17.eng.
145. Dovinova I, Cacanyiova S, Faberova V, Kristek F. The effect of an NO donor, pentaerythrityl tetranitrate, on biochemical, functional, and morphological attributes of cardiovascular system of spontaneously hypertensive rats. Gen Physiol Biophys. 2009;28(1):86–93. Epub 2009/04/25.eng.
146. Oppermann M, Balz V, Adams V, Dao VT, Bas M, Suvorava T, et al. Pharmacological induction of vascular extracellular superoxide dismutase expression in vivo. J Cell Mol Med. 2009;13(7):1271–8. Epub 2009/03/27.eng.
147. Hacker A, Muller S, Meyer W, Kojda G. The nitric oxide donor pentaerythritol tetranitrate can preserve endothelial function in established atherosclerosis. Br J Pharmacol. 2001;132(8):1707–14.
148. Kojda G, Stein D, Kottenberg E, Schnaith EM, Noack E. In vivo effects of pentaerythrityl-tetranitrate and isosorbide-5-mononitrate on the development of atherosclerosis and endothelial dysfunction in cholesterol-fed rabbits. J Cardiovasc Pharmacol. 1995;25(5):763–73. Epub 1995/05/01.eng.
149. Wu Z, Siuda D, Xia N, Reifenberg G, Daiber A, Munzel T, et al. Maternal treatment of spontaneously hypertensive rats with pentaerythritol tetranitrate reduces blood pressure in female offspring. Hypertension. 2015;65(1):232–7.
150. Thum T, Fraccarollo D, Thum S, Schultheiss M, Daiber A, Wenzel P, et al. Differential effects of organic nitrates on endothelial progenitor cells are determined by oxidative stress. Arterioscler Thromb Vasc Biol. 2007;27:748–54.
151. DiFabio JM, Thomas GR, Zucco L, Kuliszewski MA, Bennett BM, Kutryk MJ, et al. Nitroglycerin attenuates human endothelial progenitor cell differentiation, function, and survival. J Pharmacol Exp Ther. 2006;318(1):117–23.
152. Sugamura K, Keaney Jr JF. Reactive oxygen species in cardiovascular disease. Free Radic Biol Med. 2011;51(5):978–92. Pubmed Central PMCID: 3156326.
153. Chen K, Keaney Jr JF. Evolving concepts of oxidative stress and reactive oxygen species in cardiovascular disease. Curr Atheroscler Rep. 2012;14(5):476–83. Pubmed Central PMCID: 3872835.
154. Warnholtz A, Mollnau H, Heitzer T, Kontush A, Moller-Bertram T, Lavall D, et al. Adverse effects of nitroglycerin treatment on endothelial function, vascular nitrotyrosine levels and cGMP-dependent protein kinase activity in hyperlipidemic Watanabe rabbits. J Am Coll Cardiol. 2002;40(7):1356–63.

Chapter 16
Nitrite and Nitrate in Ischemia–Reperfusion Injury

David J. Lefer, Nathan S. Bryan, and Chelsea L. Organ

Key Points

- There is a long and rich history of the use of nitrite for medical conditions.
- Nitrite has a number of essential functions in mammalian tissues.
- Nitrite is reduced to nitric oxide in blood and tissues, particularly in hypoxic or ischemic conditions.
- Nitrite protects from ischemia–reperfusion injury due to modulation of mitochondrial respiration.
- Nitrite has been shown in animal models to be beneficial in cardiovascular disease, cerebral, hepatic and renal I–R injury, effective for pulmonary disorders, peripheral artery disease, and sickle cell disease.
- Very low levels of nitrite are needed for protection.
- There are now a number of clinical trials for the use of nitrite in a number of disease conditions.
- Nitrite-based therapies may provide a safe and cost-effective strategy for conditions of NO insufficiency.

Keywords Myocardial infarction • Hypoxia • Ischemia • Sickle cell disease • Peripheral artery disease • Cytoprotection • Kidney injury

Introduction

The therapeutic effects of nitrite and nitric oxide (NO) on the cardiovascular system were initially realized during medieval times in ancient Chinese medicine. In 1900 a Daoist monk discovered thousands of medieval Buddhist manuscripts, paintings, and other documents in a grotto in the ancient Silk Road town of Dunhuang where they were hidden for 900 years. Anthony Butler, Zhou Wuzong, and John Moffett translated a medical recipe from these manuscripts recently [1]. In this recipe, the patient is instructed to place potassium nitrate under the tongue, and then swallow the saliva to treat

D.J. Lefer, Ph.D. (✉) • N.S. Bryan, Ph.D. • C.L. Organ, B.S.
Cardiovascular Center of Excellence and Department of Pharmacology, LSU Health Center Science-New Orleans, 533 Bolivar Street, CSRB Room 408, New Orleans, LA 70112, USA

Department of Pharmacology and Experimental Therapeutics, Louisiana State University Health Sciences Center, 1901 Perdido Street, New Orleans, LA 70112, USA
e-mail: dlefe1@lsuhsc.edu; Nathan.bryan@bcm.edu; corgan@lsuhsc.edu

N.S. Bryan, J. Loscalzo (eds.), *Nitrite and Nitrate in Human Health and Disease*, Nutrition and Health, DOI 10.1007/978-3-319-46189-2_16, © Springer International Publishing AG 2017

symptoms of angina and digital ischemia. The significance of the instructions is that salivary nitrate-reducing bacteria convert the nitrate into nitrite. Therefore, if the patient followed the physician's instructions fully, he or she would, in fact, be consuming nitrite, known to be effective in alleviating pain resulting from angina. In 1859 the English chemist Fredrick Guthrie noted that the vapor from amyl nitrite on inhalation caused an "immediate flushing of the neck, temples and forehead and an acceleration in the action of the heart." These suggestive vasodilatory actions of nitrite inspired Sir Thomas Lauder Brunton to administer amyl nitrite to his patients with angina pectoris, and in what was considered to be the first report on the medical use of nitrite, he described how within a minute after inhalation, a few drops of amyl nitrite caused the anginal pain to disappear.

Close on the heels of organic nitrite, inorganic sodium nitrite was shown to increase coronary flow as well in the 1920s [2]. In comparison to organic amyl nitrite, inorganic sodium nitrite produced a more sustained vasodilatory effect [3] and hence, was considered a better drug.

Although organic nitrites were used for the treatment of angina and hypertension, they fell out of repute because of concern about multiple side effects. Amyl nitrite was reported to cause blindness, an effect not seen during the last century of its use [4]. Alkyl nitrite administered at very high doses has been associated with increased incidence of Kaposi's Sarcoma probably due to its carcinogenic potential presumably from the formation of N-nitroso compounds [5]. Nitrite therapy at extremely high dosages has the potential for inducing methemoglobinemia (for which reason it is used today as treatment for cyanide poisoning) [6]. In addition, amyl nitrite possesses a potential for abuse [7]. These side effects together with increased use of the organic nitrate nitroglycerin [8] for treatment of angina led to gradual decrease in the use of organic nitrite in the clinical setting; however, interest in inorganic nitrite resurfaced following the discovery of the physiological role of NO in both health and disease states, and with characterization of its metabolism into nitrite and nitrate in mammalian tissues.

NO generated in cells and tissues is oxidized to produce nitrite. Hence, in cells and tissues, inorganic nitrite is an oxidative metabolite of NO. In this regard, nitrite has long been considered to be a physiologically and biochemically defunct by-product of NO metabolism. Today however, nitrite is spearheading the field of NO biology with the discovery that it represents a physiologically critical storage form and also a valuable source of NO in blood and tissues. Nitrite storage pools can readily be reduced to NO under certain pathological conditions [9–12]. This intriguing series of discoveries has spawned an entirely new field of research that involves the investigation of the molecular, biochemical, and physiological activities of inorganic nitrite under a variety of physiological and pathophysiological states. Recent experimental studies clearly demonstrate that nitrite therapy is extremely cytoprotective in a number of animal models of disease. In this chapter, we will highlight the most recent evidence supporting the cytoprotective actions of nitrite in the setting of ischemia–reperfusion (I-R) injury. We will also elucidate its potential role as a therapeutic agent and as a biomarker of cardiovascular health and disease. We then offer comparison between nitrite and classical NO donors in the treatment of ischemic heart disease.

Bioconversion of Inorganic Nitrate and Nitrite to Nitric Oxide

Nitrite in the circulation and tissues is normally derived exclusively from either exogenous dietary sources and/or endogenous production, following generation of NO by endothelial nitric oxide synthase [1]. Exogenously, nitrite is obtained from ingestion of cured meats, green vegetables, and, more importantly, by conversion of dietary nitrate by commensal bacteria in the upper gastrointestinal tract and oral microbiota [13]. Nitrite, when swallowed is subsequently reduced to NO in the acidic environment in the stomach [14] and additionally nitrite and NO are generated by gut bacteria with active nitrate and nitrite reductase enzyme systems, respectively [15]. Endogenously, nitrite is formed from NO, which itself is synthesized from the amino acid, L-arginine, primarily by the action of eNOS. Nitrite levels therefore directly correlate with the activity of the eNOS enzyme [16] and have even been proposed as a biomarker of this key enzyme [17]. It is estimated that roughly 50 % of the circulating nitrite in blood is derived from nitric oxide production, while the other 50 % may be derived from dietary sources and reduction of salivary nitrate [18, 19].

Nitrite in plasma can be reduced to NO under certain conditions by a number of nitrite reductases, including deoxyhemoglobin [20], myoglobin [21], neuroglobin, cytoglobin [22], xanthine oxidase [23], mitochondrial enzymes such as cytochrome c [24], and aldehyde dehydrogenase 2 [25, 26], and also by acidic disproportionation [27, 28]. Nitrite moreover forms nitrosation products of thiols and nitrosylation products of heme with [29] or without the intermediate formation of NO [30]. Nitrite and these nitrosation products have now evolved to represent principal storage pools of NO in the body rather than be inert metabolites of NO devoid of any biological function [31].

Nitrite has numerous fundamental circulatory functions in mammalian tissues. Nitrite has been shown to mediate vasodilation during hypoxia by both NO-dependent [32] and -independent pathways [33]. Very importantly, in hypoxic or ischemic conditions, nitrite is the primary source of NO, which then leads to vasodilation and cytoprotection [34, 35]. This is because the endogenous L-arginine–eNOS–NO pathway requires a constant supply of oxygen to produce NO in tissues, and hypoxic conditions cripple this pathway and its ability to synthesize NO [36–38] (Fig. 16.1). Additionally, xanthine oxidase, a nitrite reductase, is more active under hypoxic and acidic conditions

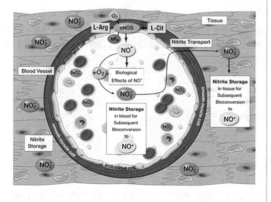
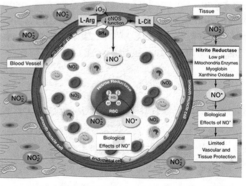

PHYSIOLOGICAL NITRIC OXIDE (NO$^\bullet$) GENERATION

Normoxia

NITRIC OXIDE (NO$^\bullet$) GENERATION DURING ISCHEMIA

Hypoxia

Fig. 16.1 During normoxia oxygen tension is adequate and endothelial nitric oxide synthase [151] functions normally to produce NO. NO acts on target cells and tissues, and any excess NO constantly replenishes nitrite stores via oxidation reactions with oxygen. Nitrite represents a critical NO storage pool. During hypoxia, eNOS is dysfunctional resulting in a fall in NO synthesis and endogenous nitrite generation. Nitrite stores are very rapidly reduced to NO via the actions of hemoglobin in the blood stream and via the actions of mitochondrial enzymes, myoglobin, and xanthine oxidase in the tissues. Thus, formation of NO from nitrite represents an NO salvage pathway that promotes cytoprotection during hypoxia and, hence, is another important source of NO (more so than eNOS) during ischemia [23, 39]. In addition to the proposed role of nitrite in regulating blood flow to organs, nitrite plays a prominent role in defending cells against I–R injury, as discussed later.

Mechanisms of Protection by Nitrite and Nitric Oxide in I–R

Ischemia–reperfusion injury is characterized by numerous detrimental cellular events, such as oxidative damage to proteins and lipids, enzyme release, and inflammatory responses, which ultimately lead to necrosis and apoptosis of tissues [40–42]. I–R injury propagates mitochondrial dysfunction (leading to reduced ATP synthesis) [43], deranges mitochondrial enzymes leading to formation of reactive oxygen species (ROS) [44, 45], and instigates opening of the mitochondrial permeability transition pore (MPTP) [46]. These changes eventually lead to increased calcium influx into the mitochondria exacerbating mitochondrial and cellular injury [47, 48]. The presence of NO abrogates mitochondrial pathology and leads to substantial protection during I–R. In this regard, NO reversibly inhibits mitochondrial enzymes [49, 50] and prevents formation of reactive oxygen species [51]. Similarly, nitrite-derived NO reduces mitochondrial ROS generation at the time of reperfusion [42, 52]. NO hinders release of cytochrome c [53] and very potently inhibits apoptosis [54]. Similarly, nitrite inhibits mitochondrial complex 1 following I–R, thereby limiting formation of ROS [42, 55]. Nitrite-derived NO further activates guanylyl cyclase, inhibits cytochrome P450, and modulates HSP 70 and HO-1 activity [30].

Vascular inflammation adversely affects blood vessels and the endothelium in the basal state and exacerbates the severity of I–R injury [56–58]. Plasma nitrite levels have a strong inverse correlation with the severity of endothelial dysfunction [17, 59–61], with decreased plasma nitrite levels observed in persons with known cardiovascular risk factors. Dietary supplementation of nitrite increases levels of the eNOS cofactor, tetrahydrobiopterin, and decreases levels of the inflammatory marker, C-reactive protein, thereby reducing endothelial dysfunction in mice fed a high cholesterol diet [62]. Thus, nitrite therapy to improve NO bioavailability can potentially be used for the numerous conditions associated with endothelial dysfunction (Fig. 16.2). Nitrite therapy also retards the progression of coronary atherosclerosis, hypertension, cardiovascular disease, and potentially attenuates the severity of any I–R injury that may arise in due course.

Nitrite-Mediated Cytoprotection in Various Organs

Nitrite in Cardiovascular Disease

Endogenous NO derived from NOS plays a vital role in imparting protection during reperfusion injury; its absence leads to increased injury in the heart [19, 63]. Administering nitrite alleviates reperfusion injury by augmenting NO levels in the coronary circulation and in the myocardium. During ischemia and reperfusion, the decrease in plasma levels of nitrite coincides with increase in levels of S-nitrosated and nitrosylated (heme) proteins [64] suggesting that NO is derived from nitrite during the conditions of ischemia and reperfusion [42]. Even brief elevation in nitrite levels modulates redox status, lowers oxidized glutathione levels, inducing significant cardioprotection in the setting of cardiac injury [65].

Nitrite administered during myocardial I–R reduces infarct size and improves left ventricular (LV) function [21], boosting ejection fraction [66]. In a canine model of myocardial I–R, apart from

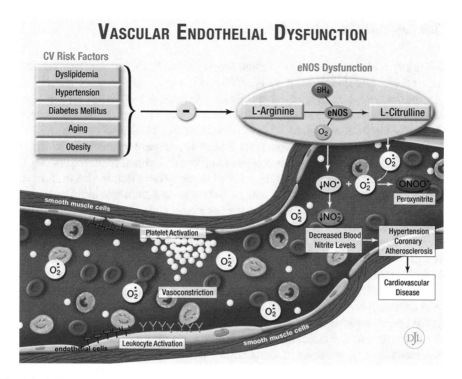

Fig. 16.2 Endothelial dysfunction is the hallmark of numerous cardiovascular risk factors, such as dyslipidemia, hypertension, diabetes, aging, and obesity. In these conditions, the L-arginine–eNOS–NO pathway activity is suppressed owing to dysfunction of eNOS, which leads to decreased nitric oxide and nitrite bioavailability in the blood and tissues. Decreased nitrite level results in abnormal platelet activation, abnormal vasoconstriction, and inability to scavenge oxygen free-radicals. In addition, dysfunctional eNOS shifts the above pathway into superoxide generation, leading to heightened oxidative stress in tissues. Ultimately, vascular endothelial dysfunction contributes to the pathology of hypertension, atherosclerosis, acute myocardial infarction, and stroke

reducing infarct size and improving LV function [66], nitrite also limits endocardial "no flow" phenomenon [66]. Nitrite administered orally is as beneficial as nitrite given intravenously. Nitrite, administered in the drinking water for 7 days prior to myocardial I–R increases plasma and myocardial nitrite levels, augments S-nitrosothiol levels, and results in decreased infarct size in mice [35]. Similarly, when mice are orally administered nitrite in their drinking water during pressure overload-induced heart failure, cardiac dysfunction is attenuated [67]. The cardioprotection observed with nitrite is independent of eNOS-derived NO [19] since eNOS knockout mice exhibit profound protection.

The mechanisms through which nitrite protects myocytes are many and varied. Nitrite protects the heart by activating K_{ATP} channels [68], reducing apoptosis [66], and decreasing the formation of ROS [21]. Of note two enzymes play a key role in nitrite-mediated protection. Xanthine oxidase-mediated NO formation is enhanced with nitrite infusions in rats during myocardial I–R injury [69]. Inhibiting xanthine oxidase abolishes the cardioprotective action of nitrite [68]. Myoglobin is another important nitrite reductase in the heart and plays an important role in the course of I–R, as well [21].

Nitrite therapy has also been shown to be highly beneficial in the setting of cardiac arrest in animal models. Cardiac arrest and subsequent resuscitation may be viewed as a situation of global reperfusion injury whereby injury occurs in organ systems throughout the body [70]. In a mouse model of asystole for 12 min followed by CPR, a single low dose of nitrite resulted in improved survival, better cardiac function, and improved neurological outcomes [71].

Nitrite Therapy and Cerebral I–R Injury

Cerebral I–R injury is characterized by increased production of ROS, cellular dysfunction, apoptosis, and cellular necrosis. NO donors such as ProliNO reduce ROS and cerebral infarct volume [72]. Similarly, nitrite therapy administered at reperfusion following middle cerebral artery occlusion (MCAO) in rats increases cerebral blood flow, decreases infarct volume, and decreases lipid peroxidation [73]. Nitrite administration within 1.5 h of permanent focal ischemia or within 3 h of transient ischemia to brain also reduces infarct volume and hastens recovery in rats [74]. Nitrite reduces the amount of micro-hypoxia in rats by minimizing the number of ischemic cells expressing HIF-1alpha [74]. Nitrite coadministered with memantine, an NMDA receptor inhibitor, also leads to reduction in oxidative stress during reperfusion [74]. It may be important to administer nitrite early in the course of reperfusion, as delayed treatment does not confer cytoprotection [75, 76].

It has been suggested that dysfunctional eNOS is responsible for the delayed vasospastic response after subarachnoid hemorrhage (SAH) [77]. Nitrite therapy in a primate model of vasospasm (induced by blood clot placement in the cerebral arteries and consequent SAH) resulted in increased nitrite and S-nitrosothiol levels in the blood and cerebrospinal fluid (CSF) with an attributable delay in cerebral vasospasm [78]. Nitrite-based therapy for cerebral vasospasm is currently being tested in humans. Phase 2 clinical trial assessing safety and pharmacokinetics of nitrite infusion in patients with subarachnoid hemorrhage has been completed and results suggest that safe and potentially therapeutic levels of nitrite can be achieved and sustained in critically ill patients after SAH from a ruptured cerebral aneurysm [79]. Hypertension increases risk of stroke in many elderly patients. Following stroke, treatment of labile hypertension is frequently challenging [80]. Nitrite treatment may help control blood pressure in hypertension and improve cerebral blood flow [81] in many of these patients.

Nitrite Therapy and Hepatic I–R Injury

Reperfusion injury in the liver leads to considerable morbidity and mortality. I–R injury commonly occurs with liver transplantation, thermal injury, and shock. During liver I–R, there is an imbalance between vasoconstrictors and vasodilators leading to disruption of microcirculation [82]. There is also a rapid decrease in NO bioavailability [83]. NO protects the liver during I–R injury by inactivation of caspase 3, opening mitochondrial K_{ATP} channels, and inhibition of complex 1 [84]. NO prevents MPTP-dependent necrosis of ischemic hepatocytes after reperfusion through a guanylyl cyclase and cGMP-dependent kinase signaling pathway [85].

In the liver, as with heart and brain, endogenous nitrite is a vital source of NO. During ischemic and hypoxic conditions, endogenous nitrite protects against development of hepatocellular injury through xanthine oxidase-mediated NO release [86]. During ischemia, NO generated from nitrite mediates signaling and formation of nitrosation and nitrosylation protein products in tissues [31]. Endogenous stores of nitrite are undoubtedly consumed to combat the oxidative stress associated with I–R injury. Exogenous administration of nitrite, then, replenishes these depleted endogenous nitrite stores and additionally leads to formation of N-, S-, and heme-nitrosated or nitrosylated hepatic proteins [64].

As with the heart, nitrite may be administered orally to protect the liver against I–R injury. Regular dietary intake of nitrite and its precursor inorganic nitrate maintains normal levels of nitrite in plasma and deficiency in diet leads to reduced levels in the blood and therefore exacerbated I–R injury [87]. The effect of one time administration of nitrite lasts up to 24 h. Nitrite given orally 24 h prior to ischemia–reperfusion inhibits complex 1 in mitochondria, thus reducing the formation of reactive oxygen species and attenuating both hepatic and myocardial I–R injury [55].

Nitrite Therapy and Renal I–R Injury

NO donors, including those with structures similar to S-nitrosothiols, effectively combat inflammatory responses that occur during I–R injury [79]. NO donor drugs, such as molsidomine [88] and sodium nitroprusside [89], attenuate I–R injury in kidneys; NO derived from these drugs diminishes apoptosis [90], suppresses tubular necrosis [91], and inhibits ET-1 expression [89].

In the kidney, nitrite is an important source of NO during ischemic syndromes [92]. eNOS has also been proposed to be an important nitrite reductase contributing to protection during renal I–R in ischemic conditions [93]. Additionally, xanthine oxidase mediates the reduction of nitrite to NO while the presence of a xanthine oxidase inhibitor prevents NO formation leading to enhanced renal injury [94]. However, nitrite administration must precede renal ischemic events to offer optimal protection since some studies revealed failure of nitrite treatment administered during ischemia [95], but beneficial effects when administered prior to onset of ischemia [96]. Clinically, this pharmacologic "preconditioning" may be applicable only to such scenarios as renal transplantation. Nitrite is also a potential antihypertensive. Administering nitrite orally may replete circulating NO levels, and thus protect the kidney during hypertension [97]. It is possible that the hypotensive effect of a diet rich in fruits and vegetables is due to release of NO [98, 99]. Nitrite therapy may additionally benefit individuals with renal insufficiency by controlling blood pressure and slowing the progression of hypertension-induced renal injury.

Nitrite Therapy for Pulmonary Disorders

Nitrite therapy is beneficial in several pulmonary disorders. Nitrite given by inhalation in newborn lambs with normoxic pulmonary hypertension has been shown to cause pulmonary vasodilation. This inhalational route has the advantage of limiting the systemic appearance of methemoglobin levels [100] and hence, can be used in neonates. Pulmonary hypoxia leads to depletion of RSNO levels and leads to reflex constriction of the pulmonary arteries [101]. Nitrite attenuates this resultant acute pulmonary vasoconstriction during hypoxia [102]. Nitrite treatment prevents smooth muscle proliferation in pulmonary arteries and attenuates the pathological right ventricular hypertrophy seen in pulmonary hypertension [103]. The beneficial reduction of pulmonary arterial pressure is sustained since it lasts for approximately 1 h after cessation of nitrite infusion [104]. This sustained pulmonary vasodilation despite plasma nitrite levels returning to baseline value suggests that nitrite has prolonged vasodilatory effects after one time administration. As with other organs, xanthine oxidase reduces nitrite to NO, especially under hypoxic conditions in the lung [105].

Additionally, nitrite treatment conferred benefits during hemolysis and pulmonary embolism. Pulmonary hypertension accompanies hemolysis [106] and nitrite infusion counteracts this elevation of pulmonary pressure [107]. Nitrite treatment has been beneficial for pulmonary thromboembolism where it improved cardiac index and pulmonary and systemic vascular resistance indices [108]. High altitude pulmonary hypertension is associated with reduced nitrite levels [109]. Higher blood flow and circulating NO products offset high-altitude hypoxia among Tibetans [110]. Replenishing nitrite levels in people with high altitude pulmonary edema (HAPE) or pulmonary hypertension may provide benefit for these patients, as well.

Nitrite Therapy for Peripheral Arterial Disease

Plasma nitrite levels correlate with vasodilatory responses in normal individuals [111]. Patients with peripheral arterial disease (PAD) are affected by abnormal vasoreactivity and reduced peripheral vascular perfusion that correlates with deranged nitrite levels [112, 113]. Nitrite-derived NO is known to vasodilate blood vessels during hypoxia and directly stimulates angiogenesis and vasculogenesis [75, 114, 115]. Not surprisingly, NO-based therapies have been shown to induce angiogenesis in the setting of the chronic ischemia that results from PAD [116]. Chronic nitrite therapy leads to endothelial cell proliferation and neovascularization in ischemic limbs [117]. In a clinically relevant porcine model of critical limb ischemia, a novel sustained release formulation of sodium nitrite (SR-nitrite) increased ischemic tissue VEGF signaling and vascular density of microvessels through endothelial cell proliferation as early as 3 days after ischemia. Therapy with SR-nitrite also significantly increased levels of circulating tissue nitrite and NO metabolites in the ischemic limb [118]. Interestingly, the levels of these metabolites were lower in the nonischemic limb [118, 119]. Thus, nitrite therapy leads to an effective and highly tissue-selective angiogenic response in a short period of time following limb ischemia. However, it is important to note that in this model of limb ischemia in the setting of obesity and metabolic syndrome (i.e., Ossabaw Swine), SR-nitrite failed to significantly improve blood flow to the ischemic limb compared to non-ischemic limb despite increases in pro-angiogenic signaling in the ischemic limb [118]. However, in the same study it was also reported [119] that SR-nitrite significantly restored vascular reactivity in metabolic syndrome providing further evidence for potent endocrine effects of nitrite under pathological conditions.

Nitrite Therapy for Sickle Cell Disease

Sickle cell disease is the most common genetic disorder in the African American population and is characterized by oxidative stress and endothelial dysfunction [120]. Multiple microvascular vaso-occlusive episodes lead to insufficient perfusion and ultimately to multiorgan injury. Reduced NO bioavailability is now thought to play an important role in pathophysiology of this disease [121]. Hydroxyurea (an NO donor) represents the only FDA-approved drug currently for the treatment of this crippling condition. A reduction of nitrite levels has been observed in patients with sickle cell crisis [122]. A human trial testing the ability of nitrite infusions to improve blood flow showed that sickle cell patients did indeed achieve a dose-dependent increase in regional blood flow though to a lesser extent compared to healthy individuals [123]. This suggests that sickle patients are more resistant to NO-based therapy than healthy counterparts. However, nitrite therapy proved to be safe in this study and represents a potential therapeutic drug.

Benefits of Low Dose Nitrite Therapy

In most of the studies mentioned in this chapter, a low dose of nitrite was employed to attenuate I–R injury or demonstrate other beneficial effects (Table 16.1).

Clinical Trials on Nitrite Therapy

It is surprisingly evident that studies in human subjects on the use of nitrite in health and disease have been scarce, and this void is now being filled with a spate of clinical trials being conducted by the NHLBI in the last few years. Pharmacokinetics of systemic nitrite infusion and NO formation are being evaluated in a phase 1 clinical trial in healthy volunteers aged 21–40. The effects of nitrite on blood pressure and the mechanism by which it increases blood flow are also being studied in this unpublished trial. Many other phase 1 clinical trials in healthy volunteers have been recently completed: studying the effects of nitrite on exercise tolerance, determining the optimum dose of nitrite for infusion, determining whether nitrite can dilate blood vessels through formation of NO, confirming the role of nitrite in ischemic preconditioning, and whether the amount of nitrites and nitrates in the diet correlates with the amount of NO exhaled. The results of these studies in healthy volunteers will greatly augment the understanding of role of nitrite in physiological regulation of blood pressure, exercise tolerance, ischemic preconditioning, and its pharmacokinetics.

Apart from studies on healthy volunteers, clinical trials of possible benefits of nitrite in disease have also been conducted. In a double-blind randomized placebo-controlled parallel-group trial, nitrite was administered immediately prior to reperfusion in patients with acute ST-elevation myocardial infarction (STEMI). A total of 229 patients presenting with acute STEMI were randomized to receive either an i.v. infusion of 70 μmol sodium nitrite (n=118) or matching placebo (n=111) over 5 min immediately before primary percutaneous intervention (PPCI). Patients underwent cardiac magnetic resonance imaging (CMR) at 6–8 days and at 6 months and serial blood sampling was performed over 72 h for the measurement of plasma creatine kinase (CK) and troponin I. Myocardial infarct size (extent of late gadolinium enhancement at 6–8 days by CMR—the primary endpoint) did not differ between nitrite and placebo groups after adjustment for area at risk, diabetes status, and center. There were no significant differences in any of the secondary endpoints, including plasma troponin I and CK area under the curve, left ventricular volumes (LV), and ejection fraction (EF) measured at 6–8 days and at 6 months and final infarct size (FIS) measured at 6 months [124]. Unfortunately in this trial, sodium nitrite administered intravenously immediately prior to reperfusion in patients with acute STEMI did not reduce infarct size. However, in 10 patients with inducible myocardial ischemia, saline and low-dose sodium nitrite (1.5 μmol/min for 20 min) were administered in a double-blind fashion during dobutamine stress echocardiography, at separate visits and in a random order; long-axis myocardial function was quantified by peak systolic velocity and strain rate responses. In 19 healthy subjects, flow-mediated dilation was assessed before and after whole-arm ischemia–reperfusion; nitrite was given before ischemia or during reperfusion. Peak systolic velocities and strain rate at peak dobutamine increased in regions exhibiting ischemia, whereas they did not change in normally functioning regions. With nitrite treatment the increment of peak systolic velocity (normalized for increase in heart rate) increased only in poorly functioning myocardial regions. Peak flow-mediated dilation decreased by 43 % after ischemia–reperfusion when subjects received only saline whereas administration of nitrite before ischemia prevented this decrease in flow-mediated dilation [125]. In this trial, low-dose nitrite improved functional responses in ischemic myocardium but had no effect on normal regions. These authors reported that infarct size in their patients was relatively large compared to placebo-treated patients in a remote conditioning study from 2010, yet there was no relationship between patients with smaller or larger infarcts, varying risk areas, or chest pain duration.

Table 16.1 Summary of the concentration and dose of nitrite solutions used in various experiments

Authors	Nitrite dose	References
Webb et al.	10–100 μM infusion 15 min prior to I–R	[69]
Hendgen-Cotta et al.	0–50 μM 5 min before R	[21]
Gonzalez et al.	0.17–0.20 μM/kg/min infusion during last 60 min and last 5 min of ischemia	[66]
Cosby et al.	36 μM/mL infusion in healthy individuals	[20]
Duranski et al.	48 nmol blood conc. (165 μg/kg i.p.) during hepatic and myocardial I–R	[64]
Jung et al.	48 and 480 nmol in 500 μL as infusion at R	[73]
Kumar et al.	165 μg/kg via i.p. injection twice daily for 3–7 days after limb ischemia	[117]
Stokes et al.	33 or 99 mg/L elemental nitrite in drinking water with high-cholesterol-rich diet for 3 weeks	[62]
Bryan et al.	50 mg/L nitrite in drinking water for 1 week prior to I–R	[19]
Bryan et al.	50 mg/L nitrite in drinking water	[35]
Dezfulian et al.	0.13 mg/kg iv at the time of CPR	[71]

The majority of studies have utilized nitrite in μM concentrations

A more recent trial investigated the use of nitrite in patients undergoing primary percutaneous coronary intervention. Eighty patients were randomized to receive intracoronary (10 mL) sodium nitrite (1.8 μmol) or NaCl (placebo) before balloon inflation. The primary end point was infarct size assessed by measuring creatine kinase release. Secondary outcomes included infarct size assessed by troponin T release and by cardiac MRI on day 2. There was no evidence of differences in creatine kinase release ($P=0.92$), troponin T ($P=0.85$), or cardiac MRI-assessed infarct size ($P=0.254$). In contrast, there was an improvement in myocardial salvage index ($P=0.05$) and reduction in major adverse cardiac event at 1 year (2.6% versus 15.8%; $P=0.04$) in the nitrite group. In a 66-patient subgroup with thrombolysis in myocardial infarction ≤ 1 flow, there was reduced serum creatine kinase ($P=0.03$) and a 19% reduction in cardiac MRI-determined infarct size ($P=0.034$) with nitrite. No adverse effects of nitrite were detected. In this phase II study, intracoronary nitrite infusion did not alter infarct size, although a trend to improved myocardial salvage index and a significant reduction in major adverse cardiac event was evident. In a subgroup of patients with thrombolysis in myocardial infarction flow ≤ 1, nitrite reduced infarct size and major adverse cardiac event and improved myocardial salvage index, indicating that a phase III clinical trial assessing intracoronary nitrite administration as an adjunct to percutaneous coronary intervention in ST-elevated myocardial infarction patients is warranted [126].

Studies using a patented nitrite lozenge (US patents 8,303,995; 8,298,589; 84,355,708; 8,962,038; 9,119,823; and 9,241,999) using 15–20 mg of sodium nitrite in the form of an orally disintegrating tablet found that it could modify cardiovascular risk factors in patients over the age of 40, significantly reduce triglycerides, and reduce blood pressure [127]. This same lozenge was used in a pediatric patient with argininosuccinic aciduria and significantly reduced his blood pressure when prescription medications were ineffective [128]. A more recent clinical trial using the nitrite lozenge reveals that a single lozenge can significantly reduce blood pressure, dilate blood vessels, improve endothelial function and arterial compliance in hypertensive patients [129]. Furthermore in a study of prehypertensive patients (BP >120/80 <139/89), administration of one lozenge twice daily leads to a significant reduction in blood pressure (12 mmHg systolic and 6 mmHg diastolic) after 30 days [130]. The same lozenge was used in an exercise study and was found to lead to a significant improvement in exercise performance [131]. Most recently, use of the nitrite lozenge without the use of a statin shows to reduce plaque as measured by carotid intima media thickness (CIMT) by 10.2% in the course of approximately 6 months. Longer term, larger trials are needed to confirm these findings. These acute studies clearly demonstrate the safety and efficacy of low doses of nitrite in humans [132].

Implications of Dietary Nitrite and Nitrate Supplementation: Safety and Cost-Effectiveness Analysis

Fruits and vegetables have numerous protective components, which include nitrates and nitrites [133]. Nitrate contained in vegetables has excellent bioavailability [134]. Similarly sodium nitrite ingested orally is absorbed well and has low first-pass metabolism in the liver [135]. It is now known that a diet rich in fruits and vegetable reduces the incidence of coronary heart disease [136] and stroke [137]. Fruit and vegetable consumption in women is associated with a decreased incidence of diabetes [138]. A diet rich in fruits, vegetables, and low-fat dairy products along with a low daily intake of total and saturated fats has been shown to reduce blood pressure comparable to a single antihypertensive medication [139, 140]. The blood pressure-lowering effects of ingesting nitrate interestingly correlate not with nitrate levels in plasma but with nitrite levels, suggesting that nitrite is the final effector of blood pressure reduction [141]. In fact, in a 60 kg adult, high-nitrate DASH diet pattern has been shown to exceed the World Health Organization's Acceptable Daily Intake for nitrate by 550% [98]. Even in normal individuals, diets rich in nitrate have been shown to reduce diastolic blood pressure [142]. Nitrate ingestion has additional benefits, such as inhibition of platelet aggregation [143] and reduced oxygen demand. Nitrate utilization by skeletal muscle thereby augments exercise capacity [144–147]. It is highly likely that nitrate in all of these instances is converted to nitrite in the body, which may account for the benefits demonstrated in these studies.

Another source of nitrite is the food processing industry. Sodium nitrite is added during the curing of meats to produce the characteristic dark color and to preserve freshness. There have been concerns of nitrite combining with proteins in meat to form N-nitrosamines, which are carcinogenic. This led to the food industry using antioxidants, such as vitamin C and vitamin E, to suppress the formation of N-nitrosamines. Today, governments of many countries regulate nitrate and nitrite levels in beverages and drinking water; however, the evidence is not convincing about causation of cancer by nitrites and nitrates [148]. The risk of development of cancer is far outweighed by the health benefits of dietary nitrites and nitrates [98, 149].

Conclusions

Nitrite has now emerged as a critically important signaling molecule involved in maintaining perfusion and redox status in tissues, not solely a metabolic product of NO in tissues. It is the principal source of NO in hypoxic conditions and oxidative stress states (Fig. 16.3). Nitrite has several disparate

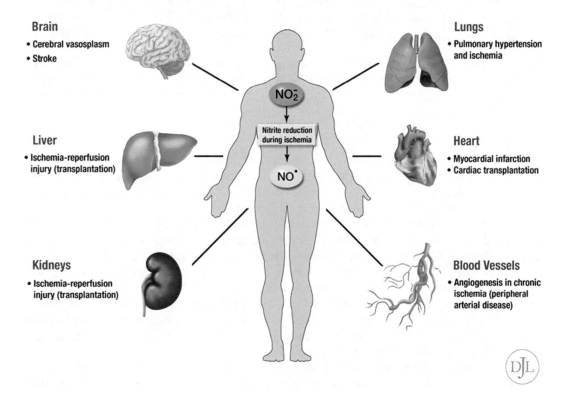

Fig. 16.3 Nitrite is an important source of NO during both acute and chronic ischemic disease states. NO derived from nitrite has been shown to limit both ischemic and reperfusion injury in the brain, lungs, heart, peripheral circulation, kidneys, and liver in a number of preclinical studies. At present, clinical trials are underway to evaluate both the efficacy and safety of nitrite therapy in a variety of ischemic syndromes

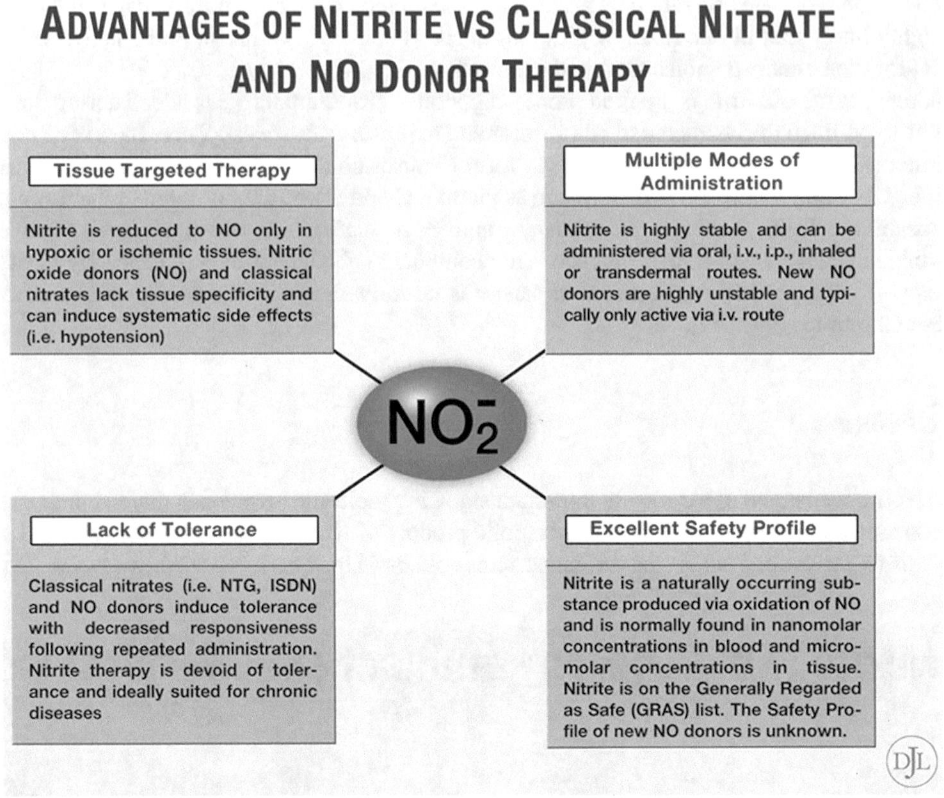

ADVANTAGES OF NITRITE VS CLASSICAL NITRATE AND NO DONOR THERAPY

Tissue Targeted Therapy

Nitrite is reduced to NO only in hypoxic or ischemic tissues. Nitric oxide donors (NO) and classical nitrates lack tissue specificity and can induce systematic side effects (i.e. hypotension)

Multiple Modes of Administration

Nitrite is highly stable and can be administered via oral, i.v., i.p., inhaled or transdermal routes. New NO donors are highly unstable and typically only active via i.v. route

NO_2^-

Lack of Tolerance

Classical nitrates (i.e. NTG, ISDN) and NO donors induce tolerance with decreased responsiveness following repeated administration. Nitrite therapy is devoid of tolerance and ideally suited for chronic diseases

Excellent Safety Profile

Nitrite is a naturally occurring substance produced via oxidation of NO and is normally found in nanomolar concentrations in blood and micromolar concentrations in tissue. Nitrite is on the Generally Regarded as Safe (GRAS) list. The Safety Profile of new NO donors is unknown.

Fig. 16.4 Nitrite has numerous advantages over classical organic nitrates and NO donors. Nitrite can be administered by multiple routes and has proven thus far to be very safe. Specific tissue targeting by nitrite is possible since hypoxic tissues selectively utilize nitrite to generate NO. Repeated administration of nitrite does not induce tolerance unlike classical nitrates

advantages as a therapeutic agent in that it is an extremely stable compound, relatively inexpensive, and can be easily administered as shown by its demonstrable efficacy as an oral, inhalational, and intravenous agent (Fig. 16.4). Nitrite, unlike the current NO donors in clinical use, does not lead to tolerance [150] and, hence, can be used for chronic therapy.

Nitrite offers an immense therapeutic potential for treatment of the most prevalent diseases and their complications, as they impose huge societal burdens in terms of healthcare resources. Conditions such as ischemic heart disease, stroke, and respiratory disorders represent three of the top five killer diseases in the US, and thankfully, may be amenable to nitrite therapy. Additionally, nitrite treatment may be beneficial in chronic cardiovascular disease associated with central obesity and diabetes. The future translation of nitrite from bench to bedside will undeniably prove to be an exciting time in the treatment of ischemic states, such as myocardial infarction, stroke, and solid organ transplantation. Further insights gleaned from preclinical investigations will likely provide evidence for the investigation of various forms of sodium nitrite for other clinical conditions.

Despite the overwhelming enthusiasm to develop sodium nitrite therapy for the treatment of cardiovascular disease, it is very important to proceed with caution. There is general agreement that very high levels of NO can promote tissue injury, decrease blood pressure, and reduce cardiac performance. Thus, it is extremely important to consider the timing of administration as well as the dosing when dealing with nitrite or any other agents that augment NO bioavailability. Finally, one must also consider possible drug interactions between nitrite-based therapies and other agents to avoid further potential toxic effects. When these pharmacokinetic and pharmacodynamic issues have been addressed, the potential clinical benefits of nitrite therapy may be realized.

Disclosures David J. Lefer (D.J.L.) is a participant on two pending United States Patents (patents no. 60/511244 and 61/003150) on the use of sodium nitrite in cardiovascular disease. Chelsea L. Organ has no disclosures. Nathan S. Bryan is an inventor on multiple issued US patents, receives royalties from patents from the University of Texas, is a Founder and Chief Science Officer for Neogenis Labs, Advisor and Stock owner for SAJE Pharma.

References

1. Lundberg JO, Weitzberg E, Gladwin MT. The nitrate-nitrite-nitric oxide pathway in physiology and therapeutics. Nat Rev Drug Discov. 2008;7(8):156–67.
2. Bodo R. The effect of the "heart-tonics" and other drugs upon the heart-tone and coronary circulation. J Physiol. 1928;64(4):365–87.
3. Nossaman VE, Nossaman BD, Kadowitz PJ. Nitrates and nitrites in the treatment of ischemic cardiac disease. Cardiol Rev. 2010;18(4):190–7.
4. Fledelius HC. Irreversible blindness after amyl nitrite inhalation. Acta Ophthalmol Scand. 1999;77(6):719–21.
5. Haverkos HW, Dougherty J. Health hazards of nitrite inhalants. Am J Med. 1988;84(3 Pt 1):479–82.
6. Gracia R, Shepherd G. Cyanide poisoning and its treatment. Pharmacotherapy. 2004;24(10):1358–65.
7. Wu LT, Schlenger WE, Ringwalt CL. Use of nitrite inhalants ("poppers") among American youth. J Adolesc Health. 2005;37(1):52–60.
8. Dezfulian C, Raat N, Shiva S, Gladwin MT. Role of the anion nitrite in ischemia-reperfusion cytoprotection and therapeutics. Cardiovasc Res. 2007;75(2):327–38.
9. Feelisch M, Fernandez BO, Bryan NS, Garcia-Saura MF, Bauer S, Whitlock DR, et al. Tissue processing of nitrite in hypoxia: an intricate interplay of nitric oxide-generating and -scavenging systems. J Biol Chem. 2008;283(49):33927–34.
10. Bryan NS. Nitrite in nitric oxide biology: cause or consequence? A systems-based review. Free Radic Biol Med. 2006;41(5):691–701.
11. Jensen FB. The role of nitrite in nitric oxide homeostasis: a comparative perspective. Biochim Biophys Acta. 2009;1787(7):841–8.
12. Lefer DJ. Emerging role of nitrite in myocardial protection. Arch Pharm Res. 2009;32(8):1127–38.
13. Lundberg JO, Weitzberg E, Cole JA, Benjamin N. Nitrate, bacteria and human health. Nat Rev Microbiol. 2004;2(7):593–602.
14. Benjamin N, O'Driscoll F, Dougall H, Duncan C, Smith L, Golden M, et al. Stomach NO synthesis. Nature. 1994;368(6471):502.
15. Sobko T, Reinders CI, Jansson E, Norin E, Midtvedt T, Lundberg JO. Gastrointestinal bacteria generate nitric oxide from nitrate and nitrite. Nitric Oxide. 2005;13(4):272–8.
16. Kleinbongard P, Dejam A, Lauer T, Rassaf T, Schindler A, Picker O, et al. Plasma nitrite reflects constitutive nitric oxide synthase activity in mammals. Free Radic Biol Med. 2003;35(7):790–6.
17. Kleinbongard P, Dejam A, Lauer T, Jax T, Kerber S, Gharini P, et al. Plasma nitrite concentrations reflect the degree of endothelial dysfunction in humans. Free Radic Biol Med. 2006;40(2):295–302.
18. Bryan NS. Cardioprotective actions of nitrite therapy and dietary considerations. Front Biosci. 2009;14:4793–808.
19. Bryan NS, Calvert JW, Gundewar S, Lefer DJ. Dietary nitrite restores NO homeostasis and is cardioprotective in endothelial nitric oxide synthase-deficient mice. Free Radic Biol Med. 2008;45(4):468–74.
20. Cosby K, Partovi KS, Crawford JH, Patel RK, Reiter CD, Martyr S, et al. Nitrite reduction to nitric oxide by deoxyhemoglobin vasodilates the human circulation. Nat Med. 2003;9:1498–505.
21. Hendgen-Cotta UB, Merx MW, Shiva S, Schmitz J, Becher S, Klare JP, et al. Nitrite reductase activity of myoglobin regulates respiration and cellular viability in myocardial ischemia-reperfusion injury. Proc Natl Acad Sci U S A. 2008;105(29):10256–61.
22. Petersen MG, Dewilde S, Fago A. Reactions of ferrous neuroglobin and cytoglobin with nitrite under anaerobic conditions. J Inorg Biochem. 2008;102(9):1777–82.
23. Li H, Samouilov A, Liu X, Zweier JL. Characterization of the effects of oxygen on xanthine oxidase-mediated nitric oxide formation. J Biol Chem. 2004;279:16939–46.
24. Basu S, Azarova NA, Font MD, King SB, Hogg N, Gladwin MT, et al. Nitrite reductase activity of cytochrome c. J Biol Chem. 2008;283(47):32590–7.

25. Daiber A, Munzel T. Nitrate reductase activity of mitochondrial aldehyde dehydrogenase (ALDH-2) as a redox sensor for cardiovascular oxidative stress. Methods Mol Biol. 2010;594:43–55.
26. Golwala NH, Hodenette C, Murthy SN, Nossaman BD, Kadowitz PJ. Vascular responses to nitrite are mediated by xanthine oxidoreductase and mitochondrial aldehyde dehydrogenase in the rat. Can J Physiol Pharmacol. 2009;87(12):1095–101.
27. Weitzberg E, Lundberg JO. Nonenzymatic nitric oxide production in humans. Nitric Oxide. 1998;2(1):1–7.
28. Zweier JL, Wang P, Samouilov A, Kuppusamy P. Enzyme-independent formation of nitric oxide in biological tissues. Nat Med. 1995;1(8):804–9.
29. Tiravanti E, Samouilov A, Zweier JL. Nitrosyl-heme complexes are formed in the ischemic heart: evidence of nitrite-derived nitric oxide formation, storage, and signaling in post-ischemic tissues. J Biol Chem. 2004;279(12):11065–73.
30. Bryan NS, Fernandez BO, Bauer SM, Garcia-Saura MF, Milsom AB, Rassaf T, et al. Nitrite is a signaling molecule and regulator of gene expression in mammalian tissues. Nat Chem Biol. 2005;1(5):290–7.
31. Bryan NS, Rassaf T, Maloney RE, Rodriguez CM, Saijo F, Rodriguez JR, et al. Cellular targets and mechanisms of nitros(yl)ation: an insight into their nature and kinetics in vivo. Proc Natl Acad Sci U S A. 2004;101(12):4308–13.
32. Dalsgaard T, Simonsen U, Fago A. Nitrite-dependent vasodilation is facilitated by hypoxia and is independent of known NO-generating nitrite reductase activities. Am J Physiol Heart Circ Physiol. 2007;292(6):H3072–8.
33. Lundberg JO, Weitzberg E. NO-synthase independent NO generation in mammals. Biochem Biophys Res Commun. 2010;396(1):39–45.
34. Gladwin MT, Raat NJ, Shiva S, Dezfulian C, Hogg N, Kim-Shapiro DB, et al. Nitrite as a vascular endocrine nitric oxide reservoir that contributes to hypoxic signaling, cytoprotection, and vasodilation. Am J Physiol Heart Circ Physiol. 2006;291(5):H2026–35.
35. Bryan NS, Calvert JW, Elrod JW, Gundewar S, Ji SY, Lefer DJ. Dietary nitrite supplementation protects against myocardial ischemia-reperfusion injury. Proc Natl Acad Sci U S A. 2007;104(48):19144–9.
36. Fish JE, Yan MS, Matouk CC, St Bernard R, Ho JJ, Gavryushova A, et al. Hypoxic repression of endothelial nitric-oxide synthase transcription is coupled with eviction of promoter histones. J Biol Chem. 2010;285(2):810–26.
37. McQuillan LP, Leung GK, Marsden PA, Kostyk SK, Kourembanas S. Hypoxia inhibits expression of eNOS via transcriptional and posttranscriptional mechanisms. Am J Physiol. 1994;267(5 Pt 2):H1921–7.
38. Tai SC, Robb GB, Marsden PA. Endothelial nitric oxide synthase: a new paradigm for gene regulation in the injured blood vessel. Arterioscler Thromb Vasc Biol. 2004;24(3):405–12.
39. Webb AJ, Milsom AB, Rathod KS, Chu WL, Qureshi S, Lovell MJ, et al. Mechanisms underlying erythrocyte and endothelial nitrite reduction to nitric oxide in hypoxia: role for xanthine oxidoreductase and endothelial nitric oxide synthase. Circ Res. 2008;103(9):957–64.
40. Braunwald E, Kloner RA. Myocardial reperfusion: a double-edged sword? J Clin Invest. 1985;76(5):1713–9.
41. Lefer AM, Lefer DJ. The role of nitric oxide and cell adhesion molecules on the microcirculation in ischaemia-reperfusion. Cardiovasc Res. 1996;32(4):743–51.
42. Calvert JW, Lefer DJ. Myocardial protection by nitrite. Cardiovasc Res. 2009;83(2):195–203.
43. Sims NR, Muyderman H. Mitochondria, oxidative metabolism and cell death in stroke. Biochim Biophys Acta. 2010;1802(1):80–91.
44. Chen Q, Camara AK, Stowe DF, Hoppel CL, Lesnefsky EJ. Modulation of electron transport protects cardiac mitochondria and decreases myocardial injury during ischemia and reperfusion. Am J Physiol Cell Physiol. 2007;292(1):C137–47.
45. Sack MN. Mitochondrial depolarization and the role of uncoupling proteins in ischemia tolerance. Cardiovasc Res. 2006;72(2):210–9.
46. Kim JS, He L, Qian T, Lemasters JJ. Role of the mitochondrial permeability transition in apoptotic and necrotic death after ischemia/reperfusion injury to hepatocytes. Curr Mol Med. 2003;3(6):527–35.
47. Garcia-Rivas GJ, Torre-Amione G. Abnormal mitochondrial function during ischemia reperfusion provides targets for pharmacological therapy. Methodist Debakey Cardiovasc J. 2009;5(3):2–7.
48. Murphy E, Steenbergen C. Mechanisms underlying acute protection from cardiac ischemia-reperfusion injury. Physiol Rev. 2008;88(2):581–609.
49. Brown GC, Cooper CE. Nanomolar concentrations of nitric oxide reversibly inhibit synaptosomal respiration by competing with oxygen at cytochrome oxidase. FEBS Lett. 1994;356:295–8.
50. Cleeter MW, Cooper JM, Darley-Usmar VM, Moncada S, Schapira AH. Reversible inhibition of cytochrome c oxidase, the terminal enzyme of the mitochondrial respiratory chain, by nitric oxide. Implications for neurodegenerative diseases. FEBS Lett. 1994;345:50–4.
51. Brookes P, Darley-Usmar VM. Hypothesis: the mitochondrial NO(*) signaling pathway, and the transduction of nitrosative to oxidative cell signals: an alternative function for cytochrome C oxidase. Free Radic Biol Med. 2002;32(4):370–4.
52. Raat NJ, Shiva S, Gladwin MT. Effects of nitrite on modulating ROS generation following ischemia and reperfusion. Adv Drug Deliv Rev. 2009;61(4):339–50.

53. Brookes PS, Salinas EP, Darley-Usmar K, Eiserich JP, Freeman BA, Darley-Usmar VM, et al. Concentration-dependent effects of nitric oxide on mitochondrial permeability transition and cytochrome c release. J Biol Chem. 2000;275(27):20474–9.

54. Kim YM, Kim TH, Seol DW, Talanian RV, Billiar TR. Nitric oxide suppression of apoptosis occurs in association with an inhibition of Bcl-2 cleavage and cytochrome c release. J Biol Chem. 1998;273(47):31437–41.

55. Shiva S, Sack MN, Greer JJ, Duranski MR, Ringwood LA, Burwell L, et al. Nitrite augments tolerance to ischemia/reperfusion injury via the modulation of mitochondrial electron transfer. J Exp Med. 2007;204(9):2089–102.

56. Carden DL, Granger DN. Pathophysiology of ischaemia-reperfusion injury. J Pathol. 2000;190(3):255–66.

57. Granger DN, Rodrigues SF, Yildirim A, Senchenkova EY. Microvascular responses to cardiovascular risk factors. Microcirculation. 2010;17(3):192–205.

58. Qi XL, Nguyen TL, Andries L, Sys SU, Rouleau JL. Vascular endothelial dysfunction contributes to myocardial depression in ischemia-reperfusion in the rat. Can J Physiol Pharmacol. 1998;76(1):35–45.

59. Lauer T, Heiss C, Balzer J, Kehmeier E, Mangold S, Leyendecker T, et al. Age-dependent endothelial dysfunction is associated with failure to increase plasma nitrite in response to exercise. Basic Res Cardiol. 2008;103(3):291–7.

60. Rassaf T, Heiss C, Hendgen-Cotta U, Balzer J, Matern S, Kleinbongard P, et al. Plasma nitrite reserve and endothelial function in the human forearm circulation. Free Radic Biol Med. 2006;41(2):295–301.

61. Rassaf T, Heiss C, Mangold S, Leyendecker T, Kehmeier ES, Kelm M, et al. Vascular formation of nitrite after exercise is abolished in patients with cardiovascular risk factors and coronary artery disease. J Am Coll Cardiol. 2010;55(14):1502–3.

62. Stokes KY, Dugas TR, Tang Y, Garg H, Guidry E, Bryan NS. Dietary nitrite prevents hypercholesterolemic microvascular inflammation and reverses endothelial dysfunction. Am J Physiol Heart Circ Physiol. 2009;296(5):H1281–8.

63. Jones SP, Greer JJ, Kakkar AK, Ware PD, Turnage RH, Hicks M, et al. Endothelial nitric oxide synthase overexpression attenuates myocardial reperfusion injury. Am J Physiol Heart Circ Physiol. 2004;286(1):H276–82.

64. Duranski MR, Greer JJ, Dejam A, Jaganmohan S, Hogg N, Langston W, et al. Cytoprotective effects of nitrite during in vivo ischemia-reperfusion of the heart and liver. J Clin Invest. 2005;115(5):1232–40.

65. Perlman DH, Bauer SM, Ashrafian H, Bryan NS, Garcia-Saura MF, Lim CC, et al. Mechanistic insights into nitrite-induced cardioprotection using an integrated metabonomic/proteomic approach. Circ Res. 2009; 104(6):796–804.

66. Gonzalez FM, Shiva S, Vincent PS, Ringwood LA, Hsu LY, Hon YY, et al. Nitrite anion provides potent cytoprotective and antiapoptotic effects as adjunctive therapy to reperfusion for acute myocardial infarction. Circulation. 2008;117(23):2986–94.

67. Bhushan S, Kondo K, Polhemus DJ, Otsuka H, Nicholson CK, Tao YX, et al. Nitrite therapy improves left ventricular function during heart failure via restoration of nitric oxide-mediated cytoprotective signaling. Circ Res. 2014;114(8):1281–91.

68. Baker JE, Su J, Fu X, Hsu A, Gross GJ, Tweddell JS, et al. Nitrite confers protection against myocardial infarction: role of xanthine oxidoreductase, NADPH oxidase and K(ATP) channels. J Mol Cell Cardiol. 2007;43(4):437–44.

69. Webb A, Bond R, McLean P, Uppal R, Benjamin N, Ahluwalia A. Reduction of nitrite to nitric oxide during ischemia protects against myocardial ischemia-reperfusion damage. Proc Natl Acad Sci U S A. 2004;101:13683–8.

70. Binks A, Nolan JP. Post-cardiac arrest syndrome. Minerva Anestesiol. 2010;76(5):362–8.

71. Dezfulian C, Shiva S, Alekseyenko A, Pendyal A, Beiser DG, Munasinghe JP, et al. Nitrite therapy after cardiac arrest reduces reactive oxygen species generation, improves cardiac and neurological function, and enhances survival via reversible inhibition of mitochondrial complex I. Circulation. 2009;120(10):897–905.

72. Pluta RM, Rak R, Wink DA, Woodward JJ, Khaldi A, Oldfield EH, et al. Effects of nitric oxide on reactive oxygen species production and infarction size after brain reperfusion injury. Neurosurgery. 2001;48(4):884–92; discussion 92–3.

73. Jung KH, Chu K, Ko SY, Lee ST, Sinn DI, Park DK, et al. Early intravenous infusion of sodium nitrite protects brain against in vivo ischemia-reperfusion injury. Stroke. 2006;37(11):2744–50.

74. Jung KH, Chu K, Lee ST, Park HK, Kim JH, Kang KM, et al. Augmentation of nitrite therapy in cerebral ischemia by NMDA receptor inhibition. Biochem Biophys Res Commun. 2009;378(3):507–12.

75. Calvert JW, Lefer DJ. Clinical translation of nitrite therapy for cardiovascular diseases. Nitric Oxide. 2010;22(2):91–7.

76. Schatlo B, Henning EC, Pluta RM, Latour LL, Golpayegani N, Merrill MJ, et al. Nitrite does not provide additional protection to thrombolysis in a rat model of stroke with delayed reperfusion. J Cereb Blood Flow Metab. 2008;28(3):482–9.

77. Pluta RM. Dysfunction of nitric oxide synthases as a cause and therapeutic target in delayed cerebral vasospasm after SAH. Acta Neurochir Suppl. 2008;104:139–47.

78. Pluta RM, Dejam A, Grimes G, Gladwin MT, Oldfield EH. Nitrite infusions to prevent delayed cerebral vasospasm in a primate model of subarachnoid hemorrhage. JAMA. 2005;293(12):1477–84.

79. Garcia-Criado FJ, Rodriguez-Barca P, Garcia-Cenador MB, Rivas-Elena JV, Grande MT, Lopez-Marcos JF, et al. Protective effect of new nitrosothiols on the early inflammatory response to kidney ischemia/reperfusion and transplantation in rats. J Interferon Cytokine Res. 2009;29(8):441–50.

80. Varon J. Diagnosis and management of labile blood pressure during acute cerebrovascular accidents and other hypertensive crises. Am J Emerg Med. 2007;25(8):949–59.

81. Rifkind JM, Nagababu E, Barbiro-Michaely E, Ramasamy S, Pluta RM, Mayevsky A. Nitrite infusion increases cerebral blood flow and decreases mean arterial blood pressure in rats: a role for red cell NO. Nitric Oxide. 2007;16(4):448–56.

82. Teoh NC, Farrell GC. Hepatic ischemia reperfusion injury: pathogenic mechanisms and basis for hepatoprotection. J Gastroenterol Hepatol. 2003;18(8):891–902.

83. Abe Y, Hines IN, Zibari G, Pavlick K, Gray L, Kitagawa Y, et al. Mouse model of liver ischemia and reperfusion injury: method for studying reactive oxygen and nitrogen metabolites in vivo. Free Radic Biol Med. 2009;46(1):1–7.

84. Abe Y, Hines I, Zibari G, Grisham MB. Hepatocellular protection by nitric oxide or nitrite in ischemia and reperfusion injury. Arch Biochem Biophys. 2009;484(2):232–7.

85. Kim JS, Ohshima S, Pediaditakis P, Lemasters JJ. Nitric oxide protects rat hepatocytes against reperfusion injury mediated by the mitochondrial permeability transition. Hepatology. 2004;39(6):1533–43.

86. Lu P, Liu F, Yao Z, Wang CY, Chen DD, Tian Y, et al. Nitrite-derived nitric oxide by xanthine oxidoreductase protects the liver against ischemia-reperfusion injury. Hepatobiliary Pancreat Dis Int. 2005;4(3):350–5.

87. Raat NJ, Noguchi AC, Liu VB, Raghavachari N, Liu D, Xu X, et al. Dietary nitrate and nitrite modulate blood and organ nitrite and the cellular ischemic stress response. Free Radic Biol Med. 2009;47(5):510–7.

88. Garcia-Criado FJ, Eleno N, Santos-Benito F, Valdunciel JJ, Reverte M, Lozano-Sanchez FS, et al. Protective effect of exogenous nitric oxide on the renal function and inflammatory response in a model of ischemia-reperfusion. Transplantation. 1998;66(8):982–90.

89. Jeong GY, Chung KY, Lee WJ, Kim YS, Sung SH. The effect of a nitric oxide donor on endogenous endothelin-1 expression in renal ischemia/reperfusion injury. Transplant Proc. 2004;36(7):1943–5.

90. Martinez-Mier G, Toledo-Pereyra LH, Bussell S, Gauvin J, Vercruysse G, Arab A, et al. Nitric oxide diminishes apoptosis and p53 gene expression after renal ischemia and reperfusion injury. Transplantation. 2000;70(10):1431–7.

91. Kucuk HF, Kaptanoglu L, Ozalp F, Kurt N, Bingul S, Torlak OA, et al. Role of glyceryl trinitrate, a nitric oxide donor, in the renal ischemia-reperfusion injury of rats. Eur Surg Res. 2006;38(5):431–7.

92. Okamoto M, Tsuchiya K, Kanematsu Y, Izawa Y, Yoshizumi M, Kagawa S, et al. Nitrite-derived nitric oxide formation following ischemia-reperfusion injury in kidney. Am J Physiol Renal Physiol. 2005;288(1):F182–7.

93. Milsom AB, Patel NS, Mazzon E, Tripatara P, Storey A, Mota-Filipe H, et al. Role for endothelial nitric oxide synthase in nitrite-induced protection against renal ischemia-reperfusion injury in mice. Nitric Oxide. 2010;22(2):141–8.

94. Tripatara P, Patel NS, Webb A, Rathod K, Lecomte FM, Mazzon E, et al. Nitrite-derived nitric oxide protects the rat kidney against ischemia/reperfusion injury in vivo: role for xanthine oxidoreductase. J Am Soc Nephrol. 2007;18(2):570–80.

95. Basireddy M, Isbell TS, Teng X, Patel RP, Agarwal A. Effects of sodium nitrite on ischemia-reperfusion injury in the rat kidney. Am J Physiol Renal Physiol. 2006;290(4):F779–86.

96. Nakajima A, Ueda K, Takaoka M, Kurata H, Takayama J, Ohkita M, et al. Effects of pre- and post-ischemic treatments with FK409, a nitric oxide donor, on ischemia/reperfusion-induced renal injury and endothelin-1 production in rats. Biol Pharm Bull. 2006;29(3):577–9.

97. Tsuchiya K, Tomita S, Ishizawa K, Abe S, Ikeda Y, Kihira Y, et al. Dietary nitrite ameliorates renal injury in L-NAME-induced hypertensive rats. Nitric Oxide. 2010;22(2):98–103.

98. Hord NG, Tang Y, Bryan NS. Food sources of nitrates and nitrites: the physiologic context for potential health benefits. Am J Clin Nutr. 2009;90(1):1–10.

99. Milkowski A, Garg HK, Coughlin JR, Bryan NS. Nutritional epidemiology in the context of nitric oxide biology: a risk-benefit evaluation for dietary nitrite and nitrate. Nitric Oxide. 2010;22(2):110–9.

100. Hunter CJ, Dejam A, Blood AB, Shields H, Kim-Shapiro DB, Machado R, et al. Inhaled nebulized nitrite is a hypoxia-sensitive NO-dependent selective pulmonary vasodilator. Nat Med. 2004;10:1122–7.

101. Wu X, Du L, Xu X, Tan L, Li R. Increased nitrosoglutathione reductase activity in hypoxic pulmonary hypertension in mice. J Pharmacol Sci. 2010;113(1):32–40.

102. Egemnazarov B, Schermuly RT, Dahal BK, Elliott GT, Hoglen NC, Surber MW, et al. Nebulization of the acidified sodium nitrite formulation attenuates acute hypoxic pulmonary vasoconstriction. Respir Res. 2010;11:81.

103. Zuckerbraun BS, Shiva S, Ifedigbo E, Mathier MA, Mollen KP, Rao J, et al. Nitrite potently inhibits hypoxic and inflammatory pulmonary arterial hypertension and smooth muscle proliferation via xanthine oxidoreductase-dependent nitric oxide generation. Circulation. 2010;121(1):98–109.

104. Ingram TE, Pinder AG, Bailey DM, Fraser AG, James PE. Low-dose sodium nitrite vasodilates hypoxic human pulmonary vasculature by a means that is not dependent on a simultaneous elevation in plasma nitrite. Am J Physiol Heart Circ Physiol. 2010;298(2):H331–9.

105. Casey DB, Badejo Jr AM, Dhaliwal JS, Murthy SN, Hyman AL, Nossaman BD, et al. Pulmonary vasodilator responses to sodium nitrite are mediated by an allopurinol-sensitive mechanism in the rat. Am J Physiol Heart Circ Physiol. 2009;296(2):H524–33.
106. Hsu LL, Champion HC, Campbell-Lee SA, Bivalacqua TJ, Manci EA, Diwan BA, et al. Hemolysis in sickle cell mice causes pulmonary hypertension due to global impairment in nitric oxide bioavailability. Blood. 2007;109(7):3088–98.
107. Minneci PC, Deans KJ, Shiva S, Zhi H, Banks SM, Kern S, et al. Nitrite reductase activity of hemoglobin as a systemic nitric oxide generator mechanism to detoxify plasma hemoglobin produced during hemolysis. Am J Physiol Heart Circ Physiol. 2008;295(2):H743–54.
108. Dias-Junior CA, Gladwin MT, Tanus-Santos JE. Low-dose intravenous nitrite improves hemodynamics in a canine model of acute pulmonary thromboembolism. Free Radic Biol Med. 2006;41(12):1764–70.
109. Berger MM, Dehnert C, Bailey DM, Luks AM, Menold E, Castell C, et al. Transpulmonary plasma ET-1 and nitrite differences in high altitude pulmonary hypertension. High Alt Med Biol. 2009;10(1):17–24.
110. Erzurum SC, Ghosh S, Janocha AJ, Xu W, Bauer S, Bryan NS, et al. Higher blood flow and circulating NO products offset high-altitude hypoxia among Tibetans. Proc Natl Acad Sci U S A. 2007;104(45):17593–8.
111. Casey DP, Beck DT, Braith RW. Systemic plasma levels of nitrite/nitrate (NOx) reflect brachial flow-mediated dilation responses in young men and women. Clin Exp Pharmacol Physiol. 2007;34(12):1291–3.
112. Allen JD, Miller EM, Schwark E, Robbins JL, Duscha BD, Annex BH. Plasma nitrite response and arterial reactivity differentiate vascular health and performance. Nitric Oxide. 2009;20(4):231–7.
113. Casey DP, Nichols WW, Conti CR, Braith RW. Relationship between endogenous concentrations of vasoactive substances and measures of peripheral vasodilator function in patients with coronary artery disease. Clin Exp Pharmacol Physiol. 2010;37(1):24–8.
114. Gladwin MT. Evidence mounts that nitrite contributes to hypoxic vasodilation in the human circulation. Circulation. 2008;117(5):594–7.
115. Maher AR, Milsom AB, Gunaruwan P, Abozguia K, Ahmed I, Weaver RA, et al. Hypoxic modulation of exogenous nitrite-induced vasodilation in humans. Circulation. 2008;117(5):670–7.
116. Mendoza MG, Robles HV, Romo E, Rios A, Escalante B. Nitric oxide-dependent neovascularization role in the lower extremity disease. Curr Pharm Des. 2007;13(35):3591–6.
117. Kumar D, Branch BG, Pattillo CB, Hood J, Thoma S, Simpson S, et al. Chronic sodium nitrite therapy augments ischemia-induced angiogenesis and arteriogenesis. Proc Natl Acad Sci U S A. 2008;105(21):7540–5.
118. Polhemus DJ, Bradley JM, Islam KN, Brewster LP, Calvert JW, Tao YX, et al. Therapeutic potential of sustained-release sodium nitrite for critical limb ischemia in the setting of metabolic syndrome. Am J Physiol Heart Circ Physiol. 2015;309(1):H82–92.
119. Bradley JM, Islam KN, Polhemus DJ, Donnarumma E, Brewster LP, Tao YX, et al. Sustained release nitrite therapy results in myocardial protection in a porcine model of metabolic syndrome with peripheral vascular disease. Am J Physiol Heart Circ Physiol. 2015;309(2):H305–17.
120. Wood KC, Granger DN. Sickle cell disease: role of reactive oxygen and nitrogen metabolites. Clin Exp Pharmacol Physiol. 2007;34(9):926–32.
121. Wood KC, Hsu LL, Gladwin MT. Sickle cell disease vasculopathy: a state of nitric oxide resistance. Free Radic Biol Med. 2008;44(8):1506–28.
122. Lopez BL, Barnett J, Ballas SK, Christopher TA, Davis-Moon L, Ma X. Nitric oxide metabolite levels in acute vaso-occlusive sickle-cell crisis. Acad Emerg Med. 1996;3(12):1098–103.
123. Mack AK, McGowan Ii VR, Tremonti CK, Ackah D, Barnett C, Machado RF, et al. Sodium nitrite promotes regional blood flow in patients with sickle cell disease: a phase I/II study. Br J Haematol. 2008;142(6):971–8.
124. Siddiqi N, Neil C, Bruce M, MacLennan G, Cotton S, Papadopoulou S, et al. Intravenous sodium nitrite in acute ST-elevation myocardial infarction: a randomized controlled trial (NIAMI). Eur Heart J. 2014;35(19):1255–62.
125. Ingram TE, Fraser AG, Bleasdale RA, Ellins EA, Margulescu AD, Halcox JP, et al. Low-dose sodium nitrite attenuates myocardial ischemia and vascular ischemia-reperfusion injury in human models. J Am Coll Cardiol. 2013;61(25):2534–41.
126. Jones DA, Pellaton C, Velmurugan S, Rathod KS, Andiapen M, Antoniou S, et al. Randomized phase 2 trial of intracoronary nitrite during acute myocardial infarction. Circ Res. 2015;116(3):437–47.
127. Zand J, Lanza F, Garg HK, Bryan NS. All-natural nitrite and nitrate containing dietary supplement promotes nitric oxide production and reduces triglycerides in humans. Nutr Res. 2011;31(4):262–9.
128. Nagamani SC, Campeau PM, Shchelochkov OA, Premkumar MH, Guse K, Brunetti-Pierri N, et al. Nitric-oxide supplementation for treatment of long-term complications in argininosuccinic aciduria. Am J Hum Genet. 2012;90(5):836–46.
129. Houston M, Hays L. Acute effects of an oral nitric oxide supplement on blood pressure, endothelial function, and vascular compliance in hypertensive patients. J Clin Hypertens (Greenwich). 2014;16(7):524–9.
130. Biswas OS, Gonzalez VR, Schwarz ER. Effects of an oral nitric oxide supplement on functional capacity and blood pressure in adults with prehypertension. J Cardiovasc Pharmacol Ther. 2014;20(1):52–8.

131. Lee J, Kim HT, Solares GJ, Kim K, Ding Z, Ivy JL. Caffeinated nitric oxide-releasing lozenge improves cycling time trial performance. Int J Sports Med. 2015;36(2):107–12.
132. Lee E. Effect of nitric oxide on carotid intima media thickness: a pilot study. Altern Ther Health Med. 2016;22 Suppl 2:32–4.
133. Lundberg JO, Feelisch M, Bjorne H, Jansson EA, Weitzberg E. Cardioprotective effects of vegetables: is nitrate the answer? Nitric Oxide. 2006;15(4):359–62.
134. van Velzen AG, Sips AJ, Schothorst RC, Lambers AC, Meulenbelt J. The oral bioavailability of nitrate from nitrate-rich vegetables in humans. Toxicol Lett. 2008;181(3):177–81.
135. Hunault CC, van Velzen AG, Sips AJ, Schothorst RC, Meulenbelt J. Bioavailability of sodium nitrite from an aqueous solution in healthy adults. Toxicol Lett. 2009;190(1):48–53.
136. Dauchet L, Amouyel P, Hercberg S, Dallongeville J. Fruit and vegetable consumption and risk of coronary heart disease: a meta-analysis of cohort studies. J Nutr. 2006;136(10):2588–93.
137. Joshipura KJ, Ascherio A, Manson JE, Stampfer MJ, Rimm EB, Speizer FE, et al. Fruit and vegetable intake in relation to risk of ischemic stroke. JAMA. 1999;282(13):1233–9.
138. Bazzano LA, Li TY, Joshipura KJ, Hu FB. Intake of fruit, vegetables, and fruit juices and risk of diabetes in women. Diabetes Care. 2008;31(7):1311–7.
139. Appel LJ, Moore TJ, Obarzanek E, Vollmer WM, Svetkey LP, Sacks FM, et al. A clinical trial of the effects of dietary patterns on blood pressure. DASH Collaborative Research Group. N Engl J Med. 1997;336(16):1117–24.
140. Sacks FM, Svetkey LP, Vollmer WM, Appel LJ, Bray GA, Harsha D, et al. Effects on blood pressure of reduced dietary sodium and the Dietary Approaches to Stop Hypertension (DASH) diet. DASH-Sodium Collaborative Research Group. N Engl J Med. 2001;344(1):3–10.
141. Kapil V, Milsom AB, Okorie M, Maleki-Toyserkani S, Akram F, Rehman F, et al. Inorganic nitrate supplementation lowers blood pressure in humans: role for nitrite-derived NO. Hypertension. 2010;56(2):274–81.
142. Sobko T, Marcus C, Govoni M, Kamiya S. Dietary nitrate in Japanese traditional foods lowers diastolic blood pressure in healthy volunteers. Nitric Oxide. 2010;22(2):136–40.
143. Webb AJ, Patel N, Loukogeorgakis S, Okorie M, Aboud Z, Misra S, et al. Acute blood pressure lowering, vasoprotective, and antiplatelet properties of dietary nitrate via bioconversion to nitrite. Hypertension. 2008;51(3):784–90.
144. Bailey SJ, Fulford J, Vanhatalo A, Winyard PG, Blackwell JR, DiMenna FJ, et al. Dietary nitrate supplementation enhances muscle contractile efficiency during knee-extensor exercise in humans. J Appl Physiol. 2010;109(1):135–48.
145. Bailey SJ, Winyard P, Vanhatalo A, Blackwell JR, Dimenna FJ, Wilkerson DP, et al. Dietary nitrate supplementation reduces the O_2 cost of low-intensity exercise and enhances tolerance to high-intensity exercise in humans. J Appl Physiol. 2009;107(4):1144–55.
146. Larsen FJ, Weitzberg E, Lundberg JO, Ekblom B. Effects of dietary nitrate on oxygen cost during exercise. Acta Physiol (Oxf). 2007;191(1):59–66.
147. Larsen FJ, Weitzberg E, Lundberg JO, Ekblom B. Dietary nitrate reduces maximal oxygen consumption while maintaining work performance in maximal exercise. Free Radic Biol Med. 2010;48(2):342–7.
148. Bryan NS, Alexander DD, Coughlin JR, Milkowski AL, Boffetta P. Ingested nitrate and nitrite and stomach cancer risk: an updated review. Food Chem Toxicol. 2012;50(10):3646–65.
149. Bryan NS, Ivy JL. Inorganic nitrite and nitrate: evidence to support consideration as dietary nutrients. Nutr Res. 2015;35(8):643–54.
150. Dejam A, Hunter CJ, Tremonti C, Pluta RM, Hon YY, Grimes G, et al. Nitrite infusion in humans and nonhuman primates: endocrine effects, pharmacokinetics, and tolerance formation. Circulation. 2007;116(16):1821–31.
151. Antonakoudis G, Poulimenos I, Kifnidis K, Zouras C, Antonakoudis H. Blood pressure control and cardiovascular risk reduction. Hippokratia. 2007;11(3):114–9.

Chapter 17
Nitrite and Nitrate as a Treatment for Hypertension

Vikas Kapil

Key Points

- Lack of endothelial nitric oxide production has been implicated in the pathogenesis of hypertension.
- Supplementation of inorganic nitrate lowers blood pressure in a dose-dependent manner in healthy volunteers.
- The nitrate-nitrite-NO pathway is functional in hypertensive patients and supplementation with dietary nitrate in hypertensive patients lowers BP by equivalent amounts to medication at standard doses, with no tolerance in medium-term studies.
- Oral hygiene products interfere with nitrate reduction in the oral cavity and may increase blood pressure by reducing circulating nitrite concentration.
- Basal nitrite levels may be responsible in part for setting blood pressure and vascular regulation.

Keywords Nitrite • Nitrate • Nitric oxide • Blood pressure • Hypertension • Antihypertensive • Cardiovascular risk • Microbiota

Introduction

The Burden of Raised Blood Pressure

Raised blood pressure (BP) is the largest attributable risk factor for mortality worldwide [1], and this enormous burden is ever increasing with predictions of 1-in-3 adults worldwide having hypertension within the next decade [2]. Indeed, over the past 30 years, the absolute number of patients with uncontrolled hypertension has inexorably risen [3, 4]. These are concerning statistics since raised BP is responsible for 50 % of all coronary heart disease events and more than 60 % of all strokes [5, 6]. Furthermore,

V. Kapil MD MRCP PhD (✉)
William Harvey Research Institute, Centre for Clinical Pharmacology, NIHR Biomedical Research Unit in Cardiovascular Disease at Barts, Queen Mary University of London, Charterhouse Square, London, UK, EC1M 6BQ

Barts Heart Centre, Barts BP Centre of Excellence, St Bartholomew's Hospital, Barts Health NHS Trust, West Smithfield, London, UK, EC1A 7BE
e-mail: v.kapil@qmul.ac.uk

N.S. Bryan, J. Loscalzo (eds.), *Nitrite and Nitrate in Human Health and Disease*, Nutrition and Health, DOI 10.1007/978-3-319-46189-2_17, © Springer International Publishing AG 2017

even within a range of BP considered *normal* (systolic blood pressure (SBP) 115–140 mmHg) that would not qualify for treatment according to current international guidelines [7–11], there is a positive linear relationship between usual BP levels and cardiovascular (CV) morbidity and mortality [12, 13]. Recent clinical trial evidence may also support the concept of driving usual BP to lower levels to reduce CV risk [14]. However in *real-world* conditions outside of clinical trials, only 50–60 % of treated hypertensive patients achieve currently recommended (laxer) BP targets [3, 15, 16]. Such statistics clearly support the rationale for identification of novel therapeutic approaches for hypertension and one strategy that has been explored over the past 30 years is delivery of nitric oxide (NO) to the CV system [17].

Nitric Oxide and the Cardiovascular System

The discovery of an endothelium-derived relaxing factor (EDRF) [18] and its eventual elucidation as the diatomic, amphipathic, free-radical *gas* NO [19–21] quickly led to the recognition of its fundamental role in maintaining CV homeostasis through its vasodilatory action on both arterial [22–24] and venous [25, 26] sides of the circulation. The development of the endothelial nitric oxide synthase (eNOS) knockout mouse demonstrated that the dependence of the vasculature on eNOS-derived NO ultimately impacts BP, with the demonstration that mean arterial pressure (MAP) is 20 mmHg higher in knockout mice compared to littermate controls [27]. Conversely, mice with eNOS overexpression have constitutively low BP [28]. Importantly, along the whole spectrum of the CV continuum [29], it is evident that there is reduced bioavailability of endothelial-derived NO from simple, classical CV risk factors, including systemic hypertension [30–32], to established CV disease phenotypes [33].

The therapeutic potential of NO in CV disease has been trialed over a number of years through modulation of the L-arginine:NO pathway to augment vasoprotective NO generation but supplementation with substrate, L-arginine, and also eNOS cofactors have yielded equivocal results [34]. Strategies that provide exogenous, synthesized sources of NO have been used for medical purposes for more than 120 years before the discovery of endogenous NO synthesis. The first *NO compounds* used for medical purposes were the organic nitrites and nitrates, amyl nitrite and nitroglycerin [35–40]. However, despite these early successes, the utility of organic nitrites and nitrates has been limited to some degree by their chemistry: tolerance [41] and inception of endothelial dysfunction [42–44] which perhaps explains in part why these drugs have failed in large-scale clinical trials [45]. While these limiting effects of organic nitrites and nitrates have been disappointing, recent proposals suggest that exploiting NO donors in CVD may still be an option using inorganic nitrite (NO_2^-) and nitrate (NO_3^-).

The Nitrate-Nitrite-NO Pathway

Until recently, a catholic view of mammalian NO biology would include the production of NO uniquely from the five electron *oxidation* of the amino acid, L-arginine, by NOS enzymes [46]. In recent years, there have been increasing evidences of a complementary system for NO production from the one electron *reduction* of inorganic NO_2^- [47, 48]. Numerous mammalian NO_2^- reductases have now been identified, containing different transition metals at their active center, such as carbonic anhydrase (Cu) [49]; xanthine oxidoreductase [50–53], aldehyde oxidase [54], and sulfite oxidase (Mo) [55]; deoxyhemoglobin [56–60], deoxymyoglobin [61, 62], neuroglobin [63, 64], cytochrome C [65] and eNOS (Fe) [52, 66]. The relative contribution of each of these pathways may vary dependent upon the prevalent conditions [67–69].

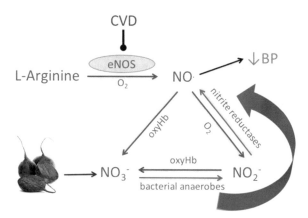

Fig. 17.1 The L-arginine: NO and nitrate-nitrite-NO pathways and hypothesis for BP lowering. The ability of endothelial nitric oxide synthase to catalyze the formation of nitric oxide from L-arginine is diminished in cardiovascular disease. Nitrate and nitrite, oxidative metabolites of nitric oxide, can be sequentially reduced involving both microbial and mammalian reductive pathways, to produce bioactive NO. Augmentation of the nitrate-nitrite-NO pathway by the provision of additional nitrate (including via dietary supplementation) may produce sufficient NO to lower BP (*BP* blood pressure, *CVD* cardiovascular disease, *NO* nitric oxide, *NO3−* nitrate, *NO2−* nitrite, *NOS* nitric oxide synthase, *O2* oxygen, *oxyHb* oxyhemoglobin)

Furthermore, the entero-salivary recirculation of NO_3^- from the circulation to the oral cavity allows the action of specific NO_3^- reductase-containing oral microbiota in a NO_3^--rich environment to facilitate the reduction of inorganic NO_3^- to NO_2^- [70–74]. Thus, there is a full reductive pathway from NO_3^- to NO_2^-, and thence NO. This paradigm shift in our understanding of NO and its related metabolites (that were previously thought of as one-way, linear termination end products of free-radical NO biology [75, 76]), reveals these chemical species to be in a *NO cycle* [77] and as such the NO_3^--NO_2^--NO pathway can be considered perhaps to be a primordial *back-up* system for NO generation [78, 79] that is particularly relevant in situations of acidotic and hypoxic [51, 54, 57, 80] environments in which the classical L-arginine/eNOS pathway is dysfunctional (Fig. 17.1) [81].

Historical Uses of Inorganic Nitrite and Nitrate

The antianginal properties of amyl nitrite reported above led others to explore related chemicals, including inorganic NO_2^- salts. Reichert and Mitchell published an extensive monograph on the physiological actions of potassium nitrite (KNO_2) in which Mitchell's human experiments revealed that the effects of KNO_2 were of similar effect to that of amyl nitrite, while Reichert's canine and feline experiments revealed that large doses of KNO_2 could cause death through profound hypotension [82]. Further comparative analysis between the pharmacodynamics of $NaNO_2$ and organic NO donors in humans demonstrated more prolonged BP reduction following inorganic NO_2^- dosing, with greater effect (25 % reduction in SBP) in patients with severely elevated BP (SBP>170 mmHg) compared to healthy individuals (13 % reduction in SBP) and maximal SBP reductions in hypertensive patients of up to 50 mmHg [83, 84]. The use of inorganic NO_2^- for the treatment of BP was established in the early part of the twentieth century and appeared in *materiae medicae* and were produced by several pharmaceutical suppliers for use in hypertension (in Butler and Feelisch [85]).

Inorganic NO_3^- may have been used to treat CVD for over a millennium in traditional Chinese medicine according to an eighth century CE manuscript discovered in the Mogao caves in Gansu province of China [86] but it was not popularized until more recently. Stieglitz produced a body of work in the 1920–1930s relating to his studies and treatments of patients with bismuth subnitrate (chemical formula: $Bi_5O(OH)_9(NO_3)_4$). Bismuth subnitrate was already an established and recommended treatment for peptic ulcer disease and diarrhea but prolonged use was cautioned against due to the risk of hypotension [87]. Stieglitz was aware of the discovery that symbiotic colonic bacteria, such as *E. coli*, were able to metabolize inorganic NO_3^- to NO_2^- [88, 89]. He thus surmised the following (Stieglitz, 1927):

> "*Therefore, theoretically, small frequent doses* [of bismuth subnitrate] *should lead to the liberation of small amounts of nitrite, uniformly and continuously absorbed. The effect of this is quite different from the violent, very transient vasodilatory effect of other forms of nitrite, such as nitroglycerol, amyl nitrite and sodium nitrite. The action to be expected is a gradually increasing vascular relaxation, with localized physiological rest to the arteriolar musculature.*"

His reports of the sustained hypotensive effects in almost 1000 patients [90–93] were followed by methodological improvements to the colorimetric techniques available at the time to measure NO_2^- levels in biological fluids [94]. He was then able to determine that the basal levels of NO_2^- in whole blood as 0.5–1.0 µg NO_2^-/100 mL blood (~110–220 nM) and that NO_2^- was not a normal constituent of fresh urine but was detectable after an inorganic NO_3^- load [95], in keeping with significant elevations in blood NO_2^- after ingestion of inorganic NO_3^- [96]. Lastly, he expended much effort into in vitro determination of the rate of NO_2^- formation from NO_3^- under the action of NO_3^--reducing bacteria to hypothesize that bacterial reduction of NO_3^- could be responsible for significant physiological effects in vivo [96].

Interest in Stieglitz's approach waned in the following decades due to concerns regarding methemoglobinemia in particular [97, 98]. However support for this approach has come recently from the recognition from large cohort studies that dietary patterns rich in green leafy vegetables, that are particularly abundant in inorganic (dietary) NO_3^- [99], reduce the risk of incident hypertension [100], ischemic heart disease [101], ischemic stroke [102], and total CV disease [103]. These and other observations led some authors to consider the possibility that inorganic NO_3^- may have cardioprotective effects [104] and may explain the beneficial effects of a vegetable-rich diet [105–107] that was recognized from small-scale clinical trials [108–111]. This chapter will detail the progress of this research to date, focusing on effects on vascular tone and BP.

Nitrite Is a Vasodilator

Despite these early studies, in the years following the discovery of EDRF as NO, the view of NO_2^- and NO_3^- was of chemically inert metabolites of endogenous NO metabolism that does not participate in any important chemical reactions [75, 76]. Although it had been recognized that supraphysiological concentrations of acidified inorganic NO_2^- had vasodilatory [112] and cardioprotective effects [113], it was not until the last 15 years that the view that NO_2^- was inert at physiological levels within the CV system was disproved. In rat aortic rings, application of low µM NO_2^- (2.5 µM), while inactive under physiological pH, relaxed contracted rat aorta under acidic (pH 6.6) conditions [114]. Following this, intra-arterial (brachial) infusion of $NaNO_2$ (36 µmol/min for 15 min) of 18 healthy subjects achieved local intravascular NO_2^- concentrations of ~200 µM. This resulted in NO-dependent vasodilation and an increase of ~175 % in forearm blood flow (a measure of arteriolar dilatation), associated with a 7 mmHg drop in MAP [57]. Infusion of approximately 100-fold less $NaNO_2$ (400 nmol/min) to achieve lower, physiologically relevant concentrations of NO_2^- (local venous concentrations of ~2.5 µM) in ten healthy subjects increased forearm blood flow by ~30 % [57]. This publication was followed by another from the same group demonstrating that infusion into the brachial artery of increasing doses of sodium nitrite ($NaNO_2$) (0.1–1.6 µmol/kg/min) resulted in local venous plasma NO_2^- levels of 25–30 μM and caused a ~10 mmHg drop in MAP that persisted for up to 3 h [115].

The effect of exercise to increase NO_2^--induced vasodilation [57] suggested that relative hypoxia might increase NO_2^--derived NO production and/or consumption. Utilizing radiolabelled, autologous blood and standard forearm plethysmographic techniques to determine arterial and venous vasodilation separately, intra-arterial NO_2^- (314 nmol/min–7.84 μmol/min) under normoxia decreased venous tone 20–35 % in a dose-dependent manner [116]. Arterial dilation was only apparent (increasing FBF by 60–80 %) at much higher doses (3.14–7.84 μmol/min) [116] and these findings are commensurate with earlier works demonstrating that dilation of the venous circulation (and not arterial circulation) was responsible for NO_2^--induced CV collapse [117, 118].

However, under hypoxia (inspired O_2 12 %) to get the arterial side of the circulation *hypoxic*, there was no augmentation of the effects of NO_2^- on venodilation but infusion of low dose NO_2^- (314 nmol/min, which had no effect in normoxic conditions) increased FBF by ~40 % [116]. Such a finding is consistent with the idea that the vasodilator potential of NO_2^- is proportional to the extent of *hypoxia* in the tissues and blood, rather than only in extreme anoxia and/or acidosis [80] as it deoxygenates from arterial to venous sides. More recently, the effect of NO_2^- to vasodilate muscular, conduit blood vessels (similarly to the actions of nitroglycerin) has been demonstrated. Infusion of $NaNO_2$ (8.7 μmol/min) into the brachial artery caused 200 % increase in radial artery diameter with no increase in radial artery flow, suggesting this was not a flow-mediated phenomenon [119]. Intriguingly, this effect was most pronounced in normoxia and was inhibited by hypoxia [119], suggesting that there is potential selectivity of NO_2^- to dilate different parts of the circulation dependent on prevalent conditions and possibly due to differing mechanisms.

These reported effects of NO_2^- on vascular tone raise the possibility that systemic NO_2^- administration or dietary NO_3^- supplementation may be useful in the treatment of hypertension.

Animal Studies with Nitrite and Nitrate for Hypertension

These demonstrations of the vasodilator potential of NO_2^- are also concordant with the effects of NO_2^- supplementation in animals. In anesthetized wild-type rats, i.v. NO_2^- (10–1000 μmol/kg for 5 min) caused dose-dependent reductions in MAP (measured by cannulation) lasting 30 min [120]. In free-moving wild-type rats, supplementation of drinking water with NO_2^- (36 mM) reduced BP measured by telemetric probes [120]. Similarly, in spontaneously hypertensive rats, prolonged (up to 1 year) oral administration of large amounts of NO_2^- in the drinking water (50–100 mM) was associated with dose-dependent improvement in BP [121, 122]. In addition, the utility of NO_2^- as an antihypertensive either supplemented in water or by daily oral garage has been confirmed in both the DOCA-salt and 2-kidney/1-clip models of hypertension [123–125]. The authors of the earlier studies considered whether elevation of systemic NO_2^- levels from NO_3^- (via bacterial reduction) would reduce BP [104, 120]. Indeed, there is now a substantial body of evidence in both animal models and humans to support this proposal.

BP was reduced in both anesthetized (by cannulation, ~20 mmHg) and conscious (telemetrically, ~5 mmHg) wild-type rats supplemented for 1 week in their drinking with NO_3^- (10 mM) compared to matched control [126]. The importance of bacterial NO_3^- reduction to NO_2^- was confirmed by use of an antibacterial mouthwash (chlorhexidine 0.2 % twice daily), to disrupt the entero-salivary recirculation of NO_3^- to NO_2^-, concomitantly with NO_3^- supplementation that abrogated the BP reduction apparent in the control state, but had no effect on rats given NO_2^- (1 mM) [126]. Further evidence in animal models of hypertension (uni-nephrectomized high-salt diet) suggests there may be clinical utility of supplementation with NO_3^- (0.1–1 mmol/kg/day) [127].

Furthermore, a series of publications have sought to explore the utility of NO_2^- or NO_3^- to restore NO homeostasis and lower BP in different models of reduced eNOS activity, which may better reflect the alteration to NO biology found in human hypertension and CV disease in general as discussed above. In the L-NAME (eNOS inhibitor)-induced model of hypertension, coadministration of dietary NO_2^- in drinking water (20–200 μM) or by oral gavage (0.2 mmol/kg/day) for 3–4 weeks prevented NOS inhibition-induced hypertension [128, 129]. Similarly, oral gavage with 0.1 mmol/kg/day inorganic NO_3^- reduces BP in both eNOS knockout mice and L-NAME treated rats [130, 131], suggesting that NO_2^-, whether directly or via dietary NO_3^- supplementation, can compensate for the depletion of eNOS-derived NO in situations such as found in human CV disease.

Human Studies with Nitrite and Nitrate

Such effects of NO_2^- and NO_3^- have been translated into the clinical setting in recent years in healthy subjects and patients.

Healthy Subjects

While much of the preclinical data suggests that inorganic NO_2^- and NO_3^- to be attractive approaches to lower BP, most of the clinical studies to date have used NO_3^- for reasons that are expounded later in this chapter. Recently, a double-blind placebo-controlled trial was conducted in older adults (aged 50–79) free of all cardiometabolic disease (but with diminished endothelial function at baseline) with three parallel groups randomized to receive either 40 mg (0.6 mmol), 80 mg (1.2 mmol) $NaNO_2$ or placebo, all twice daily for 10 weeks ($n = 10$–11), which represents the longest intervention period to date studied in a robust clinical trial design of NO_2^- or NO_3^- supplementation. Although trough plasma NO_2^- levels were significantly elevated (~2-fold) in both intervention groups after 10 weeks, there was no discernible change in BP (baseline BP ~119/73 mmHg in intervention cohorts) [132]. The lack of BP lowering in subjects with optimal/normal BP may be considered to be ideal to prevent symptomatic hypotension and/or hypotension-related syncope. This probably reflects the fact that the magnitude of BP reduction expected from any intervention is proportional to magnitude of basal BP level [133]. While the same phenomenon is observed following the elevation of plasma NO_2^- levels from inorganic NO_3^- supplementation [134], most studies using a NO_3^- supplementation approach to elevate plasma NO_2^- levels have demonstrated small yet significant reductions in BP at similar basal BP levels and the controlled clinical studies in healthy subjects to date that report change in BP (clinic unless stated otherwise) are summarized in Table 17.1 [107, 134–160] and some of the important studies are expanded on below.

The first modern demonstration of the beneficial effect of inorganic NO_3^-, via conversion to NO_2^-, was using supplementation with sodium nitrate ($NaNO_3$) (0.1 mmol/kg [6.2 mg NO_3^-/kg] daily for 3 days) compared to matched NaCl control in 17 healthy subjects. Plasma NO_2^- levels increased by ~1.5-fold in NO_3^--supplemented subjects with associated reduction in diastolic blood pressure (DBP) of 3.7 mmHg compared to placebo [154]. Following on from this, the equivalency of providing NO_3^- in salt form (as potassium nitrate (KNO_3) capsules) compared to dietary form (as NO_3^--rich (red) beet juice) was established. In a series of papers, the acute, single administration of 22.5 mmol (1395 mg) NO_3^- as beet juice (compared to water control) and an approximate equivalent of KNO_3 (24 mmol; 1488 mg NO_3^-) compared to KCl control in separate cohorts of healthy subjects had similar time courses over 24 h (peak NO_2^- levels and BP reduction at 2.5–3 h) and magnitudes of peak BP reduction from baseline (~10/7 mmHg), with no changes in either placebo group [107, 134]. Furthermore,

Table 17.1 Effect of inorganic nitrate on blood pressure in healthy subjects

ref	subjects	design	intervention	control	timeframe	baseline BP (mmHg)	BP effect (mmHg)
(154)	HV age=24	cross-over n=17	NaNO$_3$ 0.1 mmol/kg/d	NaCl 0.1 mmol/kg/d	3 days	*clinic BP not stated*	ns/-3.7
(107)	HV age=26	cross-over n=14	beet juice 22.5 mmol NO$_3^-$	matched volume water	1 day	clinic 108/70	-10.4/-8.0 (peak 2.5-3h)
(138)	HV age=26	cross-over n=8	beet juice 11.2 mmol NO$_3^-$	matched volume blackcurrant juice	6 days	*clinic BP not stated*	-6/ns
(134)	HV age=23	cross-over n=20	KNO$_3$ 24 mmol	KCl 24 mmol	1 day	clinic 110/70	-9.4/-6.0 (peak 2.5-6h)
(134)	HV age=25	cross-over n=9	beet juice 5.6 mmol NO$_3^-$	matched volume water	3h	clinic 121/71	-5.4/ns
(139)	HV age=28	cross-over n=7	beet juice 5.1 mmol NO$_3^-$	matched volume blackcurrant juice	6 days	clinic 125/73	-5/-2
(159)	HV age=29	cross-over n=8	beet juice 5.2 mmol NO$_3^-$	matched volume blackcurrant juice	15 days	clinic 127/72	-7/ns
(158)	HV age=36	cross-over n=25	Japanese diet 0.3 mmol/kg NO$_3^-$ daily	low NO$_3^-$ diet *NO$_3^-$ content not stated*	10 days	*clinic BP not stated*	ns/-4.5
(152)	HV age=22	cross-over n=9	beet juice 6.2 mmol NO$_3^-$	NO$_3^-$-deplete beet juice	6 days	clinic 129/66	-5/ns
(153)	HV age=21	cross-over n=9	beet juice 6.2 mmol NO$_3^-$	NO$_3^-$-deplete beet juice	2.5h	clinic 131/72	-6/ns

(continued)

Table 17.1 (continued)

ref	subjects	design	intervention	control	timeframe	baseline BP (mmHg)	BP effect (mmHg)
(145)	HV age=31	cross-over n=12	beet juice 8 mmol NO_3^-	NO_3^--deplete beet juice	6 days	clinic 119/74	ns
(146)	HV age=43	cross-over n=30	beet juice 7.5 mmol NO_3^-	matched volume apple juice (coloured)	1 day	clinic 132/81 *24h ABP not stated*	ns ns
(137)	HV age=28	cross-over n=14	8mmol KNO_3	8 mmol KCl	3h	116/67	-4.9/ns
(142)	HV age=47	cross-over n=30	high nitrate meal (spinach) 2.9 mmol NO_3^-	low nitrate meal (apple) or drink (rice milk) <0.1 mmol each	3.3h	clinic 112/68	-2.7/ns
(151)	HV age=64	cross-over n=12	beet juice 9.6 mmol NO_3^-	NO_3^--deplete beet juice	3 days	clinic 125/74	-5/-3
(156)	HV age=59	cross-over n=26	high NO_3^- meal (spinach) 3.5 mmol NO_3^-	low NO_3^- meal + rice milk *NO_3^- content not stated*	3.5h	clinic 119/71	-7.5/ns (peak 2h)
(160)	HV age=23	cross-over (4 limbs) n=10	beet juice 4.2, 8.4 and 16.8 mmol NO_3^-	140mL water	1 day	clinic 119/68	16.8 mmol: -9/-4 8.4 mmol: -10/-3 4.2 mmol: -5/ns (peak 2-4h)
(140, 141)	HV age=21	cross-over n=12	beet juice 12.1 mmol NO_3^-	matched volume orange juice	2h	*clinic BP not stated*	-5/ns
(143)	HV age=47	cross-over n=38	high nitrate diet estimated >4.8 mmol NO_3^- daily	low nitrate diet estimated <1.6 mmol NO_3^- daily	7 days	clinic 130/76	ns

ref	subjects	design	intervention	control	timeframe	baseline BP (mmHg)	BP effect (mmHg)
(157)	HV, mod CV risk age=63	parallel (2-arm) n=10-11	NaNO$_3$ 0.15 umol/kg daily	matched molar weight NaCl	28 days	clinic 137/80	-8/ns
(135, 148)	HV, overweight age=62	parallel (2-arm) n=10-11	beet juice 2.6-6.5 mmol NO$_3^-$ daily	matched volume blackcurrant juice	21 days	home 130/77 clinic 135/77 24h ABP 126/81	home: -7.2/ns clinic: ns 24h ABP: ns
(150)	HV age=25	cross-over n=27	high NO$_3^-$ soup (spinach) 13.6 mmol NO$_3^-$ daily	low NO$_3^-$ soup 0.01 mmol daily	7 days	clinic 116/69	-4.1/-4.4
(136)	HV age=20	cross-over n=19	high NO$_3^-$ food estimated 5.5 mmol NO$_3^-$ daily	low NO$_3^-$ food estimated 0.13 mmol NO$_3^-$ daily	7 days	clinic 107/63	-4.0/ns
(144)	HV age=31	cross-over n=12	NaNO$_3$ 0.1 mmol/kg/d	NaCl 0.1 mmol/kg/d	3 days	plethysmographic finger BP 125/75	ns
(155)	HV age=22	cross-over n=14	beet juice 6.4 mmol NO$_3^-$	NO$_3^-$-deplete beet juice	15 days	clinic 116/77	-4/-4 *inferred*
(149)	HV age=28	cross-over (4-limb) n=18	NaNO$_3$ 12.9 mmol	rocket, beet and spinach juice all 12.9 mmol NO$_3^-$	5h	*clinic BP not stated*	NaNO$_3$: ns/-4 rocket: -6/-6 beet: -5/-7.5 spinach: -7/-5 (peak 2.5-5h *inferred*)
(147)	HV age=33	cross-over (7-limb) n=12	beet juice 3, 6 and 12 mmol NO$_3^-$; NaNO$_3$ 3, 6 and 12 mmol	85 mL water	3h	*clinic BP not stated*	-6/ns (beet juice 12 mmol)

List of all randomized controlled clinical studies of inorganic nitrate supplementation in healthy subjects reporting on change in blood pressure between intervention and control limb, ordered by year of publication

Blood pressure effect reported for statistically significant ($p < 0.05$) results only or as *ns* (not significant) (*ABP* ambulatory blood pressure, *BP* blood pressure, *DBP* diastolic blood pressure, *HV* healthy volunteers, *NO3−* inorganic nitrate, *ns* not significant, *SBP* systolic blood pressure) change in SBP was significantly inversely correlated to changes in plasma NO_2^- levels only, suggesting that the change in plasma NO_2^- level was the determining effect on BP [107, 134]. The importance of entero-salivary generation of NO_2^- was confirmed by asking subjects to avoid swallowing saliva post-NO_3^- ingestion. This prevented the rise in circulating plasma NO_2^- level (elevation of NO_3^- level unaffected) and abrogated BP reduction completely [107], and confirmed NO_2^- as the bioactive moiety after NO_3^- ingestion [107]. These studies additionally demonstrated a dose-dependent effect of inorganic NO_3^- supplementation on BP and modulation of the NO_3^--NO_2^--NO pathway lead to significant elevation of cyclic guanosine monophosphate (cGMP) [134] which is the canonical intracellular secondary messenger for NO [161] and is an exquisitely sensitive marker confirming bioactive NO generation [162] via this novel pathway for the first time.

Importantly, BP-lowering effects of acute (i.e., <24 h) inorganic and dietary NO_3^- supplementation are sustained over prolonged time periods (>7 days) in healthy subjects. Comparison of the effects of a traditional, NO_3^--rich Japanese diet (0.3 mmol/kg (18.6 mg/kg) NO_3^- daily), with a low NO_3^--containing control diet in a crossover study ($n=25$, 10 days per dietary intervention), demonstrated significantly lower BP (DBP −4.5 mmHg) [158]. Similarly, dietary NO_3^- as beet juice (5.2 [322 mg] or 6.4 [397 mg] mmol/day NO_3^-) compared to low NO_3^- control juice ($n=8$–14 for 15 days per intervention) effected significant sustained lowering of BP (estimated 5–6.5 mmHg) [155, 159]. More recently, in a randomized, placebo-controlled, parallel study in healthy, older persons (baseline BP 137/80 mmHg, $n=10$–11), 4 weeks of daily 150 μmol/kg (9.3 mg NO_3^-/kg) $NaNO_3$ (~10 mmol [620 mg] NO_3^- for 70-kg subject) lowered SBP by 8 mmHg compared to matched NaCl [157], though in contrast there was only an effect on home BP (not clinic or ambulatory blood pressure (ABP)) in older, overweight persons ($n=10$–11) randomized to either 2.6–6.4 mmol (161–397 mg) NO_3^- daily (from beet juice) or NO_3^--free blackcurrant juice over 3 weeks of supplementation, though the wide range in possible NO_3^- content supplied in this cohort diminishes the impact of these results [135, 148].

Patient Groups

The translation of these overwhelmingly positive effects of inorganic or dietary NO_3^- supplementation on BP to patient groups that may benefit from BP lowering have recently been published.

Hypertension

The first study in hypertensive patients replicated the single-dose design over 24 h of the early healthy volunteer studies and used water as a control limb in a crossover design. Fifteen stage 1, drug-naïve hypertensive patients (baseline daytime ABP 142/85 mmHg) supplemented with dietary NO_3^- (3.3 mmol (205 mg) as beet juice) had peak BP reduction following dietary NO_3^- ingestion of ~11/10 mmHg [163], similar to the peak reduction in healthy volunteers in whom the NO_3^- dose was ~7-fold higher [107, 134]. Furthermore, healthy subjects given a comparable 4 mmol (248 mg) NO_3^- (KNO_3 capsules) had only a nonsignificant reduction in BP of 2/4 mmHg [134]. SBP was still reduced 24 h post-supplementation by ~8 mmHg [163], similar to expected BP-lowering effect of antihypertensive medications at standard dose in mild hypertension (~9.1 mmHg) [5] and represents a peak:trough ratio of 60 % for dietary NO_3^- in this study which is consistent with an effect size suitable (>50 %) to consider once-daily dosage for an antihypertensive medication [164, 165].

This enhanced sensitivity of BP response in hypertensive subjects could simply reflect the higher baseline BP evident in the hypertensive cohort, as the effect of BP-lowering drugs is proportional to pretreatment BP [133]. However it may be more complex that this as NO_2^- has a diminished vasodilatory effect in wild-type rats given phenylephrine (to achieve the similar baseline BP) when compared to the effects in spontaneously hypertensive rats [163]. Thus there may be additional explanations, such as sensitization of downstream pathways in the classical NO-signalling paradigm. For instance, it has been demonstrated that there is an enhanced vasorelaxant response to NO donors at the level of soluble guanylyl cyclase (sGC) in wild-type mice treated with NOS inhibitors [166] and in eNOS knockout mice [167]. However, whether this is similar in human hypertension, a situation in which eNOS activity is similarly reduced [30, 31], is not known.

Importantly, this initial study has been taken forward into a prolonged study [168]. In this double-blind, randomized, placebo-controlled, parallel study ($n = 32$ in each limb) of drug-naive and treated hypertensive patients with uncontrolled BP, 4 weeks of daily dietary NO_3^- (6.4 mmol (397 mg) as beet juice) was compared to a NO_3^--free juice (originally developed to provide a suitable placebo [169]) as control intervention to maintain patient and investigator *blindness* to achieve optimal study design. In addition, BP was measured by three separate methods and all demonstrated significant reduction in BP over 4 weeks in the active intervention limb only (baseline ABP 149/89 mmHg): clinic (−8/2 mmHg), home (−8/4 mmHg), and 24 h ABP (−8/5 mmHg) with no pharmacodynamic tolerance demonstrated by weekly home BP (Fig. 17.2) [168]. In contrast, in a smaller, shorter duration ($n = 27$, 1 week) crossover study in treated, controlled hypertensive patients (baseline BP 133/76 mmHg), 7 mmol (434 mg) NO_3^- daily (as beet juice) compared to control demonstrated no effect on home or ABP [170].

Other Patient Groups

The lack of effect in treated, controlled hypertensive patients is similar to reports in diabetic patients. In a 2-week crossover study (baseline clinic BP 143/81 mmHg, $n = 27$) comparing daily NO_3^- supplementation (7.5 mmol [465 mg] as beet juice) to NO_3^--free control, there was no effect on 24 h BP [171]. Similarly, in diabetic patients with peripheral arterial disease, 10 weeks of $NaNO_2$ (1.2 or 2.4 mmol daily, compared to placebo; $n = 18$ each) was associated with no change in BP in either active intervention group [172]. The reasons for these discrepant results compared to the robust improvements in BP demonstrated in uncomplicated, hypertensive patients are not clear but may reflect the study methodology, older demographics or pertain to concomitant medications. Another possibility is that there is specific aberrant vascular physiology/biochemistry in diabetic patients. For example, in nondiabetic patients with peripheral arterial disease ($n = 8$), acute supplementation with 9.1 mmol (564 mg) NO_3^- compared to matched volume of low-NO_3^- orange juice did cause significant change in resting, clinic BP (DBP −8 mmHg) [173], perhaps suggesting therefore some interaction between diabetes *per se* and the NO_3^--NO_2^--NO pathway.

There is equivocal data in relation to BP lowering with inorganic NO_3^- supplementation in patients with chronic obstructive pulmonary disease (COPD). Several short-term (<24 h) controlled studies ($n = 11$–21) in COPD patients using inorganic NO_3^- doses (as beet juice) ranging from 7.6 to 12.9 mmol (471–800 mg NO_3^-) have demonstrated SBP reductions of 6–12 mmHg [174–176]. In contrast, COPD patients ($n = 14$ for both) taking 9.6–13.6 mmol (595–843 mg) NO_3^- (as beet juice) daily for 2–3 days resulted in no change in any BP parameter [177, 178]. Similarly, in patients with heart failure with preserved ejection fraction (HFpEF, $n = 18$), inorganic NO_3^- (as beet juice, 6.1 mmol (378 mg) NO_3^-) resulted in significant reduction in SBP with both acute (−7 mmHg) and repetitive dosing (−14 mmHg), though the repetitive dosing phase of the study was not placebo-controlled [179]. This conflicts, at least with respect to the controlled, acute effects of inorganic NO_3^- with other results in HFpEF ($n = 17$) patients given 12.9 mmol (800 mg) NO_3^- (as beet juice) or control demonstrating no effects on BP [180].

The sum of the evidence to date suggests that inorganic NO_3^- supplementation has beneficial, robust, clinically meaningful BP-lowering effects in hypertensive subjects that are persistent in the medium term at least. Whether these effects will translate to other disease phenotypes where raised BP levels may be part of the pathophysiology, such as HFpEF, is currently not certain in the long term.

Arterial Stiffness

Arterial stiffness is increasingly considered to be important as an independent predictive marker of CV risk [181] and is best measured by aortic pulse wave velocity (PWV) [182]. Rapid reflection of incident pressure waves in stiffened central arteries leads to greater augmentation of central SBP [183] which is also more predictive of CV events than brachial SBP [184]. Increased arterial stiffness precedes incident hypertension in large, prospective cohorts [185, 186] and recent computational studies suggest that large artery stiffening alone is sufficient to explain both age-related increases in BP and failure of normal regulatory mechanisms (primarily renal and baroreflex-mediated) to provide homeostasis [187].

Inhibition of endothelial-derived NO production modulates elastic artery stiffness in vivo [188–190]. Hence increasing bioavailable NO, via increment of systemic NO_2^-, may be expected to be beneficial. In this respect, NO_2^- supplementation in aged rodents ($NaNO_2$ 0.8 mM for 3 weeks) and adults (1.2 or 2.4 mmol $NaNO_2$ orally for 10 weeks) reduced PWV [132, 191, 192]. Infusion of $NaNO_2$ (8.7 μmol/min i.v.) for 60 min in healthy subjects resulted in significant, large reductions in central SBP [119] with no change in brachial SBP, similar to the effects of oral NO_2^- dosing [132].

In healthy subjects, acute (8 mmol (496 mg) KNO_3) or chronic (4 weeks of daily 150 μmol/kg $NaNO_3$; ~ 10 mmol (620 mg) NO_3^-) for 70-kg subject) supplementation with NO_3^- is associated with reduction in PWV [137, 157]. This effect of NO_3^- supplementation was accompanied by significant elevations in plasma NO_2^- and cGMP levels, confirming the production of bioactive NO [137]. These results using inorganic NO_3^- as a source NO were furthered in patients with hypertension. Acute (3.3 mmol (205 mg) NO_3^- as beet juice; $n = 15$) or chronic (daily 6.4 mmol (397 mg) NO_3^- as beet juice for 4 weeks, $n = 32$) inorganic NO_3^- supplementation resulted in significant reductions in PWV [163, 168]. These results suggest that the effects of NO_3^- supplementation observed in hypertensive cohorts are not simply a reflection of elevated arterial tone and may indeed reflect modulation of elastic distensibility of the arterial tree by NO_2^--derived NO. Whether the effects of NO_3^-- or NO_2^--derived NO are in part mediated by reduction of distending BP in the aorta and other elastic arteries or whether there are specific direct effects on arterial *destiffening* is not yet known but is an important avenue to further explore.

Entero-salivary Generation of Nitrite Regulates BP

NO_2^- measured in the circulation is likely to represent multiple sources. Seventy percent of plasma NO_2^- is likely derived from the oxidation of eNOS-derived NO [193, 194] and plasma NO_2^- concentration inversely correlates to increasing number of CV risk factors and correlates directly to measures of endothelial function, suggesting that plasma NO_2^- levels could be used as a surrogate for endothelial function in those with CVD risk factors [195].

However, in addition to eNOS activity, under fasted, basal conditions, plasma NO_2^- levels will also be determined by the entero-salivary recirculation of NO_3^- to NO_2^- in the oral cavity. It has been estimated that 1 mmol (62 mg) NO_3^- is synthesized endogenously daily [196] from NOS-derived NO and this will also undergo entero-salivary recirculation. Thus, this generation of salivary NO_2^- (derived entirely from oral microbiome-dependent NO_3^- reduction as fresh saliva taken from salivary gland ducts contains no NO_2^- [197]) when swallowed may account for the other 30 % of plasma NO_2^- concentration that cannot be accounted for by eNOS activity.

The discovery of significant human artery-vein gradients of NO_2^- in the forearm circulation of healthy subjects suggested consumption of NO_2^- within the human circulation under basal conditions might influence regulation of vascular tone [198]. The eventual confirmation that NO_2^- is bioactive in the circulation leading to vasodilatation [57] and the inverse correlation between baseline plasma

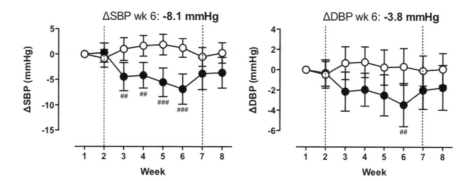

Fig. 17.2 Dietary nitrate supplementation reduces blood pressure in uncontrolled hypertensive patients, with no evidence of pharmacodynamic tolerance. The effects of 4 weeks dietary nitrate consumption (beet juice 6.4 mmol daily) or placebo (nitrate-depleted beetroot juice daily) on change in weekly systolic and diastolic blood pressure from baseline (Week 1) measured at home. Data are expressed as mean ± SD. Significant comparisons between treatment allocations for 2-way ANOVA followed by ##$p < 0.01$ and ###$p < 0.001$ for Bonferroni *post hoc* test (*BP* blood pressure, *DBP* diastolic blood pressure, *SBP* systolic blood pressure). The vertical dotted lines at 2 and 7 weeks signify the end of the 2-week run-in and the beginning of the washout period. Data adapted with permission from Kapil et al. © Wolters Kluwer [168]

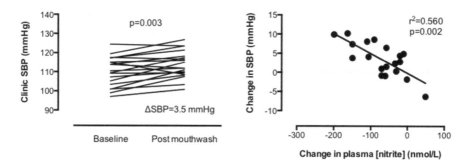

Fig. 17.3 Disturbance of nitrite homeostasis elevates basal blood pressure. Effects of a 7-day twice-daily use of antiseptic mouthwash on clinic systolic blood pressure, and the relationship between plasma nitrite concentration and clinic systolic blood pressure. Statistical significance determined using paired student's *t*-test of $n = 19$. Correlations determined using Pearson's correlation coefficient determination (*SBP* systolic blood pressure). Data adapted with permission from Kapil et al. © Elsevier [199]

NO_2^- levels and baseline BP in healthy subjects [134] raises the possibility that intrinsic plasma NO_2^- levels may be involved in *setting* the BP of healthy subjects that is dependent on a functional oral microbiome. Therefore, it could be reasonably hypothesized that interruption of the entero-salivary recirculation under basal conditions may lower basal plasma NO_2^- levels and thus affect vascular tone and increase BP. To test this hypothesis, an intervention to disrupt oral microbiome-related NO_3^- reduction (i.e., NO_2^- synthesis) was used.

In a 2-week crossover study ($n = 19$), twice-daily use of an antiseptic mouthwash for 7 days near-abolished oral conversion of NO_3^- to NO_2^-. More importantly, this was accompanied by decrease in plasma NO_2^- levels (~25%, similar to predicted [193, 194]) and a concomitant increase in BP (2–3.5 mmHg) measured by three distinct methods: clinic, home, and ABP [199]. Inspection of daily home BP measurements reveals persistence of BP effect from day 1 of instillation, with no evidence of tachyphylaxis [199]. In a similar 3-day crossover study in treated, controlled hypertensive patients ($n = 15$, baseline home BP ~ 134/79 mmHg) randomized to receive antiseptic mouthwash as above or water control, antiseptic mouthwash use led to significant increase in home SBP (2.3 mmHg) that as not apparent under control conditions [200]. The impact of different oral hygiene treatments on NO_3^- reduction and response to an inorganic NO_3^- load has been investigated and demonstrates some gradation of effect dependent on the severity of impact on oral microbiome function [201, 202].

Support for the hypothesis that these changes in BP are related to lesser NO_2^- generated from the oral microbiome comes from robust correlation of changes in plasma NO_2^- concentration and changes in BP after 7 days of antiseptic mouthwash intervention (Fig. 17.3) [199]. It is unlikely that these effects relate to other mechanisms, such as increased stress due to instillation of mouthwash, since the effects on SBP were evident in the averaged nighttime BP mean as well [199]. Although these increases in BP may seem small (~ΔSBP 2–3.5 mmHg), if persistent they could result in a 7–12% increase in the risk of mortality due to ischemic heart disease and a 10–18% increase in the risk of mortality due to stroke, using data from previous large-scale population analyses [12].

The conversion of dietary or endogenous NO_3^- to NO_2^- by the oral microbiome has long been considered to be a harmful process since salivary NO_2^-, when swallowed, can give rise to N-nitrosamines [71, 197, 203–212]. Alterations in oral microbiome species have been associated to CV disease itself [213–215] and there is widespread use of nonprescription, antiseptic, oral hygiene products in the populace [216–218]. However, the data presented above suggests that there may be a pivotal role of this pathway at least for BP. The evidence suggests a role for oral NO_3^--reducing bacteria in modulation of vascular NO_2^- and thereby NO homeostasis under basal conditions, leading to a physiological role of NO_2^-, derived from the oral reduction of endogenously (basal) generated NO_3^-, on BP regulation. Furthermore, disturbance of NO_2^- homeostasis has small, yet potentially important implications for CV health.

Benefits and Risk of Inorganic Nitrite or Nitrate Strategy

Inorganic NO_3^- is as a *prodrug* for bioactive NO_2^-, formed by the action of microbial NO_3 reductases and as such one might expect similar effectiveness from either approach, if one could match the pharmacodynamic doses. However, there are some advantages to inorganic NO_3^- approach over NO_2^-. The half-life ($t_{1/2}$) of inorganic and dietary NO_3^- is much longer (~6 h) [219] compared to NO_2^- given either by oral or *i.v.* routes (~15–45 min) [115, 220] and the peak:trough ratio [164, 165] of BP-lowering post NO_3^- supplementation in hypertension is >50% [163]. Therefore, NO_3^- can be given as an once-daily dosing regimen [168] while NO_2^- must be at least twice daily [132, 172]. It is clear that in hypertensive patients, once-daily posology improves compliance in the clinical setting [221–223] and therefore once-daily dosing is a significant advantage.

There may be an added attraction that dietary approaches may be more attractive to those patients wanting to avoid medication since they offer a *natural* approach to hypertension and other CV diseases. It may be easier/cheaper to implement such an approach in low- and middle-income health economies. However, an important limitation of the dietary approach to NO_3^- supplementation is the range of NO_3^- concentrations found naturally in batches of plants [224, 225] dependent on season of growth [226]. This produces a problem when trying to give a fixed dose of NO_3^- in the form of a vegetable or vegetable-based drink, though concentration of juice to provide a specified *dose* or use of NO_3^- salts in capsule form can overcome this.

The lack of tolerance to the BP-lowering effects of circulating NO_2^- elevation either by systemic NO_2^- administration [115] or via inorganic NO_3^- supplementation [157, 168] is encouraging and confirms much earlier reports of a lack of tolerance to prolonged NO_2^- administration. In hypertensive patients who were treated with stable doses of NO_2^-, thrice daily for periods of more than 2 weeks, there was no diminution of the hypotensive effects thereof [83]. Similarly, in healthy subjects, oral bolus doses of NO_2^- (2.2–7 mmol $NaNO_2$) produced similar BP reductions before and after a period of sustained oral administration of NO_2^- (4–5 days) [227].

The fruit- and vegetable-rich DASH diet [111] that lowers BP could be conservatively estimated to contain ~5–10 mmol (310–620 mg) NO_3^- [228], exceeding the recommended daily intake for NO_3^-, which currently is set at 3.7 mg/kg daily [229] which would be ~4 mmol in a 70-kg person. However, most of the positive studies using NO_3^- described above clearly exceed this advised limit. Strict control of levels of NO_3^- in drinking water supplies and recommended maximum NO_3^- consumption are in force in many countries based upon two main concerns: methemoglobinemia and carcinogenesis.

Initial concerns regarding the aetiological role of NO_3^- and methemoglobinemia were first reported in the early 1900s in children treated with bismuth subnitrate [230]. Surveys of NO_3^- levels in drinking well-water and associated cases of infant methemoglobinemia [97, 98] led to regulatory frameworks to control water NO_3^- level <50 mg/L [231]. However NO_3^- itself cannot cause methemoglobinemia without initial reduction to NO_2^- [59]. Several recent studies have tried to address whether elevation of circulating NO_2^- that achieves important beneficial pharmacodynamic effects leads to significant and/or clinically relevant methemoglobinemia (symptoms apparent >8 % [232]. After intra-arterial infusion of NO_2^- in healthy subjects associated with large drops in MAP (7–20 mmHg), achieved systemic NO_2^- levels were 5–16 μM and methemoglobin levels were 1–5 % [57, 233]. Chronic oral dosing of $NaNO_2$ (2.4 mmol daily for 10 weeks) or dietary NO_3^- (6.4 mmol (397 mg) daily for 4 weeks) that resulted in reductions in PWV and BP was associated with methemoglobin levels of <1.1 % [132, 168].

NO_3^- itself is not thought to be carcinogenic [229] but requires the endogenous conversion to NO_2^- [71, 197, 203–212] and the further reaction of NO_2^- with secondary amines to form potentially carcinogenic N-nitrosamines [234–238]. Similarly, comprehensive, long-term (2 years) dose-escalation studies using $NaNO_2$ in rodents conducted by the US National Toxicology Program concluded that there was no evidence of carcinogenicity [239]. Importantly, comprehensive review by WHO concluded there was no evidence NO_3^- ingestion was associated with carcinogenesis in humans [229] and in large cohorts, those with highest NO_3^- intake do not have increased cancer incidence or mortality [103, 240, 241].

Conclusion

It is now well established that the protean gasotransmitter NO can be formed from two distinct pathways in mammalian systems. The *alternative* NO_3^--NO_2^--NO pathway can be manipulated to boost NO activity in vivo and may provide a potentially easy and affordable pathway to improve both CV health and disease. It is now clear that provision of either inorganic NO_2^- or NO_3^- (the latter is salt form or as dietary intervention) is associated with robust improvements in vascular parameters including BP and measures of arterial stiffness over medium-term intervention periods. However, it is important to be aware that other interventions to *boost* endogenous NO levels, such as chronic L-arginine therapy, have not replicated the acute effects of such interventions [34] and have been associated with harm [242, 243]. Thus, it is crucial that long-term, well-conducted clinical trials powered to clinically meaningful outcomes are performed before NO_2^- or NO_3^- supplementation can be recommended in the clinical setting. What is perhaps more intriguing is the truly symbiotic role that the oral microbiome, via reduction of dietary and endogenously-derived NO_3^- to NO_2^-, may play in regulating vascular tone and therefore BP under basal conditions and the implications of this for the suitability and nonprescription provision of antiseptic oral hygiene products in the healthy and patient population. These evidences challenge dogmatic views around the harm of NO_3^- intake and on the contrary suggest that NO_3^- in sufficient supply through the diet together with the functioning, oral microbiome may be fundamental for maintaining healthy CV homeostasis.

References

1. Ezzati M, Lopez AD, Rodgers A, Vander Hoorn S, Murray CJ, Comparative Risk Assessment Collaborating Group. Selected major risk factors and global and regional burden of disease. Lancet. 2002;360:1347–60.
2. Kearney PM, Whelton M, Reynolds K, Muntner P, Whelton PK, He J. Global burden of hypertension: analysis of worldwide data. Lancet. 2005;365:217–23.
3. Egan BM, Zhao Y, Axon RN. US trends in prevalence, awareness, treatment, and control of hypertension, 1988–2008. JAMA. 2010;303:2043–50.
4. Egan BM, Zhao Y, Axon RN, Brzezinski WA, Ferdinand KC. Uncontrolled and apparent treatment resistant hypertension in the United States, 1988 to 2008. Circulation. 2011;124:1046–58.
5. Law MR, Morris JK, Wald NJ. Use of blood pressure lowering drugs in the prevention of cardiovascular disease: meta-analysis of 147 randomised trials in the context of expectations from prospective epidemiological studies. BMJ. 2009;338:b1665.
6. Rodgers A, Vaughan P, Prentice T, et al. The world health report—reducing risks, promoting healthy life. Geneva: World Health Organization; 2002.
7. James PA, Oparil S, Carter BL, et al. 2014 evidence-based guideline for the management of high blood pressure in adults: report from the panel members appointed to the Eighth Joint National Committee (JNC 8). JAMA. 2014;311:507–20.
8. Weber MA, Schiffrin EL, White WB, et al. Clinical practice guidelines for the management of hypertension in the community: a statement by the American Society of Hypertension and the International Society of Hypertension. J Hypertens. 2014;32(1):3–15.
9. National Institute of Clinical Excellence. Hypertension: clinical management of primary hypertension in adults; 2011.
10. Leung AA, Nerenberg K, Daskalopoulou SS, et al. Hypertension Canada's 2016 Canadian hypertension education program guidelines for blood pressure measurement, diagnosis, assessment of risk, prevention, and treatment of hypertension. Can J Cardiol. 2016;32:569–88.
11. Mancia G, Fagard R, Narkiewicz K, et al. 2013 ESH/ESC guidelines for the management of arterial hypertension: the task force for the management of arterial hypertension of the European Society of Hypertension (ESH) and of the European Society of Cardiology (ESC). Eur Heart J. 2013;34:2159–219.
12. Lewington S, Clarke R, Qizilbash N, Peto R, Collins R, Collaboration PS. Age-specific relevance of usual blood pressure to vascular mortality: a meta-analysis of individual data for one million adults in 61 prospective studies. Lancet. 2002;360:1903–13.
13. Rapsomaniki E, Timmis A, George J, et al. Blood pressure and incidence of twelve cardiovascular diseases: lifetime risks, healthy life-years lost, and age-specific associations in 1.25 million people. Lancet. 2014;383:1899–911.
14. Wright JTJ, Williamson JD, Whelton PK, et al. A randomized trial of intensive versus standard blood-pressure control. N Engl J Med. 2015;373:2103–16.
15. Falaschetti E, Mindell J, Knott C, Poulter N. Hypertension management in England: a serial cross-sectional study from 1994 to 2011. Lancet. 2014;383:1912–9.
16. Joffres M, Falaschetti E, Gillespie C, et al. Hypertension prevalence, awareness, treatment and control in national surveys from England, the USA and Canada, and correlation with stroke and ischaemic heart disease mortality: a cross-sectional study. BMJ Open. 2013;3, e003423.
17. Herman AG, Moncada S. Therapeutic potential of nitric oxide donors in the prevention and treatment of atherosclerosis. Eur Heart J. 2005;26:1945–55.
18. Furchgott RF, Zawadzki JV. The obligatory role of endothelial cells in the relaxation of arterial smooth muscle by acetylcholine. Nature. 1980;288:373–6.
19. Ignarro LJ, Buga GM, Wood KS, Byrns RE, Chaudhuri G. Endothelium-derived relaxing factor produced and released from artery and vein is nitric oxide. Proc Natl Acad Sci U S A. 1987;84:9265–9.
20. Palmer RM, Ferrige AG, Moncada S. Nitric oxide release accounts for the biological activity of endothelium-derived relaxing factor. Nature. 1987;327:524–6.
21. Furchgott RF. Studies on relaxation relaxation of rabbit aorta rabbit aorta by sodium nitrite: basis for the proposal that the acid-activatable component of the inhibitory factor from retractor penis is inorganic nitrite and the endothelium-derived relaxing factor is nitric oxide. In: Vanhoutte PM, editor. Vasodilatation: vascular smooth muscle, peptides, and endothelium. New York: Raven; 1988. p. 401–14.
22. Rees DD, Palmer RM, Hodson HF, Moncada S. A specific inhibitor of nitric oxide formation from L-arginine attenuates endothelium-dependent relaxation. Br J Pharmacol. 1989;96:418–24.
23. Rees DD, Palmer RM, Moncada S. Role of endothelium-derived nitric oxide in the regulation of blood pressure. Proc Natl Acad Sci U S A. 1989;86:3375–8.

24. Vallance P, Collier J, Moncada S. Effects of endothelium-derived nitric oxide on peripheral arteriolar tone in man. Lancet. 1989;334:997–1000.
25. Blackman DJ, Morris-Thurgood JA, Atherton JJ, et al. Endothelium-derived nitric oxide contributes to the regulation of venous tone in humans. Circulation. 2000;101:165–70.
26. Vallance P, Collier J, Moncada S. Nitric oxide synthesised from L-arginine mediates endothelium dependent dilatation in human veins in vivo. Cardiovasc Res. 1989;23:1053–7.
27. Huang PL, Huang Z, Mashimo H, et al. Hypertension in mice lacking the gene for endothelial nitric oxide synthase. Nature. 1995;377:239–42.
28. Ohashi Y, Kawashima S, Hirata K, et al. Hypotension and reduced nitric oxide-elicited vasorelaxation in transgenic mice overexpressing endothelial nitric oxide synthase. J Clin Invest. 1998;102:2061–71.
29. Dzau V, Braunwald E. Resolved and unresolved issues in the prevention and treatment of coronary artery disease: a workshop consensus statement. Am Heart J. 1991;121:1244–63.
30. Linder L, Kiowski W, Bühler FR, Lüscher TF. Indirect evidence for release of endothelium-derived relaxing factor in human forearm circulation in vivo. Blunted response in essential hypertension. Circulation. 1990;81:1762–7.
31. Panza JA, Casino PR, Kilcoyne CM, Quyyumi AA. Role of endothelium-derived nitric oxide in the abnormal endothelium-dependent vascular relaxation of patients with essential hypertension. Circulation. 1993;87:1468–74.
32. Forte P, Copland M, Smith LM, Milne E, Sutherland J, Benjamin N. Basal nitric oxide synthesis in essential hypertension. Lancet. 1997;349:837–42.
33. Ludmer PL, Selwyn AP, Shook TL, et al. Paradoxical vasoconstriction induced by acetylcholine in atherosclerotic coronary arteries. N Engl J Med. 1986;315:1046–51.
34. Zhang Y, Janssens SP, Wingler K, Schmidt HH, Moens AL. Modulating endothelial nitric oxide synthase: a new cardiovascular therapeutic strategy. Am J Physiol Heart Circ Physiol. 2011;301:H634–46.
35. Guthrie F. Contributions to the knowledge of the amyl group. Q J Chem Soc. 1859;11:245–52.
36. Brunton TD. On the use of nitrite of amyl in angina pectoris. Lancet. 1867;2:97–8.
37. Murrell W. Nitro-glycerine as a remedy for angina pectoris. Lancet. 1879;113:80–1.
38. Murrell W. Nitro-glycerine as a remedy for angina pectoris. Lancet. 1879;113:113–5.
39. Murrell W. Nitro-glycerine as a remedy for angina pectoris. Lancet. 1879;113:151–2.
40. Murrell W. Nitro-glycerine as a remedy for angina pectoris. Lancet. 1879;113:225–7.
41. Elkayam U, Kulick D, McIntosh N, Roth A, Hsueh W, Rahimtoola SH. Incidence of early tolerance to hemodynamic effects of continuous infusion of nitroglycerin in patients with coronary artery disease and heart failure. Circulation. 1987;76:577–84.
42. Caramori PR, Adelman AG, Azevedo ER, Newton GE, Parker AB, Parker JD. Therapy with nitroglycerin increases coronary vasoconstriction in response to acetylcholine. J Am Coll Cardiol. 1998;32:1969–74.
43. Gori T, Mak SS, Kelly S, Parker JD. Evidence supporting abnormalities in nitric oxide synthase function induced by nitroglycerin in humans. J Am Coll Cardiol. 2001;38:1096–101.
44. Heitzer T, Just H, Brockhoff C, Meinertz T, Olschewski M, Münzel T. Long-term nitroglycerin treatment is associated with supersensitivity to vasoconstrictors in men with stable coronary artery disease: prevention by concomitant treatment with captopril. J Am Coll Cardiol. 1998;31:83–8.
45. ISIS-4 Collaborative Group. ISIS-4: a randomised factorial trial assessing early oral captopril, oral mononitrate, and intravenous magnesium sulphate in 58,050 patients with suspected acute myocardial infarction. ISIS-4 (Fourth International Study of Infarct Survival) Collaborative Group. Lancet. 1995;345:669–85.
46. Stuehr DJ. Mammalian nitric oxide synthases. Biochim Biophys Acta. 1999;1411:217–30.
47. Benjamin N, O'Driscoll F, Dougall H, et al. Stomach NO synthesis. Nature. 1994;368:502.
48. Lundberg JO, Weitzberg E, Lundberg JM, Alving K. Intragastric nitric oxide production in humans: measurements in expelled air. Gut. 1994;35:1543–6.
49. Aamand R, Dalsgaard T, Jensen FB, Simonsen U, Roepstorff A, Fago A. Generation of nitric oxide from nitrite by carbonic anhydrase: a possible link between metabolic activity and vasodilation. Am J Physiol Heart Circ Physiol. 2009;297:H2068–74.
50. Li H, Samouilov A, Liu X, Zweier JL. Characterization of the effects of oxygen on xanthine oxidase-mediated nitric oxide formation. J Biol Chem. 2004;279:16939–46.
51. Webb A, Bond R, McLean P, Uppal R, Benjamin N, Ahluwalia A. Reduction of nitrite to nitric oxide during ischemia protects against myocardial ischemia-reperfusion damage. Proc Natl Acad Sci U S A. 2004;101:13683–8.
52. Webb AJ, Milsom AB, Rathod KS, et al. Mechanisms underlying erythrocyte and endothelial nitrite reduction to nitric oxide in hypoxia: role for xanthine oxidoreductase and endothelial nitric oxide synthase. Circ Res. 2008;103:957–64.
53. Zhang Z, Naughton D, Winyard PG, Benjamin N, Blake DR, Symons MC. Generation of nitric oxide by a nitrite reductase activity of xanthine oxidase: a potential pathway for nitric oxide formation in the absence of nitric oxide synthase activity. Biochem Biophys Res Commun. 1998;249:767–72.
54. Li H, Cui H, Kundu TK, Alzawahra W, Zweier JL. Nitric oxide production from nitrite occurs primarily in tissues not in the blood: critical role of xanthine oxidase and aldehyde oxidase. J Biol Chem. 2008;283:17855–63.

55. Wang J, Krizowski S, Fischer-Schrader K, et al. Sulfite oxidase catalyzes single-electron transfer at molybdenum domain to reduce nitrite to nitric oxide. Antioxid Redox Signal. 2015;23:283–94.

56. Brooks J. The action of nitrite on haemoglobin in the absence of oxygen. Proc R Soc Lond B Biol Sci. 1937;123:368–82.

57. Cosby K, Partovi KS, Crawford JH, et al. Nitrite reduction to nitric oxide by deoxyhemoglobin vasodilates the human circulation. Nat Med. 2003;9:1498–505.

58. Crawford JH, Isbell TS, Huang Z, et al. Hypoxia, red blood cells, and nitrite regulate NO-dependent hypoxic vasodilation. Blood. 2006;107:566–74.

59. Doyle MP, Pickering RA, DeWeert TM, Hoekstra JW, Pater D. Kinetics and mechanism of the oxidation of human deoxyhemoglobin by nitrites. J Biol Chem. 1981;256:12393–8.

60. Nagababu E, Ramasamy S, Abernethy DR, Rifkind JM. Active nitric oxide produced in the red cell under hypoxic conditions by deoxyhemoglobin-mediated nitrite reduction. J Biol Chem. 2003;278:46349–56.

61. Rassaf T, Flögel U, Drexhage C, Hendgen-Cotta U, Kelm M, Schrader J. Nitrite reductase function of deoxymyoglobin: oxygen sensor and regulator of cardiac energetics and function. Circ Res. 2007;100:1749–54.

62. Shiva S, Huang Z, Grubina R, et al. Deoxymyoglobin is a nitrite reductase that generates nitric oxide and regulates mitochondrial respiration. Circ Res. 2007;100:654–61.

63. Petersen MG, Dewilde S, Fago A. Reactions of ferrous neuroglobin and cytoglobin with nitrite under anaerobic conditions. J Inorg Biochem. 2008;102:1777–82.

64. Tiso M, Tejero J, Basu S, et al. Human neuroglobin functions as a redox-regulated nitrite reductase. J Biol Chem. 2011;286:18277–89.

65. Castello P, David PS, McClure T, Crook Z, Poyton R. Mitochondrial cytochrome oxidase produces nitric oxide under hypoxic conditions: implications for oxygen sensing and hypoxic signaling in eukaryotes. Cell Metab. 2006;3:277–87.

66. Gautier C, van Faassen E, Mikula I, Martasek P, Slama-Schwok A. Endothelial nitric oxide synthase reduces nitrite anions to NO under anoxia. Biochem Biophys Res Commun. 2006;341:816–21.

67. Gladwin MT, Schechter AN, Kim-Shapiro DB, et al. The emerging biology of the nitrite anion. Nat Chem Biol. 2005;1:308–14.

68. van Faassen EE, Bahrami S, Feelisch M, et al. Nitrite as regulator of hypoxic signaling in mammalian physiology. Med Res Rev. 2009;29:683–741.

69. Webb AJ, Ahluwalia A. Mechanisms of nitrite reduction in ischemia in the cardiovascular system. In: Ignarro L, editor. Nitric oxide: biology and pathobiology. Los Angeles: Elsevier; 2010. p. 555–86.

70. Maruyuma S, Murumatsu K, Shimizu S, Maki S. Reduction of nitrate with *bacillus coagulans* in human saliva. Food Hygiene and Safety Science (Shokuhin Eiseigaku Zasshi). 1976;17:19–26.

71. Tannenbaum SR, Weisman M, Fett D. The effect of nitrate intake on nitrite formation in human saliva. Food Cosmet Toxicol. 1976;14:549–52.

72. Murumatsu K, Maruyuma S, Nishizawa S. Nitrate-reducing bacterial flora and its ability to reduce nitrate in human saliva. Food Hygiene and Safety Science (Shokuhin Eiseigaku Zasshi). 1979;20:106–14.

73. Duncan C, Dougall H, Johnston P, et al. Chemical generation of nitric oxide in the mouth from the enterosalivary circulation of dietary nitrate. Nat Med. 1995;1:546–51.

74. Doel JJ, Benjamin N, Hector MP, Rogers M, Allaker RP. Evaluation of bacterial nitrate reduction in the human oral cavity. Eur J Oral Sci. 2005;113:14–9.

75. Bonner FT, Hughes MN. The aqueous solution chemistry of nitrogen in low positive oxidation states. Comm Inorg Chem. 1988;7:215–34.

76. Ignarro LJ, Fukuto JM, Griscavage JM, Rogers NE, Byrns RE. Oxidation of nitric oxide in aqueous solution to nitrite but not nitrate: comparison with enzymatically formed nitric oxide from L-arginine. Proc Natl Acad Sci U S A. 1993;90:8103–7.

77. Reutov VP, Sorokina EG. NO-synthase and nitrite-reductase components of nitric oxide cycle. Biochemistry (Mosc). 1998;63:874–84.

78. Lundberg JO, Weitzberg E, Gladwin MT. The nitrate-nitrite-nitric oxide pathway in physiology and therapeutics. Nat Rev Drug Discov. 2008;7:156–67.

79. Kapil V, Webb AJ, Ahluwalia A. Inorganic nitrate and the cardiovascular system. Heart. 2010;96:1703–9.

80. Zweier JL, Wang P, Samouilov A, Kuppusamy P. Enzyme-independent formation of nitric oxide in biological tissues. Nat Med. 1995;1:804–9.

81. Giraldez RR, Panda A, Xia Y, Sanders SP, Zweier JL. Decreased nitric-oxide synthase activity causes impaired endothelium-dependent relaxation in the postischemic heart. J Biol Chem. 1997;272:21420–6.

82. Reichert E, Mitchell SW. On the physiological action of potassium nitrite. Am J Med Sci. 1880;80:158–80.

83. Matthew E. Vaso-dilators in high blood pressure. QJM. 1909;2:261–78.

84. Wallace GB, Ringer AI. The lowering of blood-pressure by the nitrite group. JAMA. 1909;LIII:1629–30.

85. Butler AR, Feelisch M. Therapeutic uses of inorganic nitrite and nitrate: from the past to the future. Circulation. 2008;117:2151–9.

86. Butler A, Moffett J. A treatment for cardiovascular dysfunction in a Dunhuang medical manuscript. In: Lo EY, Cullen C, editors. Medieval Chinese medicine: the Dunhuang medical manuscripts. London: RoutledgeCurzon; 2005. p. 363–8.
87. Frick A. Medical treatment of peptic ulcer without alkalis. JAMA. 1924;82:595–9.
88. Salen EB. On the incidence and clinical significance of nitrites in the urine of humans. Acta Med Scand. 1925;63:369–424.
89. Zobell CE. Factors influencing the reduction of nitrates and nitrites by bacteria in semisolid media. J Bacteriol. 1932;24:273–81.
90. Stieglitz EJ. Bismuth Subnitrate in the therapy of hypertension. J Pharmacol Exp Ther. 1927;32:23–35.
91. Stieglitz EJ. The pharmacodynamics and value of bismuth subnitrate in hypertension. J Pharmacol Exp Ther. 1928;34:407–23.
92. Stieglitz EJ. Bismuth subnitrate in the treatment of arterial hypertension. JAMA. 1930;95:842–6.
93. Stieglitz EJ. Therapeutic results with bismuth subnitrate in hypertensive arterial disease. J Pharmacol Exp Ther. 1932;46:343–56.
94. Stieglitz EJ, Palmer AE. A colorimetric method for the determination of nitrite in blood. J Pharmacol Exp Ther. 1934;51:398–410.
95. Stieglitz EJ, Palmer AE. The blood nitrite. Arch Intern Med. 1937;59:620–30.
96. Stieglitz EJ, Palmer AE. Studies on the pharmacology of the nitrite effect of bismuth subnitrate. J Pharmacol Exp Ther. 1936;56:216–22.
97. Comly HH. Cyanosis in infants caused by nitrates in well water. JAMA. 1945;129:112–6.
98. Walton G. Survey of literature relating to infant methemoglobinemia due to nitrate-contaminated water. Am J Public Health Nations Health. 1951;41:986–96.
99. Santamaria P. Nitrate in vegetables: toxicity, content, intake and EC regulation. J Sci Food Agric. 2006;86:10–7.
100. Golzarand M, Bahadoran Z, Mirmiran P, Zadeh-Vakili A, Azizi F. Consumption of nitrate-containing vegetables is inversely associated with hypertension in adults: a prospective investigation from the Tehran Lipid and Glucose Study. J Nephrol. 2015.
101. Joshipura KJ, Hu FB, Manson JE, et al. The effect of fruit and vegetable intake on risk for coronary heart disease. Ann Intern Med. 2001;134:1106–14.
102. Joshipura KJ, Ascherio A, Manson JE, et al. Fruit and vegetable intake in relation to risk of ischemic stroke. JAMA. 1999;282:1233–9.
103. Hung HC, Joshipura KJ, Jiang R, et al. Fruit and vegetable intake and risk of major chronic disease. J Natl Cancer Inst. 2004;96:1577–84.
104. Classen HG, Stein-Hammer C, Thöni H. Hypothesis: the effect of oral nitrite on blood pressure in the spontaneously hypertensive rat. Does dietary nitrate mitigate hypertension after conversion to nitrite? J Am Coll Nutr. 1990;9:500–2.
105. Lundberg JO, Feelisch M, Björne H, Jansson EA, Weitzberg E. Cardioprotective effects of vegetables: is nitrate the answer? Nitric Oxide. 2006;15:359–62.
106. Ralt D. Does NO, metabolism play a role in the effects of vegetables in health? Nitric oxide formation via the reduction of nitrites and nitrates. Med Hypotheses. 2009;73:794–6.
107. Webb AJ, Patel N, Loukogeorgakis S, et al. Acute blood pressure lowering, vasoprotective, and antiplatelet properties of dietary nitrate via bioconversion to nitrite. Hypertension. 2008;51:784–90.
108. Rouse IL, Beilin LJ, Armstrong BK, Vandongen R. Blood-pressure-lowering effect of a vegetarian diet: controlled trial in normotensive subjects. Lancet. 1983;322:742–3.
109. Margetts BM, Beilin LJ, Vandongen R, Armstrong BK. Vegetarian diet in mild hypertension: a randomised controlled trial. BMJ. 1986;293:1468–71.
110. John JH, Ziebland S, Yudkin P, Roe LS, Neil HA. Effects of fruit and vegetable consumption on plasma antioxidant concentrations and blood pressure: a randomised controlled trial. Lancet. 2002;359:1969–74.
111. Appel LJ, Moore TJ, Obarzanek E, et al. A clinical trial of the effects of dietary patterns on blood pressure. DASH Collaborative Research Group. N Engl J Med. 1997;336:1117–24.
112. Furchgott RF, Bhadrakom S. Reactions of strips of rabbit aorta to epinephrine, isopropylarterenol, sodium nitrite and other drugs. J Pharmacol Exp Ther. 1953;108:129–43.
113. Johnson G, Tsao PS, Mulloy D, Lefer AM. Cardioprotective effects of acidified sodium nitrite in myocardial ischemia with reperfusion. J Pharmacol Exp Ther. 1990;252:35–41.
114. Modin A, Björne H, Herulf M, Alving K, Weitzberg E, Lundberg JO. Nitrite-derived nitric oxide: a possible mediator of 'acidic-metabolic' vasodilation. Acta Physiol Scand. 2001;171:9–16.
115. Dejam A, Hunter CJ, Tremonti C, et al. Nitrite infusion in humans and nonhuman primates: endocrine effects, pharmacokinetics, and tolerance formation. Circulation. 2007;116:1821–31.
116. Maher AR, Milsom AB, Gunaruwan P, et al. Hypoxic modulation of exogenous nitrite-induced vasodilation in humans. Circulation. 2008;117:670–7.

117. Weiss S, Wilkins RW, Haynes FW. The nature of circulatory collapse induced by sodium nitrite. J Clin Invest. 1937;16:73–84.
118. Wilkins RW, Haynes FW, Weiss S. The role of the venous system in circulatory collapse induced by sodium nitrite. J Clin Invest. 1937;16:85–91.
119. Omar SA, Fok H, Tilgner KD, et al. Paradoxical normoxia-dependent selective actions of inorganic nitrite in human muscular conduit arteries and related selective actions on central blood pressures. Circulation. 2015;131:381–9; discussion 389.
120. Vleeming W, van de Kuil A, te Biesebeek JD, Meulenbelt J, Boink AB. Effect of nitrite on blood pressure in anaesthetized and free-moving rats. Food Chem Toxicol. 1997;35:615–9.
121. Beier S, Classen HG, Loeffler K, Schumacher E, Thöni H. Antihypertensive effect of oral nitrite uptake in the spontaneously hypertensive rat. Arzneimittelforschung. 1995;45:258–61.
122. Haas M, Classen HG, Thöni H, Classen UG, Drescher B. Persistent antihypertensive effect of oral nitrite supplied up to one year via the drinking water in spontaneously hypertensive rats. Arzneimittelforschung. 1999;49:318–23.
123. Amaral JH, Ferreira GC, Pinheiro LC, Montenegro MF, Tanus-Santos JE. Consistent antioxidant and antihypertensive effects of oral sodium nitrite in DOCA-salt hypertension. Redox Biol. 2015;5:340–6.
124. Montenegro MF, Amaral JH, Pinheiro LC, et al. Sodium nitrite downregulates vascular NADPH oxidase and exerts antihypertensive effects in hypertension. Free Radic Biol Med. 2011;51:144–52.
125. Montenegro MF, Pinheiro LC, Amaral JH, et al. Antihypertensive and antioxidant effects of a single daily dose of sodium nitrite in a model of renovascular hypertension. Naunyn Schmiedebergs Arch Pharmacol. 2012;385:509–17.
126. Petersson J, Carlström M, Schreiber O, et al. Gastroprotective and blood pressure lowering effects of dietary nitrate are abolished by an antiseptic mouthwash. Free Radic Biol Med. 2009;46:1068–75.
127. Carlström M, Persson AE, Larsson E, et al. Dietary nitrate attenuates oxidative stress, prevents cardiac and renal injuries, and reduces blood pressure in salt-induced hypertension. Cardiovasc Res. 2011;89:574–85.
128. Montenegro MF, Pinheiro LC, Amaral JH, Ferreira GC, Portella RL, Tanus-Santos JE. Vascular xanthine oxidoreductase contributes to the antihypertensive effects of sodium nitrite in L-NAME hypertension. Naunyn Schmiedebergs Arch Pharmacol. 2014;387:591–8.
129. Tsuchiya K, Kanematsu Y, Yoshizumi M, et al. Nitrite is an alternative source of NO in vivo. Am J Physiol Heart Circ Physiol. 2005;288:H2163–70.
130. Carlstrom M, Liu M, Yang T, et al. Cross-talk between nitrate-nitrite-NO and NO synthase pathways in control of vascular NO homeostasis. Antioxid Redox Signal. 2015;23:295–306.
131. Carlström M, Larsen FJ, Nyström T, et al. Dietary inorganic nitrate reverses features of metabolic syndrome in endothelial nitric oxide synthase-deficient mice. Proc Natl Acad Sci U S A. 2010;107:17716–20.
132. DeVan AE, Johnson LC, Brooks FA, et al. Effects of sodium nitrite supplementation on vascular function and related small metabolite signatures in middle-aged and older adults. J Appl Physiol (1985). 2016;120:416–25.
133. Law MR, Wald NJ, Morris JK, Jordan RE. Value of low dose combination treatment with blood pressure lowering drugs: analysis of 354 randomised trials. BMJ. 2003;326:1427.
134. Kapil V, Milsom AB, Okorie M, et al. Inorganic nitrate supplementation lowers blood pressure in humans: role for nitrite-derived NO. Hypertension. 2010;56:274–81.
135. Ashor AW, Jajja A, Sutyarjoko A, et al. Effects of beetroot juice supplementation on microvascular blood flow in older overweight and obese subjects: a pilot randomised controlled study [letter]. J Hum Hypertens. 2015;29(8):511–3.
136. Ashworth A, Mitchell K, Blackwell JR, Vanhatalo A, Jones AM. High-nitrate vegetable diet increases plasma nitrate and nitrite concentrations and reduces blood pressure in healthy women. Public Health Nutr. 2015;18:2669–78.
137. Bahra M, Kapil V, Pearl V, Ghosh S, Ahluwalia A. Inorganic nitrate ingestion improves vascular compliance but does not alter flow-mediated dilatation in healthy volunteers. Nitric Oxide. 2012;26:197–202.
138. Bailey SJ, Winyard P, Vanhatalo A, et al. Dietary nitrate supplementation reduces the O_2 cost of low-intensity exercise and enhances tolerance to high-intensity exercise in humans. J Appl Physiol. 2009;107:1144–55.
139. Bailey SJ, Fulford J, Vanhatalo A, et al. Dietary nitrate supplementation enhances muscle contractile efficiency during knee-extensor exercise in humans. J Appl Physiol. 2010;109:135–48.
140. Bond VJ, Curry BH, Adams RG, Asadi MS, Millis RM, Haddad GE. Effects of dietary nitrates on systemic and cerebrovascular hemodynamics. Cardiol Res Pract. 2013;2013:435629.
141. Bond VJ, Curry BH, Adams RG, Millis RM, Haddad GE. Cardiorespiratory function associated with dietary nitrate supplementation. Appl Physiol Nutr Metab. 2014;39:168–72.
142. Bondonno CP, Yang X, Croft KD, et al. Flavonoid-rich apples and nitrate-rich spinach augment nitric oxide status and improve endothelial function in healthy men and women: a randomized controlled trial. Free Radic Biol Med. 2012;52:95–102.
143. Bondonno CP, Liu AH, Croft KD, et al. Short-term effects of nitrate-rich green leafy vegetables on blood pressure and arterial stiffness in individuals with high-normal blood pressure. Free Radic Biol Med. 2014;77:353–62.

144. Bourdillon N, Fan JL, Uva B, Muller H, Meyer P, Kayser B. Effect of oral nitrate supplementation on pulmonary hemodynamics during exercise and time trial performance in normoxia and hypoxia: a randomized controlled trial. Front Physiol. 2015;6:288.
145. Cermak NM, Gibala MJ, van Loon LJC. Nitrate supplementation's improvement of 10-km time-trial performance in trained cyclists. Int J Sport Nutr Exerc Metab. 2012;22:64–71.
146. Coles LT, Clifton PM. Effect of beetroot juice on lowering blood pressure in free-living, disease-free adults: a randomized, placebo-controlled trial. Nutr J. 2012;11:106.
147. Flueck JL, Bogdanova A, Mettler S, Perret C. Is beetroot juice more effective than sodium nitrate? The effects of equimolar nitrate dosages of nitrate-rich beetroot juice and sodium nitrate on oxygen consumption during exercise. Appl Physiol Nutr Metab. 2016;41:421–9.
148. Jajja A, Sutyarjoko A, Lara J, et al. Beetroot supplementation lowers daily systolic blood pressure in older, overweight subjects. Nutr Res. 2014;34:868–75.
149. Jonvik KL, Nyakayiru J, Pinckaers PJ, Senden JM, van Loon LJ, Verdijk LB. Nitrate-rich vegetables increase plasma nitrate and nitrite concentrations and lower blood pressure in healthy adults. J Nutr. 2016;146(5):986–93.
150. Jovanovski E, Bosco L, Khan K, et al. Effect of spinach, a high dietary nitrate source, on arterial stiffness and related hemodynamic measures: a randomized, controlled trial in healthy adults. Clin Nutr Res. 2015;4:160–7.
151. Kelly J, Fulford J, Vanhatalo A, et al. Effects of short-term dietary nitrate supplementation on blood pressure, O_2 uptake kinetics, and muscle and cognitive function in older adults. Am J Physiol Regul Integr Comp Physiol. 2013;304:R73–83.
152. Lansley KE, Winyard PG, Bailey SJ, et al. Acute dietary nitrate supplementation improves cycling time trial performance. Med Sci Sports Exerc. 2011;43:1125–31.
153. Lansley KE, Winyard PG, Fulford J, et al. Dietary nitrate supplementation reduces the O_2 cost of walking and running: a placebo-controlled study. J Appl Physiol. 2011;110:591–600.
154. Larsen FJ, Ekblom B, Sahlin K, Lundberg JO, Weitzberg E. Effects of dietary nitrate on blood pressure in healthy volunteers. N Engl J Med. 2006;355:2792–3.
155. Lee JS, Stebbins CL, Jung E, et al. Effects of chronic dietary nitrate supplementation on the hemodynamic response to dynamic exercise. Am J Physiol Regul Integr Comp Physiol. 2015;309:R459–66.
156. Liu AH, Bondonno CP, Croft KD, et al. Effects of a nitrate-rich meal on arterial stiffness and blood pressure in healthy volunteers. Nitric Oxide. 2013;35:123–30.
157. Rammos C, Hendgen-Cotta UB, Sobierajski J, Bernard A, Kelm M, Rassaf T. Dietary nitrate reverses vascular dysfunction in older adults with moderately increased cardiovascular risk [letter]. J Am Coll Cardiol. 2014;63(15):1584–5.
158. Sobko T, Marcus C, Govoni M, Kamiya S. Dietary nitrate in Japanese traditional foods lowers diastolic blood pressure in healthy volunteers. Nitric Oxide. 2010;22:136–40.
159. Vanhatalo A, Bailey SJ, Blackwell JR, et al. Acute and chronic effects of dietary nitrate supplementation on blood pressure and the physiological responses to moderate-intensity and incremental exercise. Am J Physiol Regul Integr Comp Physiol. 2010;299:R1121–31.
160. Wylie LJ, Kelly J, Bailey SJ, et al. Beetroot juice and exercise: pharmacodynamic and dose-response relationships. J Appl Physiol (1985). 2013;115:325–36.
161. Hobbs AJ, Stasch J-P. Soluble guanylate cyclase: allosteric activation and redox regulation. In: Ignarro L, editor. Nitric oxide: biology and pathobiology. New York: Academic; 2010. p. 301–26.
162. Batchelor AM, Bartus K, Reynell C, et al. Exquisite sensitivity to subsecond, picomolar nitric oxide transients conferred on cells by guanylyl cyclase-coupled receptors. Proc Natl Acad Sci U S A. 2010;107:22060–5.
163. Ghosh SM, Kapil V, Fuentes-Calvo I, et al. Enhanced vasodilator activity of nitrite in hypertension: critical role for erythrocytic xanthine oxidoreductase and translational potential. Hypertension. 2013;61:1091–102.
164. Lipicky RJ. Trough: peak ratio: the rationale behind the United States Food and Drug Administration recommendations. J Hypertens Suppl. 1994;12:S17–9.
165. Meredith PA. New FDA, guidelines on the treatment of hypertension: comparison of different therapeutic classes according to trough/peak blood pressure responses. Arch Mal Coeur Vaiss. 1994;87:1423–9.
166. Moncada S, Palmer RM, Higgs EA. Nitric oxide: physiology, pathophysiology, and pharmacology. Pharmacol Rev. 1991;43:109–42.
167. Brandes RP, Kim D, Schmitz-Winnenthal FH, et al. Increased nitrovasodilator sensitivity in endothelial nitric oxide synthase knockout mice: role of soluble guanylyl cyclase. Hypertension. 2000;35:231–6.
168. Kapil V, Khambata RS, Robertson A, Caulfield MJ, Ahluwalia A. Dietary nitrate provides sustained blood pressure lowering in hypertensive patients: a randomized, phase 2, double-blind, placebo-controlled study. Hypertension. 2015;65:320–7.
169. Gilchrist M, Winyard PG, Fulford J, Anning C, Shore AC, Benjamin N. Dietary nitrate supplementation improves reaction time in type 2 diabetes: development and application of a novel nitrate-depleted beetroot juice placebo. Nitric Oxide. 2014;40:67–74.

170. Bondonno CP, Liu AH, Croft KD, et al. Absence of an effect of high nitrate intake from beetroot juice on blood pressure in treated hypertensive individuals: a randomized controlled trial. Am J Clin Nutr. 2015;102:368–75.
171. Gilchrist M, Winyard P, Aizawa K, Anning C, Shore A, Benjamin N. Effect of dietary nitrate on blood pressure, endothelial function, and insulin sensitivity in type 2 diabetes. Free Radic Biol Med. 2013;60:89–97.
172. Mohler ER, Hiatt WR, Gornik HL, et al. Sodium nitrite in patients with peripheral artery disease and diabetes mellitus: safety, walking distance and endothelial function. Vasc Med. 2014;19:9–17.
173. Kenjale AA, Ham KL, Stabler T, et al. Dietary nitrate supplementation enhances exercise performance in peripheral arterial disease. J Appl Physiol. 2011;110:1582–91.
174. Curtis KJ, O'Brien KA, Tanner RJ, et al. Acute dietary nitrate supplementation and exercise performance in COPD: a double-blind, placebo-controlled. Randomised controlled pilot study. PLoS One. 2015;10:e0144504.
175. Berry MJ, Justus NW, Hauser JI, et al. Dietary nitrate supplementation improves exercise performance and decreases blood pressure in COPD patients. Nitric Oxide. 2015;48:22–30.
176. Kerley CP, Cahill K, Bolger K, et al. Dietary nitrate supplementation in COPD: an acute, double-blind, randomized, placebo-controlled, crossover trial. Nitric Oxide. 2015;44:105–11.
177. Leong P, Basham JE, Yong T, et al. A double blind randomized placebo control crossover trial on the effect of dietary nitrate supplementation on exercise tolerance in stable moderate chronic obstructive pulmonary disease. BMC Pulm Med. 2015;15:52.
178. Shepherd AI, Wilkerson DP, Dobson L, et al. The effect of dietary nitrate supplementation on the oxygen cost of cycling, walking performance and resting blood pressure in individuals with chronic obstructive pulmonary disease: a double blind placebo controlled, randomised control trial. Nitric Oxide. 2015;48:31–7.
179. Eggebeen J, Kim-Shapiro DB, Haykowsky M, et al. One week of daily dosing with beetroot juice improves submaximal endurance and blood pressure in older patients with heart failure and preserved ejection fraction. JACC Heart Fail. 2016;4(6):428–37.
180. Zamani P, Rawat D, Shiva-Kumar P, et al. Effect of inorganic nitrate on exercise capacity in heart failure with preserved ejection fraction. Circulation. 2015;131:371–80; discussion 380.
181. Vlachopoulos C, Aznaouridis K, Stefanadis C. Prediction of cardiovascular events and all-cause mortality with arterial stiffness: a systematic review and meta-analysis. J Am Coll Cardiol. 2010;55:1318–27.
182. Laurent S, Cockcroft JR, Van Bortel LM, et al. Expert consensus document on arterial stiffness: methodological issues and clinical applications. Eur Heart J. 2006;27:2588–605.
183. Safar ME. Arterial aging—hemodynamic changes and therapeutic options. Nat Rev Cardiol. 2010;7:442–9.
184. Vlachopoulos C, Aznaouridis K, O'Rourke MF, Safar ME, Baou K, Stefanadis C. Prediction of cardiovascular events and all-cause mortality with central haemodynamics: a systematic review and meta-analysis. Eur Heart J. 2010;31:1865–71.
185. Kaess BM, Rong J, Larson MG, et al. Aortic stiffness, blood pressure progression, and incident hypertension. JAMA. 2012;308:875–81.
186. Zheng X, Jin C, Liu Y, et al. Arterial stiffness as a predictor of clinical hypertension. J Clin Hypertens (Greenwich). 2015;17:582–91.
187. Pettersen KH, Bugenhagen SM, Nauman J, Beard DA, Omholt SW. Arterial stiffening provides sufficient explanation for primary hypertension. PLoS Comput Biol. 2014;10, e1003634.
188. Bellien J, Favre J, Iacob M, et al. Arterial stiffness is regulated by nitric oxide and endothelium-derived hyperpolarizing factor during changes in blood flow in humans. Hypertension. 2010;55:674–80.
189. Schmitt M, Avolio A, Qasem A, et al. Basal NO locally modulates human iliac artery function in vivo. Hypertension. 2005;46:227–31.
190. Wilkinson IB, Qasem A, McEniery CM, Webb DJ, Avolio AP, Cockcroft JR. Nitric oxide regulates local arterial distensibility in vivo. Circulation. 2002;105:213–7.
191. Fleenor BS, Sindler AL, Eng JS, Nair DP, Dodson RB, Seals DR. Sodium nitrite de-stiffening of large elastic arteries with aging: role of normalization of advanced glycation end-products. Exp Geront. 2012;47:588–94.
192. Sindler AL, Fleenor BS, Calvert JW, et al. Nitrite supplementation reverses vascular endothelial dysfunction and large elastic artery stiffness with aging. Aging Cell. 2011;10:429–37.
193. Lauer T, Preik M, Rassaf T, et al. Plasma nitrite rather than nitrate reflects regional endothelial nitric oxide synthase activity but lacks intrinsic vasodilator action. Proc Natl Acad Sci U S A. 2001;98:12814–9.
194. Kleinbongard P, Dejam A, Lauer T, et al. Plasma nitrite reflects constitutive nitric oxide synthase activity in mammals. Free Radic Biol Med. 2003;35:790–6.
195. Kleinbongard P, Dejam A, Lauer T, et al. Plasma nitrite concentrations reflect the degree of endothelial dysfunction in humans. Free Radic Biol Med. 2006;40:295–302.
196. Wagner DA, Schultz DS, Deen WM, Young VR, Tannenbaum SR. Metabolic fate of an oral dose of ^{15}N-labeled nitrate in humans: effect of diet supplementation with ascorbic acid. Cancer Res. 1983;43:1921–5.
197. Ishiwata H, Tanimura A, Ishidate M. Studies on in vivo formation of nitroso compounds (V). Food Hygiene and Safety Science (Shokuhin Eiseigaku Zasshi). 1975;16:234–9.

198. Gladwin MT, Shelhamer JH, Schechter AN, et al. Role of circulating nitrite and S-nitrosohemoglobin in the regulation of regional blood flow in humans. Proc Natl Acad Sci U S A. 2000;97:11482–7.
199. Kapil V, Haydar SMA, Pearl V, Lundberg JO, Weitzberg E, Ahluwalia A. Physiological role for nitrate-reducing oral bacteria in blood pressure control. Free Radic Biol Med. 2013;55:93–100.
200. Bondonno CP, Liu AH, Croft KD, et al. Antibacterial mouthwash blunts oral nitrate reduction and increases blood pressure in treated hypertensive men and women. Am J Hypertens. 2015;28:572–5.
201. Woessner M, Smoliga JM, Tarzia B, Stabler T, Van Bruggen M, Allen JD. A stepwise reduction in plasma and salivary nitrite with increasing strengths of mouthwash following a dietary nitrate load. Nitric Oxide. 2016;54:1–7.
202. McDonagh ST, Wylie LJ, Winyard PG, Vanhatalo A, Jones AM. The effects of chronic nitrate supplementation and the use of strong and weak antibacterial agents on plasma nitrite concentration and exercise blood pressure. Int J Sports Med. 2015;36:1177–85.
203. Harada M, Ishiwata H, Nakamura Y, Tanimura A, Ishidate M. Studies on in vivo formation of nitroso compounds. Food Hygiene and Safety Science (Shokuhin Eiseigaku Zasshi). 1974;15:206–7.
204. Ishiwata H. Studies on in vivo formation of nitroso compounds (VII). Food Hygiene and Safety Science (Shokuhin Eiseigaku Zasshi). 1976;17:369–73.
205. Ishiwata H, Boriboon P, Nakamura Y, Harada M, Tanimura A, Ishidate M. Studies on in vivo formation of nitroso compounds (II). Food Hygiene and Safety Science (Shokuhin Eiseigaku Zasshi). 1975;16:19–24.
206. Ishiwata H. Studies on in vivo formation of nitroso compounds (VIII). Food Hygiene and Safety Science (Shokuhin Eiseigaku Zasshi). 1976;17:423–7.
207. Ishiwata H, Tanimura A, Ishidate M. Studies on in vivo formation of nitroso compounds (III). Food Hygiene and Safety Science (Shokuhin Eiseigaku Zasshi). 1975;16:89–92.
208. Ishiwata H, Boriboon P, Harada M, Tanimura A, Ishidate M. Studies on in vivo formation of nitroso compounds (IV). Food Hygiene and Safety Science (Shokuhin Eiseigaku Zasshi). 1975;16:93–8.
209. Tannenbaum SR, Correa P. Nitrate and gastric cancer risks. Nature. 1985;317:675–6.
210. Eisenbrand G, Spiegelhalder B, Preussmann R. Nitrate and nitrite in saliva. Oncology. 1980;37:227–31.
211. Tannenbaum SR, Sinskey AJ, Weisman M, Bishop W. Nitrite in human saliva. Its possible relationship to nitrosamine formation. J Natl Cancer Inst. 1974;53:79–84.
212. Spiegelhalder B, Eisenbrand G, Preussmann R. Influence of dietary nitrate on nitrite content of human saliva: possible relevance to in vivo formation of N-nitroso compounds. Food Cosmet Toxicol. 1976;14:545–8.
213. Desvarieux M, Demmer RT, Rundek T, et al. Relationship between periodontal disease, tooth loss, and carotid artery plaque: the Oral Infections and Vascular Disease Epidemiology Study (INVEST). Stroke. 2003;34:2120–5.
214. Desvarieux M, Demmer RT, Rundek T, et al. Periodontal microbiota and carotid intima-media thickness: the Oral Infections and Vascular Disease Epidemiology Study (INVEST). Circulation. 2005;111:576–82.
215. Desvarieux M, Demmer RT, Jacobs DR, et al. Periodontal bacteria and hypertension: the oral infections and vascular disease epidemiology study (INVEST). J Hypertens. 2010;28:1413–21.
216. Elmore JG, Horwitz RI. Oral cancer and mouthwash use: evaluation of the epidemiologic evidence. Otolaryngol Head Neck Surg. 1995;113:253–61.
217. Fedorowicz Z, Aljufairi H, Nasser M, Outhouse TL, Pedrazzi V. Mouthrinses for the treatment of halitosis. Cochrane Database Syst Rev. 2008;(4):CD006701.
218. Chadwick B, White D, Lader D, Pitts N. Preventative behaviour and risks to oral health. In: O'Sullivan I, editor. Adult dental health survey 2009. Leeds: The Health and Social Care Information Centre; 2011. p. 1–44.
219. van Velzen AG, Sips AJ, Schothorst RC, Lambers AC, Meulenbelt J. The oral bioavailability of nitrate from nitrate-rich vegetables in humans. Toxicol Lett. 2008;181:177–81.
220. Hunault CC, van Velzen AG, Sips AJ, Schothorst RC, Meulenbelt J. Bioavailability of sodium nitrite from an aqueous solution in healthy adults. Toxicol Lett. 2009;190:48–53.
221. Iskedjian M, Einarson TR, MacKeigan LD, et al. Relationship between daily dose frequency and adherence to antihypertensive pharmacotherapy: evidence from a meta-analysis. Clin Ther. 2002;24:302–16.
222. Schroeder K, Fahey T, Ebrahim S. How can we improve adherence to blood pressure-lowering medication in ambulatory care? Systematic review of randomized controlled trials. Arch Intern Med. 2004;164:722–32.
223. Claxton AJ, Cramer J, Pierce C. A systematic review of the associations between dose regimens and medication compliance. Clin Ther. 2001;23:1296–310.
224. Martinoia E, Maeshima M, Neuhaus HE. Vacuolar transporters and their essential role in plant metabolism. J Exp Bot. 2007;58:83–102.
225. Nunez de Gonzalez MT, Osburn WN, Hardin MD, et al. A survey of nitrate and nitrite concentrations in conventional and organic-labeled raw vegetables at retail. J Food Sci. 2015;80:C942–9.
226. Food Standards Agency. 2004 UK monitoring programme for nitrate in lettuce and spinach [online]. 2004.
227. Crandall Jr LA, Leake CD, Loevenhart AS. Acquired tolerance to and cross tolerance between the nitrous and nitric acid esters and sodium nitrite in man. J Pharmacol Exp Ther. 1931;41:103–19.
228. Hord NG, Tang Y, Bryan NS. Food sources of nitrates and nitrites: the physiologic context for potential health benefits. Am J Clin Nutr. 2009;90:1–10.

229. Food and Agriculture Organization of the United Nations/World Health Organization. Nitrate (and potential endogenous formation of N-nitroso compounds). Geneva: World Health Organization; 2003.
230. Beck EG. Toxic effects from bismuth subnitrate with reports of cases to date. JAMA. 1909;LII:14–8.
231. U.S. Public Health Service. Public health service drinking water standards. Washington: United States Government Printing Office; 1962.
232. Ash-Bernal R, Wise R, Wright SM. Acquired methemoglobinemia: a retrospective series of 138 cases at 2 teaching hospitals. Medicine (Baltimore). 2004;83:265–73.
233. Pluta RM, Oldfield EH, Bakhtian KD, et al. Safety and feasibility of long-term intravenous sodium nitrite infusion in healthy volunteers. PLoS One. 2011;6, e14504.
234. Magee PN, Barnes JM. The production of malignant primary hepatic tumours in the rat by feeding dimethylnitrosamine. Br J Cancer. 1956;10:114–22.
235. Magee PN, Barnes JM. Carcinogenic nitroso compounds. Adv Cancer Res. 1967;10:163–246.
236. Fine DH, Ross R, Rounbehler DP, Silvergleid A, Song L. Formation in vivo of volatile N-nitrosamines in man after ingestion of cooked bacon and spinach. Nature. 1977;265:753–5.
237. Sen NP, Smith DC, Schwinghamer L. Formation of N-nitrosamines from secondary amines and nitrite in human and animal gastric juice. Food Cosmet Toxicol. 1969;7:301–7.
238. Bogovski P, Bogovski S. Animal species in which N-nitroso compounds induce cancer. Int J Cancer. 1981;27:471–4.
239. National Toxicology Program. NTP technical report on toxicology and carcinogenesis studies of sodium nitrite in F344/N rats and B6C3F1 mice (drinking water studies). Durham: National Institutes of Health; 2001.
240. World Cancer Research Fund/American Institute for Cancer Research. Food, nutrition, physical activity, and the prevention of cancer: a global perspective. Washington: AICR; 2007.
241. Boffetta P, Couto E, Wichmann J, et al. Fruit and vegetable intake and overall cancer risk in the European Prospective Investigation into Cancer and Nutrition (EPIC). J Natl Cancer Inst. 2010;102:529–37.
242. Wilson AM, Harada R, Nair N, Balasubramanian N, Cooke JP. L-arginine supplementation in peripheral arterial disease: no benefit and possible harm. Circulation. 2007;116:188–95.
243. Schulman SP, Becker LC, Kass DA, et al. L-arginine therapy in acute myocardial infarction: the Vascular Interaction with Age in Myocardial Infarction (VINTAGE MI) randomized clinical trial. JAMA. 2006;295:58–64.

Chapter 18
Nitrate and Nitrite in Aging and Age-Related Disease

Lawrence C. Johnson, Allison E. DeVan, Jamie N. Justice, and Douglas R. Seals

Key Points

- Advancing age is associated with declines in physiological function and is the leading risk factor for the majority of chronic degenerative diseases in modern societies.
- Central to age-associated declines in physiological function is a reduction in the endogenous production and bioavailability of the ubiquitous signaling molecule nitric oxide.
- Supplementation with nitrate and/or nitrite boosts nitric oxide bioavailability and improves physiological function in the presence of aging.
- Evidence suggests that nitrate and/or nitrite supplementation may hold promise as therapies to preserve and/or improve physiological function and reduce the risk of chronic degenerative diseases in middle-aged and older populations.

Keywords Vascular function • Physical function • Cognitive function • Oxidative stress • Inflammation

Introduction

Advancing age is associated with declines in physiological function and is the leading risk factor for the majority of chronic degenerative diseases in modern societies [1]. This fact, combined with the rapidly changing demographics of aging and record number of older adults, projects unprecedented levels of clinical disease, disability, and health care burden in the near future [2–4]. As such, it is imperative to identify the mechanisms underlying age-related physiological dysfunction and to

L.C. Johnson, M.S.
Department of Integrative Physiology, University of Colorado Boulder, 354 UCB, Boulder, CO, 80309, USA

A.E. DeVan, Ph.D.
Medical College of Wisconsin, Cardiovascular Center, 8701 Watertown Plank Road, Milwaukee, WI, 53226, USA

J.N. Justice, Ph.D.
Department of Internal Medicine—Geriatrics, Wake Forest School of Medicine,
1 Medical Center Blvd., Winston-Salem, NC, 27012, USA

D.R. Seals, Ph.D. (✉)
Department of Integrative Physiology, University of Colorado Boulder, 354 UCB, Boulder, CO, 80309, USA
e-mail: seals@colorado.edu

N.S. Bryan, J. Loscalzo (eds.), *Nitrite and Nitrate in Human Health and Disease*, Nutrition and Health,
DOI 10.1007/978-3-319-46189-2_18, © Springer International Publishing AG 2017

develop therapeutic treatments that can prevent or slow these processes; delay the onset of functional limitations, disability, and chronic disease; and extend quality of life to later ages. Indeed, increasing health span, or the number of years free of major functional impairment and clinical disease, is now recognized as a top priority in biomedical research [4–6].

Age-Associated Declines in Nitric Oxide

Nitric oxide (NO) is a gaseous signaling molecule critical to the regulation of numerous physiological processes, and its presence is essential to the preservation of physiological function and health with advancing age. NO is produced via two main biological pathways in vivo. First discovered was the L-arginine–NO-synthase pathway in which a stimulus causes nitric oxide synthase (NOS) enzymes to catalyze the reaction of L-arginine with oxygen to produce NO [7]. More recently, a second pathway has been elucidated by which NO can be produced through the reduction of nitrate and nitrite by various reducing mechanisms, including nitrate and nitrite reductases [8, 9]. NO has a very short half-life, as the molecule is quickly oxidized to nitrite and, subsequently, nitrate. Both nitrite and nitrate are unique in that they can act as stable circulating storage forms of NO that can be rapidly reduced to restore NO levels in vivo.

NO bioavailability decreases with age, incurring deleterious effects on the systems that require NO to maintain proper signaling and function. The reasons for the age-associated decrease in NO bioavailability are multifactorial, but evidence suggests that an age-related increase in oxidative stress is the primary cause [8, 10]. Oxidative stress is defined as an imbalance in reactive oxygen species (ROS) relative to antioxidant defenses (Fig. 18.1) [11]. Oxidative stress reduces NO bioavailability with aging via several mechanisms. Excessive age-related production of superoxide, a prominent ROS, reacts readily with NO to form peroxynitrite and, in so doing, directly reduces the abundance of

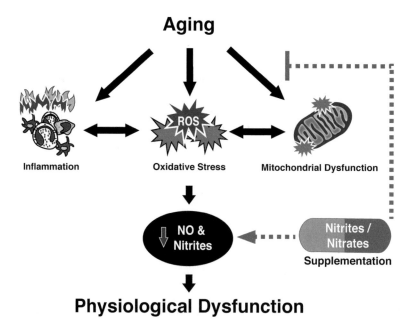

Fig. 18.1 Potential mechanisms by which advancing age leads to physiological dysfunction and the beneficial effects of nitrate and nitrite supplementation on these processes. *ROS* reactive oxygen species, *NO* nitric oxide

NO [12]. Excessive superoxide also dysregulates ("uncouples") NOS enzymes by oxidizing the essential cofactor for NO production, tetrahydrobiopterin (BH_4), resulting in further superoxide production and reduced NO synthesis [13].

In addition, age-associated oxidative stress is created and maintained by increased expression and activity of pro-oxidant enzymes (e.g., NADPH oxidase) in the absence of compensatory up-regulation of endogenous antioxidant enzymes [14–16]. More recently, dysregulated mitochondria have been implicated in excess production of ROS with aging [17, 18]. Chronic, low-grade inflammation also develops with advancing age, reinforces oxidative stress, and contributes to systemic physiological dysfunction [19–22] even in the absence of clinical disease. Given the earlier, lifestyle and pharmacological strategies that restore NO bioavailability with aging, perhaps in part by inhibiting these pro-oxidant processes and inflammation, hold the promise of enhancing and maintaining physiological function with aging..

Nitrate and Nitrite Supplementation Improves Physiological Function

Nitrate and nitrite supplementation-based therapies are effective in increasing the concentrations of these NO precursors in vivo [23–25]. Such treatments can be successfully administered via several different methods, including inhalation, intravenous or intra-arterial infusion, topical application, or most commonly, oral ingestion. Circulating and tissue concentrations of nitrate and nitrite have been boosted by oral consumption of salts (e.g., $NaNO_3-$, KNO_3, $NaNO_2^-$) or green leafy or root vegetables (e.g., spinach, beetroot juice). These modes of delivery provide a pool of systemically available nitrate and nitrite, substrates with which to produce NO independent of the endogenous L-arginine–NO-synthase pathway in both healthy individuals and those suffering from age-related disease states associated with decreased NO bioavailability. The following sections review changes in selective physiological functions with aging, the potential mechanisms by which age-related changes in NO bioavailability or signaling may affect these functions, and the current evidence from preclinical and clinical trials that suggest efficacy of nitrate or nitrite supplementation on age-related physiological dysfunction (Fig. 18.2).

Cardiovascular Diseases

Age is the major risk factor for cardiovascular diseases (CVD), as >90 % of all deaths from CVD occur in older adults above 55 years of age [26]. The increased risk of CVD with aging is due in large part to adverse changes occurring in arteries associated with vascular dysfunction [10, 27]. Among these changes, a decline in NO bioavailability is a critical event [10, 28]. NO is produced within the vascular endothelium and diffuses to the vascular smooth muscle, providing a powerful vasodilatory signal that adjusts the diameter of the vessel to accommodate changes in blood flow. In aging as well as pathological states in which NO bioavailability is low, this signaling pattern is disrupted, creating an environment conducive to endothelial dysfunction and stiffening of the large elastic arteries, two primary contributors to the increased risk of CVD in middle-aged and older adults [29, 30]. Resulting from these changes to the large elastic arteries, sensitive organ systems such as the kidney, brain, and liver can be damaged owing to alterations in blood flow or the elevated pulsatile hemodynamics created and directed to delicate microvascular systems in these organs [31, 32]. Therefore, restoring the age-related decline in NO bioavailability and thereby decreasing large elastic artery stiffness and improving endothelial function may serve to reduce the prevalence of CVD and preserve function in multiple organ systems.

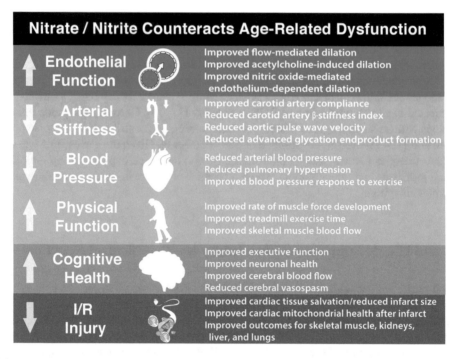

Fig. 18.2 Nitrate and nitrite supplementation counteract age-associated functional declines and improve health outcomes. *I/R* ischemia/reperfusion

Endothelial Dysfunction

The endothelium is a single cell layer that lines the vasculature at the interface of the arterial wall and the lumen of the vessel, in direct contact with the flow of blood. Via release of NO, the endothelium exerts a major influence on vasodilation, inhibition or activation of smooth muscle proliferation, and inflammatory processes [33]. As such, the endothelium plays an important role in maintaining vascular health [34].

Advancing age is associated with endothelial dysfunction, a pathological state of the endothelium in which there is an imbalance between vasodilating and vasoconstricting factors [33]. Additionally, a dysfunctional endothelium induces smooth muscle cell proliferation, platelet activation and adhesion, and inflammation, resulting in a phenotype that further drives vascular pathology and declines in function [35]. A primary hallmark of age-related endothelial dysfunction is the decreased synthesis and/or bioavailability of NO [28].

The ability of nitrate and nitrite therapies to produce NO through a pathway independent of eNOS makes them attractive therapies for restoring vascular function [36, 37]. Indeed, the vasodilating abilities of nitrate and nitrite have been known for decades [38–41], but due to safety concerns and misconceptions regarding their use, these compounds have not been tested for efficacy in treating vascular aging until recently. Early studies in humans focused on the acute effects of infused nitrate or nitrite on vascular function as measured by changes in forearm blood flow. Intravascular infusions of sodium nitrite acutely increased forearm blood flow in both young and middle-aged healthy adults [42–44], older adults [45], and some clinical populations [46, 47]. Other studies directly assessed endothelial-dependent dilation (EDD), a well-established measure of endothelial function, and demonstrated the acute beneficial effects of nitrate or nitrite administration [48–51].

To address the chronic effects of sodium nitrite, vascular endothelial function was assessed via EDD in response to increasing doses of acetylcholine, a compound used for pharmacological activation of eNOS and production of NO in mouse models of primary aging. Compared with young animals, arteries from old mice displayed impaired vasodilation in response to acetylcholine. However, 3 weeks of sodium nitrite [52] or 8 weeks of sodium nitrate [53] administered in the drinking water restored EDD in old mice to levels seen in young animals. Pharmacological investigation of the mechanisms by which sodium nitrite improves EDD revealed that the effects on vascular function were mediated by increased NO bioavailability [52]. The effects of sodium nitrate were confirmed in a human trial in which 4 weeks of sodium nitrate supplementation improved EDD in healthy older adults with elevated risk factors for CVD [54]. Results of a recent study in which sodium nitrite was orally administered for 10 weeks also showed improvements in EDD in healthy middle-aged and older adults [55]. These findings indicate that acute and chronic nitrate and nitrite supplementation is well tolerated, and that the effects of these therapies are sufficient to improve age-associated declines in NO bioavailability and resulting impairments in endothelial function.

Large Elastic Artery Stiffness

The large elastic arteries play an important role in regulating blood flow and pressure throughout the systemic circulation. The elastic properties of the large arteries work to dampen the pressure waves associated with the ejection of blood from the heart and protect sensitive organs from high pulsatile flows. Permitting this phenomenon in healthy arteries are elastin, the protein component of arterial walls that allows for increased elasticity and compliance; and collagen, a nonelastic protein component utilized for structural integrity under high-pressure loads. A proper ratio of structural components contributes to a healthy, compliant vessel.

With primary aging, as well as in the presence of certain disease states, there is a gradual increase in large artery stiffness. The stiffening of the large elastic arteries is facilitated through the structural remodeling of the wall of the large arteries [56]. Fragmentation of elastin and an increase in collagen deposition contribute to reduced arterial compliance and stiffen the large elastic arteries [57]. Lastly, the formation of advanced glycation end-products (AGEs) also contributes to the stiffening process of large elastic arteries by cross-linking structural proteins in the extracellular matrix [58]. These changes occur in settings of both healthy aging and disease and have been associated with increased risk of cardiovascular events in humans [30, 56, 59–61].

Trials to test the efficacy of using nitrate and nitrite therapies as interventions to improve large elastic artery stiffness are relatively recent. In a preclinical study, increased arterial stiffness was observed in healthy old mice compared to young controls. Although there were no effects on elastin or collagen content, treatment with sodium nitrite reversed the age-associated increase in stiffness through reduced abundance of AGEs in the arterial wall [52, 62]. Supporting the effects in mice, acute and chronic supplementation with nitrates reduces arterial stiffness in healthy and diseased adults. Specifically, central arterial stiffness was reduced in response to an acute dose of potassium nitrate [63], intra-arterial infusion of sodium nitrite [64], or dietary nitrate from beetroot juice [65] in young healthy adults, and/or healthy controls and patients with hypertension. In addition, within 30 min, arterial compliance was improved in hypertensive patients after a single acute dose of a nitrite containing lozenge [66]. Consistent with these observations, chronic supplementation of nitrite and nitrate also improved measures of arterial stiffness. Healthy middle-aged and older adults given sodium nitrite supplementation for 10 weeks demonstrated increased carotid artery compliance and decreased β-stiffness index without alterations in central or peripheral blood pressure [55]. Additionally, 4 weeks of beetroot juice or sodium nitrate dissolved in water decreased aortic pulse wave velocity, the gold-standard measure of aortic stiffness, in patients with hypertension [67] and older adults at risk for developing CVD [54]. Overall, there is some evidence that nitrate and nitrite may have beneficial effects in populations with increased baseline large elastic artery stiffness, but much more work is needed to demonstrate this benefit definitively.

Blood Pressure

Older adults are at high risk for the development of hypertension [68], and the literature currently suggests that nitrate and nitrite treatment may be effective at reducing blood pressure. Several studies have reported that acute nitrite administration reduces blood pressure in animal models of hypertension [69], in clinical studies of healthy adults [70], in addition to patients with diabetes [71], pre-hypertension [72], untreated hypertension [66], and urea cycle disorders [73]. Studies involving nitrate administration have yielded more varied results. On the one hand, a meta-analysis found that nitrate supplementation for 7 to 21 days did not have a pooled systemic effect on ambulatory blood pressures [74]. Additionally, patients with diabetes and those previously treated for hypertension displayed no further reduction in blood pressures after at least 1 week of nitrate supplementation, despite experiencing increases in circulating nitrate and nitrite [75, 76].On the other hand, both acute and chronic nitrate (up to 15 days) supplementation has been reported to reduce blood pressures in healthy adults and those with moderate risk factors for, but no history of, CVD [50, 77–79]. Moreover, in a recent double-blinded, placebo-controlled randomized trial, dietary nitrate was shown to provide sustained blood pressure lowering in hypertensive patients [67]. Daily dietary nitrate supplementation was associated with both systolic and diastolic blood pressure reductions as measured by three different methods. Blood pressure measured at home was significantly reduced by 8.1/3.8 mmHg (P<0.001 and P<0.01), while blood pressure measured in the clinic experienced a mean reduction of 7.7/2.4 mmHg (P<0.001 and P=0.050). Importantly, 24-h ambulatory blood pressure declined by 7.7/5.2 mmHg (P<0.001 for both) during the 4 week intervention period in the absence of tachyphylaxis. Vascular function also improved after dietary nitrate consumption, as arterial stiffness was reduced by 0.59 m/s (P < 0.01) while endothelial function was significantly improved by ≈ 20 % (P < 0.001) with no changes in the placebo group. Supporting these findings is evidence that 4 weeks of daily dietary inorganic nitrate supplementation lowers systolic blood pressure, reduces vascular stiffness, and improves endothelial function in an elderly population with moderately increased cardiovascular disease risk [57]. More studies will be needed to determine how effective inorganic nitrite and nitrate truly are in the treatment of hypertension. Thus, presently there is no scientific consensus as to the effects of nitrate and nitrite on age-associated hypertension, but this remains a promising avenue of investigation.

Myocardium

With advancing age, key structural changes, such as myocyte hypertrophy, diminished myocyte number, and increased connective tissue, contribute to higher levels of myocardial stiffness, impaired contractile response, and other alterations in excitation–contraction coupling that diminish myocardial function [27]. The effects of nitrate or nitrite therapy on primary cardiac aging are currently unknown, but results from numerous studies have demonstrated the beneficial effects of nitrite or nitrate administration for reducing myocardial ischemia/reperfusion injury in preclinical models [23, 80–83]. The protection afforded by nitrate and nitrite is attributed, at least in part, to their roles as physiological stores of NO and modulators of mitochondrial function. Confirmation of these findings in humans is limited at this time. Currently, only two clinical trials have addressed this issue in patients with ST-segment elevation myocardial infarction (STEMI) undergoing percutaneous coronary intervention, with conflicting results that could be due, in part, to differing doses and routes of administration. Intravenous infusion of sodium nitrite did not reduce myocardial infarct size or alter secondary end-points in one study [84], whereas infusion of a higher dose of nitrite directly into the coronary arteries was associated with a lower number of major adverse cardiac events and a higher myocardial salvage index at 1-year follow-up [85]. In a subgroup analysis of patients treated with thrombolysis, nitrite also reduced infarct size [85], and a phase III trial is presently underway.

Another target for therapy with nitrate or nitrite administration is heart failure. Patients with heart failure have low cyclic guanosine monophosphate (cGMP) content and low cGMP-dependent protein kinase (PKG) activity in their myocardium, which contribute to the development of myocardial hypertrophy, increased passive stiffness, and delayed myocardial relaxation [86]. Enhanced NO bioavailability improves cGMP/PKG signaling as well as vasodilation in the coronary and peripheral circulations, augmenting perfusion of the heart and sub-endocardium [47, 87]. Consistent with such benefits, a recent pilot study found that a single dose of nitrate-rich beetroot juice improved exercise capacity in patients with heart failure [88]. Similarly, infusion of sodium nitrite markedly improved forearm blood flow in patients with congestive heart failure [47] and improved functional cardiac responses during a dobutamine stress echocardiogram in patients with inducible myocardial ischemia [49]. Taken together, the results from these initial studies support the continued investigation of nitrate and nitrite for the management of heart failure, and perhaps myocardial infarction.

Dysfunction of Other Physiological Systems

In addition to their promise as therapies for CVD, nitrate and nitrite may also have beneficial effects on other domains of physiological function and health with aging. Because NO is a universal signaling molecule, it affects multiple organ systems and integrative functions. By increasing NO bioavailability and its other NO-independent actions, nitrate and nitrite therapies may attenuate age-related declines in cognitive and physical functions.

Cognitive Function

Cognitive decline and dementias directly related to age have been well documented. Advancing age preferentially impairs select domains of cognitive function, most notably memory and executive functioning, i.e., the processes that support strategic organization required for complex, goal-oriented tasks [89]. A number of pathophysiological changes occur in aging that are candidates for causing age-associated executive and memory difficulties, namely, neuronal atrophy, white matter abnormalities, and neurochemical changes within the brain [90, 91]. Indeed, 65 % of nondemented older adults (>75 years) show white matter abnormalities [92], which are consistent with general atrophy and loss in brain volume [93]. Small infarcts are also prevalent, as are white matter lesions thought to result from vascular disorders, such as small vessel disease [94, 95]. Importantly, these white matter lesions are strongly associated with, and predictive of, declines in both executive function and memory [90], even in nondemented older adults [96, 97]. Several lines of evidence suggest that age-associated vascular dysfunction may be an important mechanism underlying executive and memory impairment [98], and that oxidative stress and NO deficiency are key underlying factors [98, 99].

Accumulating findings indicate that NO plays an important role in the preservation of cognitive health with aging. As a multifunctional messenger molecule, NO has a prominent role in both regulation of cerebral blood flow and cell-to-cell communication in the brain. Through its vasodilatory effects, NO contributes to the regulation of cerebral perfusion [100]. Numerous theories now posit that a reduction in NO bioavailability, whether from advanced age or a perfusion-lowering disease condition, results in hemodynamic microcirculatory insufficiency [98]. If impaired perfusion persists below a key threshold, referred to as the "critically attained threshold of cerebral hypoperfusion" (CATCH), it can lead to a restricted energy state that may destabilize neurons, synapses, and neurotransmission, and ultimately affect cognitive function [101, 102]. An additional mechanism by which NO may link vascular function to cognition is through regulation of neurovascular coupling, which is the time- and regional-dependent connectivity between local neural activity and subsequent changes in cerebral blood flow. For example, when neurons and glia generate signals, this initiates a coordinated cascade of vascular events, ultimately dependent on NO to produce vasodilation to the specific area of activation in a timely manner. Reduced NO bioavailability and signaling has been implicated in diffuse and disrupted coupling in which the cerebral blood flow is no longer matched to the metabolic requirements of the tissue [103].

Evidence supports the potential for administration of nitrates, nitrites, and NO donors to improve NO bioavailability and improve neuronal/cognitive function with age, although this has not been thoroughly vetted in clinical trials of older adults. A preclinical trial established that old rats display impaired retention, object recognition, and discrimination capabilities compared to young animals; yet, old animals administered molsidomine, a direct NO donor, showed complete restoration of function to that of young rats in retention and discrimination abilities [104]. The effects of boosting NO on learning and memory in rodents have been reviewed extensively elsewhere [105], and the general consensus from these investigations is that proper NO signaling improves behaviors reliant on cognition, while the inhibition of NO synthesis induces cognitive impairment. In contrast to this preclinical evidence, little work has been performed examining the effects of NO boosting agents in older men and women. In one of only few available trials performed in healthy middle-aged and older adults, short-term (3-day) supplementation with dietary nitrate in the form of beetroot juice failed to induce improvements in cognition as determined by a computerized battery of tests, although coadministration of sodium nitrate with the carbonic anhydrase inhibitor acetazolamide did improve cerebral blood flow to visual stimuli in healthy males [106]. However, nonhuman primate models of stroke have shown nitrite can cross the blood–brain barrier and inhibit cerebral vasospasm, supporting its possible efficacy for improving age-associated brain health [107]. Moreover, 10 weeks of sodium nitrite supplementation improved performance on Trail Making Tests A and B, measures of executive function, in healthy middle-aged and older adults [108]. Overall, these results suggest that NO plays a significant role in learning and memory mechanisms affected by increasing age, and demonstrate the necessity of NO as a signaling molecule and vasoactive regulator in the domain of cognitive function, although the length and type of supplementation may be key to inducing clinically important improvements.

Physical Function

The ability to perform physical tasks is critically important to maintaining overall functional capacity [109–111], and physical function has emerged as a predictor of morbidity and mortality in older adults [111–113]. Although no single cause has emerged as being responsible for the onset of deficits in physical function with advanced age, many interconnected factors contribute to this inevitable decline [114]. The age-related physical disablement process begins with physiological impairments such as motor neuron loss and subsequent remodeling, impaired transmission at the neuromuscular junction, increased skeletal muscle excitation–contraction uncoupling, loss of mitochondrial efficiency, impaired vascular coupling, and eventual skeletal muscle atrophy [109, 114–117]. These physiological impairments lead first to observable deficits in muscle power and strength, then functional limitations such as reduced walking speed or ability to rise from a chair, and eventually culminate in disability and loss of independence [109, 118].

Many points in this sequence of physiological events contributing to functional limitation and age-related physical disability may be mediated by NO bioavailability and signaling, although little work has been performed linking NO to physical function in primary aging. First, as described previously, NO has beneficial effects contributing to neuroprotection which could theoretically inhibit the morphological loss of motor neurons and loss of axonal transmission of neural signals that are a hallmark of primary aging and a leading contributor to subsequent impairment, though this has not been tested empirically. Second, NO has known antioxidant properties that could limit excitation–contraction uncoupling, which is largely mediated by the oxidative modification of dihydropyridine receptor (DHPR) in the electromechanical transduction step linking neural input to the release of Ca_2^+ intracellularly to cause cross-bridge binding and force production [116]. Previous work in young animals supports this mechanistic action as boosting NO through dietary nitrate increases DHPR expression, intracellular Ca_2^+ release, and force production in skeletal muscle [119]. Third, as reviewed elsewhere in this book, NO is critically important in mitochondrial regulation and improves mitochondrial efficiency in young adults, which is accompanied by increased work rate and exercise tolerance [120]. Finally, vascular function is significantly related to physical function in older adults, including muscle fiber type and morphology, muscle power and performance in activities of daily living that require balance, upper and lower body strength, flexibility, balance, and coordination [121, 122].

The ability of nitrate and nitrite to modulate vascular function beneficially, along with its positive antioxidant and mitochondrial effects, provides optimism that nitrate and nitrite may restore physical function in middle-aged and older adults; yet, few studies have tested the hypothesis that nitrate and nitrite can improve measures of physical function with advanced age. In one such study, young and old mice were assessed for grip strength, open-field distance, and rota-rod endurance [123]. Results showed that old mice had deficits in these functional measures compared to young, and that 8 weeks of sodium nitrite supplementation improved grip strength and open field distance while completely restoring rota-rod endurance to that of young animals. Preclinical and clinical studies have also confirmed nitrate as an effective means to improve measures of physical performance, including measures of strength, exercise capacity, and endurance in young individuals as a result of increased NO bioavailability [124–126]. Trials in older adults are few, but an early study investigating the role of NO in physical function found that administration of L-arginine, a precursor of NO, improved measures of force production in postmenopausal women [127]. Although increasing NO bioavailability with 3 days of oral nitrate supplementation through beetroot juice consumption was insufficient to improve performance in a 6-min walk test in older adults [72], utilizing acute doses of nitrate has been found to be beneficial for certain domains of physical function. Acute administration of beetroot juice was shown to increase circulating nitrite levels in healthy older adults and significantly improve contracting skeletal muscle blood flow during handgrip exercise under hypoxic conditions [128]. Furthermore, older patients with peripheral artery disease experience a decrease in blood pressure and an improvement in exercise time prior to claudication pain after an acute dose of nitrate [129]. These results support the potentially beneficial role of nitrate on exercise capacity and motor performance in older populations, and particularly in groups with impaired skeletal muscle blood flow. In agreement with these findings administering nitrate, a recent study demonstrated that 10 weeks of sodium nitrite supplementation improves indices of balance, endurance, and muscle power in healthy middle-aged and older adults free of disease and disability [108]. Further clinical trials with nitrate or nitrite are needed to assess efficacy in a more comprehensive battery of physical function assessments in both healthy older adults and patients with clinical disease.

Injury and Disease

Research into the possible health benefits of nitrate and nitrite includes numerous other systems, tissues, and conditions related to aging, including pulmonary and renal function and ischemia–reperfusion injury. Preclinical models of pulmonary function show beneficial effects of nitrate and nitrite therapies. Specifically, pulmonary hypertension is improved with nitrate and nitrite supplementation in preclinical models [130], and clinical investigations confirm these effects in humans, with an acute nitrite infusion reducing pulmonary pressures in states of hypoxia [44], and a single dose of dietary nitrate improving exercise performance and blood pressure responses in patients with chronic obtrusive pulmonary disease [131].

Numerous studies have established nitrite as a protective treatment in the setting of ischemia–reperfusion injury in multiple tissues and organs [23, 80, 132, 133]. Nitrite is effective at improving outcomes after ischemia in skeletal muscle [134], kidneys [135], liver [136], and lungs [137]. Furthermore, renal function is favorably affected by nitrite in other compromised states, such as renal injury under conditions of eNOS inhibition [138] and brain death-mediated renal injury [139]. Lastly, nitrite is effective at improving function and reducing adverse outcomes in models of heart transplantation [140]. Taken together, these investigations confirm that nitrate and nitrite therapies may have efficacy in numerous pathophysiological states that affect older adults.

Mechanisms by Which Nitrate and Nitrite Improve or Preserve Function with Advancing Age

Oxidative Stress

Aging is associated with increases in oxidative stress that can damage cellular components and induce dysfunction in organs and systems, promoting disease [141–143]. For example, increased oxidative stress within arteries has been shown to be a primary contributor to the development of arterial dysfunction with age [144–146]. Administration of sodium nitrite to old mice normalized nitrotyrosine levels (a cellular marker of oxidative damage to proteins) to that of young animals, indicating a reduction in age-associated oxidative stress [52, 62]. Age-associated oxidative stress is driven by excessively high levels of superoxide production, which was attenuated with sodium nitrite [52]. Nitrate and nitrite supplementation also reduce oxidative stress in preclinical models of injury and disease. Hypertension in both the presence and absence of compromised renal function, ischemia–reperfusion injury, and cardiomyopathy are all associated with increased oxidative stress that is ameliorated with nitrate or nitrite supplementation [65, 147–149].

Increasing age is associated with elevated concentrations and activity of the superoxide-generating enzyme NADPH oxidase in multiple organ systems, including the vasculature [150–152]. NADPH oxidase protein expression is higher in old compared to young aortas of mice, suggesting an age-associated increase in NADPH oxidase abundance. Three weeks of sodium nitrite supplementation reduced the abundance of NADPH oxidase in old mice to levels seen in the young, demonstrating the ability of sodium nitrite to down-regulate the expression of this pro-oxidant enzyme [52]. To determine if functional changes are associated with altered enzymatic activity of NADPH oxidase, isolated carotid arteries from aged animals were incubated with the NADPH oxidase inhibitor apocynin. Old animals treated with apocynin had improved endothelial function, while no effect was seen in young controls or old animals treated with sodium nitrite [52], indicating that age-associated increases of NADPH oxidase and its activity inhibit EDD, and that sodium nitrite is successful in reversing these effects. Similarly, pentaerythritol tetranitrate, an organic nitrate, reduces NADPH oxidase activity in the cardiac tissues of diabetic rats [153]. Collectively, these results suggest that nitrate and nitrite are effective in lowering oxidative stress by reducing the expression and activity of the pro-oxidant enzyme NADPH oxidase.

Antioxidant Defenses

Endogenous antioxidant enzymes are primarily responsible for combating the increase in oxidative stress. Antioxidant enzymes, such as superoxide dismutase (SOD), scavenge ROS in an attempt to maintain oxidative homeostasis. Declines in antioxidant enzyme expression and activity contribute to the development of oxidative stress with aging [154, 155]. SOD activity declines with aging in the aortas of mice, and sodium nitrite supplementation restored SOD activity to levels observed in the young animals [52]. Subsequent trials have confirmed the ability of nitrite to increase antioxidant defenses in preclinical models of vascular hypertension, ischemia–reperfusion injury, renal injury, and alcohol-induced liver injury [69, 139, 156–159]. A recent study in humans showed that sodium nitrite can improve oxidative states associated with peripheral artery disease and diabetes by improving the GSH:GSSG ratio [160]. These investigations suggest nitrite supplementation may be an effective means to restore antioxidant defenses in aging and disease.

eNOS Uncoupling

An increase in eNOS uncoupling via decreased BH_4 has been implicated in the decrease in NO bioavailability. To understand the contribution of eNOS uncoupling to reduced EDD, carotid arteries from young and old mice were incubated ex vivo with sepiapterin, a compound that restores BH_4 bioavailability and re-couples eNOS, dramatically increasing NO production. In old mice, sepiapterin improved EDD to levels similar to young animals while having no effect in old mice administered sodium nitrite, suggesting that eNOS uncoupling via BH_4 deficiency is at least partially responsible for impaired EDD in old mice [161]. Additionally, nitrite has been shown to improve EDD in hypercholesterolemic mice through maintenance of BH_4/BH_2 ratios [162]. Importantly, these results also support nitrite supplementation as a possible treatment to increase BH_4 bioavailability, recouple eNOS, and increase the bioavailability of NO in states characterized by eNOS uncoupling, including advanced age, and some diseases.

Mitochondrial Dysfunction

Recently, mitochondrial dysfunction has been identified as a potential mechanism underlying several chronic diseases associated with aging, including CVD and diabetes [25, 163–165]. Age-associated declines in mitochondrial function have been established in the vasculature of old mice and have been implicated in the development of vascular dysfunction [166]. Studies investigating the role of NO on mitochondrial function have found that low NO concentrations can cause impaired mitochondrial fitness, including unfavorable mitochondrial remodeling and a decline in ATP production. Conversely, evidence suggests that mitochondrial biogenesis and mitochondrial antioxidant enzyme expression can be up-regulated in conditions of sufficient NO bioavailability [167–171].

The effects of nitrate supplementation on mitochondrial function were evaluated in a placebo-controlled crossover study involving healthy humans. Independent of increases in mitochondrial content, nitrate supplementation increased the capacity to produce ATP, with enriched mitochondrial coupling efficiency identified as a key mechanism [120]. Similar to nitrate, nitrite treatment of rat aortic smooth muscle cells increased mitochondrial quantity via enhanced mitochondrial biogenesis by altering the expression and activity of AMP kinase and its downstream target, peroxisome proliferator-activated receptor-γ coactivator 1α (PGC1α) [170]. Nitrite also protects mitochondria in models of hypoxia [172] and after lethal doses of the inflammatory cytokine tumor necrosis factor-α (TNF-α) [173]. Collectively, these results offer compelling evidence that nitrate and nitrite supplementation may be effective for normalizing and/or enhancing mitochondrial function in aging and disease.

Inflammation

Chronic low-grade inflammation increases with advancing age and contributes to several expressions of physiological dysfunction [174, 175]. As an example, upregulation of pro-inflammatory cytokines, such as IL-1β, IL-6, TNF-α, and interferon-γ, has been observed in the aorta of old mice when compared to young controls. Sodium nitrite supplementation reversed levels of these inflammatory cytokines to those seen in young animals, indicating a potent anti-inflammatory action [52]. The anti-inflammatory effects of nitrite have also been shown in hypercholesterolemic mice, which experienced an improvement in vascular function associated with a decline in C-reactive protein and leukocyte markers of inflammation [162]. Additionally, nitrate and nitrite reduce inflammation in the microvasculature. Pretreatment with nitrate and nitrite inhibited the migration of leukocytes after myeloperoxidase-2 administration in the microvessels of mice. Furthermore, nitrate was shown to have robust effects in reducing systemic inflammation in the presence of intestinal damage [176]. The anti-inflammatory effects of nitrate have also been demonstrated. Four weeks of nitrate supplementation reduced age-associated increases in inflammatory macrophage migration inhibitory factor [79]. Moreover, nitrite decreases inflammation in models of endotoxemic shock [173, 177, 178], crush injury [179], and ischemia–reperfusion injury [137–140, 180–182]. Taken together, these findings support the possible use of nitrate and nitrite treatments for reducing inflammation with aging and disease without compromising important aspects of immune function.

Metabolic Signaling

Recent evidence demonstrates that supplementation with sodium nitrite induces systemic changes in multiple metabolic pathways in healthy older adults, as indicated by numerous alterations in the concentrations of small metabolites assessed via untargeted metabolomics analysis [55, 108]. Importantly, many of these changes to the plasma metabolome in response to oral sodium nitrite are associated with improvements of physiological function [55, 108]. Finally, baseline metabolic signatures can be used to predict responsiveness (changes in physiological function) to a sodium nitrite intervention [55, 108]. Taken together, these new observations support the idea that nitrite/nitrate supplementation may produce functional and health benefits through broad activation of metabolic signaling pathways.

Conclusions

Accumulating evidence suggests that nitrate and nitrite therapies hold promise for increasing NO bioavailability independent of the endogenous L-arginine–NO-synthase pathway. In the setting of aging, restoring NO bioavailability decreases oxidative stress via reduced free radical production and increased antioxidant defenses, decreases inflammation, and normalizes mitochondrial function with associated improvements in physiological outcomes. As the majority of work to date has been performed in preclinical models or with acute dosing in humans, longer term studies with nitrate and nitrite supplementation in humans are needed to establish efficacy of these strategies for preserving physiological function and optimal health with advancing age.

References

1. Niccoli T, Partridge L. Ageing as a risk factor for disease. Curr Biol. 2012;22(17):R741–52.
2. Heidenreich PA, Trogdon JG, Khavjou OA, Butler J, Dracup K, Ezekowitz MD, et al. Forecasting the future of cardiovascular disease in the United States: a policy statement from the American Heart Association. Circulation. 2011;123(8):933–44.
3. U.S. Census Bureau Population Division Table 12. Projections of the population by age and sex for the United States 2010-2050; 2008.
4. Lunenfeld B, Stratton P. The clinical consequences of an ageing world and preventive strategies. Best Pract Res Clin Obstet Gynaecol. 2013;27(5):643–59.
5. Kirkland JL. Translating advances from the basic biology of aging into clinical application. Exp Gerontol. 2013;48(1):1–5.
6. Seals DR, Justice JN, LaRocca TJ. Physiological geroscience: targeting function to increase healthspan and achieve optimal longevity. J Physiol. 2016;594(8):2001–24.
7. Reutov VP, Sorokina EG. NO-synthase and nitrite-reductase components of nitric oxide cycle. Biochemistry. 1998;63(7):874–84.
8. Torregrossa AC, Aranke M, Bryan NS. Nitric oxide and geriatrics: implications in diagnostics and treatment of the elderly. J Geriatr Cardiol. 2011;8(4):230–42.
9. Lundberg JO, Gladwin MT, Ahluwalia A, Benjamin N, Bryan NS, Butler A, et al. Nitrate and nitrite in biology, nutrition and therapeutics. Nat Chem Biol. 2009;5(12):865–9.
10. Seals DR, Jablonski KL, Donato AJ. Aging and vascular endothelial function in humans. Clin Sci. 2011;120(9):357–75.
11. Kregel KC, Zhang HJ. An integrated view of oxidative stress in aging: basic mechanisms, functional effects, and pathological considerations. Am J Physiol Regul Integr Comp Physiol. 2007;292(1):R18–36.
12. Ferrer-Sueta G, Radi R. Chemical biology of peroxynitrite: kinetics, diffusion, and radicals. ACS Chem Biol. 2009;4(3):161–77.
13. Luo S, Lei H, Qin H, Xia Y. Molecular mechanisms of endothelial NO synthase uncoupling. Curr Pharm Des. 2014;20(22):3548–53.
14. Bernard K, Hecker L, Luckhardt TR, Cheng G, Thannickal VJ. NADPH oxidases in lung health and disease. Antioxid Redox Signal. 2014;20(17):2838–53.
15. Montezano AC, Touyz RM. Reactive oxygen species, vascular Noxs, and hypertension: focus on translational and clinical research. Antioxid Redox Signal. 2014;20(1):164–82.
16. Lambeth JD. Nox enzymes, ROS, and chronic disease: an example of antagonistic pleiotropy. Free Radic Biol Med. 2007;43(3):332–47.
17. Dai DF, Rabinovitch PS, Ungvari Z. Mitochondria and cardiovascular aging. Circ Res. 2012;110(8):1109–24.
18. Schulz E, Wenzel P, Munzel T, Daiber A. Mitochondrial redox signaling: interaction of mitochondrial reactive oxygen species with other sources of oxidative stress. Antioxid Redox Signal. 2014;20(2):308–24.
19. Chung JH, Seo AY, Chung SW, Kim MK, Leeuwenburgh C, Yu BP, et al. Molecular mechanism of PPAR in the regulation of age-related inflammation. Ageing Res Rev. 2008;7(2):126–36.
20. Csiszar A, Wang M, Lakatta EG, Ungvari Z. Inflammation and endothelial dysfunction during aging: role of NF-kappaB. J Appl Physiol. 2008;105(4):1333–41.
21. Sarkar D, Fisher PB. Molecular mechanisms of aging-associated inflammation. Cancer Lett. 2006;236(1):13–23.
22. Maggio M, Basaria S, Ble A, Lauretani F, Bandinelli S, Ceda GP, et al. Correlation between testosterone and the inflammatory marker soluble interleukin-6 receptor in older men. J Clin Endocrinol Metab. 2006;91(1):345–7.
23. Bryan NS, Calvert JW, Gundewar S, Lefer DJ. Dietary nitrite restores NO homeostasis and is cardioprotective in endothelial nitric oxide synthase-deficient mice. Free Radic Biol Med. 2008;45(4):468–74.
24. Lundberg JO, Carlstrom M, Larsen FJ, Weitzberg E. Roles of dietary inorganic nitrate in cardiovascular health and disease. Cardiovasc Res. 2011;89(3):525–32.
25. Rocha BS, Gago B, Pereira C, Barbosa RM, Bartesaghi S, Lundberg JO, et al. Dietary nitrite in nitric oxide biology: a redox interplay with implications for pathophysiology and therapeutics. Curr Drug Targets. 2011;12(9):1351–63.

26. Mozaffarian D, Benjamin EJ, Go AS, Arnett DK, Blaha MJ, Cushman M, et al. Heart disease and stroke statistics—2015 update: a report from the American Heart Association. Circulation. 2015;131(4):e29–322.

27. Lakatta EG. Arterial and cardiac aging: major shareholders in cardiovascular disease enterprises: part III: cellular and molecular clues to heart and arterial aging. Circulation. 2003;107(3):490–7.

28. Cau SB, Carneiro FS, Tostes RC. Differential modulation of nitric oxide synthases in aging: therapeutic opportunities. Front Physiol. 2012;3:218.

29. Najjar SS, Scuteri A, Lakatta EG. Arterial aging: is it an immutable cardiovascular risk factor? Hypertension. 2005;46(3):454–62.

30. Mitchell GF, Hwang SJ, Vasan RS, Larson MG, Pencina MJ, Hamburg NM, et al. Arterial stiffness and cardiovascular events: the Framingham Heart Study. Circulation. 2010;121(4):505–11.

31. Mitchell GF. Effects of central arterial aging on the structure and function of the peripheral vasculature: implications for end-organ damage. J Appl Physiol. 2008;105(5):1652–60.

32. Safar ME, Nilsson PM, Blacher J, Mimran A. Pulse pressure, arterial stiffness, and end-organ damage. Curr Hypertens Rep. 2012;14(4):339–44.

33. Davignon J, Ganz P. Role of endothelial dysfunction in atherosclerosis. Circulation. 2004;109(23 Suppl 1):III27–32.

34. Donato AJ, Morgan RG, Walker AE, Lesniewski LA. Cellular and molecular biology of aging endothelial cells. J Mol Cell Cardiol. 2015;89(Pt B):122–35.

35. Luscher TF, Barton M. Biology of the endothelium. Clin Cardiol. 1997;20(11 Suppl 2):II-3–10.

36. Sindler AL, Devan AE, Fleenor BS, Seals DR. Inorganic nitrite supplementation for healthy arterial aging. J Appl Physiol. 2014;116(5):463–77.

37. Lara J, Ashor AW, Oggioni C, Ahluwalia A, Mathers JC, Siervo M. Effects of inorganic nitrate and beetroot supplementation on endothelial function: a systematic review and meta-analysis. Eur J Nutr. 2015;55(2):451–9.

38. Ignarro LJ, Gruetter CA. Requirement of thiols for activation of coronary arterial guanylate cyclase by glyceryl trinitrate and sodium nitrite: possible involvement of S-nitrosothiols. Biochim Biophys Acta. 1980;631(2):221–31.

39. Gruetter CA, Gruetter DY, Lyon JE, Kadowitz PJ, Ignarro LJ. Relationship between cyclic guanosine 3′:5′-monophosphate formation and relaxation of coronary arterial smooth muscle by glyceryl trinitrate, nitroprusside, nitrite and nitric oxide: effects of methylene blue and methemoglobin. J Pharmacol Exp Ther. 1981;219(1):181–6.

40. Ignarro LJ, Lippton H, Edwards JC, Baricos WH, Hyman AL, Kadowitz PJ, et al. Mechanism of vascular smooth muscle relaxation by organic nitrates, nitrites, nitroprusside and nitric oxide: evidence for the involvement of S-nitrosothiols as active intermediates. J Pharmacol Exp Ther. 1981;218(3):739–49.

41. Moulds RF, Jauernig RA, Shaw J. A comparison of the effects of hydrallazine, diazoxide, sodium nitrite and sodium nitroprusside on human isolated arteries and veins. Br J Clin Pharmacol. 1981;11(1):57–61.

42. Dejam A, Hunter CJ, Tremonti C, Pluta RM, Hon YY, Grimes G, et al. Nitrite infusion in humans and nonhuman primates: endocrine effects, pharmacokinetics, and tolerance formation. Circulation. 2007;116(16):1821–31.

43. Cosby K, Partovi KS, Crawford JH, Patel RP, Reiter CD, Martyr S, et al. Nitrite reduction to nitric oxide by deoxyhemoglobin vasodilates the human circulation. Nat Med. 2003;9(12):1498–505.

44. Ingram TE, Pinder AG, Bailey DM, Fraser AG, James PE. Low-dose sodium nitrite vasodilates hypoxic human pulmonary vasculature by a means that is not dependent on a simultaneous elevation in plasma nitrite. Am J Physiol Heart Circ Physiol. 2010;298(2):H331–9.

45. Maher AR, Milsom AB, Gunaruwan P, Abozguia K, Ahmed I, Weaver RA, et al. Hypoxic modulation of exogenous nitrite-induced vasodilation in humans. Circulation. 2008;117(5):670–7.

46. Mack AK, McGowan Ii VR, Tremonti CK, Ackah D, Barnett C, Machado RF, et al. Sodium nitrite promotes regional blood flow in patients with sickle cell disease: a phase I/II study. Br J Haematol. 2008;142(6):971–8.

47. Maher AR, Arif S, Madhani M, Abozguia K, Ahmed I, Fernandez BO, et al. Impact of chronic congestive heart failure on pharmacokinetics and vasomotor effects of infused nitrite. Br J Pharmacol. 2013;169(3):659–70.

48. Heiss C, Meyer C, Totzeck M, Hendgen-Cotta UB, Heinen Y, Luedike P, et al. Dietary inorganic nitrate mobilizes circulating angiogenic cells. Free Radic Biol Med. 2012;52(9):1767–72.

49. Ingram TE, Fraser AG, Bleasdale RA, Ellins EA, Margulescu AD, Halcox JP, et al. Low-dose sodium nitrite attenuates myocardial ischemia and vascular ischemia-reperfusion injury in human models. J Am Coll Cardiol. 2013;61(25):2534–41.

50. Kapil V, Milsom AB, Okorie M, Maleki-Toyserkani S, Akram F, Rehman F, et al. Inorganic nitrate supplementation lowers blood pressure in humans: role for nitrite-derived NO. Hypertension. 2010;56(2):274–81.

51. Joris PJ, Mensink RP. Beetroot juice improves in overweight and slightly obese men postprandial endothelial function after consumption of a mixed meal. Atherosclerosis. 2013;231(1):78–83.

52. Sindler AL, Fleenor BS, Calvert JW, Marshall KD, Zigler ML, Lefer DJ, et al. Nitrite supplementation reverses vascular endothelial dysfunction and large elastic artery stiffness with aging. Aging Cell. 2011;10(3):429–37.

53. Rammos C, Totzeck M, Deenen R, Kohrer K, Kelm M, Rassaf T and Hendgen-Cotta UB. Dietary nitrate is a modifier of vascular gene expression in old male mice. Oxid Med Cell Longev 2015; 2015:658264 [Epub ahead of print].

54. Rammos C, Hendgen-Cotta UB, Sobierajski J, Bernard A, Kelm M, Rassaf T. Dietary nitrate reverses vascular dysfunction in older adults with moderately increased cardiovascular risk. Journal of the American College of Cardiology. 2014;63(15):1584–5.

55. DeVan AE, Johnson LC, Brooks FA, Evans TD, Justice JN, Cruickshank-Quinn C, et al. Effects of sodium nitrite supplementation on vascular function and related small metabolite signatures in middle-aged and older adults. J Appl Physiol. 2015;120(4):416–25.
56. Zieman SJ, Melenovsky V, Kass DA. Mechanisms, pathophysiology, and therapy of arterial stiffness. Arterioscler Thromb Vasc Biol. 2005;25(5):932–43.
57. Kliche K, Jeggle P, Pavenstadt H, Oberleithner H. Role of cellular mechanics in the function and life span of vascular endothelium. Pflugers Arch. 2011;462(2):209–17.
58. Soldatos G, Cooper ME. Advanced glycation end products and vascular structure and function. Curr Hypertens Rep. 2006;8(6):472–8.
59. Katsuda S, Okada Y, Minamoto T, Oda Y, Matsui Y, Nakanishi I. Collagens in human atherosclerosis. Immunohistochemical analysis using collagen type-specific antibodies. Arteriscler Thromb. 1992;12(4):494–502.
60. Semba RD, Sun K, Schwartz AV, Varadhan R, Harris TB, Satterfield S, et al. Serum carboxymethyl-lysine, an advanced glycation end product, is associated with arterial stiffness in older adults. J Hypertens. 2015;33(4):797–803; discussion.
61. Yoon SJ, Park S, Park C, Chang W, Cho DK, Ko YG, et al. Association of soluble receptor for advanced glycation end-product with increasing central aortic stiffness in hypertensive patients. Coron Artery Dis. 2012;23(2):85–90.
62. Fleenor BS, Sindler AL, Eng JS, Nair DP, Dodson RB, Seals DR. Sodium nitrite de-stiffening of large elastic arteries with aging: role of normalization of advanced glycation end-products. Exp Gerontol. 2012;47(8):588–94.
63. Bahra M, Kapil V, Pearl V, Ghosh S, Ahluwalia A. Inorganic nitrate ingestion improves vascular compliance but does not alter flow-mediated dilatation in healthy volunteers. Nitric Oxide. 2012;26(4):197–202.
64. Omar SA, Fok H, Tilgner KD, Nair A, Hunt J, Jiang B, et al. Paradoxical normoxia-dependent selective actions of inorganic nitrite in human muscular conduit arteries and related selective actions on central blood pressures. Circulation. 2015;131(4):381–9, discussion 389.
65. Ghosh SM, Kapil V, Fuentes-Calvo I, Bubb KJ, Pearl V, Milsom AB, et al. Enhanced vasodilator activity of nitrite in hypertension: critical role for erythrocytic xanthine oxidoreductase and translational potential. Hypertension. 2013;61(5):1091–102.
66. Houston M, Hay J. Acute effects of an oral nitric oxide supplement on blood pressure, endothelial function, and vascular compliance in hypertensive patients. J Clin Hypertens (Greenwich). 2014;16(7):524–9.
67. Kapil V, Khambata RS, Robertson A, Caulfield MJ, Ahluwalia A. Dietary nitrate provides sustained blood pressure lowering in hypertensive patients: a randomized, phase 2, double-blind, placebo-controlled study. Hypertension. 2015;65(2):320–7.
68. Rigaud A-S, Forette B. Hypertension in older adults. J Gerontol Ser A Biol Sci Med Sci. 2001;56(4):M217–25.
69. Montenegro MF, Pinheiro LC, Amaral JH, Marcal DM, Palei AC, Costa-Filho AJ, et al. Antihypertensive and antioxidant effects of a single daily dose of sodium nitrite in a model of renovascular hypertension. Naunyn Schmiedebergs Arch Pharmacol. 2012;385(5):509–17.
70. Pluta RM, Oldfield EH, Bakhtian KD, Fathi AR, Smith RK, Devroom HL, et al. Safety and feasibility of long-term intravenous sodium nitrite infusion in healthy volunteers. PLoS One. 2011;6(1), e14504.
71. Greenway FL, Predmore BL, Flanagan DR, Giordano T, Qiu Y, Brandon A, et al. Single-dose pharmacokinetics of different oral sodium nitrite formulations in diabetes patients. Diabetes Technol Ther. 2012;14(7):552–60.
72. Biswas OS, Gonzalez VR, Schwarz ER. Effects of an oral nitric oxide supplement on functional capacity and blood pressure in adults with prehypertension. J Cardiovasc Pharmacol Ther. 2015;20(1):52–8.
73. Nagamani SC, Campeau PM, Shchelochkov OA. Nitric-oxide supplementation for treatment of long-term complications in argininosuccinic aciduria. Am J Hum Genet. 2012;90(5):836–46.
74. Siervo M, Lara J, Jajja A, Sutyarjoko A, Ashor A, Brandt K, et al. Ageing modifies the effects of beetroot juice supplementation on 24-hour blood pressure variability: an individual participant meta-analysis. Nitric Oxide. 2015;47:97–105.
75. Gilchrist M, Winyard PG, Aizawa K, Anning C, Shore A, Benjamin N. Effect of dietary nitrate on blood pressure, endothelial function, and insulin sensitivity in type 2 diabetes. Free Radic Biol Med. 2013;60:89–97.
76. Bondonno CP, Liu AH, Croft KD, Ward NC, Shinde S, Moodley Y, et al. Absence of an effect of high nitrate intake from beetroot juice on blood pressure in treated hypertensive individuals: a randomized controlled trial. Am J Clin Nutr. 2015;102:368–75.
77. Siervo M, Lara J, Ogbonmwan I, Mathers JC. Inorganic nitrate and beetroot juice supplementation reduces blood pressure in adults: a systematic review and meta-analysis. J Nutr. 2013;143(6):818–26.
78. Kelly J, Fulford J, Vanhatalo A, Blackwell JR, French O, Bailey SJ, et al. Effects of short-term dietary nitrate supplementation on blood pressure, O2 uptake kinetics, and muscle and cognitive function in older adults. Am J Physiol Regul Integr Comp Physiol. 2013;304(2):R73–83.
79. Rammos C, Hendgen-Cotta UB, Pohl J, Totzeck M, Luedike P, Schulze VT, et al. Modulation of circulating macrophage migration inhibitory factor in the elderly. Biomed Res Int. 2014;2014:582586.
80. Bryan NS, Calvert JW, Elrod JW, Gundewar S, Ji SY, Lefer DJ. Dietary nitrite supplementation protects against myocardial ischemia-reperfusion injury. Proc Natl Acad Sci U S A. 2007;104(48):19144–9.

81. Calvert JW, Lefer DJ. Myocardial protection by nitrite. Cardiovasc Res. 2009;83(2):195–203.
82. Webb A, Bond R, McLean P, Uppal R, Benjamin N, Ahluwalia A. Reduction of nitrite to nitric oxide during ischemia protects against myocardial ischemia-reperfusion damage. Proc Natl Acad Sci U S A. 2004;101(37):13683–8.
83. Salloum FN, Sturz GR, Yin C, Rehman S, Hoke NN, Kukreja RC, et al. Beetroot juice reduces infarct size and improves cardiac function following ischemia-reperfusion injury: possible involvement of endogenous H2S. Exp Biol Med. 2015;240(5):669–81.
84. Siddiqi N, Neil C, Bruce M, MacLennan G, Cotton S, Papadopoulou S, et al. Intravenous sodium nitrite in acute ST-elevation myocardial infarction: a randomized controlled trial (NIAMI). Eur Heart J. 2014;35(19):1255–62.
85. Jones DA, Pellaton C, Velmurugan S, Rathod KS, Andiapen M, Antoniou S, et al. Randomized phase 2 trial of intra-coronary nitrite during acute myocardial infarction. Circ Res. 2015;116(3):437–47.
86. Paulus WJ, Tschope C. A novel paradigm for heart failure with preserved ejection fraction: comorbidities drive myo-cardial dysfunction and remodeling through coronary microvascular endothelial inflammation. J Am Coll Cardiol. 2013;62(4):263–71.
87. Zakeri R, Levine JA, Koepp GA, Borlaug BA, Chirinos JA, LeWinter M, et al. Nitrate's effect on activity tolerance in heart failure with preserved ejection fraction trial: rationale and design. Circ Heart Fail. 2015;8:221–8.
88. Zamani P, Rawat D, Shiva-Kumar P, Geraci S, Bhuva R, Konda P, et al. Effect of inorganic nitrate on exercise capacity in heart failure with preserved ejection fraction. Circulation. 2015;131(4):371–80, discussion 380.
89. Buckner RL. Memory and executive function in aging and AD: multiple factors that cause decline and reserve factors that compensate. Neuron. 2004;44(1):195–208.
90. Gunning-Dixon FM, Raz N. The cognitive correlates of white matter abnormalities in normal aging: a quantitative review. Neuropsychology. 2000;14(2):224–32.
91. Fjell AM, Walhovd KB, Fennema-Notestine C, McEvoy LK, Hagler DJ, Holland D, et al. One-year brain atrophy evident in healthy aging. J Neurosci. 2009;29(48):15223–31.
92. Ylikoski A, Erkinjuntti T, Raininko R, Sarna S, Sulkava R, Tilvis R. White matter hyperintensities on MRI in the neurologically nondiseased elderly. Analysis of cohorts of consecutive subjects aged 55 to 85 years living at home. Stroke. 1995;26(7):1171–7.
93. Raz N. Aging of the brain and its impact on cognitive performance: integration of structural and functional findings. In: Craik FIM, Salthouse TA, editors. Handbook of aging and cognition. 2nd ed. Mahwah: Erlbaum; 2000. p. 1–90.
94. Pantoni L, Garcia JH. Cognitive impairment and cellular/vascular changes in the cerebral white matter. Ann N Y Acad Sci. 1997;826:92–102.
95. Pugh KG, Lipsitz LA. The microvascular frontal-subcortical syndrome of aging. Neurobiol Aging. 2002;23(3):421–31.
96. de Groot JC, de Leeuw FE, Oudkerk M, van Gijn J, Hofman A, Jolles J, et al. Cerebral white matter lesions and cogni-tive function: the Rotterdam Scan Study. Ann Neurol. 2000;47(2):145–51.
97. de Groot JC, De Leeuw FE, Oudkerk M, Van Gijn J, Hofman A, Jolles J, et al. Periventricular cerebral white matter lesions predict rate of cognitive decline. Ann Neurol. 2002;52(3):335–41.
98. de la Torre JC, Stefano GB. Evidence that Alzheimer's disease is a microvascular disorder: the role of constitutive nitric oxide. Brain Res Brain Res Rev. 2000;34(3):119–36.
99. Floyd RA, Hensley K. Oxidative stress in brain aging. Implications for therapeutics of neurodegenerative diseases. Neurobiol Aging. 2002;23(5):795–807.
100. Faraci FM. Protecting against vascular disease in brain. Am J Physiol Heart Circ Physiol. 2011;300(5):H1566–82.
101. Breteler MM, Bots ML, Ott A, Hofman A. Risk factors for vascular disease and dementia. Haemostasis. 1998;28(3–4):167–73.
102. Celsis P, Agniel A, Cardebat D, Demonet JF, Ousset PJ, Puel M. Age related cognitive decline: a clinical entity? A longitudinal study of cerebral blood flow and memory performance. J Neurol Neurosurg Psychiatry. 1997;62(6):601–8.
103. Girouard H, Iadecola C. Neurovascular coupling in the normal brain and in hypertension, stroke, and Alzheimer dis-ease. J Appl Physiol. 2006;100(1):328–35.
104. Pitsikas N, Rigamonti AE, Cella SG, Sakellaridis N, Muller EE. The nitric oxide donor molsidomine antagonizes age-related memory deficits in the rat. Neurobiol Aging. 2005;26(2):259–64.
105. Paul V, Ekambaram P. Involvement of nitric oxide in learning & memory processes. Indian J Med Res. 2011;133(5):471.
106. Aamand R, Ho YC, Dalsgaard T, Roepstorff A, Lund TE. Dietary nitrate facilitates an acetazolamide-induced increase in cerebral blood flow during visual stimulation. J Appl Physiol. 2014;116(3):267–73.
107. Pluta RM, Dejam A, Grimes G, Gladwin MT, Oldfield EH. Nitrite infusions to prevent delayed cerebral vasospasm in a primate model of subarachnoid hemorrhage. JAMA. 2005;293(12):1477–84.
108. Justice JN, Johnson LC, DeVan AE, Cruickshank-Quinn C, Reisdorph N, Bassett CJ, et al. Improved motor and cogni-tive performance with sodium nitrite supplementation is related to small metabolite signatures: a pilot trial in middle-aged and older adults. Aging. 2015;7(11):1004–21.
109. Reid KF, Fielding RA. Skeletal muscle power: a critical determinant of physical functioning in older adults. Exerc Sport Sci Rev. 2012;40(1):4.

110. Cooper R, Kuh D, Cooper C, Gale CR, Lawlor DA, Matthews F, et al. Objective measures of physical capability and subsequent health: a systematic review. Age Ageing. 2011;40(1):14–23.
111. Studenski S, Perera S, Patel K, Rosano C, Faulkner K, Inzitari M, et al. Gait speed and survival in older adults. JAMA. 2011;305(1):50–8.
112. Fried LP, Guralnik JM. Disability in older adults: evidence regarding significance, etiology, and risk. J Am Geriatr Soc. 1997;45(1):92–100.
113. Rantanen T, Guralnik JM, Sakari-Rantala R, Leveille S, Simonsick EM, Ling S, et al. Disability, physical activity, and muscle strength in older women: the Women's Health and Aging Study. Arch Phys Med Rehabil. 1999;80(2):130–5.
114. Doherty TJ. Invited review: aging and sarcopenia. J Appl Physiol. 2003;95(4):1717–27.
115. Vandervoort AA. Aging of the human neuromuscular system. Muscle Nerve. 2002;25(1):17–25.
116. Payne AM, Delbono O. Neurogenesis of excitation-contraction uncoupling in aging skeletal muscle. Exerc Sport Sci Rev. 2004;32(1):36–40.
117. Conley KE, Amara CE, Jubrias SA, Marcinek DJ. Mitochondrial function, fibre types and ageing: new insights from human muscle in vivo. Exp Physiol. 2007;92(2):333–9.
118. Verbrugge LM, Jette AM. The disablement process. Social Sci Med. 1994;38(1):1–14.
119. Hernández A, Schiffer TA, Ivarsson N, Cheng AJ, Bruton JD, Lundberg JO, et al. Dietary nitrate increases tetanic [Ca2+] i and contractile force in mouse fast-twitch muscle. J Physiol. 2012;590(15):3575–83.
120. Larsen FJ, Schiffer TA, Borniquel S, Sahlin K, Ekblom B, Lundberg JO, et al. Dietary inorganic nitrate improves mitochondrial efficiency in humans. Cell Metab. 2011;13(2):149–59.
121. Heffernan KS, Chale A, Hau C, Cloutier GJ, Phillips EM, Warner P, et al. Systemic vascular function is associated with muscular power in older adults. J Aging Res. 2012;2012:386387.
122. Ronnback M, Hernelahti M, Hamalainen E, Groop PH, Tikkanen H. Effect of physical activity and muscle morphology on endothelial function and arterial stiffness. Scand J Med Sci Sports. 2007;17(5):573–9.
123. Justice JN, Gioscia-Ryan RA, Johnson LC, Battson ML, de Picciotto NE, Beck HJ, et al. Sodium nitrite supplementation improves motor function and skeletal muscle inflammatory profile in old male mice. J Appl Physiol (1985). 2014:jap.00608.2014.
124. Cermak NM, Gibala MJ, van Loon LJ. Nitrate supplementation's improvement of 10-km time-trial performance in trained cyclists. Int J Sport Nutr Exerc Metab. 2012;22(1):64–71.
125. Lansley KE, Winyard PG, Fulford J, Vanhatalo A, Bailey SJ, Blackwell JR, et al. Dietary nitrate supplementation reduces the O2 cost of walking and running: a placebo-controlled study. J Appl Physiol. 2011;110(3):591–600.
126. Larsen FJ, Weitzberg E, Lundberg JO, Ekblom B. Effects of dietary nitrate on oxygen cost during exercise. Acta Physiol (Oxf). 2007;191(1):59–66.
127. Fricke O, Baecker N, Heer M, Tutlewski B, Schoenau E. The effect of l-arginine administration on muscle force and power in postmenopausal women. Clin Physiol Funct Imaging. 2008;28(5):307–11.
128. Casey DP, Treichler DP, Ganger CT, Schneider AC, Ueda K. Acute dietary nitrate supplementation enhances compensatory vasodilation during hypoxic exercise in older adults. J Appl Physiol. 2015;118(2):178–86.
129. Kenjale AA, Ham KL, Stabler T, Robbins JL, Johnson JL, Vanbruggen M, et al. Dietary nitrate supplementation enhances exercise performance in peripheral arterial disease. J Appl Physiol. 2011;110(6):1582–91.
130. Baliga RS, Milsom AB, Ghosh SM, Trinder SL, Macallister RJ, Ahluwalia A, et al. Dietary nitrate ameliorates pulmonary hypertension: cytoprotective role for endothelial nitric oxide synthase and xanthine oxidoreductase. Circulation. 2012;125(23):2922–32.
131. Berry MJ, Justus NW, Hauser JI, Case AH, Helms CC, Basu S, et al. Dietary nitrate supplementation improves exercise performance and decreases blood pressure in COPD patients. Nitric Oxide. 2014;48:22–30.
132. Dezfulian C, Alekseyenko A, Dave KR, Raval AP, Do R, Kim F, et al. Nitrite therapy is neuroprotective and safe in cardiac arrest survivors. Nitric Oxide. 2012;26(4):241–50.
133. Shiva S, Sack MN, Greer JJ, Duranski M, Ringwood LA, Burwell L, et al. Nitrite augments tolerance to ischemia/reperfusion injury via the modulation of mitochondrial electron transfer. J Exp Med. 2007;204(9):2089–102.
134. Wang WZ, Fang XH, Stephenson LL, Zhang X, Williams SJ, Baynosa RC, et al. Nitrite attenuates ischemia-reperfusion-induced microcirculatory alterations and mitochondrial dysfunction in the microvasculature of skeletal muscle. Plast Reconstr Surg. 2011;128(4):279e–87.
135. Tripatara P, Patel NS, Webb A, Rathod K, Lecomte FM, Mazzon E, et al. Nitrite-derived nitric oxide protects the rat kidney against ischemia/reperfusion injury in vivo: role for xanthine oxidoreductase. J Am Soc Nephrol. 2007;18(2):570–80.
136. Duranski MR, Greer JJ, Dejam A, Jaganmohan S, Hogg N, Langston W, et al. Cytoprotective effects of nitrite during in vivo ischemia-reperfusion of the heart and liver. J Clin Invest. 2005;115(5):1232–40.
137. Sugimoto R, Okamoto T, Nakao A, Zhan J, Wang Y, Kohmoto J, et al. Nitrite reduces acute lung injury and improves survival in a rat lung transplantation model. Am J Transplant. 2012;12(11):2938–48.
138. Milsom AB, Patel NS, Mazzon E, Tripatara P, Storey A, Mota-Filipe H, et al. Role for endothelial nitric oxide synthase in nitrite-induced protection against renal ischemia-reperfusion injury in mice. Nitric Oxide. 2010;22(2):141–8.

139. Kelpke SS, Chen B, Bradley KM, Teng X, Chumley P, Brandon A, et al. Sodium nitrite protects against kidney injury induced by brain death and improves post-transplant function. Kidney Int. 2012;82(3):304–13.

140. Zhan J, Nakao A, Sugimoto R, Dhupar R, Wang Y, Wang Z, et al. Orally administered nitrite attenuates cardiac allograft rejection in rats. Surgery. 2009;146(2):155–65.

141. Madamanchi NR, Vendrov A, Runge MS. Oxidative stress and vascular disease. Arterioscler Thromb Vasc Biol. 2005;25(1):29–38.

142. Venkataraman K, Khurana S, Tai TC. Oxidative stress in aging—matters of the heart and mind. Int J Mol Sci. 2013;14(9):17897–925.

143. Romano AD, Serviddio G, de Matthaeis A, Bellanti F, Vendemiale G. Oxidative stress and aging. J Nephrol. 2010;23 Suppl 15:S29–36.

144. Bachschmid MM, Schildknecht S, Matsui R, Zee R, Haeussler D, Cohen RA, et al. Vascular aging: chronic oxidative stress and impairment of redox signaling-consequences for vascular homeostasis and disease. Ann Med. 2013;45(1):17–36.

145. Paneni F, Costantino S, Cosentino F. Molecular pathways of arterial aging. Clin Sci (Lond). 2015;128(2):69–79.

146. Ungvari Z, Kaley G, de Cabo R, Sonntag WE, Csiszar A. Mechanisms of vascular aging: new perspectives. J Gerontol A Biol Sci Med Sci. 2010;65(10):1028–41.

147. Zhu SG, Kukreja RC, Das A, Chen Q, Lesnefsky EJ, Xi L. Dietary nitrate supplementation protects against Doxorubicin-induced cardiomyopathy by improving mitochondrial function. J Am Coll Cardiol. 2011;57(21):2181–9.

148. Carlstrom M, Persson AE, Larsson E, Hezel M, Scheffer PG, Teerlink T, et al. Dietary nitrate attenuates oxidative stress, prevents cardiac and renal injuries, and reduces blood pressure in salt-induced hypertension. Cardiovasc Res. 2011;89(3):574–85.

149. Bir SC, Pattillo CB, Pardue S, Kolluru GK, Docherty J, Goyette D, et al. Nitrite anion stimulates ischemic arteriogenesis involving NO metabolism. Am J Physiol Heart Circ Physiol. 2012;303(2):H178–88.

150. Oudot A, Martin C, Busseuil D, Vergely C, Demaison L, Rochette L. NADPH oxidases are in part responsible for increased cardiovascular superoxide production during aging. Free Radic Biol Med. 2006;40(12):2214–22.

151. Wang M, Zhang J, Walker SJ, Dworakowski R, Lakatta EG, Shah AM. Involvement of NADPH oxidase in age-associated cardiac remodeling. J Mol Cell Cardiol. 2010;48(4):765–72.

152. Krause K-H. Aging: a revisited theory based on free radicals generated by NOX family NADPH oxidases. Exp Gerontol. 2007;42(4):256–62.

153. Schuhmacher S, Oelze M, Bollmann F, Kleinert H, Otto C, Heeren T, et al. Vascular dysfunction in experimental diabetes is improved by pentaerithrityl tetranitrate but not isosorbide-5-mononitrate therapy. Diabetes. 2011;60(10):2608–16.

154. İnal ME, Kanbak G, Sunal E. Antioxidant enzyme activities and malondialdehyde levels related to aging. Clin Chim Acta. 2001;305(1–2):75–80.

155. Corbi G, Conti V, Russomanno G, Rengo G, Vitulli P, Ciccarelli AL, et al. Is physical activity able to modify oxidative damage in cardiovascular aging? Oxid Med Cell Longev. 2012;2012:728547.

156. Singh M, Arya A, Kumar R, Bhargava K, Sethy NK. Dietary nitrite attenuates oxidative stress and activates antioxidant genes in rat heart during hypobaric hypoxia. Nitric oxide : biology and chemistry/official journal of the Nitric Oxide Society. 2012;26(1):61–73.

157. Lu XX, Wang SQ, Zhang Z, Xu HR, Liu B, Huangfu CS. [Protective effects of sodium nitrite preconditioning against alcohol-induced acute liver injury in mice]. Sheng Li Xue Bao [Acta Physiologica Sinica]. 2012;64(3):313–20.

158. Doganci S, Yildirim V, Bolcal C, Korkusuz P, Gumusel B, Demirkilic U, et al. Sodium nitrite and cardioprotective effect in pig regional myocardial ischemia-reperfusion injury model. Adv Clin Exp Med. 2012;21(6):713–26.

159. Perlman DH, Bauer SM, Ashrafian H, Bryan NS, Garcia-Saura MF, Lim CC, et al. Mechanistic insights into nitrite-induced cardioprotection using an integrated metabolomic/proteomic approach. Circ Res. 2009;104(6):796–804.

160. Mohler III ER, Hiatt WR, Gornik HL, Kevil CG, Quyyumi A, Haynes WG, et al. Sodium nitrite in patients with peripheral artery disease and diabetes mellitus: safety, walking distance and endothelial function. Vasc Med. 2014;19(1):9–17.

161. Delp MD, Behnke BJ, Spier SA, Wu G, Muller-Delp JM. Ageing diminishes endothelium-dependent vasodilatation and tetrahydrobiopterin content in rat skeletal muscle arterioles. J Physiol. 2008;586(4):1161–8.

162. Stokes KY, Dugas TR, Tang Y, Garg H, Guidry E, Bryan NS. Dietary nitrite prevents hypercholesterolemic microvascular inflammation and reverses endothelial dysfunction. Am J Physiol Heart Circ Physiol. 2009;296(5):H1281–8.

163. Ungvari Z, Sonntag WE, Csiszar A. Mitochondria and aging in the vascular system. J Mol Med (Berl). 2010;88(10):1021–7.

164. Kluge MA, Fetterman JL, Vita JA. Mitochondria and endothelial function. Circ Res. 2013;112(8):1171–88.

165. Shenouda SM, Widlansky ME, Chen K, Xu G, Holbrook M, Tabit CE, et al. Altered mitochondrial dynamics contributes to endothelial dysfunction in diabetes mellitus. Circulation. 2011;124(4):444–53.

166. Gioscia-Ryan RA, LaRocca TJ, Sindler AL, Zigler MC, Murphy MP, Seals DR. Mitochondria-targeted antioxidant (MitoQ) ameliorates age-related arterial endothelial dysfunction in mice. J Physiol. 2014;592(Pt 12):2549–61.

167. Miller MW, Knaub LA, Olivera-Fragoso LF, Keller AC, Balasubramaniam V, Watson PA, et al. Nitric oxide regulates vascular adaptive mitochondrial dynamics. Am J Physiol Heart Circ Physiol. 2013;304(12):H1624–33.
168. Shiva S. Nitrite: A physiological store of nitric oxide and modulator of mitochondrial function. Redox Biol. 2013;1(1):40–4.
169. Shiva S, Rassaf T, Patel RP, Gladwin MT. The detection of the nitrite reductase and NO-generating properties of haemoglobin by mitochondrial inhibition. Cardiovasc Res. 2011;89(3):566–73.
170. Mo L, Wang Y, Geary L, Corey C, Alef MJ, Beer-Stolz D, et al. Nitrite activates AMP kinase to stimulate mitochondrial biogenesis independent of soluble guanylate cyclase. Free Radic Biol Med. 2012;53(7):1440–50.
171. Murillo D, Kamga C, Mo L, Shiva S. Nitrite as a mediator of ischemic preconditioning and cytoprotection. Nitric Oxide. 2011;25(2):70–80.
172. Kamga Pride C, Mo L, Quesnelle K, Dagda RK, Murillo D, Geary L, et al. Nitrite activates protein kinase A in normoxia to mediate mitochondrial fusion and tolerance to ischaemia/reperfusion. Cardiovasc Res. 2014;101(1):57–68.
173. Cauwels A, Buys ES, Thoonen R, Geary L, Delanghe J, Shiva S, et al. Nitrite protects against morbidity and mortality associated with TNF- or LPS-induced shock in a soluble guanylate cyclase-dependent manner. J Exp Med. 2009;206(13):2915–24.
174. Michaud M, Balardy L, Moulis G, Gaudin C, Peyrot C, Vellas B, et al. Proinflammatory cytokines, aging, and age-related diseases. J Am Med Dir Assoc. 2013;14(12):877–82.
175. Wang M, Jiang L, Monticone RE, Lakatta EG. Proinflammation: the key to arterial aging. Trends Endocrinol Metab. 2014;25(2):72–9.
176. Jadert C, Petersson J, Massena S, Ahl D, Grapensparr L, Holm L, et al. Decreased leukocyte recruitment by inorganic nitrate and nitrite in microvascular inflammation and NSAID-induced intestinal injury. Free Radic Biol Med. 2012;52(3):683–92.
177. Cauwels A, Brouckaert P. Nitrite regulation of shock. Cardiovasc Res. 2011;89(3):553–9.
178. Hamburger T, Broecker-Preuss M, Hartmann M, Schade FU, de Groot H, Petrat F. Effects of glycine, pyruvate, resveratrol, and nitrite on tissue injury and cytokine response in endotoxemic rats. J Surg Res. 2013;183(1):e7–21.
179. Murata I, Nozaki R, Ooi K, Ohtake K, Kimura S, Ueda H, et al. Nitrite reduces ischemia/reperfusion-induced muscle damage and improves survival rates in rat crush injury model. J Trauma Acute Care Surg. 2012;72(6):1548–54.
180. Okamoto T, Tang X, Janocha A, Farver CF, Gladwin MT, McCurry KR. Nebulized nitrite protects rat lung grafts from ischemia reperfusion injury. J Thorac Cardiovasc Surg. 2013;145(4):1108–16.
181. Pattillo CB, Fang K, Terracciano J, Kevil CG. Reperfusion of chronic tissue ischemia: nitrite and dipyridamole regulation of innate immune responses. Ann N Y Acad Sci. 2011;1207:83–8.
182. Pattillo CB, Fang K, Pardue S, Kevil CG. Genome expression profiling and network analysis of nitrite therapy during chronic ischemia: possible mechanisms and interesting molecules. Nitric Oxide. 2011;22(2):168–79.

Chapter 19
The Nitrate–Nitrite–Nitric Oxide Pathway in Traditional Herbal Medicine and Dietary Supplements with Potential Benefits for Cardiovascular Diseases

Yong-Jian Geng

Key Points

- The inorganic anions nitrate and nitrite serve key intermediates in the nitrogen cycle.
- The nitrate–nitrite–NO pathway contributes to therapeutic activities of herbal medicines or dietary supplements known to have cardiovascular benefits.
- Herbal medicines contain predominantly nitrate and to a less extent nitrite.
- Herbal medicines contain reductases that are capable of reducing nitrate and nitrite into NO
- The activities of NO synthase and nitrate/nitrite reductases can be regulated by herbal medicine and dietary supplements, and thus maintain the "Yin-Yang" balance of NO production
- The ability of herbal medicines to reduce nitrate and nitrite to NO effectively may account for some of their beneficial impact on cardiovascular disorders
- Understanding the mechanism of action for herbal medicines will allow for better combinations of herbs for natural disease combat as a key part of alternative Western medicine

Keywords Nitrate Nitrite Nitric oxide • Alternative Medicine • Dietary supplements • Herb • Heart diseases

Introduction

For thousands of years, traditional herbal medicine has been serving as a major health care tool for disease prevention and treatment in many countries or communities. For instance, the herbal medicine (HM) has been an important part of the health care system in China and India, as well as many other countries in Asia and Africa. In some Asian and African countries, 80 % of the population are using traditional medicine for primary health care. HM is considered a complementary or alternative medical system in much of the Western world, mostly due to a lack of understanding of their mechanism of action and/or the active compound(s). However, the use of traditional or alternative medicine is increasing in Western nations, especially in the communities of people with Asian origin. Patients who use alternative or herbal medicine in Western countries report that the main reason for using it is

Y.-J. Geng, M.D., Ph.D. (✉)
The Center for Cardiovascular Biology and Atherosclerosis Research, University of Texas McGovern School of Medicine at Houston, 6431 Fannin Street, MSB 1.246, Houston, TX 77030, USA
e-mail: yong-jian.geng@uth.tmc.edu

N.S. Bryan, J. Loscalzo (eds.), *Nitrite and Nitrate in Human Health and Disease*, Nutrition and Health, DOI 10.1007/978-3-319-46189-2_19, © Springer International Publishing AG 2017

due to a "more natural" and potentially safer alternative for the treatment of chronic illness than pharmaceutical drugs or surgery [1]. In many developed countries, including the U.S., as much as 70–80 % of the population has used some form of alternative or complementary medicine (e.g., acupuncture or herbal supplements). Herbal treatments are the most popular form of traditional medicine and are highly lucrative in the international marketplace. The size of the global market for herbal medicinal products was estimated at $33 billion in 2013 (close to 24.5€ billion), increasing 11 % annually. A recent research by Global Industry Analysts Inc. has projected that the global herbal supplements and remedies market will reach US $115 billion (almost 105€ billion) by 2020 (http://www.strategyr.com/MarketResearch/Herbal_Supplements_and_Remedies_Market_Trends.asp). This research also takes other herbal remedies into account next to herbal medicinal products (e.g., herbal food supplements). In the US, herbal/traditional products showed steady growth over the years, with current value sales rising at a compounded annual growth rate (CAGR) of 4 % over the 2010–2015 review period. Rising awareness of some botanicals' long tradition of use in other cultures as well as growing demand for effective and natural alternatives in consumer health products supported the growth of herbal/traditional products in the US. Although the US has the largest consumer health market, herbal/traditional products held a relatively small market share in the country, as it accounted for only 7 % of total consumer health retail value sales in 2015. The market size of US (US $4.6 billion) is more than half the market size of herbal/traditional products in China (US $10.8 billion). Annual revenues in Western Europe reached $5 billion in 2003–2004. Herbal medicine revenue in Brazil was $160 million in 2007, according to WHO Fact Sheet No. 134, 2008. Other reports reveal herbal/traditional products recorded current value growth of 7 % in 2015 in Brazil, reaching sales worth R$1.7 billion. Category growth continued to be mainly driven by increasing demand for less aggressive medications to treat minor pains and illnesses. Consumers in Brazil see herbal/traditional products as a healthier and more natural way to treat minor ailments, often favoring these over regular OTC products.

Traditional herbal medicine has its origin in ancient Taoist philosophy, which views a person as an energy system in which the body, mind, and spirit are unified into one when in harmony and disrupted in disease. Traditional medicine practice treats the patient as a whole, not as a part, and it emphasizes a holistic approach that attempts to bring the mind, body, and spirit into harmony. Traditional medicine theory is extremely complex and originated thousands of years ago through meticulous observation of nature, the cosmos, and the human body. There is a growing and sustained interest in alternative medicines fueled by a combination of factors including recognition of benefits, dissatisfaction with and ineffectiveness of traditional Western medicines, an increasing commitment to holistic care, skepticism regarding adverse side effects of drug therapy, and increasing evidence for the personalized nature of different combination of herbs for specific disorders [1]. Of the approximately 500 herbs that are in use today, 50 or so are very commonly used alone or in combination. Rather than being prescribed individually, single herbs are combined into formulas that are designed to adapt to the specific needs of individual patients. An herbal formula can contain from 3 to 25 herbs. As with diet therapy, each herb has one or more of the four flavors/functions (bitter, sweet, pungent, salty, and sour) and one of the five "temperatures" "氣" (pronounced "Chi") (hot, warm, neutral, cool, cold). Herbal formulations work to balance the body from the inside out. Traditional herbal medicines include herbs, herbal materials, herbal preparations, and processed herbal products that contain parts of plants or other plant materials as active ingredients which assist with strengthening the vital energies (Chi), blood, and fluids internally. They are typically administered as tablets, tea pills, elixirs, soups, liquid extracts, and teas.

With traditional medicine practices being adopted by new populations in the United States and Europe as well as many other countries around the world, challenges have been emerging recently. To date, most countries have established national policies for traditional medicine. Regulating traditional medicine products, practices, and practitioners is difficult due to variations in definitions and categorizations of traditional medicine therapies. A single herbal product could be defined as either a food, dietary supplement, or an herbal medicine, depending on the country. This disparity in regulations at the national levels has implications for international access and distribution of products. Scientific evidence from tests done to evaluate the safety and effectiveness of traditional medicine products and practices is limited. Despite the fact that alternative treatments, such as acupuncture, some herbal medicines, and some manual therapies (e.g., massage), have been shown to be effective for specific conditions, no proven underlying mechanism is available. A further study of the products and practices is needed. Many people, including most professionals in the field of biomedicine, believe that because herbal medicines are natural and have been used traditionally they are safe and carry no risk for harm. However, traditional medicines and practices can cause harmful, adverse reactions if the product or therapy is of poor quality, or is taken inappropriately, or in conjunction with other medicines. Increased patient awareness about safe usage is important, as well as more training, collaboration, and communication among providers of traditional and other medicines. Requirements and methods for research and evaluation are complex. It is often difficult to assess the quality of processed herbal products. The safety, effectiveness, and quality of finished herbal medicine products depend on the quality of their source materials (which can include hundreds of natural constituents), and how elements are handled through production processes.

Recent research has documented the association of traditional medicines and nitric oxide (NO)-related effects [2, 3]. Inorganic nitrite and nitrate have been reported widely as alternative sources of NO production independent of the enzymatic synthesis of NO from L-arginine [4–6]; however, their mechanism of action is far from clear due to the complexity and heterogeneity of the herbal therapies as well as a lack of knowledge of content and ratio or identification of the active compound or compounds. Recent discovery of the nitrate–nitrite–nitric oxide pathway in plants and animals [7, 8] offers a new avenue by which to define the biological action as well as to assess the safety of certain herbal medicines commonly used for the treatment of cardiovascular disease. By establishing a common molecule of interest acting similarly in many of these concoctions, a purified and concentrated formulation may be developed with even greater efficacy, which may eventually penetrate the pharmaceutical markets in the countries with stricter regulations, such as the United States and European countries. This knowledge is equally important for legal and regulatory reasons, as well as for patient safety related to any contraindications with their concomitant use with other medications. In this chapter, the production of nitrate, nitrite, and NO, and the molecular basis of their biological action will be discussed. Evidence will be presented to support the notion that the nitrite, nitrate, and nitrite reductase activity ingested as herbal medicines provides a robust and natural system for NO generation to overcome pathological conditions associated with NO insufficiency and combat cardiovascular disease.

NO Production and Consequences of NO Insufficiency

Previous chapters have established the essential role of NO in the maintenance of cardiovascular health as well as healthy immune or neuronal status. Appropriate levels of NO production are critical in tissue blood perfusion and protection of cardiovascular tissues against ischemia and infarction. Sustained levels of enhanced NO production may result in direct tissue toxicity and contribute to the vascular collapse associated with various pathological conditions. Being a simple molecule, NO is a fundamental player in many different fields of biomedicine, and it was proclaimed the "Molecule of the Year" in 1992 [9]. Nitric oxide is one of the few gaseous signaling molecules known to play a role in a variety of biological processes. Acting as the 'endothelium-derived relaxing factor,' or 'EDRF' [10], NO is biosynthesized endogenously from L-arginine and oxygen by various nitric oxide synthase (NOS) enzymes. The endothelium of blood vessels uses NO to signal the surrounding smooth muscle to relax, thus resulting in vasodilation and increasing blood flow. Because it has an extremely high reactivity (having a lifetime of a few seconds), NO diffuses freely across membranes over a limited distance. These attributes make NO ideal for a transient paracrine (between adjacent cells) and autocrine (within a single cell) signaling molecule. The production of NO is elevated in populations living at high altitudes [11], which help these individuals avoid hypoxia by aiding in pulmonary vascular vasodilation. Nitroglycerin and amyl nitrite serve as vasodilators because they are converted to NO in the body. Sildenafil citrate, popularly known by the trade name *Viagra*, stimulates erections primarily by enhancing signaling through the NO pathway and, thus, dilating the arteries of the penis. NO contributes to vessel homeostasis by inhibiting vascular smooth muscle contraction and growth, platelet aggregation, and leukocyte adhesion to the endothelium. Humans with atherosclerosis, diabetes, or hypertension often show impaired NO pathways. Alternative source of NO production from diets may help compensate for the reduction in NO synthesis catalyzed by NOS [12, 13]. For years, physicians and scientists assumed simply feeding more substrate L-arginine would be sufficient to enhance NO production. It is becoming increasingly clear that this may not be the most effective strategy, especially in patients who are insufficient in NO due to endothelial dysfunction.

Nitrate and Nitrite in the Nitrogen Cycle

Since herbal medicines are mostly plant based or extracted from plants, they contain inorganic nitrogen oxides generated through the environmental nitrogen cycle. Nitrogen constitutes 78 % of earth's atmosphere and serves as a major constituent of all living tissues. As an essential element for life, the nitrogen atom is incorporated into proteins and nucleic acids. In nature, nitrogen molecules (N_2) exist mainly in air. In water and soil, nitrogen can be found predominantly in the form of nitrate and nitrite.

Nitrate and nitrite are naturally occurring anions. Compared to NO, they are chemically stable, and they can be found virtually anywhere in the environment. Nitrite is produced by nitrification or oxidation of ammonia, especially by the action of the nitrifying bacterium, *Nitrosomas*. The nitrite will then be oxidized to nitrate by the bacterium *Nitrobacter*. Nitrate is less toxic than nitrite and is used as a food source by live plants. The process of converting ammonia to nitrate is diagrammed in the *nitrogen cycle* (Fig. 19.1). Nitrification is most rapid at pH of 7–8 and at temperatures of 25–30 °C. Nitrification leads to acidification of water.

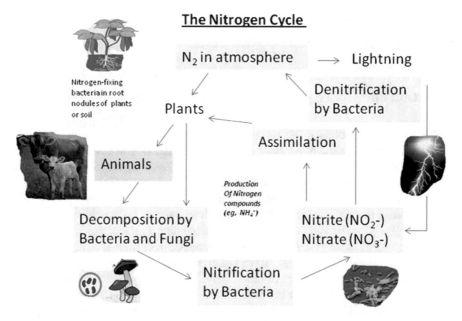

Fig. 19.1 Schematic representation of the nitrogen cycle. All nitrogen obtained by animals and humans can be traced back to ingestion of plants at some stage of the food chain as part of the nitrogen cycle

The inorganic anions, nitrate (NO_3^-) and nitrite (NO_2^-), were traditionally considered inert end-products of endogenous NO metabolism. However, recent studies show that these supposedly inert anions can be recycled in vivo to form NO, representing an important alternative source of NO to the classical L-arginine–NO-synthase pathway, in particular in hypoxic states. The emerging important biological functions of the nitrate–nitrite–NO pathway implicate the therapeutic potential of many herbal medicines traditionally used for cardiovascular diseases because herbs can provide significant amounts of nitrate and nitrite as well as an inherent ability to reduce nitrite to NO.

Medicinal Herbs: An Alternative Source of Exogenous Nitrate and Nitrite

Many herbal medicines are made from the roots and leaves of certain plants that contain abundant nitrate/nitrite or related compounds. In these plants, NO can be produced by at least four routes:

1. L-arginine-dependent or independent NO-synthesizing enzymes. Several studies [14–17] suggest that as part of the innate defense response, certain plants may express high levels of enzymes that can synthesize NO using L-arginine and other nitro compounds.
2. Plasma membrane-bound nitrate/nitrite reductase. Many research groups have demonstrated that plant cells contain nitrate or nitrite reductase [18, 19]. Their reduction may convert them into NO under conditions of low pH or hypoxia.
3. Mitochondrial electron transport chain. Mitochondrial compartments are rich in metal enzymes involved actively in the regulation of energy production [20]. The active metabolism in mitochondria may generate nitro compounds which, in turn, are converted into NO.
4. Nonenzymatic reactions. The root of certain plants may contain high levels of unstable nitro compounds, which may release NO spontaneously [21, 22]. Plants exposed to nitro types of fertilizers typically generate more NO than those grown in the wild.

Similar to that in mammal tissues, plant-derived NO functions as a signaling molecule. It acts mainly against oxidative stress and plays a role in plant–pathogen interactions. Treating cut flowers and other plants with NO has been shown to lengthen the time before wilting. Nitric oxide can bind to mitochondrial enzymes containing iron-II leading to attenuation of respiration [23]. An important biological reaction of NO is S-nitrosation [24–27] the conversion of thiol groups, including cysteine residues in proteins, to form *S*-nitrosothiols (RSNOs). S-Nitrosation is a mechanism for dynamic, posttranslational regulation of most or all major classes of protein in both plants and animals, which may also account for some of the NO activity. Inorganic nitrite and nitrate are emerging as key players in NO biology. They are important sources of NO and can be recycled in vivo under specific conditions [4, 5, 28]. There are a number of endogenous systems in mammals capable of reducing nitrite to NO [20]. Dietary nitrite has been shown to protect from tissue injury and restore NO homeostasis in eNOS$^{-/-}$ mice [29–32]. Furthermore, nitrate has also been shown to reduce blood pressure [33–36], inhibit platelet aggregation [34], and protect the heart from reperfusion injury [30]. There are a number of herbal preparations with primary indications for cardiovascular disease. There are increased interests in the role for the nitrate–nitrite–NO system in regulation of cardiovascular function by traditional herbal medicine used commonly for cardiovascular disorders, including coronary artery disease (CAD) and heart failure.

Recently, certain herbal medicines or dietary supplements have been reported to generate significant amounts of nitrate, nitrite, and NO bioactivity, which may account for some of their effects on the heart and vessels [21]. These herbs are commercially available in regular dietary supplement stores. Analysis of nitrite, nitrate, NO-modified proteins, and their NO-generating capacity as well as effects on relaxation of isolated aortic strips have provided convincing evidence that abundant nitrate and nitrate reductase activity are present in certain herbs known to have protective or therapeutic benefits to patients with CAD (Table 19.1), such as the Danshen Root (radix salviae miltiorrhizae),

Table 19.1 The measurement of nitrite, nitroso, nitrite reductase activity, and the recommended daily dose in several herbs commonly used in Chinese traditional medicine

Latin name	English name	Indications	Nitrite (ng/g)	Nitrate (mg/g)	Nitroso (nmol/g)	NO production (pmol/mg)
Radix Salviae Miltiorrhizae	Danshen Root	CAD	330	12,000	120	7
Fructus Trichosanthis	Snakegourd Fruit	CAD, acute MI, hyperlipidemia	260	278	120	46
Bulbus Allii Macrostemi	Longstamen Onion Bulb	CAD, acute MI, hyperlipidemia	150	530	842	134
Radix Notoginseng	Sanchi	CAD	210	2069	73	13
Resina Olibani	Frankincense	Hypertension	980	61	3210	72
Radix Paeonia Rubra	Red Peony Root	CAD	120	37	450	255
Radix Ginseng	Ginseng	Heart failure, CAD	300	243	76	360
Borneolum Syntheticum	Borneol	Increase other herb's function for CAD or brain disease	120	2.99	6	45
Cinnamomum	Borneol	Brain disease	160	2.39	0	875

Sanchi (radix notoginseng), and Hongshen (radix ginseng). These herbs, with specific indications for cardiovascular disease, contain high concentrations of nitrite and nitrate, generate NO from nitrite, and relax blood vessels. The therapeutic benefits of some herbal medicines occur via their ability to generate NO from nitrite providing an alternative source of NO to patients that may be unable to make NO from L-arginine owing to endothelial dysfunction.

In the field of NO biology, the inorganic anions nitrite and nitrate are known as NO donors which act as essential natural regulators of cardiovascular function [5, 37, 38]. Both nitrite and nitrate have been tested in animals and humans, and they show a good effectiveness in attenuating inflammation, reversing endothelial dysfunction, and reducing damage due to ischemia–reperfusion injury [30, 39, 40]. There is an endogenous nitrate/nitrite reductase activity in animal tissues, such as the liver and aorta, but this inherent biological capacity is low (around 1 pmol/mg protein). The reductase activity in some of these herbal medicines may exceed that detected in the animal tissues [21]. It is estimated that the increased reductase activity may occur by orders of magnitude, almost 1000 times higher than endogenous production of NO. This would equate to 300 nmoles/day of NO from a single herbal preparation. The average NO production of NO in the human body (70 kg) is 1.68 mmol NO per day (based on an NO production rate of 1 μmol/kg/h). By supplying the exogenous nitrate/nitrite and reductase activities, herbal medicines offer an alternative therapeutic strategy to combat or treat heart disease, a condition frequently associated with a deficiency in NO. Maintaining NO homeostasis requires the repletion of nitrite and nitrate through which the ability to generate NO can be restored to compensate for the inability of the endothelium to convert L-arginine to NO in coronary heart disease.

Therapeutic Potential of Herbal Medicine with Active Nitrate–Nitrite–NO Pathway

In China, physicians who practice traditional Chinese medicine routinely take the pulse (Fig. 19.2) and examine the rates, rhythms, and strength of arterial pulse. They also inspect the tongue to determine any changes in shape, color, and smell. Based on the patient's symptoms and physical examination, physicians determine whether or not there is a functional abnormality of the cardiovascular system. NO-generating herbs are considered antianginal as they can cause vasodilation, which can help reduce ischemic pain known as angina by improving perfusion and decreasing the cardiac workload as a result of venodilation and a decrease in venous return to the heart per cycle. Nitroglycerin (an organic

診 (Diagnosis)

舌象
Inspect tongue

脉象
Take Pulse

疗 (Treatment)

活血化瘀
Improve blood flow
Overcome stasis

理气和中
Justify energy
Maintain balance

Fig. 19.2 An ancient Chinese physician practicing traditional medicine and performing diagnosis and treatment of a heart disorder, such as angina or chest pain. The primary physical examination was to inspect the tongue's shape and color, and measure the pulse rate and strength. The traditional therapy was, in principle, based on improving blood flow, thereby overcoming stasis, and justifying the balance in energy

nitrate), taken sublingually, is used to prevent or treat acute chest pain. The nitroglycerin reacts with a sulfhydryl group (–SH) to produce nitric oxide, which eases the pain by causing vasodilation. Inorganic nitrite or nitrate may be beneficial for the treatment of angina due to reduced myocardial oxygen consumption, both by decreasing preload and afterload and by some direct vasodilation of coronary vessels [38, 41].

Traditional herbal medicines used for thousands of years in Asia and other regions have been proven effective in certain cardiovascular disorders. Some of the herbal medicines have profound NO bioactivity primarily due to the nitrate–nitrite–NO reduction pathway. They contain very large amounts of nitrate/nitrite in the extracts given to patients [21]. The described benefits of these ancient medications may be attributed to their inherent nitrate/nitrite contents and nitrate/nitrite reductase activity to generate NO independent of the L-arginine–NO pathway [20, 21, 31, 38]. The first use of nitrate for the treatment of patients with symptoms of angina was described in an eighth-century Chinese manuscript uncovered at the Buddhist grotto of Dunhuang. Chinese physicians in traditional medicine have tested the therapeutic effects of Xiao Shi Xiong Huang San (the Nitrum and Realgar Powder), one of the Dunhuang prescriptions, on angina pectoris caused by coronary heart disease. The patients were instructed to take Xiao Shi Xiong Huang San, hold it under the tongue for a time, and then swallow the saliva. Compared to nitroglycerin, Xiao Shi Xiong Huang San showed much higher efficacy and improvement in a clinical trial of 61 patients [42]. The significance of the instructions is that under the tongue, even in a healthy mouth, nitrate-reducing bacteria convert some of the nitrate into nitrite. Therefore, if the patient follows the physician's instructions fully, he or she will be taking nitrite, known to be effective in alleviating pain resulting from angina.

Recent clinical studies have provided new evidence on Danshen, a commonly used herbal medicine in China, which may have a similar efficacy to the known NO donor nitroglycerin [43–45]. The extract of *Salviae Miltiorrhizae*, or Danshen in Chinese, contains large amounts of nitrate [21]. The demonstration of beneficial effects of this herb on ischemic diseases offers an alternative avenue for the management of angina pectoris, myocardial infarction, or stroke [46, 47]. Danshen-related Chinese herbal medicines have been widely used for treatment of coronary heart disease in the East, and a clinical trial is on the way in the United States. Dansen is a routine herbal medication for acute angina pectoris. In addition, it may be effective for dyslipidemia, blood hyperviscosity syndrome, peripheral angiopathy (superficial thrombophlebitis, venous thrombosis, allergic arteriolitis), diabetes mellitus, and cirrhosis, and is also used for altitude sickness. Experimental studies have shown that Danshen dilates coronary arteries, increases coronary blood flow, and scavenges free radicals in ischemic diseases, reducing cellular damage from ischemia and improved heart functions, remarkably similar to known effects of NO and nitrite. However, the nitrate/nitrite reductase activity in Danshen is relatively low. Often, Danshen is mixed with other herbal products, one of which is an extract of cinnamon or borneol. Borneol is consumed excessively in China and Southeast Asian countries, particularly in a combined formula for preventing cardiovascular disease. Borneol exerts a concentration-dependent inhibitory effect on arterio-venous shunting and venous thrombosis [48]. The antithrombotic activity of borneol contributes to its action in combined formula for preventing cardiovascular diseases. We have shown that although the natural form of borneol itself contains very little nitrite and nitrate, it displays a potent nitrite/nitrate reductase activity [21]. Another herbal medicine made from the root of *Radix ginseng* may have synergistic effects with Danshen. Ginseng contains modest amounts of nitrate but has stronger reductase activity.

Certain herbal medicines may also exert regulatory effects on expression and activation of endogenous NO-synthesizing enzymes, including eNOS. Recently, we studied whether pretreatment with Tongxinluo (TXL), a mixture of traditional Chinese medicines, can attenuate the no-reflow and ischemia–reperfusion injury in an infarct animal model [49]. TXL is composed of *Radix ginseng*, *Buthus martensi*, Hirudo, *Eupolyphaga seu steleophage*, *Scolopendra subspinipes*, *Periostracum cicadae*, *Radix paeoniae rubra*, *Semen ziziphi spinosae*, *Lignum dalbergiae odoriferae*, *Lignum santali albi*, and *Borneolum syntheticum*. Owing to its efficacy and minimal adverse effects, TXL has been widely used in China to treat patients with acute coronary syndrome. Pretreatment with TXL at low to high dose for 3 days or with a high loading dose 3 h before ischemia can reduce myocardial no-reflow and infarction in a swine ischemia model. However, it is not clinically practical to pretreat patients undergoing to acute postinfarct percutaneous coronary intervention with high doses of TXL several hours or days before the operation. Therefore, whether or not low loading dose of TXL just before acute percutaneous coronary intervention can also ameliorate the no-reflow phenomenon and ischemia–reperfusion injury need to be clarified. In a 90-min ischemia and 3-h reperfusion model, miniature pigs were randomly assigned to treatment with TXL (0.05 g/kg, gavaged 1 h prior to ischemia; TXL plus H-89 (a protein kinase-A (PKA) inhibitor), intravenously infused at a dose of 1.0 µg/kg/min 30 min before ischemia); or TXL plus N(omega)-nitro-L-arginine (L-NNA, an eNOS inhibitor, intravenously administered at a dose of 10 mg/kg 30 min prior to ischemia). The results show that TXL decreased creatine kinase (CK) elevation ($P<0.05$), reduced the no-reflow area from 48.6 to 9.5 %, and reduced infarct size from 78.5 to 59.2 % ($P<0.05$); these effects of TXL were partially abolished by H-89 and completely reversed by L-NNA. TXL elevated the PKA activity and the expression of PKA, Thr^{198} phosphorylated-PKA, Ser^{1179} phosphorylated-eNOS (p-eNOS), and Ser^{635} p-eNOS in the ischemic myocardium. H-89 repressed the TXL-induced enhancement of PKA activity and the phosphorylation of Ser^{635} p-eNOS, and L-NNA counteracted the phosphorylation of eNOS at Ser^{1179} and Ser^{635} without apparent influence on the PKA activity. Thus, pretreatment with a single low loading dose of TXL 1 h before ischemia reduces the myocardial no-reflow phenomenon and ischemia–reperfusion injury by upregulating the phosphorylation of eNOS at Ser^{1179} and Ser^{635}, and this effect is partially mediated by the PKA pathway.

Biosafety of Herbal Medicine Enriching Nitrate and Nitrite

Nitrate and nitrite have long been used for food processing industry for preservation. There are two major health concerns about the safety of food or food supplements that contain abundant inorganic nitrite and nitrate, formation of methemoglobin by nitrate ingestion and nitrosamines by nitrite and secondary amines. Previous reports suggest the risk of nitrite uptake for the development of methemoglobinemia and potential carcinogenic effects of *N*-nitrosamines. Compared to nitrite, nitrate is relatively safer, and, in fact, the potential toxicity of the nitrate ion is thought to occur after its bioconversion to nitrite, which is considerably more reactive. Hemoglobin (Hb) in red blood cells plays a key role in preventing the accumulation of nitrite in blood. The formation of methemoglobin is due to oxidation of the oxygen-carrying ferrous ion (Fe^{2+}) of the heme group in the hemoglobin molecule by nitrite to the ferric state (Fe^{3+}). Methemoglobin cannot bind oxygen effectively. In the clinic, significant methemoglobinemia with cyanosis occurs when the levels increase above approximately 5%. In animal studies, intravenous administration of nitrite at doses causing vasodilation increases methemoglobin to very minor levels. This suggests that methemoglobin is not the major obstacle for nitrite administration at these dose ranges. Furthermore, the presence of antioxidants and other organic compounds in herbal formulations may further prevent oxidation of hemoglobin by nitrite and nitrate also contained in the formulations.

As published in 2001 by the US Department of Health and Human Services National Toxicology Program [50] extensive studies in toxicology and carcinogenesis have shown no significant evidence for carcinogenic activity of nitrite, despite dose escalations sufficient to produce profound methemoglobinemia and weight loss in rodents. All the rats and mice were given sodium nitrite in the drinking water for 14-week and 2-year periods. The low dose and short duration of treatment with nitrate/nitrite-containing herbal medicines has a low risk of any carcinogenic effects. Previous reports suggest that the metabolism of dietary nitrate can result in intragastric formation of nitrosamines, a compound that may be carcinogenic. However, after more than 40 years of extensive research, there is still no convincing evidence for a link between nitrate intake and gastric cancer in humans. Furthermore, many herbal medicines are enriched in antioxidants and polyphenols that have been shown to be potent inhibitors of nitrosation reactions. The context of nitrite and nitrate delivery along with systems for reduction and inhibition of nitrosation may account for the benefit of herbal medicines. Moreover, it is well known that a diet rich in vegetables is associated with a lower blood pressure and a reduced long-term risk for the development of cardiovascular disease. Recently, Larsen and colleagues [51] performed a double-blind placebo-controlled crossover evaluation of dietary nitrate supplementation in healthy young volunteers and found a significant reduction in resting blood pressure with a nitrate dose corresponding to the amount found in 150–250 g of green leafy vegetables. Remarkably, the reduction in blood pressure was similar to that described in healthy controls consuming a diet rich in fruits and vegetables in the Dietary Approaches to Stop Hypertension (DASH) trial [8, 52]. Nevertheless, further studies on safety with long-term, high-dose administration of nitrite/nitrate-rich herbal medicines are required.

Dietary nitrate found in many Chinese traditional medicines with cardiovascular benefits and in some green leafy vegetables (e.g., rocket and in beetroot) may serve as an external source of NO, via the nitrate–nitrite–NO pathway (Fig. 19.3). The herbal medicine-associated biological activities of the nitrate–nitrite–nitric oxide (NO) pathway represent a biological process in which a series of oxygen-independent and NO synthase–independent single-electron transfer reactions that ultimately facilitate vasodilation. Nitrate-rich herb and vegetable therapeutics or dietary supplements help reverse pathological features of heart failure with preserved ejection fraction (HFpEF). Recent findings from Zamani et al. [53] have indicated that the nitrate improves exercise capacity in HFpEF by reducing systemic vascular resistance (SVR) and left ventricular (LV) pulsatile afterload and enhancing cardiac output and oxygen delivery to exercising muscles. They enrolled 17 subjects with HFpEF in a double-blind, placebo-controlled, crossover trial of a single dose of 12.9 mmol nitrate in 140 mL beet root juice or nitrate-depleted placebo beet root juice. The nitrate/NO therapeutics may improve diastolic left ventricular function, exercise tolerance, vasodilatory reserve during exercise, and skeletal muscle perfusion and function. However, there was no observed significant change in exercise efficiency, the primary end point of the trial. Nitrite administration can achieve a similar effect as nitrate. Omar et al. [54] investigated the pharmacological properties of nitrite in the human circulation, showing that nitrite vasodilates not only the arteriolar and venous

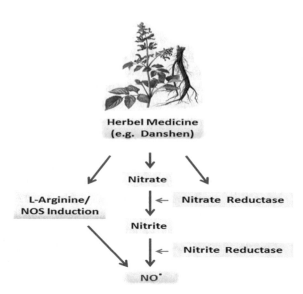

Fig. 19.3 Potential pathways leading to NO production in herbal medicine and dietary supplements. Traditional herbal medicines used for cardiovascular disorders, such as Danshen (*Radix salviae miltiorrhizae*) and certain dietary supplement, such as *Rocket/beetroot* vegetables, contain large quantities of nitrate and nitrite, and some also express high levels of nitrate reductase activities. They may also enhance eNOS expression and activation

circulation, as previously described, but also the conduit blood vessels, an effect similar to that observed with nitroglycerin. This effect was associated with a reduction in central systolic blood pressure, augmentation index, and pulsed-wave velocity, which represent hemodynamic effects that might show therapeutic promise in the setting of HFpEF. The mechanism behind the biological benefits of dietary nitrate and nitrite is not completely understood. The iron-containing enzymes of the electron transport chain, nitrate/nitrite reductases, and oxyhemoglobins are believed to involve in the effects of the nitrate/nitrite/NO donors.

Summary

Taken together, the nitrate–nitrite–NO pathway plays an important role in the therapeutic effect of many herbal medicines used for cardiovascular disease. The NO donor-like herbs may serve as an alternative source of nitrate and nitrite or other NO donors, which target various iron-containing enzymes or proteins to assert their biological effects. They contain high activities of nitrate–nitrite reductase that converts the inorganic anions into NO, which, in turn, relaxes blood vessels and prevents thrombosis. Herbs may contain regulatory factors that influence the expression and activity of NOS, and conduct a "Ying and Yang" balance in cardiovascular physiology and pathophysiology in a NOS-dependent or independent fashion (Fig. 19.4). The NOS-regulatory herbs may interact with endogenous factors that modulate NOS function. Currently, there is no clear evidence showing that the nitrate/nitrite-rich herbal medicines have any major adverse side effects, but further investigation is warranted in terms of long-term, high-dose administration of these herbal medicines.

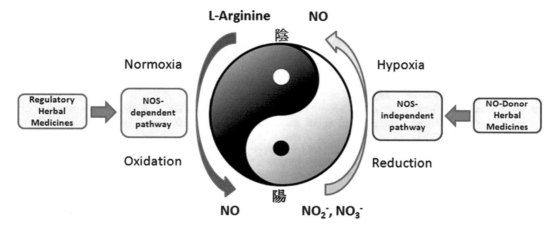

Fig. 19.4 The "Yin-Yang" (陰-陽) balance of NO production and function in association with different types of herbal medicines. As a signaling molecule well established in biology, NO plays a key role in the regulation of the function of the cardiovascular system by herbal medicines. Increasing evidence indicates that herbal medicines can help maintain the "Yin-Yang" balance through regulating the two mechanisms of NO production: oxidation of L-arginine and reduction of nitrate/nitrite

References

1. Davidson P, Hancock K, Leung D, Ang E, Chang E, Thompson DR, et al. Traditional Chinese Medicine and heart disease: what does Western medicine and nursing science know about it? Eur J Cardiovasc Nurs. 2003;2(3):171–81.
2. Gillis CN. Panax ginseng pharmacology: a nitric oxide link? Biochem Pharmacol. 1997;54(1):1–8.
3. Chen X. Cardiovascular protection by ginsenosides and their nitric oxide releasing action. Clin Exp Pharmacol Physiol. 1996;23(8):728–32.
4. Lundberg JO, Weitzberg E. NO generation from inorganic nitrate and nitrite: role in physiology, nutrition and therapeutics. Arch Pharm Res. 2009;32(8):1119–26.
5. Lundberg JO, Weitzberg E. NO-synthase independent NO generation in mammals. Biochem Biophys Res Commun. 2010;396(1):39–45.
6. Lundberg JO, Weitzberg E, Gladwin MT. The nitrate-nitrite-nitric oxide pathway in physiology and therapeutics. Nat Rev Drug Discov. 2008;7(2):156–67.
7. Garg HK, Bryan NS. Dietary sources of nitrite as a modulator of ischemia/reperfusion injury. Kidney Int. 2009;75(11):1140–4.
8. Hord NG, Tang Y, Bryan NS. Food sources of nitrates and nitrites: the physiologic context for potential health benefits. Am J Clin Nutr. 2009;90(1):1–10.
9. Koshland Jr DE. The molecule of the year. Science. 1992;258(5090):1861.
10. Griffith TM, Edwards DH, Davies RL, Harrison TJ, Evans KT. EDRF coordinates the behaviour of vascular resistance vessels. Nature. 1987;329(6138):442–5.
11. Erzurum SC, Ghosh S, Janocha AJ, Xu W, Bauer S, Bryan NS, et al. Higher blood flow and circulating NO products offset high-altitude hypoxia among Tibetans. Proc Natl Acad Sci U S A. 2007;104(45):17593–8.
12. Lundberg JO. Cardiovascular prevention by dietary nitrate and nitrite. Am J Physiol Heart Circ Physiol. 2009;296(5):H1221–3.
13. Lundberg JO, Weitzberg E. Nitrite reduction to nitric oxide in the vasculature. Am J Physiol Heart Circ Physiol. 2008;295(2):H477–8.
14. Cueto M, Hernandez-Perera O, Martin R, Bentura ML, Rodrigo J, Lamas S, et al. Presence of nitric oxide synthase activity in roots and nodules of Lupinus albus. FEBS Lett. 1996;398(2–3):159–64.
15. Ribeiro Jr EA, Cunha FQ, Tamashiro WM, Martins IS. Growth phase-dependent subcellular localization of nitric oxide synthase in maize cells. FEBS Lett. 1999;445(2–3):283–6.
16. Walker FA, Ribeiro JM, Montfort WR. Novel nitric oxide-liberating heme proteins from the saliva of bloodsucking insects. Met Ions Biol Syst. 1999;36:621–63.

17. Ninnemann H, Maier J. Indications for the occurrence of nitric oxide synthases in fungi and plants and the involvement in photoconidiation of Neurospora crassa. Photochem Photobiol. 1996;64(2):393–8.

18. Morot-Gaudry-Talarmain Y, Rockel P, Moureaux T, Quillere I, Leydecker MT, Kaiser WM, et al. Nitrite accumulation and nitric oxide emission in relation to cellular signaling in nitrite reductase antisense tobacco. Planta. 2002;215(5):708–15.

19. Rockel P, Strube F, Rockel A, Wildt J, Kaiser WM. Regulation of nitric oxide (NO) production by plant nitrate reductase in vivo and in vitro. J Exp Bot. 2002;53(366):103–10.

20. Feelisch M, Fernandez BO, Bryan NS, Garcia-Saura MF, Bauer S, Whitlock DR, et al. Tissue processing of nitrite in hypoxia: an intricate interplay of nitric oxide-generating and -scavenging systems. J Biol Chem. 2008;283(49):33927–34.

21. Tang Y, Garg H, Geng YJ, Bryan NS. Nitric oxide bioactivity of traditional Chinese medicines used for cardiovascular indications. Free Radic Biol Med. 2009;47(6):835–40.

22. Yamada K, Suzuki E, Nakaki T, Watanabe S, Kanba S. Aconiti tuber increases plasma nitrite and nitrate levels in humans. J Ethnopharmacol. 2005;96(1–2):165–9.

23. Geng YJ, Petersson AS, Wennmalm A, Hansson GK. Cytokine-induced expression of nitric oxide synthase results in nitrosylation of heme and nonheme iron proteins in vascular smooth muscle cells. Exp Cell Res. 1994;214(1):418–28.

24. Stamler JS, Jaraki O, Osborne J, Simon DI, Keaney J, Vita J, et al. Nitric oxide circulates in mammalian plasma primarily as an S-nitroso adduct of serum albumin. Proc Natl Acad Sci U S A. 1992;89(16):7674–7.

25. Stamler JS, Simon DI, Jaraki O, Osborne JA, Francis S, Mullins M, et al. S-nitrosylation of tissue-type plasminogen activator confers vasodilatory and antiplatelet properties on the enzyme. Proc Natl Acad Sci U S A. 1992;89(17):8087–91.

26. Stamler JS, Simon DI, Osborne JA, Mullins ME, Jaraki O, Michel T, et al. S-nitrosylation of proteins with nitric oxide: synthesis and characterization of biologically active compounds. Proc Natl Acad Sci U S A. 1992;89(1):444–8.

27. Janero DR, Bryan NS, Saijo F, Dhawan V, Schwalb DJ, Warren MC, et al. Differential nitros(yl)ation of blood and tissue constituents during glyceryl trinitrate biotransformation in vivo. Proc Natl Acad Sci U S A. 2004;101(48):16958–63.

28. van Faassen EE, Bahrami S, Feelisch M, Hogg N, Kelm M, Kim-Shapiro DB, et al. Nitrite as regulator of hypoxic signaling in mammalian physiology. Med Res Rev. 2009;29(5):683–741.

29. Bryan NS. Nitrite in nitric oxide biology: cause or consequence? A systems-based review. Free Radic Biol Med. 2006;41(5):691–701.

30. Bryan NS, Calvert JW, Elrod JW, Gundewar S, Ji SY, Lefer DJ. Dietary nitrite supplementation protects against myocardial ischemia-reperfusion injury. Proc Natl Acad Sci U S A. 2007;104(48):19144–9.

31. Bryan NS, Calvert JW, Gundewar S, Lefer DJ. Dietary nitrite restores NO homeostasis and is cardioprotective in endothelial nitric oxide synthase-deficient mice. Free Radic Biol Med. 2008;45(4):468–74.

32. Bryan NS, Grisham MB. Methods to detect nitric oxide and its metabolites in biological samples. Free Radic Biol Med. 2007;43(5):645–57.

33. Webb AJ, Milsom AB, Rathod KS, Chu WL, Qureshi S, Lovell MJ, et al. Mechanisms underlying erythrocyte and endothelial nitrite reduction to nitric oxide in hypoxia: role for xanthine oxidoreductase and endothelial nitric oxide synthase. Circ Res. 2008;103(9):957–64.

34. Webb AJ, Patel N, Loukogeorgakis S, Okorie M, Aboud Z, Misra S, et al. Acute blood pressure lowering, vasoprotective, and antiplatelet properties of dietary nitrate via bioconversion to nitrite. Hypertension. 2008;51(3):784–90.

35. Kapil V, Milsom AB, Okorie M, Maleki-Toyserkani S, Akram F, Rehman F, et al. Inorganic nitrate supplementation lowers blood pressure in humans: role for nitrite-derived NO. Hypertension. 2010;56(2):274–81.

36. Kapil V, Webb AJ, Ahluwalia A. Inorganic nitrate and the cardiovascular system. Heart. 2010;96(21):1703–9.

37. Lundberg JO. Nitric oxide metabolites and cardiovascular disease markers, mediators, or both? J Am Coll Cardiol. 2006;47(3):580–1.

38. Lundberg JO, Govoni M. Inorganic nitrate is a possible source for systemic generation of nitric oxide. Free Radic Biol Med. 2004;37(3):395–400.

39. Stokes KY, Dugas TR, Tang Y, Garg H, Guidry E, Bryan NS. Dietary nitrite prevents hypercholesterolemic microvascular inflammation and reverses endothelial dysfunction. Am J Physiol Heart Circ Physiol. 2009;296(5):H1281–8.

40. Webb A, Bond R, McLean P, Uppal R, Benjamin N, Ahluwalia A. Reduction of nitrite to nitric oxide during ischemia protects against myocardial ischemia-reperfusion damage. Proc Natl Acad Sci U S A. 2004;101:13683–8.

41. Milkowski A, Garg HK, Coughlin JR, Bryan NS. Nutritional epidemiology in the context of nitric oxide biology: a risk-benefit evaluation for dietary nitrite and nitrate. Nitric Oxide. 2010;22(2):110–9.

42. Liu X, Ma H, Li C, Cui Q, Guo Y, Wang L, et al. Sixty-one cases of angina pectoris due to coronary heart disease treated by external use of the paste of nitrum and realgar powder on zhiyang (GV 9). J Tradit Chin Med. 2002;22(4):243–6.

43. Lam FF, Yeung JH, Chan KM, Or PM. Mechanisms of the dilator action of cryptotanshinone on rat coronary artery. Eur J Pharmacol. 2008;578(2–3):253–60.
44. Lam FY, Ng SC, Cheung JH, Yeung JH. Mechanisms of the vasorelaxant effect of Danshen (Salvia miltiorrhiza) in rat knee joints. J Ethnopharmacol. 2006;104(3):336–44.
45. Zhou L, Zuo Z, Chow MS. Danshen: an overview of its chemistry, pharmacology, pharmacokinetics, and clinical use. J Clin Pharmacol. 2005;45(12):1345–59.
46. Li MH, Chen JM, Peng Y, Wu Q, Xiao PG. Investigation of Danshen and related medicinal plants in China. J Ethnopharmacol. 2008;120(3):419–26.
47. Li W, Li ZW, Han JP, Li XX, Gao J, Liu CX. Determination and pharmacokinetics of danshensu in rat plasma after oral administration of danshen extract using liquid chromatography/tandem mass spectrometry. Eur J Drug Metab Pharmacokinet. 2008;33(1):9–16.
48. Li YH, Sun XP, Zhang YQ, Wang NS. The antithrombotic effect of borneol related to its anticoagulant property. Am J Chin Med. 2008;36(4):719–27.
49. Li XD, Yang YJ, Geng YJ, Jin C, Hu FH, Zhao JL, et al. Tongxinluo reduces myocardial no-reflow and ischemia-reperfusion injury by stimulating the phosphorylation of eNOS via the PKA pathway. Am J Physiol Heart Circ Physiol. 2010;299(4):H1255–61.
50. Olden K, Guthrie J. Genomics: implications for toxicology. Mutat Res. 2001;473(1):3–10.
51. Larsen FJ, Weitzberg E, Lundberg JO, Ekblom B. Dietary nitrate reduces maximal oxygen consumption while maintaining work performance in maximal exercise. Free Radic Biol Med. 2010;48(2):342–7.
52. Tsuchiya K, Kanematsu Y, Yoshizumi M, Ohnishi H, Kirima K, Izawa Y, et al. Nitrite is an alternative source of NO in vivo. Am J Physiol Heart Circ Physiol. 2005;288(5):H2163–70.
53. Zamani P, Rawat D, Shiva-Kumar P, Geraci S, Bhuva R, Konda P, et al. Effect of inorganic nitrate on exercise capacity in heart failure with preserved ejection fraction. Circulation. 2015;131(4):371–80, discussion 380.
54. Omar SA, Artime E, Webb AJ. A comparison of organic and inorganic nitrates/nitrites. Nitric Oxide. 2012;26(4):229–40.

Chapter 20
Nitrate and Exercise Performance

Stephen J. Bailey, Anni Vanhatalo, and Andrew M. Jones

Key Points

- Inorganic nitrate (NO_3^-) can be reduced to nitrite (NO_2^-) and subsequently nitric oxide (NO).
- Dietary nitrate NO_3^- supplementation increases circulating plasma [NO_2^-].
- The reduction of NO_2^- to NO is enhanced in hypoxic and acidic conditions.
- Skeletal muscles become relatively acidic and hypoxic during exercise.
- Acute and chronic NO_3^- supplementation can improve exercise economy and endurance performance in moderately trained subjects, but these effects are less likely in well-trained athletes.
- NO_3^- supplementation appears particularly effective at improving physiological responses in type II (fast-twitch) muscle.
- NO_3^- supplementation has been shown to improve performance in short-duration high-intensity exercise.
- NO_3^- supplementation can improve exercise performance in some patient populations.
- NO_3^- supplementation has significant potential as a nutritional aid to enhance exercise performance in a variety of exercise settings and subject groups.
- Enhanced performance after NO_3^- supplementation might be linked to improvements in skeletal muscle perfusion, metabolism and/or contractile function.

Keywords Beetroot juice • Nitrite • Nitric oxide • Skeletal muscle • Fatigue • Efficiency • Ergogenic aid • Sports performance

Introduction

The inorganic anion, nitrate (NO_3^-), has conventionally been considered an environmental pollutant and carcinogenic agent [1, 2]. Consequently, the NO_3^- content of drinking water is regulated to 50 mg L^{-1} in the European Union, and the World Health Organisation and Food and Agricultural Organisation have set an acceptable daily intake (ADI) of 3.7 mg $kgBM^{-1}$ for NO_3^- [3], in an attempt to restrict dietary NO_3^- intake. The major source of dietary NO_3^- is the consumption of certain vegetable varieties including lettuces, radishes, spinach and beetroot [2–4]. Several countries advocate

S.J. Bailey, Ph.D. • A. Vanhatalo, Ph.D. • A.M. Jones, Ph.D. (✉)
Department of Sport and Health Sciences, College of Life and Environmental Sciences, University of Exeter, St. Luke's Campus, Heavitree Road, Exeter EX1 2LU, UK
e-mail: a.m.jones@exeter.ac.uk

N.S. Bryan, J. Loscalzo (eds.), *Nitrite and Nitrate in Human Health and Disease*, Nutrition and Health,
DOI 10.1007/978-3-319-46189-2_20, © Springer International Publishing AG 2017

increased ingestion of fruit and vegetables, through dietary initiatives such as the Dietary Approaches to Stop Hypertension (DASH) diet [5], the 5-a-day diet [6] or its variants [7], given the proven cardioprotective effects of such diets [8, 9]. Therefore, adhering to dietary initiatives to increase fruit and vegetable consumption has the potential to exceed the ADI for NO_3^-. However, in contrast to the traditional interpretation that NO_3^- is a noxious chemical which is detrimental to human health, recent evidence suggests that consumption of high NO_3^- vegetables might confer cardioprotection [10–12], and that increased dietary NO_3^- intake can have a beneficial effect on aspects of human health [1, 13, 14]. This has resulted in an exponential increase in research exploring the potential for positive effects of dietary NO_3^- supplementation on human health and function.

After oral ingestion, ~25 % of NO_3^- passes into the entero-salivary circulation and becomes concentrated in the saliva [15–17]. Subsequently, certain bacteria within the oral cavity catalyse the reduction of NO_3^- to nitrite (NO_2^-; [18–21]). Upon arrival in the stomach, the acidic milieu facilitates the reduction of ingested NO_2^- to nitric oxide (NO) and other reactive nitrogen intermediates [22, 23]. However, the elevated circulating concentrations of NO_2^- and associated reactive nitrogen intermediates [16, 17] indicate that dietary NO_3^- supplementation can also increase the systemic bioavailability of these molecules. Thereafter, circulating NO_2^- can undergo an oxygen-independent, one-electron reduction to NO in a reaction catalysed by numerous NO_2^- reductases including deoxyhaemoglobin, deoxymoglobin, xanthine oxidase and the mitochondrial respiratory complexes [24]. The discovery of this so-called NO_3^-–NO_2^-–NO pathway was important as it challenged the traditional interpretation that NO_3^- and NO_2^- were physiologically inert and revealed a potential alternative source of NO generation to complement the better known, oxygen-dependent L-arginine–NO pathway [25–28].

It is well documented that the intra-myocyte and muscle microvascular oxygen tension and pH decline in the contracting skeletal muscles [29–31]. An important feature of NO synthesis from NO_2^- reduction is that this reaction is potentiated in hypoxic and acidic conditions [32, 33]. Therefore, the physiological milieu within and surrounding the contracting myocytes would be hypothesized to facilitate increased NO_2^- reduction to NO, a postulate that is supported by the negative arterial-venous plasma [NO_2^-] difference across the contracting skeletal muscle bed [34]. Since NO has been shown to influence a wide array of physiological processes pertinent to skeletal muscle function, including perfusion, metabolism and contractility [35, 36], increasing the circulating plasma [NO_2^-] through dietary NO_3^- supplementation may increase O_2-independent skeletal muscle NO generation and positively impact skeletal muscle physiology and performance during exercise. It is also important to emphasize that, since NO_2^- can influence physiological processes independent of its reduction to NO [37], dietary NO_3^- supplementation has the potential to improve exercise physiology and performance through both NO dependent and independent mechanisms. The purpose of this chapter is to provide a critical, evidenced-based review of the recent surge of publications exploring the effects of dietary NO_3^- supplementation on exercise physiology and performance. This literature review will be used to highlight the exercise settings and human populations in which NO_3^- supplementation is more, or less, likely to be performance enhancing (ergogenic) and to inform recommendations for best practice in the use of dietary NO_3^- supplementation to improve performance in the exercising human.

Effects of NO_3^- Supplementation on Exercise Economy and Endurance Performance

Moderately Trained Participants

Interest in the ergogenic potential of dietary NO_3^- supplementation emanated from the pioneering work by Larsen et al. [38]. In this study, Larsen et al. [38] supplemented the diet of moderately trained [peak oxygen uptake ($\dot{V}O_{2peak}$) of 55 mL kg^{-1} min^{-1}] adult males with 0.1 mmol sodium nitrate

$(NaNO_3) \cdot kgBM^{-1}$ day^{-1} (6.2 mg $NaNO_3$ kgBM^{-1} day^{-1}) for 3 days and measured pulmonary oxygen uptake ($\dot{V}O_2$) during a multi-stage step incremental cycling test that comprised five sub-maximal stages at work rates corresponding to 45, 60, 70, 80 and 85% of $\dot{V}O_{2peak}$. Compared to a control condition of sodium chloride, $NaNO_3$ increased plasma [NO_3^-] and [NO_2^-], lowered $\dot{V}O_2$ during the first four sub-maximal stages by ~5%, but did not alter $\dot{V}O_2$ when cycling at 85% $\dot{V}O_{2peak}$ or during a subsequent cycling bout at 100% $\dot{V}O_{2peak}$. The improved economy (lower oxygen cost of exercising at a given work rate) during sub-maximal cycling exercise after NO_3^- supplementation was striking as it suggested that a short-term dietary intervention had the potential to favourably impact a highly stable parameter of human exercise physiology [39] and an important determinant of endurance exercise performance [40]. Consequently, this seminal work prompted a number of other research groups to investigate the effects of NO_3^- supplementation on exercise economy and endurance exercise performance. In the first investigation to follow-up the exciting observations of Larsen et al. [38], Bailey et al. [41] supplemented the diet of recreationally active ($\dot{V}O_{2peak}$ of 49 mL kg^{-1} min^{-1}) adult males with 500 mL of beetroot juice per day, which provided a daily NO_3^- dose of 5.6 mmol (347 mg), for 6 days. Consistent with Larsen et al. [38], plasma [NO_2^-] was increased and $\dot{V}O_2$ was ~5% lower after NO_3^- supplementation when cycling at a sub-maximal work rate equivalent to 80% of the gas exchange threshold [41]. In addition, Bailey et al. [41] were the first to report improved tolerance to sub-maximal cycling exercise after NO_3^- supplementation, as evidenced by the 16% improvement in time to exhaustion when cycling at a severe-intensity work rate (70% of the difference between the power output at the gas exchange threshold and $\dot{V}O_{2peak}$ + the power output at the gas exchange threshold). Taken together, these preliminary observations suggested that short-term (3–6 days) supplementation with NO_3^-, as either a NO_3^--rich food or a NO_3^- salt, has the potential to increase the circulating pool of reactive nitrogen intermediates, and to improve exercise economy and tolerance, at least in moderately trained participants. In doing so, these studies laid the foundations for a new line of scientific enquiry to address the potential for NO_3^- supplementation to function as an ergogenic aid in a variety of exercise settings and populations.

Subsequent investigations explored the potential for acute and longer term NO_3^- ingestion to impact exercise economy and exercise tolerance [42–46]. These studies revealed that $\dot{V}O_2$ was lowered (by ~3–5%) during sub-maximal cycle ergometry after the acute ingestion of 0.033 mol kgBM^{-1} (2.05 mg kgBM^{-1}, [43]), 5.2 mmol (322 mg, [44]), 6 mmol (372 mg, [46]) and 16.8 mmol NO_3^- (1042 mg, [45]), but not following the acute ingestion of 3 mmol (186 mg, [46]), 4.2 mmol (260 mg, [45]) and ~8 mmol NO_3^- (496 mg, [42]), with a trend (~1.7%, $P = 0.06$) for a lower $\dot{V}O_2$ observed after the acute ingestion of 8.4 mmol NO_3^- (521 mg, [45]). With regard to exercise tolerance, it was reported that exercise tolerance was not significantly enhanced following the acute ingestion of 4.2 mmol (260 mg, [45]) and 5.2 mmol NO_3^- (322 mg, [44]), but that the acute ingestion of 8.4 (521 mg) and 16.8 (1042 mg) mmol NO_3^- was similarly effective (+14% and 12%, respectively) at improving exercise tolerance (Fig. 20.1) [45]. Therefore, in aggregate, these investigations suggest that the acute ingestion of 5–6 mmol (310–372 mg) NO_3^-, but not <5 mmol (310 mg) NO_3^-, can improve exercise economy during sub-maximal cycling; however, a dose of at least 8.4 mmol (521 mg) NO_3^- is recommended to elicit improved exercise economy and exercise tolerance. Moreover, and in contrast to some of the physiological effects observed after organic nitrate administration [47], the improvement in cycling economy reported after acute or short-term NO_3^- supplementation is maintained for at least 15 days whilst supplementing the diet with 5.2 mmol (322 mg) $NO_3^- \cdot$day^{-1} [44] and at least 28 days whilst supplementing the diet with 6 mmol (372 mg) $NO_3^- \cdot$day^{-1} [46]. On the other hand, exercise tolerance was enhanced after 15 days supplementation with 5.2 (322 mg) mmol $NO_3^- \cdot$day^{-1}, but not after 5 days supplementation with, or the acute ingestion of, 5.2 (322 mg) mmol $NO_3^- \cdot$day^{-1} [44]. Taken together, the existing literature suggests that exercise economy and tolerance are most likely to be improved after the acute ingestion of >8 mmol (496 mg) NO_3^- or chronic supplementation with >5 mmol (310 mg) $NO_3^- \cdot$day^{-1} for 6–15 days, at least in recreationally active individuals and moderately trained endurance athletes. However, whilst the majority of studies indicate improved exercise economy or performance after adopting the aforementioned

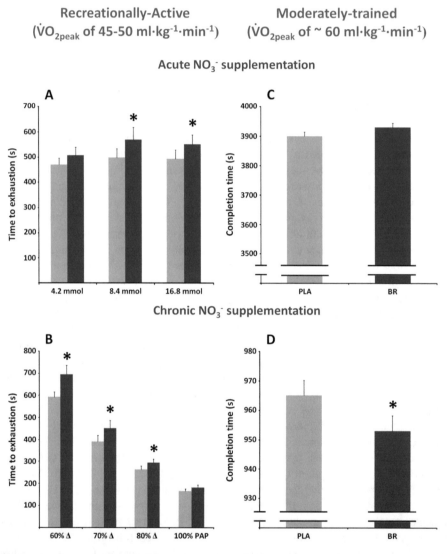

Fig. 20.1 Influence of acute and chronic dietary nitrate NO_3^- supplementation on endurance exercise performance in recreationally active and moderately trained subjects. (**a**) Illustrates the effects of different acute doses of NO_3^--rich beetroot juice (BR, *purple bars*), compared to NO_3^--depleted beetroot juice (PLA, *green bars*), on severe-intensity exercise tolerance in recreationally active subjects. Acute supplementation with 8.4 and 16.8 mmol NO_3^-, but not 4.2 mmol NO_3^-, improved exercise tolerance (redrawn from Wylie et al. 2013). (**b**) Illustrates the effects of longer-term NO_3^- supplementation (3–6 days with 8.2 mmol $NO_3^- \cdot day^{-1}$) on exercise tolerance across a series of severe-intensity work rates, where *open triangle* represents the difference between the power output at the gas exchange threshold and the peak oxygen uptake ($\dot{V}O_{2peak}$) + the power output at the gas exchange threshold, and PAP represents the peak aerobic power (the power output at $\dot{V}O_{2peak}$). Longer-term NO_3^- supplementation improved exercise tolerance at 60%Δ, 70%Δ and 80%Δ, but not 100% PAP, in recreationally active subjects (redrawn from Kelly et al. 2013). (**c**) Illustrates the effects of acute NO_3^- supplementation (8.7 mmol NO_3^-) on time trial performance in moderately trained subjects (redrawn from Cermak et al. 2012b) and (**d**) Illustrates the effects of 6 days supplementation with ~8.2 mmol $NO_3^- \cdot day^{-1}$ on time trial performance in moderately trained subjects (redrawn from Cermak et al. 2012a). Note that chronic, but not acute, NO_3^- supplementation improved endurance performance in moderately trained subjects. *Asterisk* indicates different from PLA

NO_3^- supplementation procedures, it should be acknowledged that there are examples where conforming to these supplementation procedures does not improve exercise economy or performance in recreationally active and moderately trained endurance athletes, in spite of a significant elevation in plasma $[NO_2^-]$ [48, 49]. Further research is required to resolve the mechanisms underlying responders and non-responders to NO_3^- supplementation.

In addition to NO_3^- supplementation procedures, an important consideration for researchers or practitioners wishing to utilize dietary NO_3^- supplementation as a potential ergogenic aid is to enforce abstinence from chlorohexidine-containing mouthwashes [16]. This is important because application of chlorohexidine-containing mouthwash has been shown to transiently eradicate the oral microflora that catalyse NO_3^- reduction to NO_2^- [16], a rate-limiting step in the NO_3^-–NO_2^-–NO pathway in humans [50], and therefore the rise in the circulating plasma $[NO_2^-]$ and associated physiological responses after NO_3^- ingestion [16, 51, 52]. Therefore, administering the appropriate dose and duration of NO_3^- supplementation, and avoiding chlorohexidine-containing mouthwashes, is important in order to optimize the ergogenic potential of dietary NO_3^- supplementation.

The enhanced exercise economy and tolerance after NO_3^- supplementation are not unique to cycling exercise as improvements in these variables have also been observed during knee extension [53], walking [54], desert marching [55], running [54, 56, 57], kayaking [58, 59] and rowing [60, 61] exercise after NO_3^- supplementation. Research has also been conducted to assess whether the improved exercise tolerance (i.e. time to exhaustion in a constant work rate task) reported after NO_3^- supplementation could effectively translate into improved exercise performance (i.e. a shorter time taken to complete a set distance or amount of work). In this regard, whilst the effects of acute NO_3^- ingestion on cycling performance are controversial [62, 63], chronic (6 days) NO_3^- supplementation can improve cycling performance [64] in moderately trained endurance athletes ($\dot{V}O_{2peak}$ of ≤ 60 mL kg^{-1} min^{-1}) (Fig. 20.1). Based on the existing evidence, and consistent with previous findings assessing the effects of NO_3^- supplementation on exercise tolerance [44], it appears that chronic NO_3^- supplementation holds greater potential as an ergogenic aid than acute NO_3^- ingestion.

Although dietary supplementation with NO_3^--rich beetroot juice and $NaNO_3$ has been shown to improve exercise economy, and while NO_3^--depleted beetroot juice supplementation does not improve exercise economy or performance compared to a control condition with no NO_3^- supplementation [54], it has recently been suggested that NO_3^--rich beetroot juice supplementation may be more effective at improving exercise economy than an equimolar dose of $NaNO_3$ [65]. In addition to NO_3^-, beetroot juice is rich in polyphenols and other antioxidants [66, 67], which have been shown to aid the reduction of NO_2^- to NO in the stomach [14], and might increase the circulating pool of reactive nitrogen intermediates compared to an equivalent dose of $NaNO_3$. Further research is required to determine underlying mechanisms by which NO_3^--rich beetroot juice supplementation might represent a more effective intervention to improve exercise economy than $NaNO_3$ supplementation and whether the ergogenic potential of beetroot juice is greater than that of $NaNO_3$.

Mechanisms for Improved Exercise Economy and Performance After NO_3^- Supplementation

While the preliminary studies indicated that NO_3^- supplementation, particularly chronic NO_3^- supplementation, had the potential to improve exercise economy and performance, the underlying mechanisms for these effects were obscure. Bailey et al. [53] were the first to investigate the mechanistic bases for improved exercise economy and performance after NO_3^- supplementation and used calibrated ^{31}Phosphorous magnetic resonance spectroscopy to non-invasively probe metabolic responses in contracting human skeletal muscles in vivo [53]. In accord with other studies [38, 41, 43–46, 54, 56–59], pulmonary $\dot{V}O_2$ was lower during sub-maximal two-legged knee-extensor exercise after 6 days of NO_3^- supplementation and this improved exercise economy was accompanied by a lower total ATP turnover rate through reductions in the ATP contribution from phosphocreatine (PCr) degradation and oxidative phosphorylation [53]. Therefore, these findings suggested that pulmonary $\dot{V}O_2$ was lower during exercise as a result of a lower ATP demand (i.e. energy cost) of skeletal muscle contraction and was not a function of increased anaerobic ATP turnover. The lower ATP turnover rate during exercise was reflected by lower muscle PCr utilization and lower adenosine diphosphate (ADP) and inorganic phosphate (Pi) accumulation, resulting in a physiological milieu that would be expected to attenuate the activation of oxidative phosphorylation and thus $\dot{V}O_2$ [68, 69]. In addition, PCr degradation and ADP and Pi accumulation are important contributors to skeletal muscle fatigue [70]. Therefore, the blunted perturbation of these fatigue-related muscle substrates and metabolites would be expected to contribute to the improved skeletal muscle fatigue resistance reported after dietary NO_3^- supplementation.

Although the study by Bailey et al. [53] suggested that the improved exercise economy after NO_3^- supplementation was linked to a lower ATP turnover rate, the efficiency of mitochondrial oxidative phosphorylation was not investigated in this study. Therefore, it was not possible to exclude changes in mitochondrial efficiency as a contributory factor for the improved exercise economy after NO_3^- supplementation. This possibility was explored by Larsen et al. [71] who harvested mitochondria from the *vastus lateralis* muscle of humans who had undergone 3 days of supplementation with $NaNO_3$ and measured the mitochondrial ADP to oxygen consumption ratio (P/O ratio), which indicates the oxygen cost of mitochondrial ATP resynthesis, in vitro. Larsen et al. [71] observed an increase in the mitochondrial P/O ratio (a lower oxygen consumption for a given ATP resynthesis), which was correlated to a lower pulmonary $\dot{V}O_2$ in vivo and accompanied by a lower leak respiration, after $NaNO_3$ supplementation. The lower leak respiration was attributed to increased adenine nucleotide translocate (ANT) expression and a trend for a lower uncoupling protein 3 (UCP3) content after $NaNO_3$ supplementation [71]. These findings suggested that improved efficiency of mitochondrial oxidative phosphorylation contributes to improved exercise economy after NO_3^- supplementation. However, this group also reported no improvement in the P/O ratio after acute treatment with 25 μM NO_2^- in mitochondria isolated from the human skeletal muscle [71] and human skeletal muscle myotubes [72]. Moreover, and in contrast to the observations by Larsen et al. [71], recent research has indicated that, while 5 days supplementation with NO_3^--rich beetroot juice can lower $\dot{V}O_2$, this occurs independently of changes in ANT and UCP3, mitochondrial P/O ratio and respiratory parameters in permeabilized skeletal muscle fibres in vitro [73].

Most recently it was reported that cycling economy is improved after 28 days supplementation with 6 mmol (372 mg) NO_3^-·day^{-1} and, importantly, that this improved cycling economy is still manifest 24 h after ceasing NO_3^- supplementation when plasma $[NO_2^-]$ had returned to baseline [46]. This suggests that chronic NO_3^- supplementation can improve exercise economy independent of increased plasma $[NO_2^-]$, presumably, because the increased content of proteins that influence skeletal muscle mitochondrial efficiency [71] and/or calcium (Ca^{2+}) handling ([74], see later) are still expressed for a period of time after chronic NO_3^- supplementation is ceased. However, it is unclear whether exercise performance can be improved after chronic NO_3^- supplementation when plasma $[NO_2^-]$ has returned to baseline. Therefore, the mechanisms that underlie improved exercise economy and endurance performance after NO_3^- supplementation may be to some extent dissociated and warrant further investigation.

In addition to mechanistic investigations in human skeletal muscle, results from murine models have contributed greatly to understanding of the potential mechanisms that underlie improved skeletal muscle metabolic efficiency and fatigue resistance after NO_3^- supplementation. Hernández et al. [74] supplemented the diet of mice with $NaNO_3$ for 7 days and observed increased skeletal muscle contractile force and expression of the Ca^{2+} handling proteins, calsequestrin-1 and the dihydropyridine receptor, in type II (fast-twitch) *extensor digitorum longus* muscle, but not type I (slow-twitch) *soleus* muscle.

In another set of experiments, these authors excised single skeletal muscle fibres from the *flexor digitorum brevis* muscle of these NO_3^- supplemented mice and observed increases in twitch and tetanic cytosolic $[Ca^{2+}]$ and force production [74]. Since skeletal muscle Ca^{2+} handling accounts for a significant portion of the ATP cost of skeletal muscle contraction [75], and since perturbations to skeletal muscle Ca^{2+} handling are a characteristic feature of skeletal muscle fatigue [70], improved skeletal muscle Ca^{2+} handling might account for a portion of the improvements in exercise economy and fatigue resistance that have been reported following NO_3^- supplementation. In a separate series of studies, Ferguson et al. [76] observed increased hindlimb skeletal muscle blood flow in the exercising rat after 5 days supplementation with beetroot juice, with this additional blood flow being preferentially distributed to type II skeletal muscle. These authors subsequently observed increased microvascular PO_2, which reflects the dynamic balance between skeletal muscle O_2 delivery and O_2 utilization [77], during evoked skeletal muscle contractions in type II skeletal muscle, but not type I skeletal muscle, after NO_3^- supplementation [31]. Therefore, improved skeletal muscle blood flow and blood flow distribution to type II muscle might account, at least in part, for the improved exercise performance reported after NO_3^- supplementation.

In line with findings from murine models, there is some indirect evidence of a preferential effect of NO_3^- supplementation on type II muscle in humans. The Henneman size principle of motor unit recruitment posits that skeletal muscle fibres are recruited in a hierarchical manner according to their recruitment threshold [78]. On this basis, smaller type I (slow twitch) motor units are initially recruited to meet low muscle force requirements, while larger type II (fast twitch) motor units are recruited as muscle force requirements are increased. With this in mind, Breese et al. [48] assessed the effects of 4–6 days NO_3^- supplementation on $\dot{V}O_2$ and muscle deoxyhaemoglobin (HHb; reflective of the balance between muscle O_2 utilization and muscle O_2 delivery [79]) kinetics during an initial increment in cycling work rate from a low-intensity (20 W) to a moderate-intensity work rate (where predominantly type I muscle would be recruited to produce force) and a subsequent increment to a severe-intensity work rate (where the additional force production would be predominantly achieved through the additional recruitment of type II muscle). In this study, NO_3^- supplementation speeded $\dot{V}O_2$ and HHb kinetics in the moderate–severe-intensity work rate increment, but not the low–moderate-intensity work rate increment [48], consistent with the notion of enhanced effects of NO_3^- supplementation on type II skeletal muscle. The skeletal muscle fibre populations also exhibit distinct power–velocity relationships with the contribution of type II muscle fibres to force production expected to be increased at higher contraction velocities [80]. NO_3^- supplementation has been shown to increase human peak knee extensor torque at $360°$ s^{-1}, but not $90°$, $180°$ and $270°$ s^{-1} [81], and to improve $\dot{V}O_2$ kinetics (a more rapid increase in $\dot{V}O_2$ following the onset of exercise that would be expected to increase the proportional energy contribution from oxidative metabolism), skeletal muscle oxygenation and fatigue resistance when completing a severe-intensity cycling trial at 115 rpm, but not 35 rpm [82]. Taken together, findings from human studies support observations from murine models in suggesting that NO_3^- supplementation confers greater physiological benefits to type II compared to type I skeletal muscle. These preferential effects of NO_3^- supplementation on type II skeletal muscle physiology and function might be linked to a lower microvascular PO_2 in type II muscle at rest and during contractions [83]. Given that the reduction of NO_2^- to NO is potentiated in hypoxia [32], the potential for NO generation and NO-mediated signalling after NO_3^- supplementation is greater in type II muscle and this might underlie the improved physiology and function in type II skeletal muscle after NO_3^- supplementation.

Highly Trained and Elite Participants

While the majority of studies have reported enhanced exercise economy and/or endurance performance in recreationally active and moderately trained endurance athletes after NO_3^- supplementation, the effects on these variables in highly trained and elite endurance athletes ($\dot{V}O_{2peak}$ of >60 mL kg^{-1} min^{-1})

are less encouraging. In one of the first studies to investigate the effects of NO_3^- supplementation on endurance performance in highly trained endurance athletes, Wilkerson et al. [84] reported that the acute ingestion of 500 mL of beetroot juice, which provided ~6.2 mmol (384 mg) NO_3^-, did not improve 50 mile cycling time trial performance. Interestingly, however, these authors observed a negative correlation between the change in the plasma $[NO_2^-]$ and the change in time trial completion time between the NO_3^--rich and NO_3^--depleted beetroot juice trials [84]. This suggested that highly trained endurance athletes might require a larger acute dose of dietary NO_3^- for an ergogenic response to be observed and that there are responders and non-responders to dietary NO_3^- supplementation. However, at the group mean level and consistent with the initial findings by Wilkerson et al. [84], subsequent studies have not observed improved exercise economy or performance after acute NO_3^- ingestion in well-trained endurance athletes completing running [85, 86] or cycling [87–89] exercise. Exercise economy and performance are also not improved after the acute ingestion of a large NO_3^- dose of 19.5 mmol (1209 mg, [85]). Similarly, no improvement in endurance performance has been reported in well-trained endurance athletes following chronic NO_3^- supplementation, at least up to 8 days of supplementation [57, 85, 90, 91]. The diminished effectiveness of NO_3^- supplementation to improve exercise economy and endurance performance in well-trained endurance athletes might be linked to the physiological remodelling of the skeletal muscle that occurs during chronic endurance training. Specifically, chronic endurance training has been shown to increase NOS expression in the vasculature (eNOS, [92]) and skeletal muscle (nNOS, [93]); to increase the content of Ca^{2+} handling proteins in type II muscle [94]; to lower mitochondrial UCP3 content [95], an important determinant of exercise efficiency [96]; and to promote a lower % of type IIx skeletal muscle [97]. Therefore, the divergent skeletal muscle and vascular physiology of well-trained compared to lesser trained endurance athletes might account for the blunted benefit of NO_3^- supplementation in the former compared to the latter. Nonetheless, while these studies collectively suggest that NO_3^- supplementation has limited potential as an ergogenic aid for well-trained endurance athletes, at least in cycling and running exercise, it should be highlighted that, even in well-trained endurance athletes, there are some individuals who demonstrate practically meaningful performance gains after NO_3 supplementation (e.g. [85, 91]). Therefore, NO_3^- supplementation should not be excluded as a potential ergogenic aid for some well-trained endurance athletes.

There is more encouraging evidence to suggest that NO_3^- supplementation might benefit exercise economy and endurance performance during upper body exercise in athletes who have trained the upper body musculature. Indeed, Muggeridge et al. [58] observed a lower $\dot{V}O_2$ in trained kayakers completing kayaking exercise after the acute ingestion of 70 mL of concentrated beetroot juice (providing ~5 mmol [310 mg] NO_3^-) without a change in 1 km kayaking time trial performance (~4.5 min). Peeling et al. [59] also reported a lower $\dot{V}O_2$ in trained kayakers completing kayaking exercise after acutely ingesting 70 mL of concentrated beetroot juice (providing ~4.8 mmol [298 mg] NO_3^-) without a change in 4 min kayaking time trial performance. However, in a separate arm of their study, Peeling et al. [59] observed improved 500 m kayaking time trial performance after the acute consumption of 140 mL of concentrated beetroot juice (providing ~9.6 mmol [595 mg] NO_3^-). Taken together, these findings suggest that NO_3^- supplementation can improve economy during kayaking exercise and, providing a sufficient NO_3^- dose is consumed, also improve kayaking performance. The greater potential for NO_3^- supplementation to improve economy and endurance performance in trained kayakers compared to trained cyclists and runners might be linked to the different characteristics of the skeletal muscles that are recruited in these exercise tasks. Specifically, the upper body musculature contains a greater proportion of type II muscle compared to the lower body musculature [98, 99] and, given the preferential effects of NO_3^- supplementation on improving physiological responses in type II muscle [74, 76], this might account for enhanced ergogenic effects of NO_3^- supplementation for well-trained kayakers compared to well-trained cyclists and runners. In addition to benefiting kayaking performance, acute ingestion of 140 mL concentrated beetroot juice (~8.4 mmol [521 mg] NO_3^-), but not 70 mL concentrated beetroot juice (~4.2 mmol [260 mg] NO_3^-), has been shown to improve 2 km rowing performance in trained rowers [61]. Chronic supplementation with 500 mL of standard beetroot juice for 6 days has also been shown to improve performance during 6 maximal 500 m rowing ergometer repetitions in trained rowers [60]. In summary, while NO_3^- supplementation appears to be less effective at improving economy and performance in highly trained runners and cyclists compared to moderately trained participants, supplementation with >8 mmol (496 mg) NO_3^- appears to provide performance benefits to highly trained rowers and kayakers, as illustrated in Fig. 20.2.

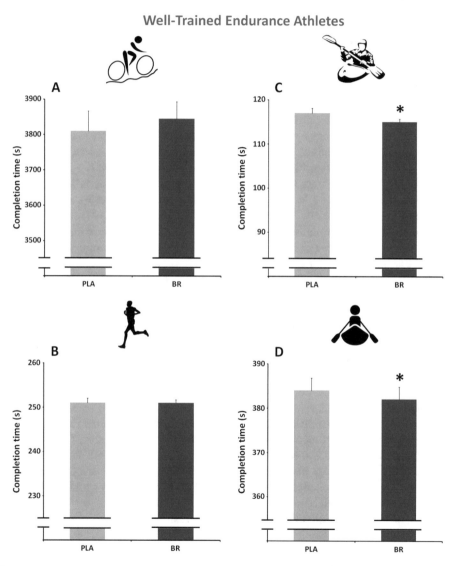

Fig. 20.2 Influence of exercise modality on the effectiveness of NO_3^- supplementation to improve endurance exercise performance in endurance-trained athletes. Time trial performance is not improved in well-trained cyclists completing a 43.83 km cycling time trial after the acute consumption of 8.4 mmol NO_3^- ((**a**), redrawn from Lane et al. 2014) or in well-trained runners completing a 1500 m running time trial after 8 days supplementation with 19.5 mmol $NO_3^- \cdot day^{-1}$ ((**b**), redrawn from Boorsma et al. 2014), but is improved in well-trained kayakers completing a 500 m kayaking time trial after the acute consumption of 9.6 mmol NO_3^- ((**c**), redrawn from Peeling et al. 2015) and well-trained rowers completing a 2000 m rowing time trial after the acute consumption of 8.4 mmol NO_3^- ((**d**), redrawn from Hoon et al. 2014). *Asterisk* indicates different from PLA

Effects of NO$_3^-$ Supplementation on Intermittent and Sprint Exercise Performance

The first studies investigating the effects of NO$_3^-$ supplementation on exercise performance focused on its potential to improve continuous endurance exercise performance. With emerging evidence that NO$_3^-$ supplementation may preferentially enhance physiological responses in type II muscle [74, 76], and given that a greater recruitment of type II muscle is mandated during high-intensity exercise [100], recent research has explored the effects of NO$_3^-$ supplementation on performance during high-intensity exercise, including intermittent exercise. Bond et al. [60] were the first to investigate the effects of NO$_3^-$ supplementation on high-intensity intermittent exercise performance and reported improved performance during 6 maximal 500 m rowing ergometer repetitions in trained rowers after 6 days supplementation with 500 mL of beetroot juice (5.5 mmol [341 mg] NO$_3^-$). Subsequent studies in recreationally active team sports athletes have reported improved performance during the Yo-Yo intermittent recovery test-1 after ingesting 29 mmol (1798 mg) NO$_3^-$ over 36 h prior to exercise [101]; two 40-min halves of repeated 2-min blocks consisting of a 6-s "all-out" sprint, 100-s active recovery and 20 s of rest after 7 days supplementation with ~12.8 mmol (794 mg) NO$_3^-$·day^{-1} (Fig. 20.3) [102]; and repeated 15-s exercise periods at 170 % of the maximal aerobic power interspersed with 30-s passive recovery periods until exhaustion after 3 days supplementation with ~5.5 (341 mg) mmol NO$_3^-$·day^{-1} [103]. However, no improvement in performance has been observed during 6×20 s maximal cycling sprints interspersed with a 100 s low-intensity recovery in well-trained cyclists after 6 days supplementation with 500 mL beetroot juice (8 mmol [496 mg] NO$_3^-$)·day^{-1} [91], and during 5×10 s maximal kayaking sprints interspersed with a 50 s low-intensity recovery in well-trained kayakers after the acute ingestion of ~5 mmol (310 mg) NO$_3^-$ [58]; with compromised performance being reported during repeated 8 s sprints separated by 30 s low-intensity recovery in recreationally active team sports athletes after the acute ingestion of ~5 mmol (310 mg) NO$_3^-$ [104]. These disparate findings might be linked to the inter-study differences in participant training status, exercise modality, NO$_3^-$ supplementation regime and/or the intermittent exercise protocol. Based on the initial evidence, it appears that NO$_3^-$ supplementation can improve high-intensity intermittent exercise performance when recreationally active team sports athletes ingest a large acute NO$_3^-$ dose or a sufficient chronic NO$_3^-$ dose (>5.5 mmol [341 mg]), but not necessarily in endurance-trained participants after the acute ingestion of ~5 mmol (310 mg) NO$_3^-$.

To address inter-study methodological differences between studies assessing the effects of NO$_3^-$ supplementation on intermittent exercise performance, Wylie et al. [105] recently investigated the effects of 3–5 days supplementation with NO$_3^-$-rich beetroot juice (8.2 mmol [508 mg] NO$_3^-$·day^{-1}) on performance during 24 6-s all-out sprints interspersed with 24-s of recovery (24×6-s), seven 30-s all-out sprints interspersed with 240-s of recovery (7×30-s) and six 60-s self-paced maximal efforts interspersed with 60-s of recovery (6×60-s) in recreational team sport players. Compared to placebo trials, performance was improved after NO$_3^-$ supplementation in the 24×6-s trial, but not the 7×30-s and 6×60-s trials. Therefore, NO$_3^-$ supplementation appears to hold promise as an ergogenic aid in exercise settings where short-duration, high-intensity intervals are interspersed with a short recovery duration, but not necessarily when longer duration intervals are interspersed with a longer recovery duration. Importantly, the interval duration and work:rest ratio in the 24×6-s trial, where performance was improved, is similar to that exhibited in many intermittent sports such as association football, rugby and field hockey [106–108]. In addition to increasing the physical work capacity during short-duration, high-intensity intermittent exercise, NO$_3^-$ supplementation has been reported to improve reaction time without compromising response accuracy during short-duration high-intensity intermittent exercise [102]. This potential for improved physical and cognitive performance after NO$_3^-$ supplementation might be expected to improve performance in intermittent team sports.

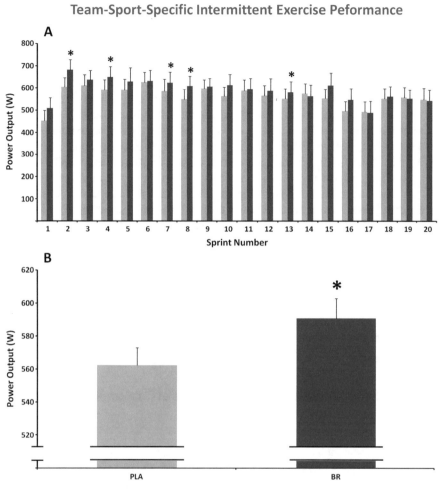

Fig. 20.3 Influence of NO_3^- supplementation on performance during a team-sport-specific intermittent exercise test. Performance was improved in the first of two 40-min halves of repeated 2-min blocks consisting of a 6-s "all-out" sprint, 100-s active recovery and 20 s of rest after 7 days supplementation with ~12.8 mmol $NO_3^- \cdot day^{-1}$ compared to PLA supplementation ((**a**), redrawn from Thompson et al. 2015) with a mean improvement of 5 % across these sprints after BR compared to PLA supplementation ((**b**), redrawn from Thompson et al. 2015). *Asterisk* indicates different from PLA

Supplementation with NO_3^- has been reported to improve: rate of force development during evoked maximal twitch contractions [109]; maximal torque production during voluntary, high-velocity, but not low-velocity, contractions [81]; and performance in a single 180 m running sprint [110]. Therefore, NO_3^- supplementation appears capable of improving speed and power during single bouts of short-duration maximal muscle contractions, in addition to short-duration, high-intensity intermittent exercise. There is also evidence that, in contrast to observations that NO_3^- supplementation is ineffective at improving endurance exercise performance in endurance-trained athletes [85–91], NO_3^- supplementation might be effective at improving speed/power performance in speed/power trained athletes [111]. These conflicting findings might be linked to the disparate muscle fibre recruitment patterns required in continuous endurance tasks compared to single or repeated sprint tasks [112] and to contrasting skeletal muscle phenotypes after endurance compared to speed/power training [97]. Collectively, these findings suggest that NO_3^- supplementation also has ergogenic potential for athletes participating in sports where speed, power and repeated sprint ability are important determinants of success.

Effects of NO_3^- Supplementation on Exercise Performance in Hypoxia

Given that the positive physiological effects observed following NO_3^- supplementation are purported to be NO mediated [50], and since the reduction of NO_2^- to NO is augmented in hypoxia [32], there has been significant interest in the ergogenic potential of NO_3^- supplementation in conditions where O_2 availability is compromised. Existing evidence indicates that NO_3^- supplementation can increase muscle oxygenation, attenuate the disturbance to muscle metabolic homeostasis (blunt PCr degradation, Pi accumulation and pH decline), lower the O_2 cost of sub-maximal exercise, extend exercise tolerance and improve endurance exercise performance in healthy subjects inhaling a gas mixture with a lower fraction of O_2 [49, 113–115]. There is also evidence to suggest that the effects of NO_3^- supplementation on exercise physiology and performance are more pronounced in hypoxia compared to normoxia [49]. In addition, it has been reported that NO_3^- supplementation can attenuate the decline in force production during evoked skeletal muscle contractions in healthy humans when muscle blood flow is impeded [116]. However, the existing evidence suggests that NO_3^- supplementation does not improve exercise economy or performance in hypoxia in well-trained endurance athletes [117–119]. Taken together, NO_3^- supplementation appears to be ergogenic in moderately trained, but not well-trained, endurance athletes when muscle O_2 delivery is impaired. Recent data also suggest that NO_3^- supplementation is more effective at improving exercise economy and endurance exercise performance after acute ultraviolet light exposure [120]. Therefore, the ergogenic potential of NO_3^- supplementation might be linked to the environmental conditions and might be particularly effective in sunny environments, where ultraviolet light exposure is greater, and at altitude, where O_2 availability is lower.

Effects of NO_3^- Supplementation on Exercise Performance in Patient Populations

Since the effects of NO_3^- supplementation are likely accentuated in individuals with hypoxemia and a greater proportion of type II muscle [49, 74, 76], which are characteristic of patients with cardiovascular, respiratory and metabolic diseases [121–126], NO_3^- supplementation might hold therapeutic potential for some patient groups. With regard to the effects of NO_3^- supplementation on functional capacity, there is evidence to suggest it does not benefit 6 min walk test (6MWT) performance inpatients with type II diabetes [66], some evidence to suggest that it can benefit exercise capacity inpatients with chronic obstructive pulmonary disease [127–130], and encouraging initial evidence that it can benefit time to claudication pain inpatients with peripheral artery disease [131] and exercise capacity and muscle contractile function in heart failure patients [132–134]. It has also been reported that NO_3^- supplementation does not improve 6MWT performance in healthy older adults [135]. However, further research is required to elucidate the ergogenic potential of NO_3^- supplementation in the elderly and different patient populations.

Conclusions

Over the last decade, NO_3^- supplementation, particularly in the form of NO_3^--rich beetroot juice, has developed into a popular sports nutrition supplement. This practice stems from initial studies showing that NO_3^- supplementation can improve exercise economy and performance during continuous exercise, and more recent evidence suggesting that NO_3^- supplementation might enhance performance in exercise tasks dependent on speed, power and repeated sprint ability. These effects can be evoked by acute and chronic consumption of 5–6 mmol (310–372 mg) NO_3^-, but chronic supplementation (>3 days) with >8 mmol (496 mg) $NO_3^- \cdot day^{-1}$ is recommended to enhance exercise performance. The mechanisms that underlie improved exercise performance after NO_3^- have yet to be completely resolved, but the latest evidence suggests that NO_3^- supplementation may preferentially enhance physiological responses in type II muscle and in conditions where muscle O_2 delivery is compromised. This might explain improved performance after NO_3^- supplementation in some patient groups and during short-duration high-intensity exercise in moderately trained and well-trained speed/power athletes, and in endurance-trained athletes completing upper body (kayaking and rowing), but not lower body (running and cycling), endurance exercise. In addition to improving exercise performance, there is also preliminary evidence to suggest that NO_3^- supplementation can aid recovery post-exercise [136]. However, further research is required to assess the impact of NO_3^- supplementation on post-exercise recovery and its implications for subsequent exercise performance; how NO_3^- supplementation interacts with other nutritional and exercise training interventions to impact exercise performance; and the mechanisms that underlie responders and non-responders to NO_3^- supplementation.

To summarize, while the efficacy of NO_3^- supplementation to positively impact physiology and performance during exercise is likely to be a function of the NO_3^- supplementation regime and the characteristics of the exercise task and the individual participant, NO_3^- supplementation has been shown to improve performance in a variety of exercise settings and functional capacity in some patient populations. Therefore, NO_3^- supplementation appears to have significant potential as an ergogenic and therapeutic aid.

References

1. Gilchrist M, Winyard PG, Benjamin N. Dietary nitrate—good or bad? Nitric Oxide. 2010;22(2):104–9.
2. Hord NG, Tang Y, Bryan NS. Food sources of nitrates and nitrites: the physiologic context for potential health benefits. Am J Clin Nutr. 2009;90(1):1–10.
3. EFSA. Nitrate in vegetables: scientific opinion of the panel on contaminants in the food chain. EFSA J. 2008;689:1–79.
4. Ysart G, Miller P, Barrett G, Farrington D, Lawrance P, Harrison N. Dietary exposures to nitrate in the UK. Food Addit Contam. 1999;16(12):521–32.
5. Appel LJ, Moore TJ, Obarzanek E, Vollmer WM, Svetkey LP, Sacks FM, Bray GA, Vogt TM, Cutler JA, Windhauser MM, Lin PH, Karanja N. A clinical trial of the effects of dietary patterns on blood pressure. DASH Collaborative Research Group. N Engl J Med. 1997;336(16):1117–24.
6. NHS Choices. Why 5 a day? 2013. http://www.nhs.uk/Livewell/5ADAY/Pages/5ADAYhome.aspx. Accessed 25 Aug 2015.
7. EUFIC. Fruit and vegetable consumption in Europe—do Europeans get enough? 2012. http://www.eufic.org/article/en/expid/Fruit-vegetable-consumption-Europe/. Accessed 25 Aug 2015.
8. Kapil V, Webb AJ, Ahluwalia A. Inorganic nitrate and the cardiovascular system. Heart. 2010;96(21):1703–9.
9. Omar SA, Webb AJ. Nitrite reduction and cardiovascular protection. J Mol Cell Cardiol. 2014;73:57–69.
10. Ashworth A, Mitchell K, Blackwell JR, Vanhatalo A, Jones AM. High-nitrate vegetable diet increases plasma nitrate and nitrite concentrations and reduces blood pressure in healthy women. Public Health Nutr. 2015;18:2669–78.
11. Joshipura KJ, Ascherio A, Manson JE, Stampfer MJ, Rimm EB, Speizer FE, et al. Fruit and vegetable intake in relation to risk of ischemic stroke. JAMA. 1999;282:1233–9.
12. Joshipura KJ, Hu FB, Manson JE, Stampfer MJ, Rimm EB, Speizer FE, et al. The effect of fruit and vegetable intake on risk for coronary heart disease. Ann Intern Med. 2001;134:1106–14.
13. Omar SA, Webb AJ, Lundberg JO, Weitzberg E. Therapeutic effects of inorganic nitrate and nitrite in cardiovascular and metabolic diseases. J Intern Med. 2016;279:315–36.
14. Weitzberg E, Lundberg JO. Novel aspects of dietary nitrate and human health. Annu Rev Nutr. 2013;33:129–59.
15. Spiegelhalder B, Eisenbrand G, Preussmann R. Influence of dietary nitrate on nitrite content of human saliva: possible relevance to in vivo formation of N-nitroso compounds. Food Cosmet Toxicol. 1976;14:545–8.
16. Govoni M, Jansson EA, Weitzberg E, Lundberg JO. The increase in plasma nitrite after a dietary nitrate load is markedly attenuated by an antibacterial mouthwash. Nitric Oxide. 2008;19:333–7.

17. Lundberg JO, Govoni M. Inorganic nitrate is a possible source for systemic generation of nitric oxide. Free Radic Biol Med. 2004;37:395–400.

18. Duncan C, Dougall H, Johnston P, Green S, Brogan R, Leifert C, et al. Chemical generation of nitric oxide in the mouth from the enterosalivary circulation of dietary nitrate. Nat Med. 1995;1:546–51.

19. Hyde ER, Andrade F, Vaksman Z, Parthasarathy K, Jiang H, Parthasarathy DK, et al. Metagenomic analysis of nitrate-reducing bacteria in the oral cavity: implications for nitric oxide homeostasis. PLoS One. 2014;9, e88645.

20. Hyde ER, Luk B, Cron S, Kusic L, McCue T, Bauch T, et al. Characterization of the rat oral microbiome and the effects of dietary nitrate. Free Radic Biol Med. 2014;77:249–57.

21. Lundberg JO, Weitzberg E, Cole JA, Benjamin N. Nitrate, bacteria and human health. Nat Rev Microbiol. 2004;2:593–602.

22. Benjamin N, O'Driscoll F, Dougall H, Duncan C, Smith L, Golden M, et al. Stomach NO synthesis. Nature. 1994;368:502.

23. Lundberg JO, Weitzberg E, Lundberg JM, Alving K. Intragastric nitric oxide production in humans: measurements in expelled air. Gut. 1994;35:1543–6.

24. van Faassen EE, Bahrami S, Feelisch M, Hogg N, Kelm M, Kim-Shapiro DB, et al. Nitrite as regulator of hypoxic signaling in mammalian physiology. Med Res Rev. 2009;29:683–741.

25. Bryan NS, Calvert JW, Gundewar S, Lefer DJ. Dietary nitrite restores NO homeostasis and is cardioprotective in endothelial nitric oxide synthase-deficient mice. Free Radic Biol Med. 2008;45:468–74.

26. Carlström M, Larsen FJ, Nyström T, Hezel M, Borniquel S, Weitzberg E, et al. Dietary inorganic nitrate reverses features of metabolic syndrome in endothelial nitric oxide synthase-deficient mice. Proc Natl Acad Sci U S A. 2010;107:17716–20.

27. Ferguson SK, Glean AA, Holdsworth CT, Wright JL, Fees AJ, Colburn TD, et al. Skeletal muscle vascular control during exercise: impact of nitrite infusion during nitric oxide synthase inhibition in healthy rats. J Cardiovasc Pharmacol Ther. 2016;21:201–8.

28. Lundberg JO, Weitzberg E, Gladwin MT. The nitrate-nitrite-nitric oxide pathway in physiology and therapeutics. Nat Rev Drug Discov. 2008;7:156–67.

29. Richardson RS, Noyszewski EA, Kendrick KF, Leigh JS, Wagner PD. Myoglobin O_2 desaturation during exercise. Evidence of limited O_2 transport. J Clin Invest. 1995;96:1916–26.

30. Tanaka Y, Poole DC, Kano Y. pH homeostasis in contracting and recovering skeletal muscle: integrated function of the microcirculation with the interstitium and intramyocyte milieu. Curr Top Med Chem. 2016;16:2656–63.

31. Ferguson SK, Holdsworth CT, Wright JL, Fees AJ, Allen JD, Jones AM, et al. Microvascular oxygen pressures in muscles comprised of different fiber types: impact of dietary nitrate supplementation. Nitric Oxide. 2015;48:38–43.

32. Castello PR, David PS, McClure T, Crook Z, Poyton RO. Mitochondrial cytochrome oxidase produces nitric oxide under hypoxic conditions: implications for oxygen sensing and hypoxic signaling in eukaryotes. Cell Metab. 2006;3:277–87.

33. Modin A, Björne H, Herulf M, Alving K, Weitzberg E, Lundberg JO. Nitrite-derived nitric oxide: a possible mediator of 'acidic-metabolic' vasodilation. Acta Physiol Scand. 2001;171:9–16.

34. Gladwin MT, Shelhamer JH, Schechter AN, Pease-Fye ME, Waclawiw MA, et al. Role of circulating nitrite and S-nitrosohemoglobin in the regulation of regional blood flow in humans. Proc Natl Acad Sci U S A. 2000;97:11482–7.

35. Stamler JS, Meissner G. Physiology of nitric oxide in skeletal muscle. Physiol Rev. 2001;81:209–37.

36. Suhr F, Gehlert S, Grau M, Bloch W. Skeletal muscle function during exercise-fine-tuning of diverse subsystems by nitric oxide. Int J Mol Sci. 2013;14:7109–39.

37. Bryan NS. Nitrite in nitric oxide biology: cause or consequence? A systems-based review. Free Radic Biol Med. 2006;41:691–701.

38. Larsen FJ, Weitzberg E, Lundberg JO, Ekblom B. Effects of dietary nitrate on oxygen cost during exercise. Acta Physiol (Oxf). 2007;191:59–66.

39. Poole DC, Richardson RS. Determinants of oxygen uptake. Implications for exercise testing. Sports Med. 1997;24:308–20.

40. Jones AM, Carter H. The effect of endurance training on parameters of aerobic fitness. Sports Med. 2000;29:373–86.

41. Bailey SJ, Winyard P, Vanhatalo A, Blackwell JR, Dimenna FJ, Wilkerson DP, et al. Dietary nitrate supplementation reduces the O_2 cost of low-intensity exercise and enhances tolerance to high-intensity exercise in humans. J Appl Physiol. 2009;107:1144–55.

42. Betteridge S, Bescós R, Martorell M, Pons A, Garnham AP, Stathis CC, et al. No effect of acute beetroot juice ingestion on oxygen consumption, glucose kinetics, or skeletal muscle metabolism during submaximal exercise in males. J Appl Physiol. 2016;120:391–8.

43. Larsen FJ, Weitzberg E, Lundberg JO, Ekblom B. Dietary nitrate reduces maximal oxygen consumption while maintaining work performance in maximal exercise. Free Radic Biol Med. 2010;48:342–7.

44. Vanhatalo A, Bailey SJ, Blackwell JR, DiMenna FJ, Pavey TG, Wilkerson DP, et al. Acute and chronic effects of dietary nitrate supplementation on blood pressure and the physiological responses to moderate-intensity and incremental exercise. Am J Physiol Regul Integr Comp Physiol. 2010;299:1121–31.
45. Wylie LJ, Kelly J, Bailey SJ, Blackwell JR, Skiba PF, Winyard PG, et al. Beetroot juice and exercise: pharmacodynamic and dose-response relationships. J Appl Physiol. 2013;115:325–36.
46. Wylie LJ, Ortiz de Zevallos J, Isidore T, Nyman L, Vanhatalo A, et al. Dose-dependent effects of dietary nitrate on the oxygen cost of moderate-intensity exercise: Acute vs. chronic supplementation. Nitric Oxide. 2016;57:30–9.
47. Omar SA, Artime E, Webb AJ. A comparison of organic and inorganic nitrates/nitrites. Nitric Oxide. 2012;26:229–40.
48. Breese BC, McNarry MA, Marwood S, Blackwell JR, Bailey SJ, Jones AM. Beetroot juice supplementation speeds O2 uptake kinetics and improves exercise tolerance during severe-intensity exercise initiated from an elevated metabolic rate. Am J Physiol Regul Integr Comp Physiol. 2013;305:1441–50.
49. Kelly J, Vanhatalo A, Bailey SJ, Wylie LJ, Tucker C, List S, et al. Dietary nitrate supplementation: effects on plasma nitrite and pulmonary O2 uptake dynamics during exercise in hypoxia and normoxia. Am J Physiol Regul Integr Comp Physiol. 2014;307:920–30.
50. Lundberg JO, Weitzberg E. NO-synthase independent NO generation in mammals. Biochem Biophys Res Commun. 2010;396:39–45.
51. McDonagh ST, Wylie LJ, Winyard PG, Vanhatalo A, Jones AM. The effects of chronic nitrate supplementation and the use of strong and weak antibacterial agents on plasma nitrite concentration and exercise blood pressure. Int J Sports Med. 2015;36:1177–85.
52. Woessner M, Smoliga JM, Tarzia B, Stabler T, Van Bruggen M, Allen JD. A stepwise reduction in plasma and salivary nitrite with increasing strengths of mouthwash following a dietary nitrate load. Nitric Oxide. 2016;54:1–7.
53. Bailey SJ, Fulford J, Vanhatalo A, Winyard PG, Blackwell JR, DiMenna FJ, et al. Dietary nitrate supplementation enhances muscle contractile efficiency during knee-extensor exercise in humans. J Appl Physiol. 2010;109:135–48.
54. Lansley KE, Winyard PG, Fulford J, Vanhatalo A, Bailey SJ, Blackwell JR, et al. Dietary nitrate supplementation reduces the O2 cost of walking and running: a placebo-controlled study. J Appl Physiol. 2011;110:591–600.
55. Kuennen M, Jansen L, Gillum T, Granados J, Castillo W, Nabiyar A, et al. Dietary nitrate reduces the O2 cost of desert marching but elevates the rise in core temperature. Eur J Appl Physiol. 2015;115:2557–69.
56. Murphy M, Eliot K, Heuertz RM, Weiss E. Whole beetroot consumption acutely improves running performance. J Acad Nutr Diet. 2012;112:548–52.
57. Porcelli S, Ramaglia M, Bellistri G, Pavei G, Pugliese L, Montorsi M, et al. Aerobic fitness affects the exercise performance responses to nitrate supplementation. Med Sci Sports Exerc. 2015;47:1643–51.
58. Muggeridge DJ, Howe CC, Spendiff O, Pedlar C, James PE, Easton C. The effects of a single dose of concentrated beetroot juice on performance in trained flatwater kayakers. Int J Sport Nutr Exerc Metab. 2013;23:498–506.
59. Peeling P, Cox GR, Bullock N, Burke LM. Beetroot juice improves on-water 500 M time-trial performance, and laboratory-based paddling economy in national and international-level kayak athletes. Int J Sport Nutr Exerc Metab. 2015;25:278–84.
60. Bond H, Morton L, Braakhuis AJ. Dietary nitrate supplementation improves rowing performance in well-trained rowers. Int J Sport Nutr Exerc Metab. 2012;22:251–6.
61. Hoon MW, Jones AM, Johnson NA, Blackwell JR, Broad EM, Lundy B, et al. The effect of variable doses of inorganic nitrate-rich beetroot juice on simulated 2,000-m rowing performance in trained athletes. Int J Sports Physiol Perform. 2014;9:615–20.
62. Cermak NM, Res P, Stinkens R, Lundberg JO, Gibala MJ, van Loon LJ. No improvement in endurance performance after a single dose of beetroot juice. Int J Sport Nutr Exerc Metab. 2012;22:470–8.
63. Lansley KE, Winyard PG, Bailey SJ, Vanhatalo A, Wilkerson DP, Blackwell JR, et al. Acute dietary nitrate supplementation improves cycling time trial performance. Med Sci Sports Exerc. 2011;43:1125–31.
64. Cermak NM, Gibala MJ, van Loon LJ. Nitrate supplementation's improvement of 10-km time-trial performance in trained cyclists. Int J Sport Nutr Exerc Metab. 2012;22:64–71.
65. Flueck JL, Bogdanova A, Mettler S, Perret C. Is beetroot juice more effective than sodium nitrate? The effects of equimolar nitrate dosages of nitrate-rich beetroot juice and sodium nitrate on oxygen consumption during exercise. Appl Physiol Nutr Metab. 2016;41:421–9.
66. Shepherd AI, Gilchrist M, Winyard PG, Jones AM, Hallmann E, Kazimierczak R, et al. Effects of dietary nitrate supplementation on the oxygen cost of exercise and walking performance in individuals with type 2 diabetes: a randomized, double-blind, placebo-controlled crossover trial. Free Radic Biol Med. 2015;86:200–8.
67. Wootton-Beard PC, Ryan L. A beetroot juice shot is a significant and convenient source of bioaccessible antioxidants. J Funct Foods. 2011;3:329–34.
68. Brown GC. Control of respiration and ATP synthesis in mammalian mitochondria and cells. Biochem J. 1992;284:1–13.
69. Chance B, Williams GR. Respiratory enzymes in oxidative phosphorylation. I. Kinetics of oxygen utilization. J Biol Chem. 1955;217:383–93.

70. Allen DG, Lamb GD, Westerblad H. Skeletal muscle fatigue: cellular mechanisms. Physiol Rev. 2008; 88:287–332.

71. Larsen FJ, Schiffer TA, Borniquel S, Sahlin K, Ekblom B, Lundberg JO, et al. Dietary inorganic nitrate improves mitochondrial efficiency in humans. Cell Metab. 2011;13:149–59.

72. Larsen FJ, Schiffer TA, Ekblom B, Mattsson MP, Checa A, Wheelock CE, et al. Dietary nitrate reduces resting metabolic rate: a randomized, crossover study in humans. Am J Clin Nutr. 2014;99:843–50.

73. Whitfield J, Ludzki A, Heigenhauser GJ, Senden JM, Verdijk LB, van Loon LJ, et al. Beetroot juice supplementation reduces whole body oxygen consumption but does not improve indices of mitochondrial efficiency in human skeletal muscle. J Physiol. 2016;594:421–35.

74. Hernández A, Schiffer TA, Ivarsson N, Cheng AJ, Bruton JD, Lundberg JO, et al. Dietary nitrate increases tetanic $[Ca^{2+}]i$ and contractile force in mouse fast-twitch muscle. J Physiol. 2012;590:3575–83.

75. Barclay CJ, Woledge RC, Curtin NA. Energy turnover for Ca^{2+} cycling in skeletal muscle. J Muscle Res Cell Motil. 2007;28:259–74.

76. Ferguson SK, Hirai DM, Copp SW, Holdsworth CT, Allen JD, Jones AM, et al. Impact of dietary nitrate supplementation via beetroot juice on exercising muscle vascular control in rats. J Physiol. 2013;591:547–57.

77. Poole DC, Behnke BJ, Padilla DJ. Dynamics of muscle microcirculatory oxygen exchange. Med Sci Sports Exerc. 2005;37:1559–66.

78. Henneman E, Somjen G, Carpenter DO. Excitability and inhibitability of motoneurons of different sizes. J Neurophysiol. 1965;28:599–620.

79. Koga S, Kano Y, Barstow TJ, Ferreira LF, Ohmae E, Sudo M, et al. Kinetics of muscle deoxygenation and microvascular PO_2 during contractions in rat: comparison of optical spectroscopy and phosphorescence-quenching techniques. J Appl Physiol. 2012;112:26–32.

80. Coyle EF, Costill DL, Lesmes GR. Leg extension power and muscle fiber composition. Med Sci Sports. 1979;11:12–5.

81. Coggan AR, Leibowitz JL, Kadkhodayan A, Thomas DP, Ramamurthy S, Spearie CA, et al. Effect of acute dietary nitrate intake on maximal knee extensor speed and power in healthy men and women. Nitric Oxide. 2015;48:16–21.

82. Bailey SJ, Varnham RL, DiMenna FJ, Breese BC, Wylie LJ, Jones AM. Inorganic nitrate supplementation improves muscle oxygenation, O_2 uptake kinetics, and exercise tolerance at high but not low pedal rates. J Appl Physiol. 2015;118:1396–405.

83. McDonough P, Behnke BJ, Padilla DJ, Musch TI, Poole DC. Control of microvascular oxygen pressures in rat muscles comprised of different fibre types. J Physiol. 2005;563:903–13.

84. Wilkerson DP, Hayward GM, Bailey SJ, Vanhatalo A, Blackwell JR, Jones AM. Influence of acute dietary nitrate supplementation on 50 mile time trial performance in well-trained cyclists. Eur J Appl Physiol. 2012;112:4127–34.

85. Boorsma RK, Whitfield J, Spriet LL. Beetroot juice supplementation does not improve performance of elite 1500-m runners. Med Sci Sports Exerc. 2014;46:2326–34.

86. Peacock O, Tjønna AE, James P, Wisløff U, Welde B, Böhlke N, et al. Dietary nitrate does not enhance running performance in elite cross-country skiers. Med Sci Sports Exerc. 2012;44:2213–9.

87. Hoon MW, Hopkins WG, Jones AM, Martin DT, Halson SL, West NP, et al. Nitrate supplementation and high-intensity performance in competitive cyclists. Appl Physiol Nutr Metab. 2014;39:1043–9.

88. Lane SC, Hawley JA, Desbrow B, Jones AM, Blackwell JR, Ross ML, et al. Single and combined effects of beetroot juice and caffeine supplementation on cycling time trial performance. Appl Physiol Nutr Metab. 2014;39:1050–7.

89. MacLeod KE, Nugent SF, Barr SI, Koehle MS, Sporer BC, MacInnis MJ. Acute beetroot juice supplementation does not improve cycling performance in normoxia or moderate hypoxia. Int J Sport Nutr Exerc Metab. 2015;25:359–66.

90. Bescós R, Ferrer-Roca V, Galilea PA, Roig A, Drobnic F, Sureda A, et al. Sodium nitrate supplementation does not enhance performance of endurance athletes. Med Sci Sports Exerc. 2012;44:2400–9.

91. Christensen PM, Nyberg M, Bangsbo J. Influence of nitrate supplementation on VO_2 kinetics and endurance of elite cyclists. Scand J Med Sci Sports. 2013;23:e21–31.

92. Green DJ, Maiorana A, O'Driscoll G, Taylor R. Effect of exercise training on endothelium-derived nitric oxide function in humans. J Physiol. 2004;561:1–25.

93. McConell GK, Bradley SJ, Stephens TJ, Canny BJ, Kingwell BA, Lee-Young RS. Skeletal muscle nNOS mu protein content is increased by exercise training in humans. Am J Physiol Regul Integr Comp Physiol. 2007;293:821–8.

94. Kinnunen S, Mänttäri S. Specific effects of endurance and sprint training on protein expression of calsequestrin and SERCA in mouse skeletal muscle. J Muscle Res Cell Motil. 2012;33:123–30.

95. Fernström M, Tonkonogi M, Sahlin K. Effects of acute and chronic endurance exercise on mitochondrial uncoupling in human skeletal muscle. J Physiol. 2004;554:755–63.

96. Mogensen M, Bagger M, Pedersen PK, Fernström M, Sahlin K. Cycling efficiency in humans is related to low UCP3 content and to type I fibres but not to mitochondrial efficiency. J Physiol. 2006;571:669–81.

97. Wilson JM, Loenneke JP, Jo E, Wilson GJ, Zourdos MC, Kim JS. The effects of endurance, strength, and power training on muscle fiber type shifting. J Strength Cond Res. 2012;26:1724–9.

98. Gollnick PD, Armstrong RB, Saubert IV CW, Piehl K, Saltin B. Enzyme activity and fiber composition in skeletal muscle of untrained and trained men. J Appl Physiol. 1972;33:312–9.

99. Polgar J, Johnson MA, Weightman D, Appleton D. Data on fibre size in thirty-six human muscles. An autopsy study. J Neurol Sci. 1973;19:307–18.

100. Krustrup P, Söderlund K, Mohr M, Bangsbo J. The slow component of oxygen uptake during intense, sub-maximal exercise in man is associated with additional fibre recruitment. Pflugers Arch. 2004;447:855–66.

101. Wylie LJ, Mohr M, Krustrup P, Jackman SR, Ermιdis G, Kelly J, et al. Dietary nitrate supplementation improves team sport-specific intense intermittent exercise performance. Eur J Appl Physiol. 2013;113:1673–84.

102. Thompson C, Wylie LJ, Fulford J, Kelly J, Black MI, McDonagh ST, et al. Dietary nitrate improves sprint performance and cognitive function during prolonged intermittent exercise. Eur J Appl Physiol. 2015;115:1825–34.

103. Aucouturier J, Boissière J, Pawlak-Chaouch M, Cuvelier G, Gamelin FX. Effect of dietary nitrate supplementation on tolerance to supramaximal intensity intermittent exercise. Nitric Oxide. 2015;49:16–25.

104. Martin K, Smee D, Thompson KG, Rattray B. No improvement of repeated-sprint performance with dietary nitrate. Int J Sports Physiol Perform. 2014;9:845–50.

105. Wylie LJ, Bailey SJ, Kelly J, Blackwell JR, Vanhatalo A, Jones AM. Influence of beetroot juice supplementation on intermittent exercise performance. Eur J Appl Physiol. 2016;116:415–25.

106. King T, Jenkins D, Gabbett T. A time-motion analysis of professional rugby league match-play. J Sports Sci. 2009;27:213–9.

107. Mohr M, Krustrup P, Bangsbo J. Match performance of high-standard soccer players with special reference to development of fatigue. J Sports Sci. 2003;21:519–28.

108. Spencer M, Lawrence S, Rechichi C, Bishop D, Dawson B, Goodman C. Time-motion analysis of elite field hockey, with special reference to repeated-sprint activity. J Sports Sci. 2004;22:843–50.

109. Haider G, Folland JP. Nitrate supplementation enhances the contractile properties of human skeletal muscle. Med Sci Sports Exerc. 2014;46:2234–43.

110. Sandbakk SB, Sandbakk Ø, Peacock O, James P, Welde B, Stokes K, et al. Effects of acute supplementation of L-arginine and nitrate on endurance and sprint performance in elite athletes. Nitric Oxide. 2015;48:10–5.

111. Rimer EG, Peterson LR, Coggan AR, Martin JC. Acute dietary nitrate supplementation increases maximal cycling power in athletes. Int J Sports Physiol Perform. 2016;11:715–20.

112. Abernethy PJ, Thayer R, Taylor AW. Acute and chronic responses of skeletal muscle to endurance and sprint exercise. A review. Sports Med. 1990;10:365–89.

113. Masschelein E, Van Thienen R, Wang X, Van Schepdael A, Thomis M, Hespel P. Dietary nitrate improves muscle but not cerebral oxygenation status during exercise in hypoxia. J Appl Physiol. 2012;113:736–45.

114. Muggeridge DJ, Howe CC, Spendiff O, Pedlar C, James PE, Easton C. A single dose of beetroot juice enhances cycling performance in simulated altitude. Med Sci Sports Exerc. 2014;46:143–50.

115. Vanhatalo A, Fulford J, Bailey SJ, Blackwell JR, Winyard PG, Jones AM. Dietary nitrate reduces muscle metabolic perturbation and improves exercise tolerance in hypoxia. J Physiol. 2011;589:5517–28.

116. Hoon MW, Fornusek C, Chapman PG, Johnson NA. The effect of nitrate supplementation on muscle contraction in healthy adults. Eur J Sport Sci. 2015;15:712–9.

117. Arnold JT, Oliver SJ, Lewis-Jones TM, Wylie LJ, Macdonald JH. Beetroot juice does not enhance altitude running performance in well-trained athletes. Appl Physiol Nutr Metab. 2015;40:590–5.

118. Bourdillon N, Fan JL, Uva B, Müller H, Meyer P, Kayser B. Effect of oral nitrate supplementation on pulmonary hemodynamics during exercise and time trial performance in normoxia and hypoxia: a randomized controlled trial. Front Physiol. 2015;6:288.

119. Carriker CR, Mermier CM, McLain TA, Johnson KE, Beltz NM, Vaughan RA, et al. Effect of acute dietary nitrate consumption on oxygen consumption during submaximal exercise in hypobaric hypoxia. Int J Sport Nutr Exerc Metab. 2016;26:315–22.

120. Muggeridge DJ, Sculthorpe N, Grace FM, Willis G, Thornhill L, Weller RB, et al. Acute whole body UVA irradiation combined with nitrate ingestion enhances time trial performance in trained cyclists. Nitric Oxide. 2015;48:3–9.

121. Askew CD, Green S, Walker PJ, Kerr GK, Green AA, Williams AD, et al. Skeletal muscle phenotype is associated with exercise tolerance in patients with peripheral arterial disease. J Vasc Surg. 2005;41:802–7.

122. Gosker HR, Wouters EF, van der Vusse GJ, Schols AM. Skeletal muscle dysfunction in chronic obstructive pulmonary disease and chronic heart failure: underlying mechanisms and therapy perspectives. Am J Clin Nutr. 2000;71:1033–47.

123. Mador MJ, Bozkanat E. Skeletal muscle dysfunction in chronic obstructive pulmonary disease. Respir Res. 2001;2:216–24.

124. Oberbach A, Bossenz Y, Lehmann S, Niebauer J, Adams V, Paschke R, et al. Altered fiber distribution and fiber-specific glycolytic and oxidative enzyme activity in skeletal muscle of patients with type 2 diabetes. Diabetes Care. 2006;29:895–900.

125. Raguso CA, Guinot SL, Janssens JP, Kayser B, Pichard C. Chronic hypoxia: common traits between chronic obstructive pulmonary disease and altitude. Curr Opin Clin Nutr Metab Care. 2004;7:411–7.

126. Schaufelberger M, Eriksson BO, Grimby G, Held P, Swedberg K. Skeletal muscle fiber composition and capillarization in patients with chronic heart failure: relation to exercise capacity and central hemodynamics. J Card Fail. 1995;1:267–72.

127. Kerley CP, Cahill K, Bolger K, McGowan A, Burke C, Faul J, et al. Dietary nitrate supplementation in COPD: an acute, double-blind, randomized, placebo-controlled, crossover trial. Nitric Oxide. 2015;44:105–11.

128. Berry MJ, Justus NW, Hauser JI, Case AH, Helms CC, Basu S, et al. Dietary nitrate supplementation improves exercise performance and decreases blood pressure in COPD patients. Nitric Oxide. 2015;48:22–30.

129. Leong P, Basham JE, Yong T, Chazan A, Finlay P, Barnes S, et al. A double blind randomized placebo control crossover trial on the effect of dietary nitrate supplementation on exercise tolerance in stable moderate chronic obstructive pulmonary disease. BMC Pulm Med. 2015;15:52.

130. Shepherd AI, Wilkerson DP, Dobson L, Kelly J, Winyard PG, Jones AM, et al. The effect of dietary nitrate supplementation on the oxygen cost of cycling, walking performance and resting blood pressure in individuals with chronic obstructive pulmonary disease: a double blind placebo controlled, randomised control trial. Nitric Oxide. 2015;48:31–7.

131. Kenjale AA, Ham KL, Stabler T, Robbins JL, Johnson JL, Vanbruggen M, et al. Dietary nitrate supplementation enhances exercise performance in peripheral arterial disease. J Appl Physiol. 2011;110:1582–91.

132. Coggan AR, Leibowitz JL, Spearie CA, Kadkhodayan A, Thomas DP, Ramamurthy S, et al. Acute dietary nitrate intake improves muscle contractile function in patients with heart failure: a double-blind, placebo-controlled, randomized trial. Circ Heart Fail. 2015;8:914–20.

133. Eggebeen J, Kim-Shapiro DB, Haykowsky M, Morgan TM, Basu S, Brubaker P, et al. One week of daily dosing with beetroot juice improves submaximal endurance and blood pressure in older patients with heart failure and preserved ejection fraction. JACC Heart Fail. 2016. pii: S2213-1779(15)00835-5. doi:10.1016/j.jchf.2015.12.013. [Epub ahead of print].

134. Zamani P, Rawat D, Shiva-Kumar P, Geraci S, Bhuva R, Konda P, et al. Effect of inorganic nitrate on exercise capacity in heart failure with preserved ejection fraction. Circulation. 2015;131:371–80.

135. Kelly J, Fulford J, Vanhatalo A, Blackwell JR, French O, Bailey SJ, et al. Effects of short-term dietary nitrate supplementation on blood pressure, O_2 uptake kinetics, and muscle and cognitive function in older adults. Am J Physiol Regul Integr Comp Physiol. 2013;304:73–83.

136. Clifford T, Bell O, West DJ, Howatson G, Stevenson EJ. The effects of beetroot juice supplementation on indices of muscle damage following eccentric exercise. Eur J Appl Physiol. 2016;116:353–62.

Chapter 21
Nitrite and Nitrate in Cancer

David M. Klurfield

Key Points

- Studies in animal models established the feasibility of sodium nitrite contributing to gastric carcinogenesis primarily via conversion to nitrosamines.
- Most animal studies did not corroborate this assumption.
- Exposure of humans to nitrates is primarily derived from vegetables.
- Since a fraction of nitrate is reduced to nitrite by oral bacteria, the largest source of nitrite exposure is also derived from vegetables.
- Meat processed with nitrite has reducing agents added that virtually eliminate the formation of nitrosamines.
- Existing epidemiological studies are not definitive, and it is likely that additional studies of the same type will not clarify any putative relationship owing to the inherent weaknesses in such epidemiological studies.
- Nitrate and nitrite use has clear benefits for food preservation and only theoretical long-term risk that remains unproven.

Keywords Nitrosamines • Cancer • Epidemiology • Tumor • Carcinogenesis

Introduction

The putative linkage between exposure to nitrites or nitrates and risk of developing cancer derives from studies done in vitro, in animal models, or via epidemiological observations. Much of the experimental literature suggesting increased risk is several decades old, some of which was discounted because of contaminating nitrosamines in drinking water or bedding of test animals, and predates our understanding of the endogenous synthesis and biology of nitrogen-based secondary messengers. It often relied on enormous doses of nitrites/nitrates that have no relevance for normal dietary exposure and were usually instilled directly into the stomach of animals. In addition, it was demonstrated that sodium nitrite could be a precursor to carcinogenic N-nitroso compounds. Under acidic conditions

D.M. Klurfeld, Ph.D. (✉)
National Program Leader for Human Nutrition, USDA Agricultural Research Service,
5601 Sunnyside Ave., Beltsville, MD 20705-5138, USA
e-mail: david.klurfeld@ars.usda.gov

N.S. Bryan, J. Loscalzo (eds.), *Nitrite and Nitrate in Human Health and Disease*, Nutrition and Health,
DOI 10.1007/978-3-319-46189-2_21, © Springer International Publishing AG 2017

in the stomach, some nitrites form nitrous acid, which may then nitrosate low-molecular-weight secondary amines (if present) to form genotoxic and carcinogenic N-nitrosamines. Such studies only demonstrate plausibility although it should be recognized that regulatory agencies generally rely on data from animal studies in the absence of reliable exposure/carcinogenicity information in humans. Furthermore, the effects of nitrites on experimental gastric cancer were also demonstrated before it was known that many times as much nitrate, mostly from vegetables whose intake may reduce risk of cancer, is in the food supply as nitrite and that there is significant concentration of nitrate in saliva where it is converted to nitrite by the oral microflora. Hence, the relationship between exposure to nitrates and nitrites with cancer remains controversial. Informed observers of the environment and cancer generally agree that the level of evidence is convincing when there are concordant epidemiological studies showing increased risk of sufficient magnitude supported by animal and cell/molecular studies to demonstrate mechanistic connections. Hill [1] set out nine viewpoints on the information that would contribute to a conclusion of causality from associations. The reader is referred to the original paper for details of the factors, but Hill listed strength of the association as, by far, the most important of them. He also specifically mentioned an example of needing only fair evidence for replacement of a toxic chemical with a safer one that is readily available and economical compared with the requirement of very strong evidence if people were to cease eating sugar and fat in excess.

Animal Studies

Modern standards of testing generally call for lifetime exposure of both sexes from two animal species to a chemical for assessment of carcinogenicity. The dosages tested are often chosen by first determining the maximum tolerated dose and then reducing exposure so that there are no apparent toxic effects. This paradigm has resulted in about half of all chemicals tested being deemed carcinogens. There has been an ongoing debate about whether this experimental scheme is appropriate since such high doses of many chemicals could result in increased mitosis while lower doses do not [2] for which reason carcinogenicity in such a situation may be an artifact. Two-year exposure of groups of 50 F344 rats of both sexes to either sodium nitrite added to the drinking water at 0, 0.125 % (125 mg/100 mL = 18.1 mM), or 0.25 % (250 mg/100 mL = 36.2 mM) or to sodium nitrate in the diet at 0, 2.5 % (2.5 g/100 g = 294 mM), or 5.0 % (5 g/100 g = 588 mM) resulted in two statistically significant cancer differences between treated and control groups—there was a significant decrease in all tumors in females exposed to 0.25 % nitrite and all treated groups showed reductions in mononuclear cell leukemias [3], the main cause of death in rats of this strain. These changes were seen even in the presence of a threefold increase of N-nitrosodimethylamine in the stomach contents of only male rats fed the higher dose of nitrite. Despite the absence of any carcinogenic effect of either nitrate or nitrite and a significant reduction in two measures of tumorigenesis, the authors concluded by recommending reductions of both compounds in foods as very important because they are precursors of N-nitroso compounds. Furthermore, these doses could never be achieved through dietary means. The U.S. National Toxicology Program evaluated sodium nitrite in the drinking water of both 50 rats and mice of each sex for 2 years of exposure at 0, 0.075 %, 0.15 %, and 0.3 %, and concluded there was no evidence of carcinogenicity from nitrite even though the highest concentration resulted in hyperplasia of the forestomach epithelium in both sexes of the rat and a positive trend in the incidence of squamous cell papilloma and carcinoma combined in female mice [4]; the latter was considered equivocal evidence. People do not consume sodium nitrate or nitrite alone, so Olsen et al. [5] tested in two generations of rats' diets that contained either casein or chopped pork as the sole protein source at 45 % by weight of diet; this level is more than double the typical protein content of a rodent diet. Sodium nitrite was added to the pork to achieve final diet concentrations of 200, 1000, or 4000 mg/kg. There was no effect on reproduction or on cancer, except for a nonsignificant tendency to an increased number of tumor-bearing rats fed the highest dose. Again, despite the absence of any statistically significant effect on carcinogenesis, the authors recommended a reduction in the use of nitrite for curing meat.

Human Exposure to Nitrates/Nitrites

The International Agency for Research on Cancer evaluated the carcinogenicity of ingested nitrate and nitrites in 2010 [6]. The reviewers concluded there is inadequate evidence in experimental animals for carcinogenicity of nitrate and limited evidence for nitrite per se, but sufficient evidence for nitrite in combination with amines or amides. In humans, there is inadequate evidence for carcinogenicity of nitrate in food or in drinking water but there is limited evidence for carcinogenicity of nitrite in foods, specifically for stomach cancer. The overall evaluation was that ingested nitrate or nitrite under conditions that result in endogenous nitrosation is probably carcinogenic to humans, placing both compounds in Group 2A (of probable carcinogens) that include well-established carcinogens such as benzo(a)anthracene, 1,2-dimethylhydrazine, and N-ethyl-N-nitrosourea. The IARC report states that the overall evaluation was made despite the observation that none of the human studies linking nitrite to stomach cancer had taken into account potential confounding or effect modification by *Helicobacter pylori*, an important microbial risk factor for stomach cancer.

To assess human exposure to nitrates and nitrites, four primary conditions must be met. First, we need an accurate database of sources of the compounds in food, water, and any other relevant sources of exposure. Second, we need accurate methods of determining exposure over time to foods and beverages that provide precursors to nitrosamine formation. Third, we need to know about exposure to preformed carcinogenic nitrosamines since nitrates and nitrites themselves are not harmful at usual levels of exposure. Fourth, we need to know what proportion of exogenous or endogenous nitrates or nitrites are converted to carcinogenic nitrosamines after ingestion. In addition to exposure factors, we need to know if there are other biological mechanisms by which these compounds affect carcinogenesis.

There are no official government databases that include nitrate or nitrite concentrations in food. The primary reason for this omission is the marked variability within foods and the fact that drinking water can be a major source of exposure. For example, nitrate content of seven vegetables in Denmark show 20-fold differences within each type with about threefold seasonal variation [7]. From an estimated total intake of 61 mg/day of nitrate and 0.5 mg/day of nitrite, the seven vegetables surveyed contribute 49 and 0.09 mg/day, respectively. Similar variability was found in vegetables and processed foods likely to contain nitrite (mostly meat and dairy products) in New Zealand [8]. In both countries, lettuce was the food providing the most nitrate. Based on the acceptable daily intake (ADI) values for nitrate and nitrite in New Zealand, the mean adult intake from food and water was 16 and 13 % of the ADI. A maximally exposed adult (a vegetarian consuming well water with high nitrate), however, was estimated to exceed the nitrate limit by sevenfold. Because of the endogenous conversion of nitrate to nitrite by oral microflora, 10 % of the New Zealand population with an average conversion rate of 5 % exceeds the ADI for nitrite, and among those with a high conversion rate of 20 %, half that portion of the population exceeds the ADI. Factors affecting nitrate and nitrite variability of vegetables include variety, soil conditions, growth conditions (including fertilization and water stress), transport, and storage conditions. Hord et al. [9] modeled the nitrate and nitrite content of the DASH diet (Dietary Approaches to Stop Hypertension), which stresses high fruit and vegetable intake along with three servings of low-fat dairy and a reduction in meat and eggs, that has been shown in intervention studies to reduce mild hypertension and is one of the dietary patterns recommended in the *Dietary Guidelines for Americans*. Selecting seven low-nitrate fruits and vegetables could yield 174 mg nitrate and 0.41 mg nitrite, while choosing high-nitrate plant foods would yield

1222 mg nitrate plus 0.35 mg nitrite daily, not taking into account the salivary reduction of nitrate to nitrite. Likewise, it has been pointed out that traditional Mediterranean and Japanese diets contain more nitrate than recommended by the World Health Organization [10]. From these data, it should be apparent that dietary patterns recommended for reduction of chronic disease, including various types of cancer, can be notably high in nitrate and/or nitrite content and still afford reduced risk. In a most recent survey of several vegetables sampled at retail sites across five major metropolitan US cities, it was concluded that there is as much as a 500 % difference in nitrate content between certain vegetables and even regional differences in nitrite and nitrate content of the same vegetables, by as much as 50 %. Furthermore, there were also significant differences in nitrate content in conventionally vs. organically grown vegetables of the same type [11]. This variation in nitrite and nitrate content of vegetables of different types, but also of the same vegetables based on geographical location and method of growth (organic vs. conventional), does not allow for accurate quantification of nitrite and nitrate exposure based on historical databases.

Processed meats are the major source of preformed nitrites, but given that the conversion of nitrate to nitrite occurs at 5–20 % of the total ingested, meats are not the major sources of total exposure to nitrites. Another confounder in assessing exposure is that processing of meats differs geographically. In Europe, total intake of cured meats is lower in Mediterranean countries than in central and northern countries; however, intake of nitrate/nitrite is higher in Spain than in Germany or the United Kingdom because of the differences in the amount of nitrate/nitrite added to meats, storage conditions that may not include refrigeration, or the addition of reducing agents in processing designed to reduce formation of nitrosamines [12]; these aspects call into question the validity of international comparisons of processed meats unless all such parameters are taken into account. In addition, widespread addition of reducing agents to cured meat over the past 35 years has greatly decreased intake of preformed nitrosamines from this source; thus, much of the older literature is no longer relevant.

When evaluating the potential role of nitrate or nitrite in cancer, it is essential to know what other potential carcinogens are in foods along with these anions. This was accomplished by one group through literature searches and included the type of food, cooking method, preservation method, cooking extent, temperature, time, nitrosamines, heterocyclic amines, and polycyclic aromatic hydrocarbons [13]. One limitation of this type of approach is the variation in quality of the analytical methods among published studies. Another limitation in assessing the role of nitrate/nitrite in cancer is factors that associate with higher exposures. In addition to diet, drinking water, occupation, and environment, endogenous formation contributes to the overall exposure. We must recognize that nitrite and nitrate are formed endogenously from nitric oxide production and oxidation. Among women in the National Birth Defects Prevention Study, many characteristics distinguished participants according to the level of nitrate/nitrite intake and exposure to nitrosamines that included race/ethnicity, area of residence, household income, education, fat as percent of calories, and age at conception [14]; these factors were all associated with adverse birth outcomes and, hence, it was not possible to clearly separate food components from confounders.

If exposure to nitrate/nitrite is a key factor in cancer, then workers exposed during industrial uses, particularly in years prior to the widespread use of protective breathing devices and improved air handling, should prove revealing. In a cohort of 1327 men working in a fertilizer factory since 1945 and exposed to nitrates for at least 1 year, 537 were considered heavily exposed [15]; this was determined from salivary nitrate and nitrite concentrations, with the exposed workers having 212 nmol/mL (212 μM) nitrate and 129 nmol/mL (129 μM) nitrite in their saliva, which was approximately double the levels in nonexposed employees of the same company and almost fourfold higher than salivary levels in local residents. Exposure could have been through inhalation, accidental ingestion, or dermal contact. Urinary *N*-nitroso compounds were also increased in both the total cohort and heavily exposed workers. Despite this greatly increased chronic exposure to nitrate and its metabolites, there was no significant increase in cancers or any other specific diseases except for laryngeal cancer for which four deaths were observed when 0.9 was expected in the total cohort. Of course, smoking was highly prevalent at that time and was a factor that was not controlled for. All cause mortality was lower in the exposed workers—193 deaths observed when expected deaths were 220 in the heavily exposed men and 304 observed vs. 368 expected in the total cohort. If industrial exposure that equates to fourfold higher salivary nitrate shows no increase in cancers and about a 15 % reduction in total mortality, it is difficult to conclude that there is any risk of serious chronic disease from nitrate at the levels reported.

Unresolved Issues in Epidemiology

There is no doubt that epidemiologic investigation has contributed to our understanding of linkages between lifestyle or environmental factors and chronic diseases. While this has been most useful for cardiovascular disease, the linkages with cancer are less certain, in part, because of our lack of understanding of the causes and various promoters of carcinogenesis that likely interact differently with varying genotypes. The link between cigarette smoking and lung cancer was made first by retrospective case–control studies. This should not be surprising since the relative risk (RR) of lung cancer among smokers is at least tenfold higher than in non-smokers. RR refers to the rate of disease in those exposed to the highest amount of a factor compared with the rate in those exposed to the lowest amount, which can be zero. Thus, if the absolute lifetime rate of colon cancer is 2 in 100, then an RR of 1.5 reflects a lifetime rate of 3 in 100. Therefore, if nitrate/nitrite carries a theoretical RR of 1.5 for some types of cancer among 100 people who are not exposed, 2 will get cancer, and among people who are heavily exposed, 3 will get cancer. But 97 other people will have to avoid exposure to prevent that 1 additional case.

It is rare for a diet–cancer relationship to exceed an RR of 2, and most environmental epidemiologists outside the diet field consider an RR of that magnitude as weak to the point of having little certainty. The only dietary RR exceeding the threshold of 2.0 is the relationship between consumption of moldy cereal grains contaminated with aflatoxin and liver cancer, which carries an RR of 6; this is an important risk factor in parts of Africa and Asia. Substantial controversy exists within the epidemiology community and among other researchers about how capable current epidemiological methods are for determining food intake accurately, how much natural variation occurs in the same foods from different areas or seasons, and whether relatively weak RRs really suggest causality. One of the biggest problems facing epidemiologists is the issue of confounding—many lifestyle factors associate with another variable but not all in a causative manner so that only some factors are responsible for the association [16]. These issues will be discussed in detail in the following chapter.

Some epidemiologists working in this area have proposed that the inability of nutritional epidemiology to identify active factors from the diet is not simply a problem of quantitation but an inability to identify qualitatively dietary constituents that are effective [17] because single agents are assumed to be the active factors in systems in which multiple agents or multiple interacting host factors are responsible for the observed effects. If this criticism is valid, then the converse is also true—nutritional epidemiology is not capable, with methods currently available, of identifying factors in the diet that increase the risk of for example, cancer and, as a result, larger cohorts will not provide definitive answers owing to the inherent co-linearities and biases. Willett [18] has argued that prospective studies avoid the biases of recall and selective participation, and that diet measurement error would force the RR toward the null. However, it is established that those who volunteer for health research tend to be healthier than the general population. Furthermore, if an error in estimation of nutrient intake is nonrandom, forcing the RR to the null may not occur. Willett's position was written in reaction to the EPIC study's finding that a statistically significant reduction of 4 % (RR=0.96) for all cancers was associated with an increment of 200 g/day of fruits and vegetables. Willett concluded from these data that recommending fruits and vegetables to reduce cancer is not justified, even though it is possible that some specific fruit or vegetable may have a significant benefit.

The World Cancer Research Fund (WCRF) in 2007 published what is considered by many a definitive set of systematic reviews and meta-analyses on cancers of various organs and dietary habits [19], even though there is substantial controversy about the conclusions of that report. The only two foods recommended for avoidance were moldy cereals (RR of 6.0 for liver cancer) and processed meats (RR of 1.4 for colorectal cancer). Despite the conclusion about processed meat and colorectal cancer, WCRF noted there was not enough evidence to reach any conclusion about nitrate or nitrite. Some scientists interpreted this report as based on a fragile foundation, given the uncertainties around quantifying the connection and the weak relationships between food and cancer [20]. WCRF updated the colorectal cancer report in 2011 and found convincing evidence for red meat, processed meat, and alcoholic beverages (in men only for this last category) as increasing the risk of colorectal cancer and decreasing the benefit of foods containing dietary fiber [21]. Meta-analysis calculated relative risk of 1.18 for processed meat and colorectal cancer and 1.17 for red meat and colorectal cancer, with the increased RR about half that of the report from only 4 years earlier. The update also included a section on animal fat as a risk factor, and concluded that all newly added studies showed increased risk of colorectal cancer with greater intake; however, there was the potential for residual confounding. It is unclear how the expert group concluded there was too much confounding for animal fat but not for nitrite in meats since both components are confounded by the same factors.

One of the more provocative and eloquent commentaries on this situation called on cancer epidemiologists to be more circumspect in claiming positive results from the studies of environmental (including diet) and occupational exposures with cancer [22]. The authors pointed out that the more studies conducted on a specific topic, the lower the RR over time, primarily as a result of earlier studies finding statistically significant yet false findings that had arisen by chance leading to other studies attempting to replicate the original observation. A number of commentaries were published in response to that editorial, and the same authors were compelled to point out that "committee reports and their conclusions in themselves should not be misconstrued as science; they are consensus documents and opinions with an eye toward closure. In contrast, science is inherently open-ended, provisional and tentative in its findings and conclusions" [23]. If the reader doubts that more research can result in lessening the certainty around most individual epidemiological associations, a mathematical approach has been developed that demonstrates most published research finding are false positives [24], supporting the conclusion that claimed differences may simply be a reflection of an accurate measurement of the prevailing bias. This particular author advocates larger studies that offer more statistical power, but cautions that large studies are more likely to find a statistically significant difference that is not meaningfully different from the null (small effect size), and also points out that the totality of the evidence from different types of studies must be considered, again harkening back to the 50-year-old admonition of Hill [1].

One area vital to assessing a relationship of any dietary component and cancer risk is accurate assessment of exposure. Although 24-h diet recalls are often used in small to moderate sized studies, most large prospective cohort studies, considered the strongest design of those currently in use, rely on the food frequency questionnaire (FFQ) that can be computer scored or administered online. Not only does the FFQ have the advantage of lower participant burden and lower cost, but it also should be able to ascertain foods eaten less often. One inherent problem is that an FFQ is often "validated" by determining correlation for specific nutrients with values derived from one or more 24-h recalls. This would be acceptable if the 24-h recall exhibited both high precision and accuracy; however, most recalls or diaries approximate total energy by ±25 % around the mean, or a range of 50 %. Some FFQs are validated against plasma biomarkers, such as polyunsaturated fatty acids or carotenoids, as markers of intake. However, no FFQ has ever been validated for the numerous components of foods for which they are used to associate with disease endpoints. Just because an FFQ correlates with one or a few markers (usually with modest correlations of up to ~0.5), it does not mean it accurately and precisely captures intake of nutrients calculated from it. Another issue is that subjects who eat more total energy also are exposed to more of all nutrients. One of the responses to this concern has been to adjust for energy intake. The Observing Protein and Energy Nutrition (OPEN) study examined the ability of the FFQ to estimate total energy and protein intake by measuring the attenuation resulting from dietary measurement error [25]. For both energy and protein, the attenuation was substantial enough for the investigators to conclude that the FFQ cannot be recommended. Although multiple 24-h recalls were more accurate, the authors questioned the ability of both these instruments to detect moderate RRs between 1.5 and 2.0 (which is the maximal RR for almost all diet–cancer relationships) even when dietary factors are adjusted for energy intake. These findings led to the assertion by epidemiologists who have made extensive use of the FFQ that we are not likely to learn much more about diet and cancer by continuing to use the standard FFQ [26]. However, it is not clear how an FFQ can be modified to overcome these deficiencies. It is likely the FFQ accurately separates high and low consumers of most foods. But what used to be called semiquantitative FFQs are now being used to report food intakes to the fraction of a gram and nutrients to the milligram. In addition, many studies use a single FFQ to represent intake over the full follow-up period, despite the demonstration from studies with FFQs that 75% of women change intake of dietary fiber by two or more quintiles over 6 years of follow-up [27]; it is a given that if fiber intake changes to that extent, consumption of other foods must also vary by similar amounts. A key to understanding exposure to a food or nutrient over time is whether the FFQ can capture those changes and, if so, how to integrate multiple FFQs into a composite exposure index.

Epidemiologic Associations of Nitrates/Nitrites and Cancer

Despite the limitations described in the section above, those charged with deciding about diet and cancer risk feel pressure to use the best available evidence to make a recommendation even though it is not as strong as desired and is rarely definitive. This would be more acceptable if such recommendations carried grades about the strength of evidence but that is not commonly done; in fact, sometimes taking no position may be the correct decision even though this is almost never done. Therefore, a cursory review of the state of the science is presented here, with the assumption that some of the small but statistically significant risks associated with nitrates, nitrites, or their food sources may be real.

As mentioned above, the WCRF report [19] stimulated considerable discussion and reaction from the meat industry. Some researchers called upon the industry to reduce or eliminate the use of nitrates/nitrites in processing meat based on the assumption that these additives were proven carcinogens [28]; however, if that conclusion is incorrect, then what is proposed is a major change in the production of foods that have been popular for hundreds of years and whose new formulations are potentially less appealing, less safe, and more expensive. It is, therefore, not at all trivial for the food industry to respond to the WCRF report with changes in production.

In 1998, a review of the epidemiological evidence on nitrates, nitrites, and *N*-nitroso compounds with regard to cancer came to the conclusion that *N*-nitroso compounds are potent carcinogens in animals, but the evidence in humans for all three categories of compounds was inconclusive with regard to cancers of the esophagus, stomach, brain, and nasopharynx [29]. The rationale for the specific cancers studied is that most nitrosamines that are carcinogenic (about 90 % of those tested) exhibit target-organ specificity. All of the studies summarized in this review were of the case–control variety. This design is inherently susceptible to recall bias—that is, an individual diagnosed with cancer often remembers his or her diet differently from a healthy person when that person is in a research study exploring the link between diet and cancer. Additionally, it is not clear what time period is most important in carcinogenesis other than the obvious requirement that exposure occurs before tumor development. But is diet in the year before diagnosis important, is 10 years of continuous exposure required, or is exposure in a specific period of life critical to establishing a relationship?

Furthermore, endogenous production of *N*-nitroso compounds may be more important for exposure than exogenous intake. The authors of this review also pointed out, where exposure appears to have a small effect, the amount of uncontrollable confounding inherent in the studies is about as large as the most plausible effect, and the absence of actual exposure levels may have resulted in misclassification sufficient to explain negative study results.

A systematic review of the epidemiological evidence on esophageal and gastric cancers with food sources of nitrites and nitrosamines evaluated both case–control and cohort studies [30]. Evidence from case–control studies of gastric cancer was claimed to show elevated risk from high nitrite and nitrosamine intake, but was not considered conclusive. Importantly, no formal meta-analyses were presented in that paper. The three cohort studies summarized showed no elevated risk. Studies of meat, processed meats, preserved fish, smoked foods, preserved vegetables, and beer were included in this review although no clear definitions of any food groups were presented, and it is safe to assume there was no consistency among the foods, or their nitrate/nitrite contents, making up these food groups across studies. Again, while case–control studies tended to support a correlation between stomach cancer and several of these food categories, the cohort studies neither support such a conclusion nor was there a formal meta-analysis of the data. Another review of various dietary factors and gastric cancer risk concluded that *N*-nitroso compounds, such as nitrosamines, increase cancer risk but the evidence on nitrate and nitrite are not consistent enough to reach a conclusion [31]. These authors cited a meta-analysis of six prospective studies that found an RR of 1.15 for gastric cancer among the highest consumers of processed meat and made the assumption that this habit increased exposure to *N*-nitroso compounds. However, addition of reducing agents since the 1970s has greatly decreased the formation of these compounds in processed meats, and there is little, if any, certainty in judging endogenous production of nitrosamines.

A case–control study in Mexico City examined the intake of nitrates, nitrites, and polyphenols in 257 people with stomach cancer and 478 controls via analysis of an FFQ [32]. Cases were distinguished from controls by significantly higher *H. pylori* infection, alcohol use, energy intake, added salt, and chili consumption, but lower servings of vegetables. While estimates of nitrate and nitrite consumption by cases were significantly higher, the differences were 6% and 4%, respectively, and there were no differences in these compounds derived from animal sources. Nevertheless, the RR for consumption of nitrate from animal sources was 1.87 (95% confidence interval of 1.19–2.91). Controls ate more vegetables that resulted in significantly greater intake of nitrate, nitrite, cinnamic acids, total lignans, secoisolariciresinol, and coumestrol. These increased polyphenols reflect greater consumption of pears, mangoes, legumes, carrots and squash. Since we do not consume nutrients in isolation, except in dietary supplements, perhaps it would be more productive to characterize foods, food groups, or dietary patterns associated with reduced risk of cancer as all single nutrient prescriptions for prevention of cancer seem not to hold up over time.

Another study of nitrate intake in relation to stomach cancer was a case–control study in Korea that looked at effect modification by intake of antioxidant vitamins [33]. In Korea, the gastric cancer rate has declined, but still remains the most common form of the disease and accounts for 20% of all cancers. For a perspective, Mexico also has high stomach cancer prevalence while the US has one of the world's lowest; decline in the US paralleled adoption of refrigeration leading to speculation that spoiled foods contributed to the high stomach cancer rate. Korea's experience with stomach cancer differs from most other countries in that the rate in women is almost as high as for men; in most countries, men have twice the prevalence of stomach cancer. One limitation of the Korean study is that only 136 cases and an equal number of controls were studied. *H. pylori* infection was significantly more common among cases than controls. There were nonsignificant trends for shorter usage of refrigerators among cases, and a greater history of gastric cancer among first-degree relatives. Cases consumed about the same nitrate as controls, but cases ingested significantly less beta-carotene, vitamin C, vitamin E, and folate. There were significant correlations between nitrate intake and these plant nutrients, with vitamin C and folate correlations of 0.64 and 0.70, while the correlation for beta-carotene was 0.26. The lower correlation of nitrate with vitamin E (0.11) was not significant. The RR for stomach cancer increased as the ratio of nitrate to these vitamins increased, but only the ratio of nitrate:folate remained statistically significant following multivariate adjustment for confounders identified in this study. It should be pointed out that the mean daily intake of nitrate was 534 mg and this is more than 2.4 times the ADI as a result of high vegetable intake, particularly Korean radish, cabbage, lettuce, and spinach. The authors suggest that the ratio of dietary nitrate to antioxidant vitamins may be a fruitful area of further investigation.

An updated review of nitrate and nitrite ingestion in relation to stomach cancer risk was published in 2012 [34]. The more recently published prospective cohort studies failed to find an association between stomach cancer and estimated consumption of either nitrate or nitrite. This group of authors also reviewed the animal literature and concluded that nitrite was not carcinogenic in the absence of co-administration of nitrosamine precursors. The review concluded that nitrate and nitrite are important biological compounds, that nitrosation is an important part of nitric oxide metabolism, and that carcinogenic N-nitrosation requires conditions not usually found in normal metabolism. The authors concluded that the studies supporting a carcinogenic effect of nitrate or nitrite were of lower methodological quality and that the better designed studies do not support such a relationship.

It is enlightening to examine in some detail one of the largest prospective cohort studies on diet and cancer—the NIH-AARP Diet and Health Study, being conducted among more than 567,000 individuals aged 50–71 years at the start who were administered an FFQ at baseline. In the first 6.8 years of follow-up, over 53,000 cancers were diagnosed in 21 different organs or tissues. The mean intake of processed meat by subjects in the highest quintile of consumption was 14 times the average intake in the lowest quintile. The highest quintile had significantly elevated risks of lung (RR = 1.16) and colorectal (RR = 1.20) cancers but significantly reduced risks of leukemia and melanoma, while the other cancers reported did not differ by intake of processed meats [35]. Nitrite and nitrate intakes were not reported in this study. The authors reported the p value for the trend of each cancer, and those were significant only for the elevated risks as well as for two other cancers for which the 95 % confidence interval crossed the null. It is compelling to ask if using standard criteria of 95 % confidence intervals not crossing unity (RR = 1.0) to determine significance in epidemiological studies in which the risk of two cancers was elevated, two decreased, and 17 unchanged, were the calculated significant differences in trends really meaningful?

Another report from the NIH-AARP cohort described data from 7 years of follow-up in 300,000 from the larger group, with 2719 colorectal cancers cases diagnosed subsequent to initiation of the study [36]. Subjects were divided by quintile of red meat intake, adjusted for total dietary energy; mean red meat intake in the lowest quintile was 8.9 and 66.5 g/day in the highest quintile per 1000 kcal. However, factors that strongly correlated with red meat intake were male gender, lower education, more smoking, lower physical activity, more calorie intake, and less calcium, fiber, fruits, and vegetables. Median processed meat intakes in the lowest and highest quintiles were 2.7 and 38.0 g/day/1000 kcal. RR for colorectal cancer was significantly elevated at 1.16 (95 % CI, 1.01–1.32). The same RR for nitrate from processed meats was calculated, but the risk for nitrite from processed meats (RR = 1.11, 95 % CI, 0.97–1.25) was not significant. Failure to calculate nitrite and nitrate intake from other sources makes these data much less certain. Some investigators have suggested other factors in processed meat such as preformed heterocyclic aromatic amines or the combination of nitrites and amines make nitrate/nitrite intake from meat dangerous compared with nitrate/nitrite from other foods. But if nitrate or nitrite is designated as a culprit, there should be no plausible biological difference from ingesting larger amounts of nitrogen ions from vegetables, water, or meat.

The same large cohort was studied for relation of N-nitroso compounds with adult brain cancer [37]. This time 546,000 participants' data were analyzed to identify 585 cases of glioma. No significant trends in risk were found for consumption of red or processed meats, nitrate, or vitamins C or E. Significantly elevated risks were found for total fruit and vegetable consumption (RR = 1.42, 95 % CI, 1.08–1.86) and for nitrite from plant sources (RR = 1.59, 95 % CI, 1.20–2.10). The authors argued that nitrate from grain products might be harmful because grains do not contain the nitrosation inhibitors found in fruits and vegetables. They admit they did not statistically adjust their results for multiple comparisons, so the findings remain questionable. In addition, this cohort was examined for the risk of prostate cancer in a subset of 175,343 men [38]. Over 10,000 cases of prostate cancer were diagnosed, and 1102 were in advanced disease during 9 years of follow-up. Nitrite and nitrate from meat were not associated with total prostate cancer (RR = 1.02 for both), but there were significantly elevated risks with advanced prostate cancer (RR = 1.24, 95 % CI 1.02–1.51 and 1.31, 95 % CI, 1.07–1.61 for nitrite and nitrate, respectively). One would expect if the relationship between a subset of 1102 cases of prostate cancer and these two compounds were real, then it would be seen in a ninefold larger group of total prostate cancer cases.

The final report from the NIH-AARP cohort to be discussed here reported on total mortality [39]. Red and processed meat intakes were reported to increase the risk of total death during 10 years of follow-up in which there were over 71,000 deaths, with two-thirds of the deaths in men. Men in the quintile of highest red meat intake had an RR of 1.31 (95 % CI, 1.27–1.35), and for processed meat the RR was 1.16 (95 % CI, 1.12–1.20). The information that put these values in sharpest perspective was the risk of mortality from injuries and sudden deaths. Red meat was associated with an increased RR of 1.26 (95 % CI, 1.04–1.54), and white meat associated with a decreased RR of 0.85 (95 % CI, 0.70–1.02). There is no plausible link between accidental deaths and greater intake of red meat or a protective effect of poultry/fish consumption unless these different dietary habits reflect other lifestyle decisions that lead to fatal accidents. These data may suggest that RRs in the range reported are not signals, but are noise, even when based on huge numbers compared to most dietary epidemiology studies.

A different cohort study of about 100,000 participants, in contrast to the NIH-AARP study [35], found no association of either red or processed meat consumption with lung cancer in men or women [40]. For processed meat, the RR for lung cancer risk in the highest quintile of consumption was 1.12 (95 % CI, 0.83–1.53) in men and 0.98 (95 % CI, 0.68–1.41) in women. A wide variety of cancers in other organ sites have been studied in relation to nitrate/nitrite intake and red or processed meat as a proxy for nitrates and nitrites, even though there is no strong correlation between the foodstuffs and intake of the compounds in all studies. A meta-analysis of all published studies on red or processed meat and kidney cancer, including 12 case–control studies, three cohorts, and the Pooling Project [41] in which 13 cohort studies were evaluated, found no increased risk with high consumption of meat [42]. The RRs ranged from 1.02 to 1.19, and the total findings were interpreted as not supportive of a relationship between intake and cancer risk. Another prospective cohort study in the Netherlands examined a potential relationship between nitrate intake and bladder cancer [43]. In a population of 120,852 middle-aged men and women followed for 9.3 years, 889 cases of bladder cancer were compared against cohort members without the disease. Nitrate exposures from food, drinking water, or total sources were similar and did not elevate risk—RR for nitrate from food was 1.06 (95 % CI, 0.81–1.31). Three US cohorts were combined to examine a relationship between nitrates/nitrites and glioma in adults [44]. A total of 237,794 people were followed for up to 24 years, and 335 cases of glioma were identified. Risk of glioma was not elevated for highest consumption of nitrate (RR = 1.02, 95 % CI, 0.66–1.58) or nitrite (RR = 1.26, 95 % CI, 0.89–1.79).

The final epidemiological study to be considered here exemplifies the difficulty in examining compounds found in multiple foods and water with the development of cancer. A case–control study in Iowa, where there are high levels of nitrate in the drinking water, was conducted among 458 cases of non-Hodgkin's lymphoma and 383 controls with records on exposure [45]. There was no elevated risk from high nitrate in drinking water (RR = 1.2, 95 % CI 0.6–2.2). Nitrate in diet was associated with decreased risk (RR = 0.54, 95 % CI 0.34–0.86), but nitrite was associated with increased risk (RR = 3.1, 95 % CI 1.7–5.5); however, this latter increase was attributed to nitrite in breads and cereals.

Conclusions

No consistent relationship appears to have been substantiated between nitrate/nitrite exposure and risk of cancer or other adverse health consequences. In fact, experts remain considerably divided on the risk when summarizing and interpreting the same data [46, 47]. A relatively new area of research that challenges traditional biochemical thinking on this topic is that endogenous protein S-nitrosylation may play a substantial role in pathophysiological states, including cancer, and that this phenomenon is mostly independent of exogenous nitrogen anion exposure [48]. S-nitrosylation is the process in which nitric oxide is post-translationally added to cysteine sulfhydryl groups of proteins. This process represents a form of redox modulation and is becoming increasingly recognized as a regulatory reaction perhaps equally important as phosphorylation. A partial list of proteins related to carcinogenesis shown to undergo S-nitrosylation includes Ras, COX-2, PTEN, NF-κB, p53, Bcl-2, caspases, and DNA repair enzymes. Both nitrosylation and denitrosylation of various target proteins contribute to the net biological effect. The activities of nitrite as a signaling molecule and regulator of gene expression, combined with its abundance in vivo and its stability, suggest an important role of nitrite in normal metabolism [49, 50].

As can be seen in the epidemiology section above, some groups of researchers consistently report that nitrate/nitrite increase the risk of cancer. It is not clear why only some scientists find this relationship while others consistently do not. One of the former groups recently reported that men fed 420 g/day of processed meats had a 70-fold increase in fecal excretion of N-nitroso compounds compared with those eating a vegetarian diet [51]; however, similar increases in fecal excretion of these compounds were found when subjects consumed a diet supplemented with red meat that contained no nitrite. This strongly suggests that endogenous formation of N-nitroso compounds is independent of dietary intake of nitrite. There were five male and 11 female volunteers in this study. Curiously, only the data for the men were published with the rationale that their diets contained the largest amount of meat. The authors of this report proposed that S-nitrosothiols form in the acidic conditions of the stomach following ingestion of any meat products. Furthermore, inorganic nitrite under acidic conditions as might occur in the stomach or urinary bladder actually inhibits cancer cell growth, an effect that can be enhanced by the addition of ascorbic acid [52]. This effect appeared to result from the formation of nitric oxide or other reactive nitrogen oxide species.

Several review articles assessing the putative role of nitrate, nitrite, or red meat ingestion on risk of cancer or other health endpoints have appeared in recent years. A group of 23 experts met in 2013 to evaluate the role of red and processed meat in development of colorectal cancer and produced a summary publication that concluded both the epidemiological and experimental data are internally inconsistent and the underlying mechanisms of such a relationship are unclear [53]. Another group of 22 experts met in 2012 at the request of the German Research Foundation to evaluate specifically the benefits and risks to health of dietary nitrate and nitrite [54]. This group concluded that nitrate and nitrite are not carcinogenic, but under conditions that results in endogenous nitrosation, they may possibly be carcinogenic to humans.

The reader is referred to a risk–benefit analysis of dietary nitrate and nitrite that examined the small, uncertain risks of cancer balanced against the certain reduction of food poisoning, including the elimination of botulism and marked reduction in other microbial illnesses from commercially processed meats, and balanced these observations with the potential cardiovascular benefits from the nitrogen anions [55]. It should be clear that scientists and policy makers need to understand the full biological ramifications of exogenous exposure to nitrates/nitrites from all sources, the endogenous synthesis of nitrogen metabolites, and their biological consequences in multiple target tissues before any meaningful decisions can be made. The strength of evidence to make such decisions will not be made by counting up the number of publications, or looking at the largest cohort, but by weighing a variety of studies and synthesizing a rational conclusion that does not trade proven and immediate consumer safety for a small, long-term risk that remains theoretical despite decades of research.

References

1. Hill AB. The environment and disease: association or causation? Proc R Soc Med. 1965;58:295–300.
2. Ames BN, Gold LS. Chemical carcinogenesis: too many rodent carcinogens. Proc Natl Acad Sci U S A. 1990;87:7772–6.
3. Maekawa A, Ogiu T, Onodera H, Furuta K, Matsuoka C, Ohno Y, et al. Carcinogenicity studies of sodium nitrite and sodium nitrate in F-344 rats. Food Chem Toxicol. 1982;20:25–33.
4. National Toxicology Program. NTP technical report on toxicology and carcinogenesis studies of sodium nitrite in F344/N rats and B6C3F1 mice (drinking water studies). Durham: National Institutes of Health; 2001.
5. Olsen P, Gry J, Knudsen I, Meyer O, Poulsen E. Animal feeding study with nitrite-treated meat. IARC Sci Publ. 1984;57:667–75.
6. IARC. Working Group on the evaluation of carcinogenic risks to humans. IARC Monogr Eval Carcinog Risks Hum. 2010;94:1–412.
7. Petersen A, Stoltze S. Nitrate and nitrite in vegetables on the Danish market: content and intake. Food Addit Contam. 1999;16:291–9.
8. Thomson BM, Nokes CJ, Cressey PJ. Intake and risk assessment of nitrate and nitrite form New Zealand foods and drinking water. Food Addit Contam. 2007;24:113–21.
9. Hord NG, Tang Y, Bryan NS. Food sources of nitrates and nitrites: the physiologic context for potential health benefits. Am J Clin Nutr. 2009;90:1–10.
10. Lundberg JO, Gladwin MT, Ahluwalia A, Benjamin N, Bryan NS, Butler A, et al. Nitrate and nitrite in biology, nutrition and therapeutics. Nat Chem Biol. 2009;5:865–9.
11. Nunez de Gonzalez MT, Osburn WN, Hardin MD, Longnecker M, Garg HK, Bryan NS, et al. A survey of nitrate and nitrite concentrations in conventional and organic-labeled raw vegetables at retail. J Food Sci. 2015;80:942–9.
12. Linseisen J, Rohrmann S, Norat T, Gonzalez CA, Dorronsoro Iraeta M, Morote Gomez P, et al. Dietary intake of different types and characteristics of processed meat which might be associated with cancer risk — results from the 24-hour diet recalls in the European Prospective Investigation into Cancer and Nutrition (EPIC). Public Health Nutr. 2006;9:449–64.
13. Jakszyn P, Agudo A, Ibanez R, Garcia-Closas R, Pera G, Amiano P, et al. Development of a food database of nitrosamines, heterocyclic amines, and polycyclic aromatic hydrocarbons. J Nutr. 2004;134:2011–4.
14. Griesenbeck JS, Brender JD, Sharkey JR, Steck MD, Huber Jr JC, Rene AA, et al. Maternal characteristics associated with the dietary intake of nitrates, nitrites, and nitrosamines in women of child-bearing age: a cross-sectional study. Environ Health. 2010;9:10–6.
15. Forman D, Al-Dabbagh S, Knight T, Doll R. Nitrate exposure and the carcinogenic process. Ann N Y Acad Sci. 1988;534:597–603.
16. van den Brandt P, Voorrips L, Hertz-Picciotto I, Shuker D, Boeing H, Speijers G, et al. The contribution of epidemiology. Food Chem Toxicol. 2002;40:387–424.
17. Meyskens FL, Szabo E. Diet and cancer: the disconnect between epidemiology and randomized clinical trials. Cancer Epidemiol Biomarkers Prev. 2005;14:1366–9.
18. Willett WC. Fruit, vegetables, and cancer prevention: turmoil in the produce section. J Natl Cancer Inst. 2010;102:1–2.
19. World Cancer Research Fund/American Institute for Cancer Research. Food, nutrition, physical activity, and the prevention of cancer: a global perspective. Washington, DC: AICR; 2007.
20. Boyle P, Boffetta P, Autier P. Diet, nutrition and cancer: public, media and scientific confusion. Ann Oncol. 2008;19:1665–7.
21. World Cancer Research Fund/American Institute for Cancer Research Continuous Update Project Report. Food, nutrition, physical activity, and the prevention of colorectal cancer. Washington, DC: AICR; 2011.
22. Boffetta P, McLaughlin JK, La Vecchia C, Tarone RE, Lipworth L, Blot WJ. False-positive results in cancer epidemiology: a plea for epistemological modesty. J Natl Cancer Inst. 2008;100:988–95.
23. Boffetta P, McLaughlin JK, La Vecchia C. A further plea for adherence to the principles underlying science in general and the epidemiologic enterprise in particular. Int J Epidemiol. 2009;38:678–9.
24. Ioannidis JPA. Why most published research findings are false. PLoS Med. 2005;2:0676–701.
25. Schatzkin A, Kipnis V, Carroll RJ, Midthune D, Subar AF, Bingham S, et al. A comparison of a food frequency questionnaire with a 24-hour recall for use in an epidemiological cohort study: results from the biomarker-based Observing Protein and energy Nutrition (OPEN) study. Int J Epidemiol. 2003;32:1954–62.
26. Kristal AR, Peters U, Potter JD. Is it time to abandon the food frequency questionnaire? Cancer Epidemiol Biomarkers Prev. 2005;14:2826–8.
27. Fuchs CS, Giovannucci EL, Colditz GA, Hunter DJ, Stampfer MJ, Rosner B, et al. Dietary fiber and the risk of colorectal cancer and adenoma in women. N Engl J Med. 1999;340:169–76.

28. Demeyer D, Honikel K, De Smet S. The World Cancer Research Fund report 2007: a challenge for the meat processing industry. Meat Sci. 2008;80:953–9.
29. Eichholzer M, Gutzwiller F. Dietary nitrates, nitrites, and N-nitroso compounds and cancer risk: a review of the epidemiologic evidence. Nutr Rev. 1998;56:95–105.
30. Jakszyn P, Gonzalez CA. Nitrosamine and related food intake and gastric and oesophageal cancer risk: a systematic review of the epidemiological evidence. World J Gastroenterol. 2006;12:4296–303.
31. Liu C, Russell RM. Nutrition and gastric cancer risk: an update. Nutr Rev. 2008;66:237–49.
32. Hernandez-Ramirez RU, Galvan-Portillo MV, Ward MH, Agudo A, Gonzalez CA, et al. Dietary intake of polyphenols, nitrate and nitrite and gastric cancer risk in Mexico City. Int J Cancer. 2009;125:1424–30.
33. Kim HJ, Lee SS, Choi BY, Kim MK. Nitrate intake relative to antioxidant vitamin intake affects gastric cancer risk: a case-control study in Korea. Nutr Cancer. 2007;59:185–91.
34. Bryan NS, Alexander DD, Coughlin JR, Milkowsi AL, Boffetta P. Ingested nitrate and nitrite and stomach cancer risk: an updated review. Food Chem Toxicol. 2012;50:3646–65.
35. Cross AJ, Leitzmann MF, Gail MH, Hollenbeck AR, Schatzkin A, Sinha R. A prospective study of red and processed meat intake in relation to cancer risk. PLoS Med. 2007;4, e325.
36. Cross AJ, Ferrucci LM, Risch A, Graubard BI, Ward MH, Park Y, et al. A large prospective study of meat consumption and colorectal cancer risk: an investigation of potential mechanisms underlying this association. Cancer Res. 2010;70:2406–14.
37. Dubrow R, Darefsky AS, Park Y, Mayne ST, Moore SC, Kilfoy B, et al. Dietary components related to N-nitroso compound formation: a prospective study of adult glioma. Cancer Epidemiol Biomarkers Prev. 2010;19:1709–22.
38. Sinha R, Park Y, Graubard BI, Leitzmann MF, Hollenbeck A, Schatzkin A, et al. Meat and meat-related compounds and risk of prostate cancer in a large prospective cohort study in the United State. Am J Epidemiol. 2009;170:1165–77.
39. Sinha R, Cross AJ, Graubard BI. Meat intake and mortality. A prospective study of over half a million people. Arch Intern Med. 2009;169:562–71.
40. Tasevska N, Cross AJ, Dodd KW, Ziegler RG, Caporaso NE, Sinha R. No effect of meat, meat cooking preferences, meat mutagens or heme iron on lung cancer risk in the Prostate, Lung, Colorectal, and Ovarian Cancer Screening Trial. Int J Cancer. 2011;128:402–11.
41. Lee JE, Spiegelman D, Hunter DJ, Albanes D, Bernstein L, van den Brandt PA, et al. Fat, protein, and meat consumption and renal cell cancer risk: a pooled analysis of 13 prospective studies. J Natl Cancer Inst. 2008;100:1695–706.
42. Alexander DD, Cushing CA. Quantitative assessment of red meat or processed meat consumption and kidney cancer. Cancer Detect Prev. 2009;32:340–51.
43. Zeegers MP, Selen RFM, Kleinjans JCS, Goldbohm RA, van den Brandt PA. Nitrate intake does not influence bladder cancer risk: The Netherlands Cohort Study. Environ Health Perspect. 2006;114:1527–31.
44. Michaud DS, Holick CN, Batchelor TT, Giovannucci E, Hunter DJ. Prospective study of meat intake and dietary nitrates, nitrites, and nitrosamines and risk of adult glioma. Am J Clin Nutr. 2009;90:570–7.
45. Ward MH, Cerhan JR, Colt JS, Hartge P. Risk of non-Hodgkin lymphoma and nitrate and nitrite from drinking water and diet. Epidemiology. 2006;17:375–82.
46. Powlson DS, Addiscott TM, Benjamin N, Cassman KG, de Kok TM, van Grinsven H, et al. When does nitrate become a risk for humans? J Environ Qual. 2008;37:291–5.
47. Ward MH, deKok TM, Levallois P, Brender J, Gulis G, Nolan BT, et al. Workgroup report: drinking-water nitrate and health—recent findings and research needs. Environ Health Perspect. 2005;113:1607–14.
48. Foster MW, Hess DT, Stamler JS. Protein S-nitrosylation in health and disease: a current perspective. Trends Mol Med. 2009;15:391–404.
49. Bryan NS, Fernandex BO, Bauer SM, Garcia-Saura MF, Milson AB, Rassaf T, et al. Nitrite is a signaling molecule and regulator of gene expression in mammalian tissues. Nat Chem Biol. 2005;1:290–7.
50. Angelo M, Singel DJ, Stamler JS. An S-nitrothiol (SNO) synthase function of hemoglobin that utilizes nitrite as a substrate. Proc Natl Acad Sci U S A. 2006;103:8366–71.
51. Joosen AM, Kuhnle GG, Aspinall SM, Barrow TM, Lecommandeur E, Azqueta A. Effect of processed and red meat on endogenous nitrosation and DNA damage. Carcinogenesis. 2009;30:1402–7.
52. Morcos E, Carlsson S, Weitzberg E, Wiklund NP, Lundberg JO. Inhibition of cancer cell replication by inorganic nitrite. Nutr Cancer. 2010;62:501–4.
53. Oostindjer M, Alexander J, Amdam GV, Andersen G, Bryan NS, Chen D, et al. The role of red and processed meat in colorectal cancer development: a perspective. Meat Sci. 2014;97:583–96.
54. Habermeyer M, Roth A, Guth S, Diel P, Engel K-H, Epe B, et al. Nitrate and nitrite in the diet: how to assess their benefit and risk for human health. Mol Nutr Food Res. 2015;59:106–28.
55. Milkowski A, Garg HK, Coughlin JR, Bryan NS. Nutritional epidemiology in the context of nitric oxide biology: a risk-benefit evaluation for dietary nitrite and nitrate. Nitric Oxide. 2010;22:110–9.

Chapter 22
Looking Forward

Nathan S. Bryan and Joseph Loscalzo

Keywords Safety • Dietary guidelines • Hypertension • Prevention

Introduction

Eating a well-balanced, nutritious diet and performing moderate exercise comprise the ideal model of routinely good health and disease prevention. Probably not coincidently, these two activities are known to affect positively nitric oxide production, which, one can argue, is a key molecular mediator of the salubrious effect of both interventions. What is becoming more apparent is that NO derived from dietary sources of nitrite and nitrate comprises another key pathway in NO biology. Notwithstanding the essential nature of nitrite and nitrate in the environmental nitrogen cycle, historical use of nitrates and nitrites as medicinal agents, and the fact that these anions are produced naturally in the body from the oxidation of nitric oxide, public perception remains that these are harmful substances in our food and water supply [1].

There exists a number of nitrogen-containing molecules that are essential and fundamental for all life. Nitrogen is the largest single constituent of the earth's atmosphere; it is created by fusion processes in stars, and is estimated to be the seventh most abundant chemical element by mass in the universe. Nitric oxide, nitrite, and nitrate in the environment and in the body make up part of the overall biological nitrogen cycle along with proteins and nucleic acids, and are important intermediates or mediators of a number of biological actions. The realm of nitrogen-based chemistry is historical and complex. Just as Alfred Nobel found great irony in the fact that he was prescribed nitroglycerin for his angina later in life, which was the very substance that he had patented for safe delivery of dynamite, the current view of the biological activity of nitrite and nitrate shares similar historical irony. The discovery that a poisonous and toxic gas (NO) was produced by the inner lining of our

N.S. Bryan, Ph.D.
Department of Molecular and Human Genetics, Baylor College of Medicine,
One Baylor Plaza, BCM225, Houston, TX 77030, USA
e-mail: Nathan.bryan@bcm.edu

J. Loscalzo, M.D., Ph.D. (✉)
Department of Medicine, Harvard Medical School, Boston, MA, USA

Department of Medicine, Brigham and Women's Hospital,
75 Francis Street, TR 1, Rm 210, Boston, MA 02115, USA
e-mail: jloscalzo@partners.org

N.S. Bryan, J. Loscalzo (eds.), *Nitrite and Nitrate in Human Health and Disease*, Nutrition and Health,
DOI 10.1007/978-3-319-46189-2_22, © Springer International Publishing AG 2017

NO GAS   **NITRITE**   **NITRATE**

Fig. 22.1 The labels from a cylinder of nitric oxide gas, and bottles of sodium nitrite and sodium nitrate

blood vessels was shocking and revolutionary. This is not to say that NO is not poisonous or dangerous because its chemical properties define it as a poison or toxin; however, in the right context and in an ideal environment, it is without doubt the most important molecule produced in blood vessels. The same argument can be made for nitrite and nitrate. Their inherent chemistry makes them toxic, but in the right context and the right environment, they can have extremely beneficial actions in the human body. The illustration in Fig. 22.1 shows the label of NO, sodium nitrite, and sodium nitrate as poisonous and toxic. Yet, we appreciate and understand the essential functions NO has in the body. Becoming more apparent is the same beneficial effects of nitrite and nitrate *in the proper context*. As an analogous example, water is absolutely essential to all life on earth and is completely safe and innocuous; however, if one were to drink a large excess of water within a few minutes, the cells in the body would swell and burst owing to the hypotonic effects. In the words of Paracelsus, "the dose makes the poison." With respect to nitric oxide and nitrite, one could add the duration, location, context of exposure, and method of delivery.

Despite NO being recognized by the scientific and medical community as one of the most important molecules produced within the body and being named "Molecule of the Year" by *Science* in 1992 and a Nobel Prize in Physiology or Medicine awarded for its discovery in 1998, there are currently only three products on the market directly related to NO: (1) organic nitrates, such as nitroglycerin for the treatment of acute angina (these have been used for many decades long before the discovery of NO); (2) inhaled NO therapy for neonates for treatment of pulmonary hypertension due to underdeveloped lungs; and (3) phosphodiesterase inhibitors, such as sildenafil (Viagra®), which do not directly affect NO production per se but act through affecting the downstream second messenger of NO, cyclic guanosine monophosphate (cGMP). There are a number of other NO-based therapies in development, including technologies designated to affect post-translational protein modifications through S-nitrosation. The method of delivery of NO, nitrite, and nitrate is of utmost importance. We know that delivery of NO through controlled and enzymatic metabolism of organic nitrates is safe and effective acute treatment for angina, but still not without some adverse effects as presented in Chap. 15. The safe delivery of NO gas through inhalation therapy is also now in practice. Inhaled NO is currently approved by the U.S. Food and Drug Administration for hypoxemia in term and near-term infants with pulmonary hypertension. Although more clinical trials are underway in other disease states, currently, this is the only approved use of inhaled NO. However, many physicians use this off-label in adults with pulmonary hypertension, transplantation, and cardiothoracic surgery.

We expect the same considerations for the safe and effective delivery of nitrite and/or nitrate. Understanding their underlying chemistry and metabolism is intrinsic to developing strategies for delivery. The risk-benefit spectrum from nitrite and nitrate may very well depend upon the specific metabolism and the presence of other components that may be concomitantly ingested or available at the time of administration or ingestion. The stepwise reduction of nitrate to nitrite and NO may account for the now well described clinical benefits, while pathways leading to nitrosation of low-molecular-weight amines or amides may account for the health risks of nitrite and nitrate exposure. Understanding and affecting those pathways will certainly help in mitigating the risks. The early concerns about nitric oxide reactivity and the propensity to undergo unwanted nitrosation reactions also are invariably related to nitrite and nitrate biochemistry. We know from previous chapters that these reactions can occur, but there are also very effective inhibitors of nitrosation reactions including vitamins C and E, as well as polyphenols that are present in many foods that contain nitrite and nitrate, particularly vegetables.

The first pathway to be discovered for the endogenous production of NO was that involving L-arginine. For years scientists and physicians have investigated L-arginine supplementation as a means to enhance NO production. This strategy has been shown to work effectively in young healthy individuals with functional endothelium or in older patients with high levels of asymmetric dimethyl L-arginine [2] where the supplemental L-arginine can outcompete this natural inhibitor of NO production. Patients with endothelial dysfunction, however, by definition, are unable to convert L-arginine to NO due to a dysfunctional nitric oxide synthase (NOS) enzyme and, therefore, this strategy has not proven effective consistently in clinical trials. In fact, L-arginine therapy in acute myocardial infarction: the Vascular Interaction With Age in Myocardial Infarction (VINTAGE MI) randomized clinical trial published in the *JAMA* in 2006, concluded that L-arginine, when added to standard postinfarction therapies, did not improve vascular stiffness measurements or ejection fraction and was associated with higher postinfarction mortality [3]. L-arginine should not be recommended following acute myocardial infarction (MI). Similarly, long term L-arginine supplementation in patients with peripheral artery disease (PAD) did not provide any clinical benefit and, in fact, many patients did worse [4]. However, there are also a number of studies showing benefit to healthy patients taking L-arginine and just as many showing no benefit, no harm. Collectively, the literature suggests that strategies to enhance NO production through L-arginine supplementation are equivocal at best.

NOS and Nitrite: A Concert in NO Homeostasis

The nitrate–nitrite–nitric oxide pathway may be a redundant system for overcoming the body's inability to make NO from L-arginine. The emerging literature suggests such and the information included in this book presents a strong case for such. It appears we have at least two systems for affecting NO production/homeostasis. The first is through the classical L-arginine–NO pathway. As we know from Chap. 13, this is a complex and complicated pathway, and if any of the cofactors become limiting, then NO production from NOS shuts down, and in many cases, NOS then produces superoxide instead. The enzymatic production of NO is, indeed, a very complex and coordinated effort that normally proceeds very efficiently. However, in diseases characterized by oxidative stress where essential NOS cofactors become oxidized, NOS uncoupling, or conditions of hypoxia where oxygen is limiting, this process can no longer maintain NO production. Therefore, one can argue saliently that there has to be an alternate route for NO production. It is highly unlikely that Nature devised such a sophisticated mechanism of NO production as a sole source of a critical molecule. After all, there is enormous redundancy in physiology.

This alternate route involves the provision of nitrate and nitrite reductively recycled to NO. The two-electron reduction of nitrate to nitrite requires oral commensal nitrate reducing bacteria since humans lack a functional nitrate reductase gene. Nitrite reduction to NO can occur in a much simpler mechanism than nitrate. The one-electron reduction of nitrite can occur by ferrous heme proteins (or any redox active metal) through the following reaction:

$$NO_2^- + Fe^{(II)} + H^+ \leftrightarrow NO + Fe^{(III)} + OH^-$$

This is the same biologically active NO as that produced by NOS, with nitrite rather than L-arginine as the precursor but it is a relatively inefficient process [5]. Therefore, for this reaction to occur, the tissues or biological compartment must have a sufficient pool of nitrite stored. Since plasma nitrite is a direct measure of NOS activity [6], a compromised NOS system can also affect downstream nitrite production and metabolism, which can perhaps exacerbate any condition associated with decreased NO bioavailability. Replenishing nitrate and nitrite through dietary means may then act as a protective measure to compensate for insufficient NOS activity under conditions of hypoxia or in a number of conditions characterized by NO insufficiency. It is very likely that exogenous nitrite contributes to whole body NO production and homeostasis. Considerable published data now support this redundant pathway: NO produced from nitrite in the upper intestine is up to 10,000 times the concentrations that occur in tissues from enzymatic synthesis [7], nitrite can act as a circulating NO donor [8], and nitrite can itself perform many actions previously attributable to NO [9] without the intermediacy of NO [10].

There are known strategies that will affect both pathways positively. Exercise has been shown to enhance endothelial production of L-arginine and improve endothelial function. Since the burden of exposure of nitrite and nitrate comes from the diet, eating a diet rich in green leafy vegetables or other NO_x-rich foods can fuel the second pathway. We believe that both systems complement one another. When we are young and healthy, the endothelial production of NO through L-arginine is efficient and sufficient to produce NO; however, as we age we lose our ability to synthesize endothelial derived NO. Taddei et al. [11] have shown that there is a gradual decline in endothelial function due to aging with greater than 50 % loss in endothelial function in the oldest age group tested as measured by forearm blood flow assays. Egashira et al. [12] reported more dramatic findings in the coronary circulation of aging adults whereby there was a loss of 75 % of endothelium-derived nitric oxide in 70–80-year-old patients compared to young, healthy 20-year olds. Vita and colleagues [13] demonstrated that increasing age was one predictor of abnormal endothelium-dependent vasodilation in atherosclerotic human epicardial coronary arteries. Gerhard et al. [14] concluded from their 1996 study that age was the most significant predictor of endothelium-dependent vasodilator responses by multiple stepwise regression analysis. Collectively, these important findings illustrate that endothelium NO-dependent vasodilation in resistance vessels declines progressively with increasing age. This abnormality is present in healthy adults who have no other cardiovascular risk factors, such as diabetes, hypertension, or hypercholesterolemia. Most of these studies found that impairment of endothelium-dependent vasodilation was clearly evident by the fourth decade. In contrast, endothelium-independent vasodilation does not change significantly with aging, demonstrating that the responsiveness to NO did not change, only the ability to generate it did. These observations enable us to conclude that reduced availability of endothelium-derived nitric oxide occurs as we age and to speculate that this abnormality may create an environment that is conducive to atherogenesis and other vascular disorders. It is that early event, the inability to produce sufficient NO under the right preclinical conditions that enhances the risk for a number of diseases that plague the older population. If true, then there exist an opportunity to intervene early during this process, implement strategies to restore NO homeostasis, and, perhaps, delay or prevent the onset and progression of certain diseases. This gradual loss of NO activity with age can be sped up or slowed down based on individual lifestyle and diet. This idea is illustrated in the hypothetical graphical representation in Fig. 22.2. Adopting healthy habits such as a good diet and exercise can prolong the precipitous drop in NO production with age. To the contrary, a poor diet along with physical inactivity can accelerate the process and lead to a faster decline in NO production at a younger age. Therapeutic strategies directed at improving endothelial function and/or providing an alternative source of NO should be the primary focus because they may reduce the incidence of atherosclerosis or other diseases that occur with aging, even perhaps Alzheimer's disease or at the very least vascular dementia.

Exercise training has been shown, in many animal and human studies, to augment endothelial, NO-dependent vasodilatation in both large and small vessels (reviewed in [15, 16]); however, the response to exercise is diminished in patients with age-dependent endothelial dysfunction [17, 18].

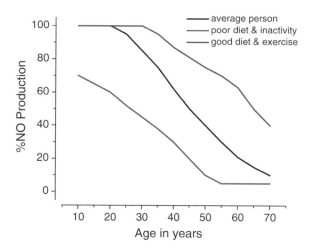

Fig. 22.2 Hypothetical representation of NO production based on diet and lifestyle

In fact, plasma nitrite has been shown to predict exercise capacity [19] and further demonstrates that endothelial production of NO declines with age and increasing risk factor burden [20]. More recent data reveal that dietary supplementation with nitrate reduces oxygen costs of low intensity exercise and enhances tolerance to high intensity exercise [21–23]. This effect is due to an increase in plasma nitrite. These data support the notion that one can compensate through dietary nitrite and nitrate for the endothelial production of NO during exercise. This dietary pathway may also extend beyond exercise. Dietary supplementation of nitrite and nitrate in animals has been shown to reverse endothelial dysfunction, suppress microvascular inflammation, and reduce levels of C-reactive protein in mice subjected to a high cholesterol diet [24] and to protect from ischemia–reperfusion injury [25–27]. This proof-of-concept in animals has now been translated and corroborated in humans. Low-dose nitrite supplementation (80–160 mg/day) has shown to improve endothelial function by 45–60 % without changes in body mass or blood lipids. Measures of carotid artery elasticity improved without changes in brachial or carotid artery blood pressure. Nitrite-induced changes in vascular measures were significantly related to 11 plasma metabolites identified by untargeted analysis [28]. In another study, nitrite supplementation improves aspects of motor and cognitive function in healthy middle aged and older adults, and that these improvements are associated with, and predicted by, the plasma metabolome [29]. Studies using a patented formulation (US patents 8,303,995, 8,298,589, 8,435,570, 8,962,038, 9,119,823 and 9,241,999) using 15–20 mg of sodium nitrite in the form of an orally disintegrating tablet found that it could modify cardiovascular risk factors in patients over the age of 40, significantly reduce triglycerides, and reduce blood pressure [30]. This same lozenge was used in a pediatric patient with argininosuccinic aciduria and significantly reduced his blood pressure when prescription medications were ineffective [31]. A more recent clinical trial using the nitrite lozenge reveals that a single lozenge can significantly reduce blood pressure, dilate blood vessels, improve endothelial function and arterial compliance in hypertensive patients [32]. Furthermore in a study of pre-hypertensive patients (BP >120/80 < 139/89), administration of one lozenge twice daily leads to a significant reduction in blood pressure (12 mmHg systolic and 6 mmHg diastolic) after 30 days [33]. The same lozenge was used in an exercise study and was found to lead to a significant improvement in exercise performance [34]. These studies provide evidence that sodium nitrite supplementation is well-tolerated, increases plasma nitrite concentrations, improves endothelial function, and lessens carotid artery stiffening in middle-aged and older adults, perhaps by altering multiple metabolic pathways, thereby warranting larger long-term clinical trials.

The Nitric Oxide Pathways

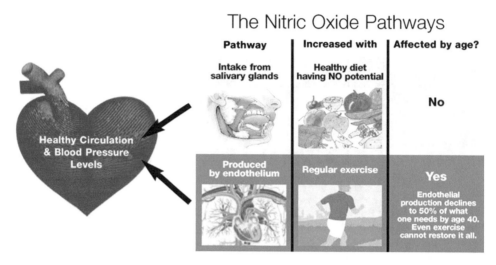

Fig. 22.3 The two pathways for NO production; one from the reduction of the nitrate and nitrite from the foods we eat and the other from the oxidation of L-arginine

Dietary nitrate has also been shown to reduce blood pressure [35–37], inhibit platelet aggregation [37], and restore endothelial function [37] in humans. What is clearly emerging is that there are two pathways for NO production, one through endothelial production via the L-arginine pathway and one through dietary sources of nitrite, nitrate, and antioxidants. This overview is illustrated in Fig. 22.3. The L-arginine pathway becomes dysfunctional with age, and we, therefore, need a backup system to compensate. Eating a diet rich in NO potential, i.e., sufficient nitrite and nitrate along with antioxidants to facilitate reduction to NO, can appear to overcome an insufficiency in endothelium-derived NO. This dietary pathway does not appear to be affected by age. However, overuse of antibiotics or antiseptic mouthwashes can affect this pathway by eliminating the commensal bacteria that are essential for the first step of nitrate reduction to nitrite. Furthermore, use of proton pump inhibitors can decrease the acid secretion in the stomach, thereby affecting the acidic disproportionation of nitrite to NO (pKa nitrite=3.4). This dietary pathway is reliant upon recognizing foods that are rich in NO potential. The inherent NO bioactivity of certain foods or diets is a delicate balance between nitrite and nitrate content as well as antioxidant capacity to facilitate reduction to NO and to inhibit any unwanted nitrosation reactions. Antioxidant status can be generically estimated by the Oxygen Radical Absorbance Capacity or ORAC. An antioxidant's strength reflects its ability to eliminate oxygen-free radicals to prevent scavenging of nitric oxide as well as provide the reductive capacity to activate dietary nitrate and nitrite to NO. Foods with a high ORAC score are thought to protect cells and their components from oxidative damage. The ORAC score combined with the inherent nitrite and nitrate content of certain foods may provide a novel scoring system for NO potential or NO index. The Bryan lab has created a nitric oxide index below by applying an algorithm considering the average nitrite/nitrate content of foods as well as their reported ORAC values. We define the NO index below:

Table 22.1 Nitric oxide index of select foods

	Nitrite + nitrate (mg/100 g)	ORAC (μmol/100 g)	NO index
High			
Kale	1950	3500	**6825**
Swiss chard	822	2500	**2055**
Arugula	612	2373	**1452**
Spinach	741	1515	**1123**
Chicory	625	1500	**938**
Wild radish	465	1750	**814**
Bok choy	310	2500	**775**
Collard greens	317	2200	**697**
Beets	174	3632	**632**
Chinese cabbage	161	3100	**499**
Lettuce	268	1447	**388**
Cabbage	125	2496	**312**
Mustard greens	116	1946	**226**
Cauliflower, raw	202	829	**167**
Parsley	115	1301	**150**
Kohlrabi	177	769	**136**
Carrot	190	666	**127**
Broccoli	39.5	3083	**122**
Medium			
Cole slaw	55.9	1500	**84**
Asparagus	50	1644	**82**
Celery	160	497	**80**
Watercress	33	2200	**73**
Artichoke	9.6	6552	**63**
Eggplant	42	933	**39**
Strawberry	9.4	3577	**34**
Potato	20	1300	**26**
Garlic	3.4	5708	**19**
Tomato	39.2	367	**14**
Vegetable soup	20.9	500	**10**
Cereal	4.9	2000	**10**
Melons	68	142	**10**
Low			
String beans	30	300	**9**
Cured, dried sausage	78.8	100	**8**
Figs	2	3383	**7**
Prunes	1	5770	**6**
Sweet potato, raw, uncooked	5.4	902	**5**
Blackberries	1	5347	**5**
Raspberries	1	4882	**5**
Raisins	1.2	3037	**4**
Banana	4.5	879	**4**
Cherries	1	3365	**3**
Cucumber	14	214	**3**
Onions	3.2	1034	**3**
Bean sprouts	1	1510	**2**
Hot dog	9	100	**1**
Kiwi	1	882	**1**
Cucumber	9.3	100	**1**

(continued)

Table 22.1 (continued)

	Nitrite + nitrate (mg/100 g)	ORAC (μmol/100 g)	NO index
Bacon	5.5	100	**1**
Apricot	1	1115	**1**
Papaya	1	500	**1**
Chickpeas	1	847	**1**
Bacon, nitrite-free	3	100	**0**
French fries	2	150	**0**
Juices	(mg/100 mL)		
Beet root	279	1727	**482**
Vegetables juice	20	548	**11**
Carrot	6.8	195	**1**
Pomegranate	1.3	2681	**3**
Cranberry	1		**0**
Red wine	1	3000	**3**
Green tea	0.02	1253	**0**
Acai	0.06	1767	**0**

$$\left[\left(\text{Nitrite} + \text{nitrate}\left(\text{mg}/100\text{g}\right) \times \text{ORAC}\left(\mu\text{mol}/100\text{g}\right)\right)\right]/1000 = \text{Nitric Oxide Index}$$

Similar to the glycemic index for diabetics, the nitric oxide index may be a useful tool for people with vascular disease or any conditions associated with NO insufficiency. A list of the NO index of select foods is included in Table 22.1. We suspect the context of a Nitric oxide Index whereby nitrite and nitrate along with antioxidants will help define the health benefits of certain foods or diets. However, the difficulty is the fact that there are highly variable differences in nitrite and nitrate of the same vegetable based on different farming practices in different geographical locations [38]. In fact, there can be more than a 100-fold difference in the nitrate content of a single vegetable depending on where it is consumed and how it was grown. Therefore, this NO index can only be used when quantities of nitrite and nitrate are quantified and known for any given vegetable. We can use averages from historical database for proof of principle but we cannot be certain that any given vegetable in the table will contain the amount of nitrite and nitrate reported.

Although dietary nitrite and nitrate have been shown to replenish blood and tissue stores of NO, it is still not clear how this pathway is regulated or controlled. The production of NO through the NOS enzymes is a precise, spatially and temporally regulated, controlled event that generates NO in a local environment upon need or specific stimulus. Gladwin and colleagues have proposed that nitrite reduction to NO is allosterically linked to oxygen saturation of hemoglobin [39, 40], thereby providing a sensing mechanism for production of NO from nitrite during hypoxic vasodilation. Myoglobin has also been implicated in serving such a function in the heart [41]. More work is needed to determine if nitrite alone is necessary and sufficient to elicit specific cellular events as has been well described for NOS-derived NO. Although it may be argued that nitrite is simply a storage pool that can be reduced to NO under appropriate conditions, we consider this an unlikely role for nitrite in normal physiology, i.e., the normoxic and neutral pH conditions under which nitrite is supposedly stable. Alternatively, nitrite acts as an important molecule in its own right but which has regulatory effects on the NO pathway. It may be that both hypoxic reduction of nitrite to NO and normoxic metabolism of nitrite represent an advantageous oxygen sensing system, which is a vestige of denitrifying microorganisms that existed long before the advent of aerobic respiration and the emergence of an NO synthase system [42]. The fact that both systems still exist today highlight the importance of nitrite in all cellular processes throughout the entire physiological oxygen gradient [9]. More research is needed in order to completely delineate the precise and regulated cell signaling aspects of nitrite, both from endogenous sources as well as from dietary sources.

The role of diet in the prevention and control of morbidity and premature mortality due to non-communicable diseases has been well established by vast population-based epidemiological studies carried out during the last decade [43]. Nothing affects our health more than what we choose to eat. Nitric oxide is essential for maintaining normal blood pressure, preventing adhesion of blood cells to the endothelium, and preventing platelet aggregation; it may, therefore, be argued that this single abnormality, the inability to generate NO, puts us at risk for diseases that plague us later in life, such as atherosclerosis, myocardial infarction, stroke, Alzheimer's, and peripheral vascular disease. Therefore, developing strategies and new technologies designed to restore NO availability is essential for inhibiting the progression of certain common chronic diseases. The provision of dietary nitrate and nitrite may allow for such a strategy. The information presented throughout this text illustrates that the beneficial effects of nitrite and nitrate are seen at doses and levels that are easily achievable by eating certain foods enriched in nitrite and nitrate, particularly green leafy vegetables. The fact that exposure to nitrite and nitrate from certain foods or diets exceeds the World Health Organizations limit for Acceptable Daily Intake (ADI) calls into question the current regulatory limits.

Hord and Bryan have shown that people following the Dietary Approaches to Stop Hypertension [44] diet exceed the ADI for nitrate by greater than 500 % [45] and infants consuming human breast milk and some formulas can also exceed the regulatory limits [46]. The DASH diet was developed by the US National Institutes of Health to lower blood pressure without medication. The DASH diet is based on the research studies and has been proven to lower blood pressure, reduce cholesterol, and improve insulin sensitivity [47]. The DASH diet provides more than just the traditional low salt or low sodium diet plans to help lower blood pressure. It is based on an eating plan proven to lower blood pressure, a plan rich in fruits, vegetables, and low-fat or nonfat dairy. DASH diet is recommended by The National Heart, Lung, and Blood Institute (one of the National Institutes of Health, of the US Department of Health and Human Services), The American Heart Association, The Dietary Guidelines for Americans, US guidelines for treatment of high blood pressure, and, the DASH diet formed the basis for the USDA MyPyramid. So how can such a diet exceed regulatory limits for the molecule that may be responsible for its blood pressure lowering effects? The same argument can be made for breast milk [46]. Human breast milk is recommended to serve as the exclusive food for the first 6 months of life and continue, along with safe, nutritious complementary foods, up to 2 years [48, 49]. Breast milk is nature's most perfect food. In fact, the U.S. Centers for Disease Controls in 2010 acknowledged, "Breast milk is widely acknowledged as the most complete form of nutrition for infants, with a range of benefits for infants' health, growth, immunity and development." Breast milk is a unique nutritional source for infants that cannot adequately be replaced by any other food, including infant formula. It remains superior to infant formula from the perspective of the overall health of both mother and child. Human milk is known to confer significant nutritional and immunological benefits for the infant [50–52]. And yet, human breast milk and colostrum is enriched in nitrite and nitrate [46, 53–55]. It may be time to reconsider new regulatory guidelines for dietary sources of nitrite and nitrate given the last two decades of research showing remarkable health benefits at levels that do not pose any significant risk.

We hope this work on nitrite and nitrate in human health and disease will provide the reader with a comprehensive body of knowledge on these molecules. The last 25 years of research has drastically altered the landscape of how we think about nitrite and nitrate. The discovery of the nitric oxide pathway revealed that these two anions are naturally produced within our bodies and are not simply synthetic food additives. Prior to this discovery, much of the scientific research focused on the toxic properties due to exposure in both industrial settings and from cured and processed foods. Surprisingly, the context for nitrite and nitrate in human health and disease has not been adequately addressed since. We feel privileged to be able to communicate this information for the Nutrition Series. Nutrition can play a key and cost-effective role in decreasing the risks of different chronic diseases. Hippocrates himself said, "let food be thy medicine and medicine be thy food." Identifying the key components of food and nutrition that may be responsible for the medicinal effects will help in refining dietary guidelines and designing optimal preventive nutrition regimens for specific disease. Future clinical studies will determine if nitrite and nitrate can provide a nutritional approach to prevention and/or treatment of specific diseases and if such positive effects will outweigh any negative health effects traditionally attributed to these anions. Nutritionists, physiologists, physicians, toxicologists, and dieticians need to converge and establish nutrient guidelines for nitrite and nitrate similar to other well recognized nutrients. The data and facts are now available for such an initiative and the ground is now fertile for such studies.

References

1. L'Hirondel JL. Nitrate and man: toxic, harmless or beneficial? Wallingford: CABI; 2001.
2. Moncada S, Higgs A. the L-Arginine-Nitric Oxide Pathway N Engl J Med. 1993 Dec 30;329(27):2002-12.
3. Schulman SP, Becker LC, Kass DA, Champion HC, Terrin ML, Forman S, et al. L-arginine therapy in acute myocardial infarction: the Vascular Interaction with Age in Myocardial Infarction (VINTAGE MI) randomized clinical trial. JAMA. 2006;295(1):58–64.
4. Wilson AM, Harada R, Nair N, Balasubramanian N, Cooke JP. L-arginine supplementation in peripheral arterial disease: no benefit and possible harm. Circulation. 2007;116(2):188–95.
5. Feelisch M, Fernandez BO, Bryan NS, Garcia-Saura MF, Bauer S, Whitlock DR, et al. Tissue processing of nitrite in hypoxia: an intricate interplay of nitric oxide-generating and -scavenging systems. J Biol Chem. 2008;283(49):33927–34.
6. Kleinbongard P, Dejam A, Lauer T, Rassaf T, Schindler A, Picker O, et al. Plasma nitrite reflects constitutive nitric oxide synthase activity in mammals. Free Rad Biol Med. 2003;35(7):790–6.
7. McKnight GM, Smith LM, Drummond RS, Duncan CW, Golden M, Benjamin N. Chemical synthesis of nitric oxide in the stomach from dietary nitrate in humans. Gut. 1997;40(2):211–4.
8. Dejam A, Hunter CJ, Schechter AN, Gladwin MT. Emerging role of nitrite in human biology. Blood Cells Mol Dis. 2004;32(3):423–9.
9. Gladwin MT, Schechter AN, Kim-Shapiro DB, Patel RP, Hogg N, Shiva S, et al. The emerging biology of the nitrite anion. Nat Chem Biol. 2005;1(6):308–14.
10. Bryan NS, Fernandez BO, Bauer SM, Garcia-Saura MF, Milsom AB, Rassaf T, et al. Nitrite is a signaling molecule and regulator of gene expression in mammalian tissues. Nature Chem Biol. 2005;1(5):290–7.
11. Taddei S, Virdis A, Ghiadoni L, Salvetti G, Bernini G, Magagna A, et al. Age-related reduction of NO availability and oxidative stress in humans. Hypertension. 2001;38(2):274–9.
12. Egashira K, Inou T, Hirooka Y, Kai H, Sugimachi M, Suzuki S, et al. Effects of age on endothelium-dependent vasodilation of resistance coronary artery by acetylcholine in humans. Circulation. 1993;88(1):77–81.
13. Vita JA, Treasure CB, Nabel EG, McLenachan JM, Fish RD, Yeung AC, et al. Coronary vasomotor response to acetylcholine relates to risk factors for coronary artery disease. Circulation. 1990;81(2):491–7.
14. Gerhard M, Roddy MA, Creager SJ, Creager MA. Aging progressively impairs endothelium-dependent vasodilation in forearm resistance vessels of humans. Hypertension. 1996;27(4):849–53.
15. Green DJ, Maiorana A, O'Driscoll G, Taylor R. Effect of exercise training on endothelium-derived nitric oxide function in humans. J Physiol. 2004;561(Pt 1):1–25.
16. Ignarro LJ, Balestrieri ML, Napoli C. Nutrition, physical activity, and cardiovascular disease: an update. Cardiovasc Res. 2007;73(2):326–40.
17. Lauer T, Heiss C, Balzer J, Kehmeier E, Mangold S, Leyendecker T, et al. Age-dependent endothelial dysfunction is associated with failure to increase plasma nitrite in response to exercise. Basic Res Cardiol. 2008;103(3):291–7.
18. Allen JD, Miller EM, Schwark E, Robbins JL, Duscha BD, Annex BH. Plasma nitrite response and arterial reactivity differentiate vascular health and performance. Nitric Oxide. 2009;20(4):231–7.
19. Rassaf T, Lauer T, Heiss C, Balzer J, Mangold S, Leyendecker T, et al. Nitric oxide synthase-derived plasma nitrite predicts exercise capacity. Br J Sports Med. 2007;41(10):669–73; discussion 73.
20. Kleinbongard P, Dejam A, Lauer T, Jax T, Kerber S, Gharini P, et al. Plasma nitrite concentrations reflect the degree of endothelial dysfunction in humans. Free Radic Biol Med. 2006;40(2):295–302.
21. Larsen FJ, Weitzberg E, Lundberg JO, Ekblom B. Effects of dietary nitrate on oxygen cost during exercise. Acta Physiol (Oxf). 2007;191(1):59–66.

22. Bailey SJ, Winyard P, Vanhatalo A, Blackwell JR, Dimenna FJ, Wilkerson DP, et al. Dietary nitrate supplementation reduces the O_2 cost of low-intensity exercise and enhances tolerance to high-intensity exercise in humans. J Appl Physiol. 2009;107(4):1144–55.

23. Larsen FJ, Weitzberg E, Lundberg JO, Ekblom B. Dietary nitrate reduces maximal oxygen consumption while maintaining work performance in maximal exercise. Free Radic Biol Med. 2010;48(2):342–7.

24. Stokes KY, Dugas TR, Tang Y, Garg H, Guidry E, Bryan NS. Dietary nitrite prevents hypercholesterolemic microvascular inflammation and reverses endothelial dysfunction. Am J Physiol Heart Circ Physiol. 2009;296(5):H1281–8.

25. Bryan NS, Calvert JW, Elrod JW, Gundewar S, Ji SY, Lefer DJ. Dietary nitrite supplementation protects against myocardial ischemia-reperfusion injury. Proc Natl Acad Sci U S A. 2007;104(48):19144–9.

26. Bryan NS, Calvert JW, Gundewar S, Lefer DJ. Dietary nitrite restores NO homeostasis and is cardioprotective in endothelial nitric oxide synthase-deficient mice. Free Radic Biol Med. 2008;45(4):468–74.

27. Shiva S, Sack MN, Greer JJ, Duranski MR, Ringwood LA, Burwell L, et al. Nitrite augments tolerance to ischemia/reperfusion injury via the modulation of mitochondrial electron transfer. J Exp Med. 2007;204(9):2089–102.

28. DeVan AE, Johnson LC, Brooks FA, Evans TD, Justice JN, Cruickshank-Quinn C, et al. Effects of sodium nitrite supplementation on vascular function and related small metabolite signatures in middle-aged and older adults. J Appl Physiol. 1985;120(4):416–25.

29. Justice JN, Gioscia-Ryan RA, Johnson LC, Battson ML, de Picciotto NE, Beck HJ, et al. Sodium nitrite supplementation improves motor function and skeletal muscle inflammatory profile in old male mice. J Appl Physiol. 2015;118(2):163–9.

30. Zand J, Lanza F, Garg HK, Bryan NS. All-natural nitrite and nitrate containing dietary supplement promotes nitric oxide production and reduces triglycerides in humans. Nutr Res. 2011;31(4):262–9.

31. Nagamani SC, Campeau PM, Shchelochkov OA, Premkumar MH, Guse K, Brunetti-Pierri N, et al. Nitric-oxide supplementation for treatment of long-term complications in argininosuccinic aciduria. Am J Hum Genet. 2012;90(5):836–46.

32. Houston M, Hays L. Acute effects of an oral nitric oxide supplement on blood pressure, endothelial function, and vascular compliance in hypertensive patients. J Clin Hypertens (Greenwich). 2014;16(7):524–9.

33. Biswas OS, Gonzalez VR, Schwarz ER. Effects of an oral nitric oxide supplement on functional capacity and blood pressure in adults with prehypertension. J Cardiovasc Pharmacol Ther. 2014;20(1):52–8.

34. Lee J, Kim HT, Solares GJ, Kim K, Ding Z, Ivy JL. Caffeinated nitric oxide-releasing lozenge improves cycling time trial performance. Int J Sports Med. 2015;36(2):107–12.

35. Larsen FJ, Ekblom B, Sahlin K, Lundberg JO, Weitzberg E. Effects of dietary nitrate on blood pressure in healthy volunteers. N Engl J Med. 2006;355(26):2792–3.

36. Kapil V, Milsom AB, Okorie M, Maleki-Toyserkani S, Akram F, Rehman F, et al. Inorganic nitrate supplementation lowers blood pressure in humans: role for nitrite-derived NO. Hypertension. 2010;56(2):274–81.

37. Webb AJ, Patel N, Loukogeorgakis S, Okorie M, Aboud Z, Misra S, et al. Acute blood pressure lowering, vasoprotective, and antiplatelet properties of dietary nitrate via bioconversion to nitrite. Hypertension. 2008;51(3):784–90.

38. Nunez de Gonzalez MT, Osburn WN, Hardin MD, Longnecker M, Garg HK, Bryan NS, et al. A survey of nitrate and nitrite concentrations in conventional and organic-labeled raw vegetables at retail. J Food Sci. 2015;80(5):C942–9.

39. Cosby K, Partovi KS, Crawford JH, Patel RK, Reiter CD, Martyr S, et al. Nitrite reduction to nitric oxide by deoxyhemoglobin vasodilates the human circulation. Nat Med. 2003;9:1498–505.

40. Huang Z, Shiva S, Kim-Shapiro DB, Patel RP, Ringwood LA, Irby CE, et al. Enzymatic function of hemoglobin as a nitrite reductase that produces NO under allosteric control. J Clin Invest. 2005;115(8):2099–107.

41. Hendgen-Cotta UB, Merx MW, Shiva S, Schmitz J, Becher S, Klare JP, et al. Nitrite reductase activity of myoglobin regulates respiration and cellular viability in myocardial ischemia-reperfusion injury. Proc Natl Acad Sci U S A. 2008;105(29):10256–61.

42. Feelisch M, Martin JF. The early role of nitric oxide in evolution. Trends Ecol Evol. 1995;10(12):496–9.

43. World Health Organization. Report on diet, nutrition and the prevention of chronic diseases; 2003.

44. Karl JM, Alaverdashvili M, Cross AR, Whishaw IQ. Thinning, movement, and volume loss of residual cortical tissue occurs after stroke in the adult rat as identified by histological and magnetic resonance imaging analysis. Neuroscience. 2010;170(1):123–37.

45. Hord NG, Tang Y, Bryan NS. Food sources of nitrates and nitrites: the physiologic context for potential health benefits. Am J Clin Nutr. 2009;90(1):1–10.

46. Hord NG, Ghannam JS, Garg HK, Berens PD, Bryan NS. Nitrate and nitrite content of human, formula, bovine and soy milks: implications for dietary nitrite and nitrate recommendations. Breastfeed Med. 2011;6(6):393–9.

47. Appel LJ, Moore TJ, Obarzanek E, Vollmer WM, Svetkey LP, Sacks FM, et al. A clinical trial of the effects of dietary patterns on blood pressure. DASH Collaborative Research Group. N Engl J Med. 1997;336(16):1117–24.

48. Heinig MJ. The American Academy of Pediatrics recommendations on breastfeeding and the use of human milk. J Hum Lact. 1998;14(1):2–3.
49. Gartner LM, Morton J, Lawrence RA, Naylor AJ, O'Hare D, Schanler RJ, et al. Breastfeeding and the use of human milk. Pediatrics. 2005;115(2):496–506.
50. Hoddinott P, Tappin D, Wright C. Breast feeding. BMJ. 2008;336(7649):881–7.
51. James DC, Lessen R. Position of the American Dietetic Association: promoting and supporting breastfeeding. J Am Dietetic Assoc. 2009;109(11):1926–42.
52. Ip S, Chung M, Raman G, Chew P, Magula N, DeVine D, et al. Breastfeeding and maternal and infant health outcomes in developed countries. Evidence Report/Technology Assessment. Rockville: Tufts-New England Medical Center Evidence-Based Practice Center, under Contract No. 290-02-00222007 Contract No.: 07-E007.
53. Iizuka T, Sasaki M, Oishi K, Uemura S, Koike M, Shinozaki M. Non-enzymatic nitric oxide generation in the stomachs of breastfed neonates. Acta Paediatr. 1999;88(10):1053–5.
54. Ohta N, Tsukahara H, Ohshima Y, Nishii M, Ogawa Y, Sekine K, et al. Nitric oxide metabolites and adrenomedullin in human breast milk. Early Hum Dev. 2004;78(1):61–5.
55. Cekmen MB, Balat A, Balat O, Aksoy F, Yurekli M, Erbagci AB, et al. Decreased adrenomedullin and total nitrite levels in breast milk of preeclamptic women. Clin Biochem. 2004;37(2):146–8.

ERRATUM TO

Chapter 13
Nitric Oxide Signaling in Health and Disease

Nathan S. Bryan and Jack R. Lancaster Jr.

N.S. Bryan, J. Loscalzo (eds.), *Nitrite and Nitrate in Human Health and Disease*, Nutrition and Health,
DOI 10.1007/978-3-319-46189-2_13, © Springer International Publishing AG 2017

DOI 10.1007/978-3-319-46189-2_23

Initially chapter 13 included a figure for which the author did not have permission. We now have permission and the chapter has been updated with additional reference.

The online version of the updated original chapter can be found at
http://dx.doi.org/10.1007/978-3-319-46189-2_13

N.S. Bryan, J. Loscalzo (eds.), *Nitrite and Nitrate in Human Health and Disease*, Nutrition and Health,
DOI 10.1007/978-3-319-46189-2_23, © Springer International Publishing AG 2017

Index

N.S. Bryan, J. Loscalzo (eds.), *Nitrite and Nitrate in Human Health and Disease*, Nutrition and Health,
DOI 10.1007/978-3-319-46189-2, © Springer International Publishing AG 2017

Printed in the United States
By Bookmasters